Global Precipitations and Climate Change

NATO ASI Series

Advanced Science Institutes Series

A series presenting the results of activities sponsored by the NATO Science Committee, which aims at the dissemination of advanced scientific and technological knowledge, with a view to strengthening links between scientific communities.

The Series is published by an international board of publishers in conjunction with the NATO Scientific Affairs Division

A Life Sciences B Physics	Plenum Publishing Corporation London and New York
C Mathematical and Physical Sciences D Behavioural and Social Sciences E Applied Sciences	Kluwer Academic Publishers Dordrecht, Boston and London
F Computer and Systems Sciences G Ecological Sciences H Cell Biology I Global Environmental Change	Springer-Verlag Berlin Heidelberg New York London Paris Tokyo Hong Kong Barcelona Budapest

NATO-PCO DATABASE

The electronic index to the NATO ASI Series provides full bibliographical references (with keywords and/or abstracts) to more than 30000 contributions from international scientists published in all sections of the NATO ASI Series. Access to the NATO-PCO DATABASE compiled by the NATO Publication Coordination Office is possible in two ways:

- via online FILE 128 (NATO-PCO DATABASE) hosted by ESRIN, Via Galileo Galilei, I-00044 Frascati, Italy.
- via CD-ROM "NATO Science & Technology Disk" with user-friendly retrieval software in English, French and German (© WTV GmbH and DATAWARE Technologies Inc. 1992).

The CD-ROM can be ordered through any member of the Board of Publishers or through NATO-PCO, Overijse, Belgium.

Series I: Global Environmental Change, Vol. 26

Global Precipitations and Climate Change

Edited by

Michel Desbois

Françoise Désalmand

Ecole Polytechnique
Laboratoire de Météorologie Dynamique
F-91128 Palaiseau, France

Springer-Verlag
Berlin Heidelberg New York London Paris Tokyo
Hong Kong Barcelona Budapest
Published in cooperation with NATO Scientific Affairs Division

Proceedings of the NATO Advanced Research Workshop on Global Precipitations and Climate Change, held at l'Agelonde, F-83250 La Londe les Maures, France, September 27 - October 1, 1993

ISBN-13: 978-3-642-79270-0 e-ISBN-13: 978-3-642-79268-7
DOI: 10.1007/978-3-642-79268-7

CIP data applied for

Softcover reprint of the hardcover 1st edition 1994

Typesetting: Camera ready by authors
SPIN 10100959 31/3130 - 5 4 3 2 1 0 - Printed on acid-free paper

FOREWORD

SUMMARY OF THE WORKSHOP AND ITS RECOMMENDATIONS

Michel Desbois, Director of the Workshop
Françoise Désalmand, co-Director

The workshop reviewed the present status of knowledge concerning the past and present evolution of the distribution of precipitations at global scale, related to climate evolution at different time scales. This review was intended to assess the availability and quality of data which could help, through validation and initialization of model studies, to improve our understanding of the processes determining these precipitation changes. On another hand, the modelling specialists presented their actual use of precipitation data. Exchanges of views between the modelling and observing communities were thus made possible, leading to a set of recommendations for future studies.

The introductory session outlined the different domains interested by improvement of precipitations knowledge and understanding, from practical concerns as droughts and floods to climate processes, through hydrology and weather forecast. T.N. Krishnamurti put emphasis on the recent Mississipi flooding, while catastrophic rains over Southeastern France were occuring not far from the workshop location, emphasizing the need of studying the extreme events.

Sessions were then devoted to specific themes : 1) paleoclimatology, 2) data collection, history and statistics, programmes, 3) methodologies and accuracy of large scale estimation of precipitation from conventional data, 4) estimation of precipitation from satellite data, 5) modelling studies.

The Paleoclimatology session had 5 papers concerning land and sea evolution of tropical precipitation related to monsoon during the last 170 000 years, tropical abrupt changes over land during the last 15 000 years, accumulation at the surface of polar ice sheets, and finally attempts to reconstitute the last 2 000 years from historic documents. It gave up to date information about the main precipitation changes related to temperature changes occuring for example during the Holocene optimum or the Younger Drias. At global scale, it was indicated that a 1°C increase in temperature corresponds to about 3 to 4 % increase in precipitation.

The session concerning accumulation and analysis of observational precipitation data had 11 papers (only 12 of which are included in this book) covering 3 domains :

- global scale assessment of precipitation, discussing for example the actual global mean of precipitation (around 1 000 mm/year) and the recent trends (decrease in the tropics, specially sharp in Sahelian Africa, slight increase in the high latitudes of Northern Hemisphere),

- current global scale programmes (GPCP and TRMM) were presented. The GPCP appeared as a focus point essential for futures studies and interaction with models, while the TRMM presentation gave views on the possible use of new techniques from space and surface,

- regional scale studies of precipitation related to other parameters, aimed to infer relationships and provide hypotheses on mechanisms, or to take into account the orography and weather regimes for a better mapping of rainfall.

The next session (5 papers) precised the methodologies and accuracy of large scale estimations from conventional data, leading to recommendations about the measurement procedures and corrections.

The session on satellite data analysis (4 papers) insisted on the passive microwave instruments, but also on the need of an appropriate space time sampling which can only be provided up to now by the IR measurements from geostationary satellites. Different kinds of approaches (physical, statistical, combined) were presented. The usefulness of these data over sea was assessed, with the need of proper validation data.

The modelling session (7 papers) adressed problems relative to forecast and climate models. Concerning forecast models, there was a debate between two approaches, one advocating the use of real time precipitation data in the model initialization, the other the improvement of the physical parameterizations, precipitation data being used only as validations. The importance of taking into account properly the microphysical processes was also outlined for climate models. Examples of validation procedures of these climate models were shown. Satisfactory validation of complete climate models running for ten years or more allows to document present phenomena, while simplified models are also useful to study more general processes.

The discussion was then organized in 4 subgroups concerned by a) measurement of precipitation over land, b) measurement over sea, c) process studies, d) model needs. These groups rised a number of detailed recommendations, among which :

a) - encourage the international data exchange, - support the international data centers, - maintain homogeneous series - produce time-dependent climatologies - enhance research on

conventional data analysis
b) - collect coastal indications on paleoclimates - use existing data from satellites over sea - develop new surface means of precipitation measurement over sea - develop and test new algorithms combining data from different sources - improve the time sampling of passive microwave measurements
c) - identify meaningful space and time scales in relation to precipitation climatology and variability - encourage paleoclimatic reconstructions to understand the range of sensitivity to external and internal forcings (glacial-interglacial and holocene changes)
d) - provide to models : daily and monthly area means (1x1°), 3 hourly in some places for diurnal cycle, real time on GTS, improved coverage, real time IR and microwave data, real time cloud data , snow depth data, variance of precipitation - develop inverse modelling for direct use of radiances - build a one month at least data set for further development of rain-rate strategies.

More general recommendations were issued by the full workshop : encourage further interaction between observation people at different time scales and modelists, develop scientific cooperation with Eastern Europe countries, pay special attention to the tropical precipitation processes, encourage international precipitation data exchange, enhance the repetition rate of microwave measurements from space.

The present book is composed from the majority of the papers which were presented to the workshop, after a careful work of reviewing and rewriting, that Françoise Désalmand tried to make as short as possible. We thank the reviewers and the authors who returned promptly their corrected papers.

Acknowledgements

In addition to NATO, which provided most of the financial support to this workshop, acknowledgements are also due to the World Meteorological Organisation (thanks to Professor G.O.P. Obasi and Professor Pierre Morel) and the French Ministery of Research for providing additional founds. We want also to thank the persons in LMD who helped for the numerous practical tasks associated to the organization of such a workshop, and specially Eliane Rier and Anny Pradel.

TABLE OF CONTENTS

Section 1 1

PRECIPITATION PALEOCLIMATOLOGY

NATURAL VARIABILITY OF THE ASIAN, INDIAN AND AFRICAN MONSOONS OVER THE LAST 130 KA

Nicole Petit-Maire
Laboratoire de Géologie du Quaternaire,
CNRS, Case 907 Luminy,
13288 Marseille Cedex 9 (France).

ABSTRACT.
A review of the geological multi-proxy records observed over the transitional area extending, in northern Africa and Asia, between the humid tropics and the desert belts, is presented for the last 130 ka. Long-term arid / humid alternations took place in rough phase with the global cold/warm variations, reflecting a coherent behaviour of the monsoon patterns, cold phases corresponding to an extension of the arid zones to the South, warm phases corresponding to their regression to the North. The total surface affected by the induced sharp environmental changes between the last two climatic extremes was about 14 000 000 km², which may have highly contributed to the variations of the methane atmospheric ratio.

KEYWORDS
Continental paleoclimates. Last climatic cycle. Paleomonsoons. Sahara / Sahel. Arabian peninsula. India. China. Global Change. Methane.

1. Introduction

The transitional zone extending between the arid belt in northern Africa, the Arabian Peninsula, northwestern India, northeastern China and the humid tropical zones is very sensitive to the irregularity of monsoonal activity and range. During the large scale changes of the last climatic cycle, this sensivity is expected to have been still more important and to have induced major environmental changes. We now dispose of a great number of dated geological records which allow to check and validate the theorical models related to this area : they can cover, at the same wide scale, the same large area liable to react to the variability of the african, indian and asian monsoons.

NATO ASI Series, Vol. I 26
Global Precipitations and Climate Change
Edited by M. Desbois and F. Désalmand

In this paper, we shall put together the dated, multi-proxy regional observations, bringing evidence to wet or dry environments over the last 130 ka, compare them and see if the results match the global deep-sea and ice records. Such an overview implies a wide space and time resolution, in order to overcome the inevitable local geographic differences, the corresponding lags in climatic responses and the statistical uncertainty of the chronology.

2. The Asian summer monsoon

The most densely studied areas in the transitional area of **China,** are the Loess Plateau and its surroundings, Beijing region, eastern Inner Mongolia and the mid-lower reach of the Yellow River. Several loess-paleosols sequences have been analyzed (Kukla 1987, Kukla et al. 1988, Kukla and An 1989, Liu, ed. 1985, 1991, Liu et al. 1993, An et al. 1990, Ding et al.1990, Clemens and Prell 1990, Maher and Thompson 1992, Rutter et al. 1990, Kemp et al. 1994, Liu T. 1994) but their chronology, based on magnetic profiles and a few TL dates, is still controversial or unprecise.

During stage 5, up to 3 paleosols provide evidence for wetter conditions dated 128-73 ka (An et al. 1990) or 140-85 ka (Li J. et al. 1990). They correspond to 3 peaks of magnetic susceptibility matching those in deep-sea isotopic stage 5 (Shackleton 1994).

Isotopic stage 4 coincides with severe aridity between 73 and 55 ka.

During stage 3, two relatively warm-wet episodes are recorded between 55 and 24 ka (An and Xiao 1990, Dong et al. 1990, Huang 1989, Li J. et al. 1990, Wu et al. 1991). Large paleolakes have just been discovered in the Tenga desert of Inner Mongolia, one of the humid phases dated 30-26 ka (Pachur et al., in prep.).

The Last Glacial Maximum is characterized by cold and dry conditions between 20 and 13 ka, as evidenced by periglacial features (Li R. 1990), pollen (Kong and Du 1980) and loess sequences (An et al. 1990). High salinity values in Daihai lake (Inner Mongolia) indicate, between 20 and 14.5 ka, a marked deterioration of the Precipitation / Evaporation ratio (Wang and Feng 1991). Abrupt disappearance of arboreal pollen takes place at 20-18 ka, temperature being estimated by Sun (1989) 12°C lower than at present, due to the longer persistence of winter conditions induced by cold polar air masses. The important increase of continentality related to the major sea-level regression certainly added to the climatic degradation : the Chinese coastline was, during the Last Glacial Maximum, 800 to 1 000 km

East of its present-day location, thus reducing further the impact of the East Asian monsoon over northern China. (An et al. 1991).

The first step of deglaciation (Duplessy et al. 1986) is marked at 14.5 ka by a sharp decrease of Lake Daihai salinity. In the Beijing area, pollen records indicate a warmer-wetter episode at about 12 ka (Kong Z. et al. 1982). A decrease of magnetic susceptibility in the loess series indicates a climatic deterioration at 11 ka, corresponding with the onset of the Younger Dryas (An et al. 1990).

The Holocene is documented by about one hundred radiocarbon dates (Shi et al. 1993). A humid phase initiates between 9.5 and 9 ka. Eolian sedimentation stops, paleosols slowly build (they are dated 7.5 ka). The optimum is dated 8.6 to 6.5 ka (Zhou et al. 1989, Zhang and Hu 1989, Li J. et al. 1990, An et al. 1990, Li B.et al. 1991), and 12-8.7 ka in the Tenga desert, Inner Mongolia (Pachur et al., in prep.). A short cooler-drier episode then interrupts the humid phase but favourable conditions again occur between 5 and 3 ka, as registered both in geological records (Li and Liang 1985, Huang 1990, Wang S. et al. 1990, Li B. et al. 1991) and in ancient chinese archives (Yan et al. 1992) giving a calendar age of 4 000 to 3 000 years ago to a temperature increase (1.5°C higher than present-day values). At 3 ka, temperature and precipitation abruptly decrease and a progressive reduction of the monsoonal rains upon the transitional zone marks the neoglacial trend.

3. The Indian summer monsoon

The environmental and climatic changes related to the large scale variability of the Indian monsoon are recorded in NW India and Tibet, the Arabian Peninsula and East Africa , by a still limited number of continental observations and by sedimentation in the Gulf of Bengal, the Arabian Sea and the Mediterranean Sea off the Nile delta.

In northeastern India, few precise data are presently available for the last interglacial and the subsequent interphase. A succession of wet / dry phases took place between 170 and 26 ka, as evidenced by an undated succession of calcareous layers, paleosols and colluvial washes into consolidated dune formations (Pant 1993). During the arid phases, the Thar desert extended 350 km to the South and East of its present-day location, into zones now receiving up to 900 mm/yr (Allchin et al. 1976). During an undetermined period of stage 5, a humid phase is registered by Didwana lake (Courty 1990). Ocean records suggest a strengthening of the summer monsoons during stage 5e (Clemens et al. 1991, Rostek et al. 1993).

During isotopic stage 3, an increase of river run-off is also registered in oceanic sedimentation at 55-30 ka (Clemens et al. 1991).

During the last glacial, a severe arid phase is dated 20-13 ka (Singhvi et al. 1982, Wasson et al. 1983, 1984), a peak of eolian sedimentation being TL dated at 14 ka (Chawla et al. 1992). Oxygen isotope records from the Indian Ocean and the Arabian Sea confirm a weakening of the summer monsoon during the Last Glacial Maximum, together with an increase of the NE winds and of the ocean salinity, due to a drastic reduction of run off from the Himalayan rivers (Prell et al. 1980, Cullen 1981, Fontugne and Duplessy 1986, Van Campo 1986, Prell and Van Campo 1986, Rostek et al. 1993, Sirocko et al. 1993).

The first effects of deglaciation are recorded at 12.8 ka, followed by a short arid episode during the Younger Dryas (Wasson et al. 1983, 1984).

The Holocene phase is characterized by an hydrological optimum dated 9.3-7.4 ka to 6 ka at Lake Didwana, followed by an increase of aridity with a new wetter episode peaking at 4.2 ka (Wasson et al. 1984). Recent work confirm permanent freshwater conditions between *c.* 9 and 6 ka (Singh et al. 1990). At 3 ka, arid conditions are again established, rapid dune building takes place (Singhvi et al. 1990, Chawla et al. 1992).

The records from the **Tibetan Plateau** match the evolution in NW India with a warm-wet phase at 123-90 (Huang C.C. 1990), high lake levels at 40-24 ka (Zheng 1990, Chen K. et al. 1990, Li B. et al. 1991), a cold-dry maximum between 22 and 15 ka, characterized by lake salinization, widespread glacial advances and vegetation degradation (Pan and Xu 1989, Chen K. et al. 1990, Hu and Chen K. 1990, Huang C.X. et al. 1983). Sedimentological studies of lake deposits date the duration of this dry stage at 25 to 9 ka, which reflects the retardation of the impact of the monsoon on the Plateau, relative to the lowland area. Gasse et al. (1991) observed a warm-wet pulse at 12.5 ka and a severe dry-cold episode during the 11-10 ka interval.

Holocene warm conditions initiated at c. 10 ka, as evidenced by the development of paleosols and forest (Kong K. et al. 1990, Chen K. et al. 1990). The beginning of the optimum is given at 7.5 ka or a little later (Huang et al. 1983, Wang 1985, Chen K. et al. 1990, Pan and Xu 1989). A mid-Holocene cold-dry episode is also recognized at 5.5 ka from pollen and lake sediments analysis (Wang 1985, Gasse et al. 1991). Geochemical and glaciological analyses (Chen K. et al. 1990) and pollen records (Kong et al. 1990) indicate the onset of neoglacial patterns at *c.* 3.5 ka, which fits the observations of a low organic carbon stage marking the end of the optimum at 3.4 ka (Gasse et al. 1991). Vegetational changes (Huang et al. 1983), lowering of lakes level (Wang 1985, Li B. et al. 1990) and glaciers advance in SE Tibet (Wang 1985, Li J. et al. 1986) are dated at about 3 ka.

The climatic changes over **the Arabian Peninsula, Sudan and Ethiopia** are recorded both in continental and marine sediments.

In Arabia, the last interglacial humid period is uncertainly dated 142 to 108 ka (Edgell 1989, Sanlaville 1992). Klein et al. (1990) describe monsoonal rains up to the Sinaï. In Oman, alteration and pedogenesis are active during stage 5e (Perthuisot 1980, Sanlaville 1992, Hoang C.H. et al. 1993). Along the coasts of Sudan and the Red Sea, stage 5e coral reefs are interbedded with sediments brought by river run-off. The three warmer peaks in oceanic stage 5 are documented by rich sapropel layers off the Nile delta, bearing evidence of enhanced tropical rains over Ethiopia and East Africa (Rossignol-Strick et al. 1982, Petit-Maire et al. 1991). During stages 5c and 5a, the sapropels off the Nile delta are again well developed, which fits Rossignol-Strick monsoon-index and the model for precipitation over East Africa and southern Asia by Prell and Kutzbach (1987).

The beginning of stage 4 is marked in coastal Sudan and Oman by active erosion and progressive aridification (Sanlaville 1992).

Stage 3 is marked, in Arabia, by plentiful proxy data (lake deposits, paleosols, fluvial terraces, hippopotamus bones) bringing evidence of two wet phases in the Rub-al-Khali, the center of Arabia, the Nefud and the region of Wadi Sirhan, at the latitude of the Sinaï, around 45 ka and between 30 and 21 ka (McClure 1976, 1988, Schulz and Whitney 1986, Sanlaville 1992). However, the monsoon index is weak on the Arabian sea, where SSTs are much colder than nowadays (Kroon et al. 1990). Fresh water inputs from the Nile into the Mediterranean are reduced, as shown by the low ratio of organic carbon off the Delta. However, at 25 ka, a slight rise in *G. ruber*'s frequency is significant (Petit-Maire et al. 1991) and active braided channels flow in the Delta (Stanley and Warne 1993).

During the Last Glacial, the area is very arid. Large dunes build in the Rub-al-Khali and the Nefud (Sanlaville 1992). Important dust fluxes reach the Arabian Sea (Shimmield et al. 1990), the peak of aridity being dated 22 to 18 ka. The volume of the White Nile is low. Evaporites are dated 20 to 12.5 ka in central Sudan (Adamson et al. 1982) and in the Nile Delta (Foucault and Stanley 1989).

The first effects of deglaciation result in a short wetter episode around 14-12 ka in Arabia (Sanlaville 1992). The *G. ruber* record off the Nile increased very rapidly between 14.4 and 11.7 ka (Petit-Maire et al. 1991), the river volume being important from 12.5 ka on (Adamson et al. 1982).

The Younger Dryas is evidenced from 11.7 to 10 ka by the *G. ruber* record and the increase in the Red Sea salinity (Almogi-Labin et al. 1991).

The Holocene is well documented. In Arabia, the humid optimum is dated 9 to 6 ka (McClure 1976, Al Sayyari and Zotl 1978, Jado and Zotl 1984, Schulz and Whitney 1986) up to the Nefud, although most of the geological evidence is located in the Southern and central areas of the Peninsula (Sanlaville 1992), as logical when dealing with a monsoon pattern. Surprisingly, in contrast with evidence from Africa, Asia, and with Global Changes, this Holocene wet phase seems to have been less important than the one in stage 3. After 6 ka, the climatic deterioration has a deep impact on the environment (Edgell 1989), although neolithic sites exist in the present-day desert until 3.5 ka, testifying to still accessible fresh-water resources. At 3 ka, aridification severly marks the neoglacial trend (Pott 1990, Fedele 1990). In central Sudan, large Nile-related swamps extend between 12.5 and 5 ka (Adamson et al. 1982). *G. ruber*'s development is maximal at 8 ka and sapropel deposits at 9-8 ka (Rossignol-Strick et al. 1982).

4. The African monsoon and the Saharo - Sahelian belt.

The existence of several wet episodes in the present-day hyperarid Sahara during the last interglacial is documented by U/Th dated records of palaeolakes in central Libya, northern Mali and the Egyptian desert. The most important record is the one observed at 27°N, in Central Libya, where the Shati palaeolake covered 2 000 km2 and contained 20 km3 of brackish water in an area presently receiving 30 mm mean annual rainfall (Petit-Maire et al. 1978, 1980, 1982). Although located to the North of the Tropic, it may be associated with monsoonal rains or tropical depressions, since it was permanent, i.e. fed by both winter and summer precipitation (Petit-Maire 1989). The extensive coquina deposits (with fresh water molluscs at the mouth of the incoming wadis and a brackish *Cardium* fauna in the opposite end of the paleolake) provided 22 U/Th ages peaking at 130-125 ka, and also indicating high levels at 92, 89 and 77 ka (Gaven et al. 1981). Despite the uncertainty of the dates (± 2 000 yrs), it seems probable that the three episodes in stage 5 are represented (or some of the numerous oscillations recorded in GRIP results). More detailed research is needed in this unique place. In northern Mali, at 23°N, thick lacustrine limestones with fresh-water molluscs have been observed at +25 m above the present-day floor of the Sbeita depression (Fabre 1983, Fabre and Petit-Maire 1983) and dated 123.1 ± 3.8 ka , 125.7 ± 4 ka and 133.1 ± 4.8 ka (Petit-Maire and Reyss, in prep). In the western desert of Egypt, at present hyperarid, stage 5 is documented by some uncertain U/Th and ESR ages at 121 ka (Szabo et al. 1989), 130 ka and 104-70 ka from calcretes and faunal remains (Close A. 1990, Wendorf et al. 1990, Miller et al. 1991). Along the northern margins of the Sahara , a hydrological optimum is dated at c. 90 ka in the Tunisian chotts (Causse et al. 1988).

No evidence of humid environments is found between 70 and 45 ka (stage 4).

Two humid episodes characterize stage 3 : one around 45-40 ka in Egypt (Szabo et al. 1989), 40 ka at the Shati lake in Libya (Petit-Maire et al. 1980), 46 and 41 ka in northern Mali (Morel et al. 1991), the other between 35 and 21 ka throughout the Sahara and Sahel (Durand 1982, Pachur and Röper 1984, Close A. 1988, Pachur et al. 1987, 1990, Petit-Maire 1989, Kröpelin 1993). The development of this last episode matches the existence of the Aterian civilisation (Tillet 1989), human presence being unlikely without availability of surface water, vegetal food and game. It is interesting to note that, in contrast with the observations in the Arabian Peninsula, those episodes in stage 3 were characterized by widespread swampy areas, related to aquifer rise, rather than by lakes as during stage 5 and the Holocene.

The Last Glacial Maximum induced a drastic reduction of the monsoonal rains over the presently semi-arid sahelian area. The 100 mm isohyet, nowadays at 17°N, shifted southwards to 13°N, the *Acacia* grassland having regressed as far as 10°N (Sarnthein et al. 1981, Rossignol-Strick and Duzer 1980, Talbot 1984, Lézine 1989, Petit-Maire 1989). This major climatic change thus resulted into the aridification of some 2 500 000 km2, relative to the present-day situation (and, as we shall see, of 6 500 000 km2 relative to the Holocene optimum !). Organized dune fields belonging to this arid phase are quite visible under the Sahelian vegetal cover, on satellite imagery. They bring evidence that the direction of the dominant winds from the North East was then similar to the present-day one (Petit-Maire and Page 1992). Those formations, now consolidated , lying under the present active dunes, were TL dated 18-9 ka at 23°N, in northern Mali.

Consequences of the first deglaciation step (Duplessy et al. 1986) only appear in the mountainous areas of Hoggar and Tibesti, where active run off is recorded at 14 ka (Faure 1969, Hagedorn and Pachur 1969, Nicholson and Flohn 1980), and along the oceanic margins of the desert, where paleosols are dated between 15 ka and 12.5 ka (Petit-Maire 1986, Petit-Maire et al. 1987). The pollen diagrams for the present Sahel (Senegal and Chad) show the reappearance of the wooded grassland between 16° and 14°N at 13-12 ka (Lézine 1989, Servant and Servant-Vildary 1980).

The Younger Dryas episode is well registered in the eastern Canary Islands, where eolian sedimentation interrupts the pedogenesis between 12.5 and 9.8 ka (Petit-Maire 1987).

About 3 000 radiocarbon dates document the Holocene wet phase from multi-proxy evidence (Commelin and Petit-Maire 1979, Vernet and Aumassip 1992, Petit-Maire 1993), synthetized in a recent paleomap (Petit-Maire et al. 1993). Two continuous palaeolacustrine records at 23°N, in the hyperarid core of western Sahara (Fabre and Petit-Maire 1988; Petit-Maire ed. 1991), provide an excellent chronology for the climatic evolution of this vast area during the Holocene. From the Nile to the Atlantic, the pattern is strikingly similar, the dates for the beginning and end of the humid phase varying by about 1 kyr according to latitude and local conditions. Precipitation progressively increased from the South to the North and then decreased progressively from the Tropic to the South, implying a classical pattern of dominant monsoonal regime. For the presently hyperarid area in the Saharan basins, with less than 20 mm mean annual rainfall and at least 5 m potential evaporation, (which represents extreme unfavourable geographical situations of continentality, low altitude and hydrological isolation), the general evolution is as follows (Eastern Sahara always remaining, as at present, a little more arid than western Sahara; Close A. 1992, Kröpelin 1993) :

- c. 10 to 9 ka : change in Precipitation / Evaporation ratio. Seasonal lakes or swamps appear in the topographical depressions.

- c. 8.5 to 7 ka : climatic optimum. Permanent fresh water lakes. Extensive swamps (Pachur and Kröpelin 1987; Ritchie and Haynes 1987; Petit-Maire ed., 1991; Pachur and Hoelzmann 1991; Kröpelin and Soulié-Märsche 1991). Sahelian vegetation and large mammalian fauna up to 22°N (Guérin and Faure 1983; Neumann 1990; Schulz 1991). Paleoprecipitation is estimated to be 50 times the present values (Haynes and Mead 1987; Petit-Maire 1989; Petit-Maire and Kröpelin 1991).The wide extension of surface water is striking during this optimum. Brown soils are evidence of a grassland cover. The Saharo-Sahelian boundary is located at 22°-23°N, i.e.some 550 km to the North of its present position (17°N) and some 1000 km to the North of its glacial maximum position (13°N).

- After 7 ka, a period of higher climatic variability is recorded both in the eastern (Pachur and Hoelzmann 1991) and western Sahara (Fabre and Petit-Maire 1988, Petit-Maire ed. 1991), as well as in the Sahel (Servant and Servant-Vildary 1980) : it is characterized by important drought features, salinization episodes in the lacustrine series and also by a decrease of temperature (Goetz and Erlenkeuser 1991). The ages of this episode vary according to latitude and geological location but it seems to have lasted about 1 kyr, roughly between c. 6.7 or 6.3 and c. 5.5 ka It was followed by a second but less humid episode. Sahelian plants and animal species migrated progressively to the South.

- 4.5 ka : both hydrological and palynological data show an abrupt decline of rainfall.

- 4.5 to 3 ka : according to latitude, disappearance of surface water and associated biotopes. Eolian erosion and dune accumulation begin. The Saharo-Sahelian boundary shifts back southwards to its present day location at 17°N, which matches the neoglacial trend.

5. General results and relationships with past Global Changes

5.1 Comparison, at wide spatial and time scales (in order to suppress the local differences due to geographical diversity), of the data from geological records observed along the arid/semi-arid zone of northern Africa and Asia disclose roughly common patterns :

The Last Interglacial is associated everywhere with a more humid climate, dates covering a wide time span, between 141 ka and 70 ka. The wet peak is at 130-125 ka, with evidence for drastic hydrological and environmental changes. Although often uncertain, the U/Th dates suggest that the beginning of the wet phase rather fits the recent Vostok chronology (transition 2 at 133 ka : Jouzel et al. 1993), rather than the classic oceanic one (transition 2 at 128 ka).

Stage 4 is arid throughout the area, which matches the sharp dust flux in the Arabian sea and in the Vostok cores (Clemens and Prell, 1990; Jouzel et al. 1993; fig. 1).

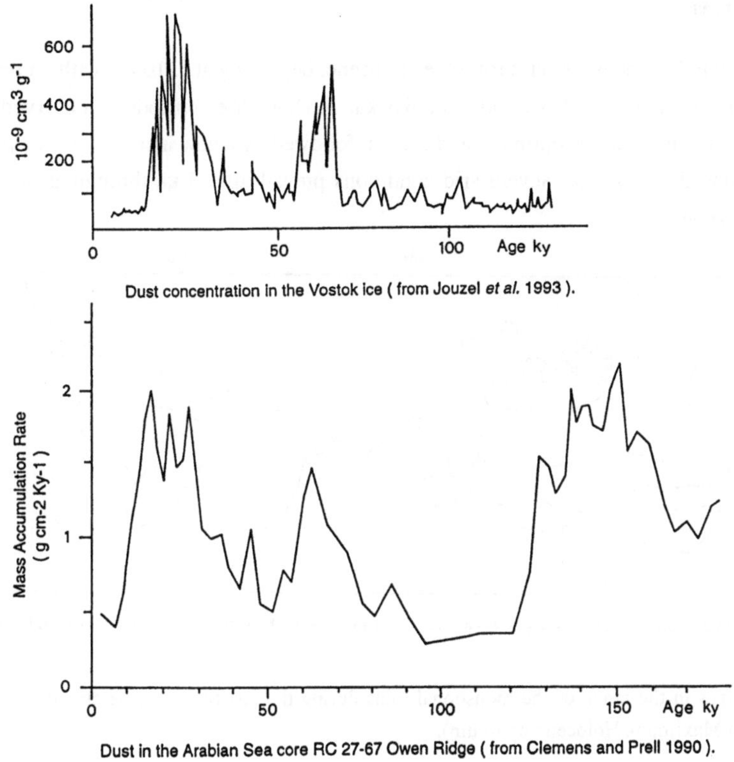

Fig.1 : Dust input into the Arabian Sea (from Clemens and Prell, 1990) and in Vostok cores (Jouzel et al. 1993).

During isotopic stage 3, two humid episodes are recorded between 55 and 42 ka and between 38 and 21 ka. They are minor relative to stage 5 and the Holocene, except in the Arabian Peninsula where they seem to have been quite marked.

This discrepancy does not contradict the dust record from the Arabian sea, since dust is brought there from India and China, during the arid phases, by the enhanced North East winds.

A severe increase of aridity (very sharp in the dust records) is associated everywhere, between 21 ka and 15 ka, with the last glacial phase. The monsoonal margins withdrew southwards in a spectacular way : the semi-arid zone in Asia and the desert in northern Africa spread to the South of their present location by 900 km in Eastern China, 350 km in northwestern India and 400 km in northern Africa (fig. 2).

The first deglaciation step is only documented in the most sensitive geographical areas by a return to wetter conditions, followed by a new onset of dry conditions during the Younger Dryas.

The whole zone is wetter during the Holocene, beginning at c. 10 ka, with an optimum generally given between 8.5 ka and 7 or 6.5 ka. A short drier episode is observed, dates varying locally, between the optimum and c. 5 ka, followed by a new increase of precipitation, ending around 3.5 ± 0.5 ka. Severe arid conditions prevail after 3 ka throughout the whole transitional zone.

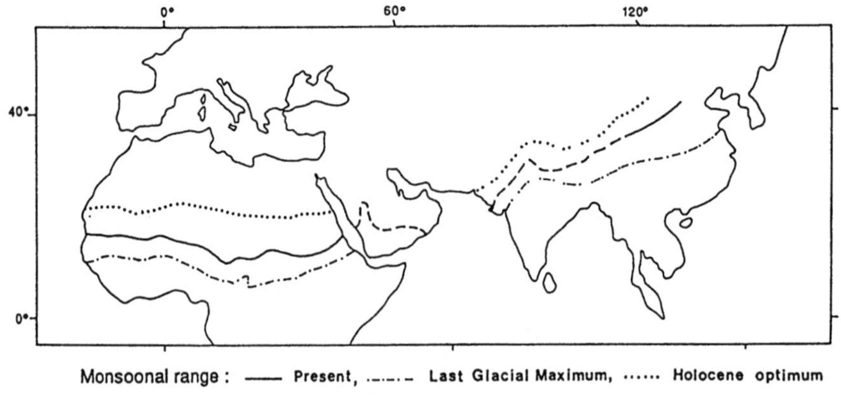

Fig.2 : Changes in the range of the monsoonal rains during the last two extreme climatic scenarios (Last Glacial Maximum, Holocene optimum).

5.2 Between the Atlantic and the coast of China, the transitional zone which underwent these humid / arid alternations covers some 14 million km² (fig. 2), which strongly comforts the proposal that the variations of the methane atmospheric ratio during the last climatic cycle were partly related with the major environmental changes in the tropical transitional zone (Chappellaz et al. 1990; Petit-Maire et al. 1991). Strong support is given to this hypothesis by the fact that the alternations recorded in this paper fit the CH_4 curve more closely than the temperature and CO_2 ones (fig. 3), in particular for isotopic stage 3 when large methanogenic swampy areas developed.

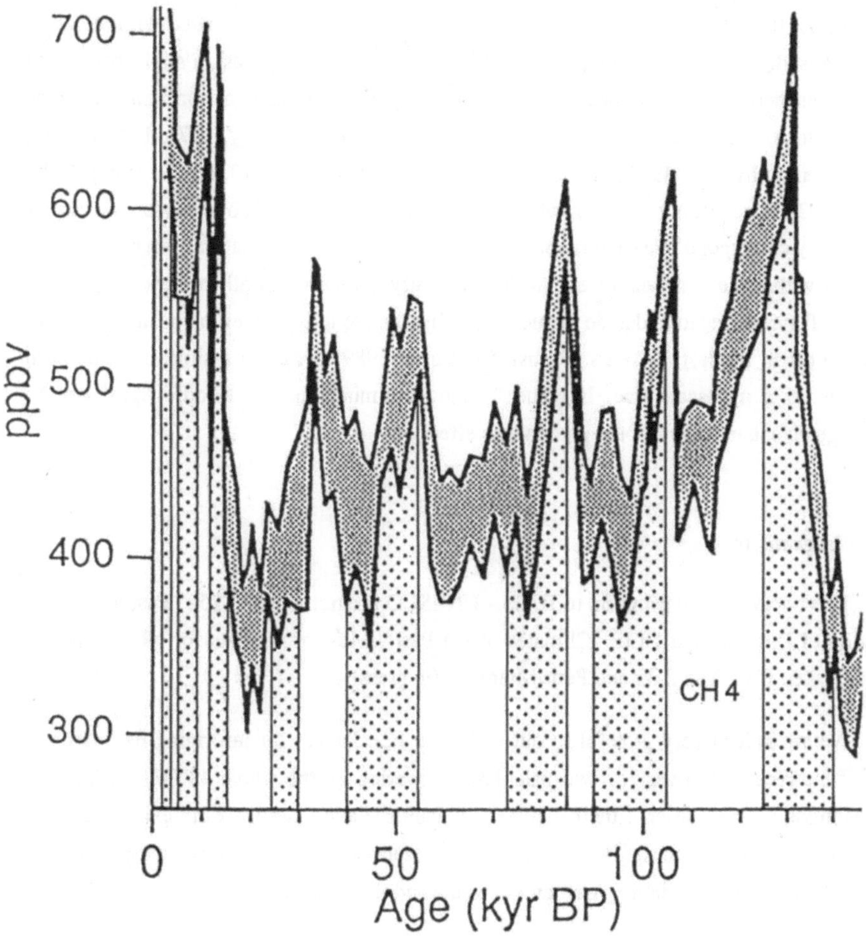

Fig.3 : Wet periods in the transitional zone of Asia and Africa, related to the variations of the CH_4 atmospheric ratio registered at Vostok (Jouzel et al. 1993).

5.3 Those large scale changes in the range of the Asian, Indian and African monsoons over the last 130 ka occurred roughly in phase and match the Global temperature changes and the astronomical forcings recorded in ice and deep-sea cores. *Global cold phases correspond to regional increases of aridity. Global warm phases correspond to regional increases of humidity.* The environmental changes toward the humid or arid scenarios have taken place according to a latitudinal gradient typical of monsoonal patterns.

This close relationship established between monsoonal activity and global temperature is possibly effective even at short term. Such a correlation has been observed in the recent past (without having any valid statistical significance) : a warming of the northern Hemisphere during the 1920s (Jones et al. 1986 , Goossens and Berger, 1987) is associated with the onset of a wet episode in northern China, India and the Sahel while, in contrast, a slight cooling of the N. Hemisphere in the 1960s was shortly followed by sharp droughts in northern China, the Tibetan plateau, India, the Arabian Peninsula and the Sahel (Wistanley, 1976; Lin and Wu, 1986, Parthasarathy et al. 1986, Zhang 1989, Demarée and Nicolis 1990, Fu and Wang 1991, Janicot 1993). One cannot yet say whether the very recent return to a somewhat wetter situation may correspond to a (natural or man-made) warming or to a (natural)very short-term oscillation within the neoglacial trend which is clearly evidenced in all the records since at least 3 ka. Time is needed to decide of such a question and to tell whether the positive trend of global run-off in the last 80 years (Probst and Tardy, 1989) is due to a short natural warm oscillation, to a mid-scale one, like the "Viking optimum" in the Middle-Ages, or to anthropogenic enhancement of the greenhouse effect.

Acknowledgements :

This paper is a contribution to IUGS - UNESCO programs IGCP 252 "Past and future Evolution of Deserts" and CLIP "Climates of the Past". Most of the data for NE China and Tibet are taken from Yan ZW. and Petit-Maire N., (in press).

We are indeed very grateful to our colleague H.J. Pachur for informing us, march 3, 1994, of the recent discovery of two paleolakes, covering on the whole 48 000 km², in the Inner Mongolia Tenga desert : their precise radiocarbon dating confirms the existence of an important holocene wet phase, but also that of an important episode during isotopic stage 3 (30-26 ka) matching those in the Arabian Peninsula and the Sahara.

REFERENCES

Adamson, D.A., R. Gillespie and M.A.J. WIlliams, 1982: Palaeogeography of the lower Blue and White Nile Valleys. In: Williams, M.A.J. and Adamson, D.A. (eds.), A land between two Niles, 165-219. Balkema, Rotterdam (NL).

Al-Sayyari, S.S. and J.G. Zötl (eds.), 1978: Quaternary period in Saudi Arabia, 1. Springer, Wien (OS), 336 p.

Allchin, B., A. Goudie and K.T.M. Hedge, 1976: The prehistory and palaeogeography of the Great Indian Desert. , 370 p.

Almogi-Labin, A., C. Hemleben, D. Meischner and H. Erlenkeuser, 1991: Palaeoenvironmental events during the last 13 000 years in the central Red Sea as recorded by Pteropods. Palaeoceanography, 6, 83-98.

An, Z., G.J. Kukla, S.C. Porter and J. Xiao, 1991: Magnetic susceptibility evidence of monsoon variation on the loess plateau of central China during the last 130000 years. Quaternary Research, 36, 29-36.

An, Z.S. and J.L. Xiao, 1990: A case study of dust deposit flux over the Loess Plateau (in Chinese). Kexue Tongbao, Science Bulletin, 35, 220-222.

Causse, C., G. Conrad, J.C. Fontes, F. Gasse, E. Gibert and A. Kassir, 1988: Le dernier 'humide' pléistocène du Sahara nord-occidental daterait de 80-100000 ans. Comptes Rendus de l'Académie des Sciences, Paris (2), 306, 1459-1464.

Chappelaz, J., J.M. Barnola, D. Raynaud, Y.S. Korotkevich and C. Lorius, 1990: Ice record of atmospheric methane over the past 160 000 years. Nature, 345, 127-131.

Chawla, S., R.P. Dhir and A.K. Singhvi, 1992: Thermoluminescence chronology of sand profiles in the Thar desert and their implications. Quaternary Science Reviews, 11, 25-32.

Chen, K.Z., J.M. Bowler and K. Kelts, 1990: Palaeoclimatic evolution within the Qinhai-Xizang (Tibet) Plateau in the last 40 000 years (in Chinese). Quaternary Sciences, 1, 21-31.

Clemens, S. and W. Prell, 1990: Late Pleistocene variability of Arabian Sea summer monsoons winds and continental aridity: eolian records from the lithogenic component of deep-sea sediments. Paleoceanography, 5, 109-145.

Clemens, S., W. Prell, D. Murray, G. Shimmield and G. Weedon, 1991: Forcing mechanisms of the Indian Monsoon. Nature, 353, 720-725.

Close, A.E., 1988: Current research and recent radiocarbon dates from northern Africa. Journal of African History, 29, 145-176.

Close, A.E., 1992: Holocene occupation of the Eastern Sahara. Africa praehistorica, 5, 137-183.

Close, A.E., F. Wendorf and R. Schild, 1990: Patterned use of a middle palaeolithic landscape: Bir Tarfawi and Bir Sahara east, eastern Sahara. Sahara, 3, 21-34.

Commelin, D., N. Petit-Maire and J. Casanova, 1979: Chronologie isotopique saharienne pour les derniers 10 000 ans. Essai d'interprétation: migrations humaines et paléoclimatologie. Bulletin du Musée d'Anthropologie préhistorique de Monaco, 23, 37-88.

Courty, M.A., 1990: Environnements géologiques dans le nord-ouest de l'Inde. Contraintes géodynamiques au peuplement protohistorique. (Bassins de la Ghaggar - Saraswati - Chautang). Université de Bordeaux 1 (FR), Thèse Doctorat Etat Sciences, 328p.

Cullen, J.L., 1981: Microfossil evidence for changing salinity patterns in the Bay of Bengal over the last 20 000 yrs. Palaeogeography, Palaeoclimatology, Palaeoecology, 35, 31-356.

Demarée, G.R. and C. Nicolis, 1990: Onset of Sahelian drought viewed as a fluctuation induced transition. Quarterly Journal of the royal meteorological Society, 16, 221-238.

Ding, Z.L., T. Liu, X.M. Liu, M.Y. Chen and Z.S. An, 1990: Thirty-seven climatic cycles in the 2.5 Ma. Chinese Science Bulletin, 34, 1494-1496.

Dong, G.R., J. Jin, S.Y. Gao, B.S. Li and H.Z. Cheng, 1990: Climatic changes since the late Pleistocene in deserts of northern China (in Chinese). Quaternary Sciences, 3, 213-222.

Duplessy, J.C., M. Arnold, P. Maurice, E. Bard, J. Duprat and J. Moyes, 1986: Direct dating of the oxygene - isotope record of the last deglaciation by ^{14}C accelerator mass spectrometry. Nature, 320, 350-352.

Durand, A., 1982: Oscillations of lake Chad over the past 50000 years: New data and new hypothesis. Palaeogeography, Palaeoclimatology, Palaeoecology, 39, 37-53.

Edgell, H.S., 1989: Evolution of the Rub' al Khali Desert. Earth Science, 3, 109-126.

Fabre, J., 1983: Le Sahara malien septentrional (22°-24° N): Taoudenni - Trhaza. Esquisse stratigraphique préliminaire des dépôts lacustres quaternaires. In: Petit-Maire, N. and Riser, J. (eds.), Sahara ou Sahel ? Quaternaire récent du Bassin de Taoudenni (Mali), 421-441. Imprimerie Lamy, Marseille (FR).

Fabre, J. and N. Petit-Maire, 1983: Lacs pléistocènes de la région de Taoudenni (Sahara malien). In: Société géologique de France (ed.), Bassins sédimentaires en Afrique. Résumés des communications, 99-100. Laboratoire des Sciences de la Terre - Faculté des Sciences St-Jérôme, Marseille (FR).

Fabre, J. and N. Petit-Maire, 1988: Holocene climatic evolution at 22-23° N from two palaeolakes in the Taoudenni area (northern Mali). Palaeogeography, Palaeoclimatology, Palaeoecology, 65, 133-148.

Faure, H., 1969: Lacs quaternaires du Sahara. Mitt. Int. Ver. Limnologie, Stuttgart, 17, 131-146.

Fedele, F.G., 1990: Man, land and Climate: emerging interactions from the Holocene of the Yemen Highlands. In: Bottema, S. et al. (eds.), Man's role in the shaping of the eastern Mediterranean landscape, 31-42. Balkema, Rotterdam (NL).

Fontugne, M.R. and J.C. Duplessy, 1986: Variations of the monsoon regime during the upper Quaternary : evidence from carbon isotopic record of organic matter in North Indian sediments cores. Palaeogeography, Palaeoclimatology, Palaeoecology, 56, 69-88.

Foulcault, A. and D.J. Stanley, 1989: Late Quaternary palaeoclimatic oscillations in East Africa recorded by heavy minerals in the Nile delta. Nature, 339, 44-46.

Fu, C.B. and Q. Wang, 1991: The synchroneousness of the 1920s' warming and the strengthening of the south Asian monsoon activity (in Chinese). Science in China B, 6.

Gasse, F., M. Arnold, J.C. Fontes, M. Fort, E. Gilbert, A. Huc, B.Y. Li, Y.F. Li, Q. Liu, F. Mélières, E. Van Campo, F.B. Wang and Q.S. Zhang, 1991: A 13 000-year climate record from western Tibet. Nature, 353, 742-745.

Gaven, C., C. Hillaire-Marcel and N. Petit-Maire, 1981: A Pleistocene lacustrine episode in southeastern Libya. Nature, 290, 131-133.

Goetz, C. and H. Erlenkeuser, 1991: Les isotopes stables de l'oxygène et du carbone. In: Petit-Maire, N. (ed.), Paléoenvironnements du Sahara. Lacs holocènes à Taoudenni (Mali), 103-106. Centre régional de Publication - CNRS, Marseille / Paris (FR).

Goossens, C. and A.L. Berger, 1987: How to recognize an abrupt climatic change. In: Berger, W.H. and Labeyrie, L.D. (eds.), Abrupt Climatic Change, 31-45. Reidel, Dordrecht (NL).

Greenland Ice-core Project (GRIP) Members. 1993: Climate instability during the last interglacial period recorded in the GRIP ice core. Nature, 364, 203-207.

Guérin, C. and M. Faure, 1983: Mammifères. In: Petit-Maire, N. and Riser, J. (eds.), Sahara ou Sahel ? Quaternaire récent du Bassin de Taoudenni (Mali), 239-272. Imprimerie Lamy, Marseille (FR).

Hagedorn, H. and H. Pachur, 1969: Observations on Climatic Geomorphology and Quaternary Evolution of Landforms in South Central Libya. 1st Symposium on the Geology of Libya 1967, 387-400.

Haynes, C.V., Jr and A.R. Mead, 1987: Radiocarbon dating and paleoclimatic significance of subfossil Limicolaria in northwestern Sudan. Quaternary Research, 28, 86-99.

Hoang, C.H., R. Dalongeville and P. Sanlaville, 1993: Uranium-series ages of corals from thc Red Sea coasts of Sudan. Clip Meeting, Barbados, sept 1993. Abstracts, 16-.

Hu, D.S. and K.Z. Chen, 1990: Recent changes of Qarhan salt lake, Tibetan Plateau (in Chinese). Kexue Tongbao, Science Bulletin, 35, 1888-1891.

Huang, C.C., 1989: Loess, paleosols and climatic changes in southern Loess Plateau during the late Pleistocene (in Chinese). Acta Geographica Sinica, 44, 1-10.

Huang, C.C., 1990: Holocene deposits and environmental changes in the plateaus and basins of northwestern China (in Chinese). Arid Land Geography, 13, 1-8.

Huang, C.X., Y.R. Wang and Y.L. Liang, 1983: Environmental changes in southern Tibet during the Holocene from pollen analysis (in Chinese). In: Series of the Scientific expedition to the Qinghai-Xizang Plateau, Quaternary geology in Xizang, 179-192. Science Press, Beijing (CN).

Jado, A.R. and J.G. Zötl (eds.), 1984: Quaternary period in Saudi Arabia, 2. Springer, Wien (OS), 361 p.

Janicot, S. and B. Fontaine, 1993: L'évolution des idées sur la variabilité interannuelle récente des précipitations en Afrique de l'Ouest. La Météorologie, 8, 28-53.

Jones, P.D., S.C.B. Raper, R.S. Bradley, H.F. Diaz, P.M. Kelly and T.M. Wigley, 1986: Northern Hemisphere surface air temperature variation: 1851-1984. Journal of Climate and Applied Meteorology, 25, 161-179.

Jouzel, J., N.I. Barkov, J.M. Barnola, M. Bender, J. Chappellaz, C. Gentho, V.M. Kotlyakov, V. Lipenkiv, C. Lorius, J.R. Petit, D. Raynaud, G. Raisbeck, C. Ritz, T. Sowers, M. Stievenard, F. Yiou and P. Yiou, 1993: Extending the Vostok ice-core record of paleoclimate to the penultimate glacial period. Nature, 364, 407-412.

Kemp, R., E. Derbyshire and X. Meng, 1994: Pedosedimentary reconstruction of a thick loess-palaeosol sequence near Lanzhou in north-central China. In: Derbyshire, E. (ed.), Wind blown sediments in the Quaternary record. Abstracts of papers and posters, Geography Department - University of London (UK).

Klein, R., Y. Loya, G. Gvirtzman, P. Isdale and M. Susic, 1990: Seasonal rainfall in the Sinai Desert during the Late Quaternary inferred from fluorescent bands in fossil corals. Nature, 345, 145-147.

Kong, Z.C., N. Du, S.F. Shan, G.B. Tong, S.J. Luo and S.X. Fan, 1990: Vegetational and climatic changes in the last 11000 years in Qinghai lake. Numerical analysis based on palynology in core QH85-14C (in Chinese). Marine Geology & Quaternary Geology, 10, 79-90.

Kong, Z.C. and N.Q. Du, 1980: Vegetational and climatic changes in Beijing in the past 30000-10000 years (in Chinese). Acta Botanica Sinica, 22, 328-338.

Kong, Z.C., N.Q. Du and Z.B. Zhang, 1982: Vegetational development and climatic changes in the last 10000 years in Beijing (in Chinese). Acta Botanica Sinica, 24, 172-181.

Kroon, D., K. Beets, S. Mowbray, G. Shimmield and T. Steens, 1991: Changes in northern Indian Ocean monsoonal wind activity during the last 500 ka. Memoirs of the geological Society of India, 44, 189-207.

Kröpelin, S., 1993: Geomorphology, landscape evolution and paleoclimates of southwest Egypt. Catena Supplement, 26, 31-65.

Kröpelin, S. and H.J. Pachur, 1991: 500 mm annual rainfall in SW Egypt during the Middle Paleolithic. Some remarks from the geoscientific point of view. Sahara, 4, 158-161.

Kukla, G., 1987: Loess stratigraphy in central China. Quaternary Science Reviews, 6, 191-219.

Kukla, G., 1989: Long continental records of climate - an introduction. Palaeogeography, Palaeoclimatology, Palaeoecology, 72, 1-9.

Kukla, G., F. Heller, Liu, X., Xu, T., Liu, T. and An, Z., 1988: Pleistocene climates in China dated by magnetic susceptilibity. Geology, 16, 811-814.

Lézine, A.M., 1989: Le Sahel: 20 000 ans d'histoire de la végétation. Bulletin de la Société géologique de France, 8, 35-42.

Li, B.Y., Q.S. Zhang and F.B. Wang, 1991: Evolution of lakes in the Karakorum-west Kunlun mountains (in Chinese). Quaternary Sciences, 1, 64-71.

Li, J.J., B.X. Zheng, X.J. Yang, Y.Q. Xie, L.Y. Zhang, Z.H. Ma and S.Y. Xu, 1986: Glaciers in Tibet (in Chinese). Science Press, Beijing (CN), 328 p.

Li, J.J., J.J. Zhu, J.C. Kang, X.M. Fang, D.F. Mu, J.X. Cao, L.Y. Tang, Y.T. Zhang and B.T. Pan, 1990: Comparison between the Lanzhou loess profiles and the Antarctic ice core records for the last interglacial-glacial cycle (in Chinese). Science in China, B 10, 1086-1094.

Li, R.Q., 1990: Periglacial stages and their division in north China (in Chinese). Quaternary Sciences, 125-136.

Li, W.Y. and Y.L. Liang, 1985: Vegetation and environment of the holocene hypsithermal interval in eastern Heibei Plain (in Chinese). Acta Botanica Sinica, 27, 640-651.

Lin, Z.Y. and X.D. Wu, 1986: A preliminary analysis of the regularity in flood, droughts and snowstorms in Tibetan Plateau during the historical times (in Chinese). Acta Meteorologica Sinica, 44, 257-264.

Liu, T., 1989: The global surface energy and the geological role of wind stress. In: Rutter, N. and Faure, H. (eds.), Global change, 43-54. Pergamon, Oxford (UK).

Liu, T. (ed.), 1991: Quaternary geology and environment in China (in Chinese). Science Press, Beijing (CN).

Liu, T., 1994: Variability of eastern Asian monsoon climate over the last climatic cycle. In: Derbyshire, E. (ed.), Wind blown sediments in the Quaternary record. Abstracts of papers and posters, 1-. Geography Department - University of London (UK).

Liu, T., Z. Ding, Z. Yu and N. Rutter, 1993: Susceptibility time series of the Baoji section and the bearings on paleoclimatic periodicities in the last 2.5 Ma. Quaternary International, 17, 33-38.

Maher, B.A. and R. Thompson, 1992: Palaeoclimatic significance of the mineral magnetic records of the Chinese loess and paleosols. Quaternary Research, 37, 155-170.

McLure, H.A., 1976: Radiocarbon chronology of late Quaternary lakes in the Arabian desert. Nature, 263, 755-756.

Miller, G., G. Wendorf, R. Ernst, R. Schild, A.E. Close, I. Friedman and H.P. Schwarcz, 1991: Dating lacustrine episodes in the eastern Sahara by the epimerization of isoleucine in ostrich egg shells. Palaeogeography, Palaeoclimatology, Palaeoecology, 84, 175-189.

Morel, A., T. Tillet and G. Poupeau, 1991: Le Bassin de Taoudenni. In: Tillet, T. (ed.), Abstracts. Colloque de Solignac (PICG 252) "Palaeoenvironments and Prehistoric Populations of the Sahara in the Upper Pleistocene".

Neumann, K., 1989: Vegetationgeschichte der Östsahara in Holozän aus prähistorischen Fundstellen. In: Küper, R. (ed.), Forschungen zur Umweltgeschichte der Östsahara, 13-181. Heinrich-Barth Institut, Köln (DE).

Nicholson, S.E. and H. Flohn, 1980: African environmental and climatic changes and the general atmospheric circulation in late Pleistocene and Holocene. Climatic Change, 2, 313-348.

Pachur, H.J. and P. Hoelzmann, 1991: Palaeoclimatic implications of Late Quaternary lacustrine sediments in Western Nubia, Sudan. Quaternary Research, 36, 257-276.

Pachur, H.J. and S. Kröpelin, 1987: Wadi Howar: paleoclimatic evidence from an extinct river system in the southeastern Sahara. Science, 237, 298-300.

Pachur, H.J., S. Kröpelin, P. Hoelzmann, M. Goschin and N. Altmann, 1990: Late quaternary fluvio-lacustrine environments of western Nubia. Berliner geowissenschaftlichen Abhandlungen (A), 120, 203-260.

Pachur, H.J. and H.P. Röper, 1984: The western Libyan desert and northern Sudan during the late Pleistocene and Holocene. Berliner geowissenschaftlichen Abhandlungen, A 50, 249-284.

Pachur, H.J., H.P. Röper, S. Kröpelin and M. Goschin, 1987: Late Quaternary hydrography of the eastern Sahara. Berliner geowissenschaftlichen Abhandlungen, A 75, 331-384.

Pachur, H.J., Wunneman and Chang, in prep.: Paleolakes in the Tenga desert, inner Mongolia.

Pan, B.T. and S.Y. Xu, 1989: Environmental changes in the eastern Qinghai-Tibet Plateau during the late Quaternary (in Chinese). Kexue Tongbao, Science Bulletin, 34, 534-536.

Pant, R.K., 1993: Spread of loess and march of desert in western India. Current Science, 64, 841-847.

Parthasarathy, B., N.A. Sontake, A.A. Monot and D.R. Kothawale, 1987: Droughts / floods in the summer monsoon season over different meteorological subdivisions of India for the period 1871-1984. Journal of Climatology, 7, 57-70.

Perthuisot, J.P., 1980: Le Quaternaire de la péninsule de Qatar. Données liminaires et hypothèses. In: Tixier, J. (ed.), Mission archéologique française à Qatar, 1, 11-31. Ministry of Information, Qatar.

Petit-Maire, N. (ed.), 1982: Le Shati, lac pléistocène du Fezzan (Libye). CNRS, Marseille / Paris (FR), 118 p.

Petit-Maire, N., 1986: Paleoclimates in the Sahara of Mali. A multidisciplinary study. Episodes, 9, 7-16.

Petit-Maire, N., 1987: Local responses to recent global climatic change: hyperarid central Sahara and coastal Sahara. In: Matheis, G. and Schandelmeier, H. (eds.), Current Research in African Earth Sciences, 431-434. Rotterdam, Balkema (NL).

Petit-Maire, N., 1989: Interglacial environments in presently hyperarid Sahara: palaeoclimatic implications. In: Leinen, M. and Sarnthein, M. (eds.), Palaeoclimatology and Palaeometeorology: Modern and Past Patterns of Global Atmospheric Transport, 637-661. Kluwer, Dordrecht (NL).

Petit-Maire, N. (ed.), 1991: Paléoenvironnements du Sahara. Lacs holocènes à Taoudenni (Mali). Centre régional de Publication - CNRS, Marseille / Paris (FR), 239 p.

Petit-Maire, N., 1992: Environnements et climats de la ceinture tropicale nord-africaine depuis 140 000 ans. In: Miskovsky, J.C. (eds.), Les applications de la géologie à la connaissance de l'environnement de l'homme fossile, 27-34. Société géologique de France, Paris (FR).

Petit-Maire, N., 1993: Past global climatic changes and the tropical arid/semi-arid belt in the North of Africa. In: Thorweihe, U. and Schandelmeier, H. (eds.), Geoscientific research in northeast Africa, 551-560. Balkema, Rotterdam (NL).

Petit-Maire, N., L. Casta, G. Delibrias, C. Gaven, with an appendix by A.M. Testud, 1980: Preliminary data on quaternary paleolacustrine deposits in the Wadi ash Shati, Libya. In: Salem, M.J. and Busrewil, M.T. (eds.), The Geology of Libya, 3, 797-807. Academic Press, London.

Petit-Maire, N., D. Commelin, J. Fabre and M. Fontugne, 1990: First evidence for Holocene rainfall in the Tanezrouft hyperdesert and its margins. Palaeogeography, Palaeoclimatology, Palaeoecology, 79, 333-338.

Petit-Maire, N., G. Delibrias and C. Gaven, 1980: Pleistocene lakes in the Shati area, Fezzan (27°30' N). In: Sarnthein, M., Seibold, E. and Rognon, P. (eds.), Sahara and the surrounding seas, 289-295. Balkema, Rotterdam (NL).

Petit-Maire, N., M. Fontugne and C. Rouland, 1991: Atmospheric methane ratio and environmental changes in the Sahara and Sahel during the last 130 000 years. Palaeogeography, Palaeoclimatology, Palaeoecology, 86, 197-204.

Petit-Maire, N., M. Fontugne and C. Rouland, 1991: Les derniers 130 000 ans: changements climatiques intertropicaux et variations du taux de méthane atmosphérique. Comptes Rendus de l'Académie des Sciences, Paris (2), 312, 1235-1241.

Petit-Maire, N. and S. Kröpelin, 1991: Les climats holocènes du Sahara le long du Tropique du Cancer. In: Petit-Maire, N. (ed.), Paléoenvironnements du Sahara. Lacs holocènes à Taoudenni (Mali), 205-210. Centre régional de Publication - CNRS, Marseille / Paris (FR).

Petit-Maire, N. and N. Page, 1992: Remote sensing and past climatic changes in tropical deserts: example of the Sahara. Episodes, 15, 113-117.

Petit-Maire, N., N. Page and J. Marchand, 1993: The Sahara in the Holocene. Unesco / CGMW., Map 1/5 000 000.

Petit-Maire, N., J. Riser and coll., 1988: Le Sahara à l'Holocène: Mali. CGMW., Map 1/1 000 000.

Potts, D.T., 1990: The Arabian Gulf in antiquity, 1: From prehistory to the fall of the Achaemenid Empire. Clarendon, Oxford (UK)

Prell, W.L., W.H. Huston, D.F. Williams, A.W. Bé, K. Geitzenauer and B. Molfinos, 1990: Surface circulation of the Indian Ocean during the Last Glacial maximum, approximately 18 000 yr B.P. Quaternary Research, 14, 309-336.

Prell, W.L. and J.E. Kutzbach, 1987: Monsoon variability over the past 150 000 years. Journal of geophysical Research, 92, 8411-8425.

Prell, W.L. and E. Van Campo, 1986: Coherent response of Arabian Sea upwelling and pollen transport to late quaternary monsoonal winds. Nature, 323, 526-528.

Probst, J.L. and Y. Tardy, 1989: Global runoff fluctuations during the last 80 years in relation to world temperature change. American Journal of Science, 289, 267-285.

Ritchie, J.C. and C.V. Haynes, 1987: Holocene vegetation zonation in the eastern Sahara. Nature, 330, 645-647.

Rossignol-Strick, M. and D. Duzer, 1980: Late Quaternary west African climate inferred from palynology of Atlantic deep sea cores. In: Sarnthein, M., Seibold, E. and Rognon, P. (eds.), Sahara and the surrounding seas, 227-228. Balkema, Rotterdam (NL).

Rossignol-Strick, M., W. Nesteroff, P. Olive and C. Vergnaud-Grazzini, 1982: After the Deluge : Mediterranean stagnation and sapropel formation. Nature, 295, 105-110.

Rostek, F., G. Ruhland, F. Bassinot, P.J. Müller, L.D. Labeyrie and E. Bard, 1993: Fluctuations of the Indian monsoon regime during the last 170,000 years: evidence from sea surface temperature, salinity and organic carbon records. NATO Workshop "Global precipitations and climate change", L'Agelonde (FR), 27 sept.- 1 oct. Abstracts, 17-.

Rutter, N., Z. Ding, M.E. Evans and Y. Wang, 1990: Magnetostratigraphy of the Baoji loess - paleosol section in the north-central China loess plateau. In: Pecsi, M. and Loczy, D. (eds.), Loess and the palaeoenvironment, 97-102. Pergamon, Oxford (UK).

Sanlaville, P., 1992: Changements climatiques dans la péninsule arabique durant le Pléistocène supérieur et l'Holocène. Paléorient, 18, 5-26.

Sarnthein, M., G. Tetzlaff, B. Koopmann, K. Wolter and U. Pflaumann, 1981: Glacial and interglacial wind regimes over the eastern subtropical Atlantic and Northwest Africa. Nature, 293, 193-196.

Schulz, E., 1991: Les pollens des lacs de Taoudenni-Agorgott. In: Petit-Maire, N. (ed.), Paléoenvironnements du Sahara. Lacs holocènes à Taoudenni (Mali), 143-162. Centre régional de Publication / CNRS, Marseille / Paris (FR).

Schulz, E. and J.W. Whitney, 1987: Upper Pleistocene and Holocene palaeoenvironments in the An Nafud, Saudi Arabia. In: Matheis, G. and Schandelmeier, H. (eds.), Current Research in African Earth Sciences, 443-446. Rotterdam, Balkema (NL).

Servant, M. and S. Servant-Vildary, 1980: L'environnement quaternaire du bassin du Tchad. In: Williams, M.A.J. and Faure, H. (eds.), The Sahara and the Nile, 133-162. Balkema, Rotterdam (NL).

Shackleton, N.J., 1994: Correlations between marine $\delta\ ^{18}O$ and loess sequences. In: Derbyshire, E. (ed.), Wind blown sediments in the Quaternary record. Abstracts of papers and posters, Geography Department - University, London (UK).

Shi, Y., Z. Kong, S. Wang, L. Tan, F. Wang, T. Yao, X. Zhao, P. Zhang and S. Shi, 1993: Mid-Holocene climates and environments in China. Global and planetary Change, 7, 219-233.

Shimmield, G.B., S.R. Mowbray and G.P. Weedon, 1990: A 350 ky history of the Indian south-west monsoon evidence from deep-sea cores, northwest Arabian Sea. Earth Science, 81, 289-299.

Singh, G., R.J. Wasson and D.P. Agrawal, 1990: Vegetational and seasonal climatic changes since the last full glacial in the Thar Desert, northwestern India. Review of Palaeobotany and Palynology, 64, 351-358.

Singhvi, A.K., Y.P. Sharama and D.P. Agrawal, 1982: Thermoluminescence dating of sand dunes in Rajahstan, India. Nature, 295, 313-315.

Sirocko, F., M. Sarnthein, H. Erlenkeuser, H. Lange, M. Arnold and J.C. Duplessy, 1993: Century-scale events in monsoonal climate over the past 24,000 years. Nature, 364, 322-324.

Sirocko, F., M. Sarnthein, H. Lange and H. Erlenkeuser, 1991: Atmospheric summer circulation and coastal upwelling in the Arabian Sea during the Holocene and the Last deglaciation. Quaternary Research, 36, 72-93.

Stanley, D.J. and A.G. Warne, 1993: Nile delta: recent geological evolution and human impact. Science, 260, 628-634.

Sun, X.J., 1989: A study of the late Pleistocene palaeovegetation at Beizhuangcun, Shanxi Province, northern China (in Chinese). Quaternary Sciences, 2, 177-189.

Szabo, B.J., W.P. McHugh, G.G. Schaber, C.V. Haynes Jr and C.S. Breed, 1989: Uranium-series dated authigenic carbonates and acheulian sites in southern Egypt. Science, 243, 1053-1056.

Talbot, M., 1984: Late Pleistocene rainfall and dune building in the Sahel. Palaeoecology of Africa, 16, 203-214.

Tillet, T., 1989: L'Atérien saharien. Essai sur le comportement d'une civilisation paléolithique face à l'accroissement de l'aridité. Bulletin de la Société géologique de France, 8, 91-97.

Van Campo, E., 1986: Monsoon fluctuations in two 20,000-yr BP oxygen isotope / pollen records off southwest India. Quaternary Research, 26, 376-388.

Vernet, R. and G. Aumassip, 1992: Le Sahara et ses marges. Paléoenvironnements et occupation humaine à l'Holocène. Inventaire des datations ^{14}C. CNRS, Meudon (FR), 107 p.

Wang, F.B., 1985: Preliminary study on climatic change and trends over the last 10 000 years on the Qinghai-Xizang Plateau. In: Yang, H.J. (ed.), Contribution to the Quaternary glaciology and Quaternary Geology 2, 153-161. Geological Publishing House, Beijing (CN).

Wang, S.M. and M. Feng, 1991: Relationship between the environmental changes at Daihai lake, Inner Mongolia, and the southeastern monsoon variations (in Chinese). Science in China, B 7, 759-768.

Wang, S.M., R.J. Wu and X.H. Jiang, 1990: Environmental evolution and palaeoclimate at Daihai lake, Inner Mongolia, since the last glacial (in Chinese). Quaternary Sciences, 3, 223-232.

Wasson, R.J., S.N. Rajaguru, V.N. Misra, D.P. Agrawal, R.P. Dhir, A.K. Singhvi and K. Kameswara Rao, 1983: Geomorphology, late Quaternary stratigraphy and paleoclimatology of the Thar dunefield. Geomorphology, 45, 117-151.

Wasson, R.J., J.I. Smith and D.P. Agrawal, 1984: Late Quaternary sediments, minerals and inferred geochemical history of Didwana Lake, Thar desert, India. Palaeogeography, Palaeoclimatology, Palaeoecology, 46, 345-372.

Wendorf, F., A.E. Close, R. Schild, A. Gautier, H. Schwarcz, G.H. Miller, K. Kowalski, H. Krolik, A. Bluuszcz, D. Robins and R. Grün, 1990: Le dernier interglaciaire dans le Sahara oriental. L'Anthropologie, 94, 361-391.

Winstanley, D., 1973: Rainfall trends in Africa, Middle-East and India. Nature, 243, 464-465.

Wu, C., X.Q. Zhu, N.H. He, Q.H. Xu, W.Y. Yuan, Z.H. Wang, D.R. Shi and M.X. Zhao, 1991: Study of the formation of palaeodrainage in north China (in Chinese). Science in China, B 2, 188-197.

Yan, Z.W., D.Z. Ye and C. Wang, 1992: Climatic jumps in flood/drought chronology of Central China. Climate Dynamics, 6, 153-160.

Yan Z.W., and N. Petit-Maire, 1994. The last 150 ka in the Afro-Asian climatic transitional zone. *Palaeogeography, Palaeoclimatology, Paleoecology.*

Zhang, M.L., 1989: Rainfall changes in East China during the recent 100 years. Kexue Tongbao, Science Bulletin, 34, 605-607.

Zhang, W.Q. and S.X. Hu, 1989: Dates and processes of the formation of paleosol in Longdong, north China (in Chinese). Kexue Tongbao, Science Bulletin, 34, 1252-1255.

Zheng, B.X., 1990: The glacial environment and its changes since the last glaciation in west China (in Chinese). Quaternary Sciences, 2, 101-110.

Zhou, W.J., M.F. Zhou and J. Hyder, 1989: ^{14}C dating of a deposit sequence at Beizhuangcun (north China) since 30 000 yr BP (in Chinese). Kexue Tongbao, Science Bulletin, 34, 1096-1099.

FLUCTUATIONS OF THE INDIAN MONSOON REGIME DURING THE LAST 170,000 YEARS: EVIDENCE FROM SEA SURFACE TEMPERATURE, SALINITY AND ORGANIC CARBON RECORDS

Frauke Rostek[1,2], Götz Ruhland[3], Franck C. Bassinot[2,4], Luc Beaufort[2], Peter J. Müller[3] & Edouard Bard[1]

1 Laboratoire de Géoscience de l'Environnement, JE DRED 192, France

2 Laboratoire de Géologie du Quaternaire, CNRS Marseille, France

3 FB Geowissenschaften, Universität Bremen, Germany

4 Centre des Faibles Radioactivités, CNRS-CEA, Gif-sur Yvette, France

ABSTRACT

We studied the variations of the Indian Monsoon regime during the last glacial cycle recorded in deep-sea sediments of Core MD900963 recovered at the junction between the Arabian Sea and the Bay of Bengal. Sea surface temperature (SST) variations were established by means of the unsaturation ratio of alkenones. The SST record allowed us to apply temperature correction to the $\delta^{18}O$ curve of Core MD900963 which was established on the planktonic foraminifera Globigerinoides ruber (white). Using a global $\delta^{18}O$ sea water curve and the $\delta^{18}O$/salinity ratios for the Arabian Sea and the Bay of Bengal we reconstructed the variability of sea surface salinity during the last 170,000 years. The salinity variations can probably be attributed to the changes in the evaporation-precipitation balance (E-P) and ultimately to the relative intensity between the monsoons over the last 170,000 years. Our results suggest that during stage 5.5 and the early Holocene, the SW monsoon was probably stronger which resulted in higher precipitation and lower surface salinity. In contrast, during the time interval between 90 and 15 kyr and in particular during the glacial stages 6, 2 and substage 5.2 the SW monsoon was probably reduced and the dry NE monsoon was the dominant feature of the seasonal climate.

The organic carbon content of Core MD900963 is strongly correlated with the $\delta^{18}O$ record: highest values are associated with glacial $\delta^{18}O$ maxima and lowest values with interglacial minima. The correlation between the organic carbon, the total C37 alkenone concentration, the mass accumulation rates of these compounds, and a coccolith productivity index suggests the occurence of important variations in carbon accumulation rate on Core site MD900963 during the last 170,000 years. The carbon and biomarker contents and accumulation rates can reflect both, changes in productivity and/or changes in preservation. In contrast, the coccolith productivity index suggests that within the glacial stages 6, 4, 2 and substages 5.4 and 5.2 the oceanic productivity was increased, probably due to the

NATO ASI Series, Vol. I 26
Global Precipitations and Climate Change
Edited by M. Desbois and F. Désalmand

injection of nutrient-rich subsurface water into the euphotic zone in response to a stronger and predominating NE monsoon. Consequently, strengthened preservation due to a oxygen minimum or anoxic conditions is not the dominant cause for high organic carbon contents.

KEYWORDS
Indian monsoon, paleo-sea surface temperature (alkenones), - salinity, - productivity.

1. Introduction

Modern surface circulation in the Northern Indian Ocean is controlled by the seasonal monsoon wind system with a strong and humid SW monsoon during summer and a moderate and dry NE monsoon during winter.

Between November and April the NE monsoon is created by the heat gradient between the Asian land mass and the ocean. Due to its lower heat capacity a high pressure area forms over the continent and dry and cool air is carried from the Asian land mass to the Indian Ocean. The NE monsoon causes the NE Monsoon Current or North Equatorial Current (Wyrtki, 1973; Bearman, 1991). From November to January a branch of the NE Monsoon Current turns north at the southern tip of India, carrying low salinity water from the Bay of Bengal into the eastern Arabian Sea (Wyrtki, 1973). Off the Somalian coast the NE Monsoon Current turns south as the Somali Current, crosses the Equator and then flows eastwards with the Equatorial Counter Current (Fig. 1a).

During April the NE monsoon wind system collapses. Due to the solar heating of the Asian continent, a low pressure field is created over the continent and warm and humid air is carried from the subtropical high pressure area over the southern subtropical Indian Ocean to India. The atmospheric changes are associated to surface circulation variations in the Indian Ocean. With the onset of the SW monsoon the surface water of the Somali Current flows north along the coast of Somalia as a western boundary current (Fig. 1a). The Somali Current, the SW Monsoon Current and the South Equatorial Current form a very strong wind driven gyre in the Equatorial Indian Ocean (Wyrtki, 1973). South of Indian subcontinent a branch of the SW Monsoon Current turns north and carries surface water from the Arabian Sea into the Bay of Bengal. During the SW monsoon season an intense upwelling occurs along the coasts of Somalia and Arabia (Wyrtki, 1973; Pickard & Emery, 1982; Bearman, 1991).

Due to these extreme monsoonal atmospheric and oceanographic changes, the Northern Indian Ocean is characterized by two main surface water masses: the high-salinity surface water (35 to 36.5 ‰) of the Arabian Sea due to strong evaporation and the low-salinity water (31 to 34 ‰) in the Bay of Bengal which is caused by high precipitation and large river runoff, particulary during the SW monsoon (Fig. 1b; Pickard & Emery, 1982).

Since sea surface salinity is strongly related to the evaporation-precipitation (E-P) balance it represents a sensitive climate indicator.

Calculations of the water vapour fluxes in the northern Indian Ocean during the SW monsoon show that about 70% of the water vapour over India comes from the Southern Hemisphere and only 30% is picked up from the Arabian Sea (Cadet & Reverdin, 1981). In the Bay of Bengal precipitation is high during the SW monsoon but in contrast to the Arabian Sea most of the water vapour picked up by evaporation is used for local precipitation (Cadet & Reverdin, 1981). The mean annual rainfall over the northern Indian Ocean obtained by microwave satellite measurements (SSM/I-Special Sensor Microwave Imager) illustrates an important gradient in precipitation from the western Arabian Sea to the Bay of Bengal (Fig.1c; Chang et al., 1993).

Deep-sea sediment records can be used to study the variations of the monsoonal wind system during the last glacial/interglacial cycle. Pollen and foraminifera data (van Campo, 1986; Cullen, 1981; Prell et al., 1980), oxygen and carbon isotope composition of planktonic foraminifera (Duplessy, 1982; Fontugne & Duplessy, 1986) and organic matter content (Fontugne & Duplessy, 1986) suggest that within the last glacial maximum (LGM, about 18,000 years ago) the SW monsoon was weaker than during the Holocene, whereas the NE monsoon had more relative importance and was even stronger than today. Magnetic susceptibility patterns recorded in paleosols (interglacial-humid climate) and loess (glacial-dry climate) successions of the loess plateau of central China represent a proxy record of the Asian summer monsoon variability for the last climatic cycle (An et al., 1991). The loess-paleosol sequences correlate with the oxygen isotope stages of deep-sea sediments (Shackleton, 1994).

Core MD900963 (05°04'N/73°53'E; 2450 m water depth; Fig. 1) was recovered on the eastern shoulder of the Maldives Islands (SEYMAMA expedition, 1990; RV *Marion Dufresne*). A detailled $\delta^{18}O$ stratigraphy is available for the 54 m long piston core (Bassinot, 1993; Bassinot et al., 1994). In order to reconstruct sea surface temperature (SST) and salinity variations for the last 170,000 years the alkenone unsaturation ratio U^{k}_{37} was measured for the uppermost 10 m of Core MD900963 (Rostek et al., 1993). In addition, we studied organic carbon content, foraminiferal fragmentation, total C_{37} alkenone content, coccolith abundance and carbon accumulation rates on this site to evaluate the productivity or preservation changes during this time interval. Situated at the junction between the Arabian Sea and the Bay of Bengal Core MD900963 is probably ideally located to monitor the changes of the Indian monsoon during the last glacial cycle.

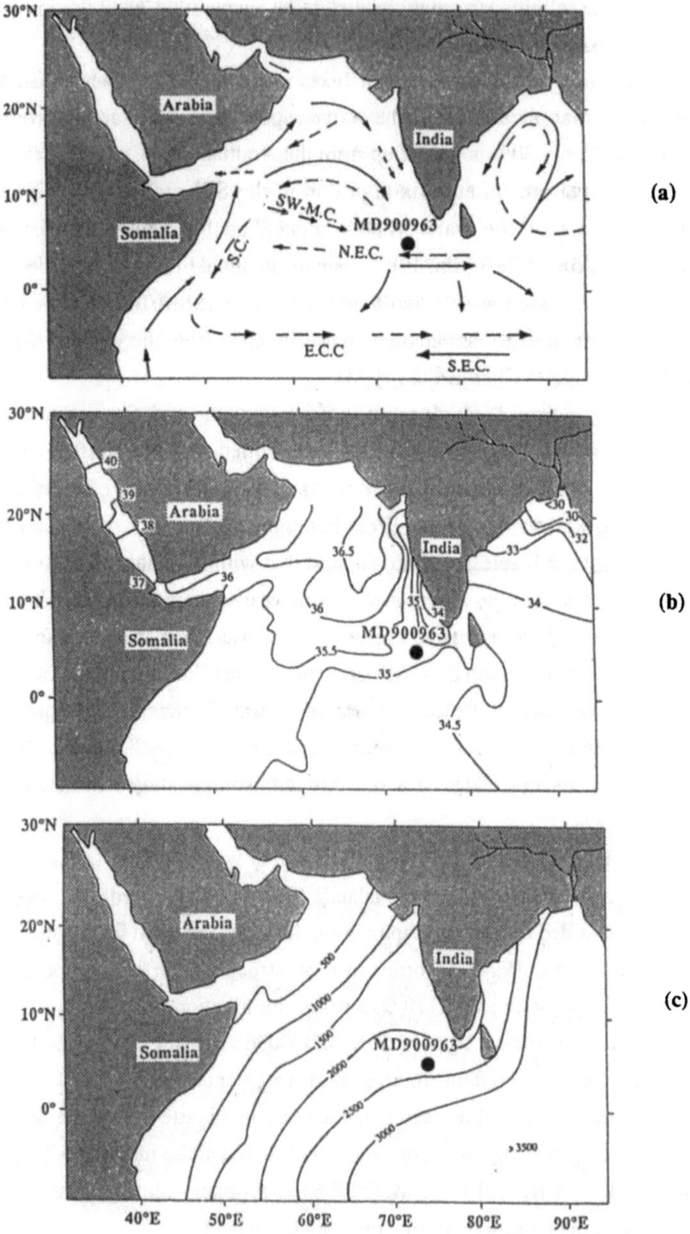

Figure 1: (a) Circulation patterns in the northern Indian Ocean and location of Core MD900963. The solid lines represent the oceanic circulation during the SW monsoon, stippled lines the circulation during the NE monsoon. S.C.=Somali Current, N.E.C.=North Equatorial Current, E.C.C.=Equatorial Counter Current, SW-M.C.=SW Monsoon Current, S.E.C.=South Equatorial Current, Conv.=Convergence. (b) Summer (July & August) surface salinity in ‰ (after Wyrtki, 1971). (c) Annual mean rainfall in mm derived from SSM/I (Special Sensor Microwave Imager) satellite measurements (after Chang et al., 1993).

2. The sea surface temperature (SST) record

Algae of the class Prymnesiophyceae in particular the coccolithophorid species *Emiliania huxleyi*, have been recognized as one source of long-chain C_{37} alkenones (Volkman et al. 1979; Prahl & Wakeham, 1987). In laboratory cultures but also in modern plankton, sediment trap material and surface sediments the alkenone unsaturation ratio (two or three double bounds, indicating the degree of unsaturation) expressed as

$$U^{k}_{37}{}' = \frac{C_{37:3} + C_{37:2}}{C_{37:2}}$$

exhibits a linear response for water temperatures between 0 and 30 °C (Fig. 2; Prahl & Wakeham, 1987; Prahl et al., 1988; Sikes et al., 1991; Sikes & Volkman, 1993). Consequently, the $U^{k}_{37}{}'$ unsaturation ratio measured in deep-sea sediments is now used as a tool to quantify SST in the past (Brassell et al., 1986; Farrington et al., 1988; Jasper & Gagosian, 1989; ten Haven & Kroon, 1991; Eglinton et al., 1992; Kennedy & Brassell, 1992; Lyle et al., 1992; Rostek et al., 1993).

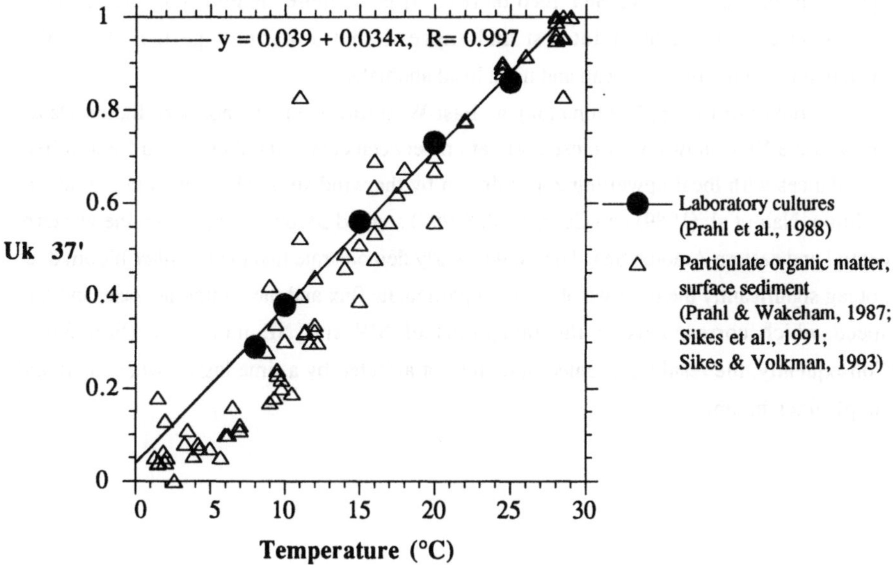

Figure 2: Compilation of $U^{k}_{37}{}'$ data versus measured sea surface temperatures. The temperatur equation of Prahl et al. (1988) is based on laboratory cultures of the algae *E. huxleyi*.

The alkenone measurements were carried out at 10 cm intervals at the Department of Geoscience of the University of Bremen. The sample preparation and measurement details are described elsewhere (Müller et al., 1993). We used the temperature equation of Prahl et al. (1988) to determine the SST variations for the last 170,000 years in the equatorial Indian Ocean. The calcareous tests of the algae*Emiliania huxleyi* and other species of the family *Gephyrocapseae* are abundant in the first 10 m of Core MD900963 (Beaufort et al., in prep.). A $\delta^{18}O$ record was established on the surface dwelling planktonic foraminifera *Globigerinoides ruber (white)* and a time scale was obtained by correlating the $\delta^{18}O$ record to the SPECMAP record (Imbrie et al., 1984).

Figure 3a, b shows the SST and $\delta^{18}O$ record as a function of age. The SST variations during the last 170,000 years are rather small between 25 and 28 °C. Low SST values are observed in stages 4 and 3 with a minimum centered within stage 3 whereas interglacial stage 5 is characterized by high SST's between 27 and 28 °C. For stage 6 the SST's are only about 0.5 °C lower than for stage 5. SST reconstructions based on foraminifera assemblages suggest a temperature drop of 1 to 2 °C during the LGM for Site MD900963 (Prell & Hutson, 1979), which is in agreement with our SST data. In particular, we compared our alkenone SST record to a foraminifer transfer function SST record of Core RC17-98 from the tropical Indian Ocean (15°S/65°E; Clemens et al., 1991). The good agreement between the two SST records (Fig. 4) gives more credibility to the alkenone record on the timescale of interest and suggests that a warm stage 6 is a genuine phenomenon in the Indian Ocean and not a local anomaly.

Studies of living plankton along an East-West traverse in the northern Indian Ocean during June 1984 show a very close correlation between coccolithophorids and foraminifers abundances with local upwelling zones driven by the wind stress (Kleijne et al., 1988). In addition, Nair et al. (1989) and Curry et al. (1992) studied sediment traps from the western, central and eastern Arabian Sea. Their data clearly demonstrate that the *G. ruber* bloom does not lag significantly the increase of the total particulate flux and the abrupt increases in wind speed which corresponds to the inceptions of SW and NE monsoons (Fig. 5a-c). Consequently, the results presented here are not affected by a time lag between algal and zooplankton blooms.

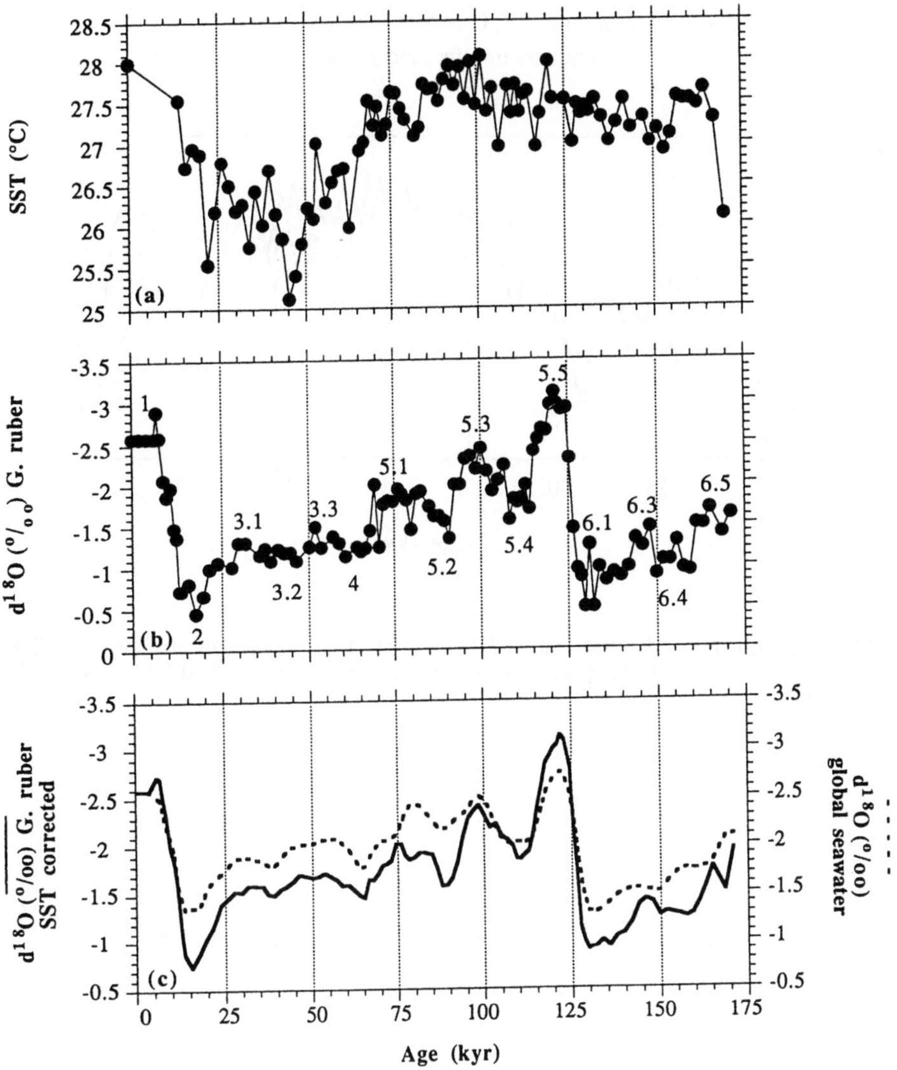

Figure 3: SST (a) and $\delta^{18}O_F$ record (b) for Core MD900963. For missing core top SST and $\delta^{18}O$ values a constant modern annual mean SST of 28 °C (Wyrtki, 1971) and a $\delta^{18}O$ value of -2.58 ‰ was used, assuming that sea level did not change during the last 6 kyr (Fairbanks, 1989). (c) SST corrected $\delta^{18}O_F$ record in comparison with the global seawater $\delta^{18}O_{WG}$ record (Labeyrie et al., 1987; Vogelsang, 1990) correlated to the timescale of Core MD900963. The SST values are interpolated to the same depth points as the $\delta^{18}O$ measurements.

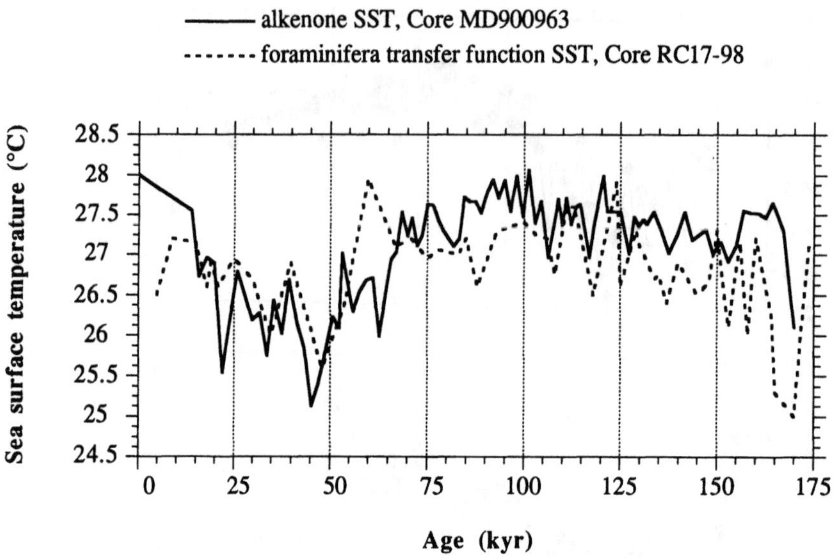

Figure 4: Comparison between the SST record obtained by means of alkenones for Core MD900963 (5°N/74°E) and by the foraminifera transfer function method for Core RC17-98 (15°S/65°E).

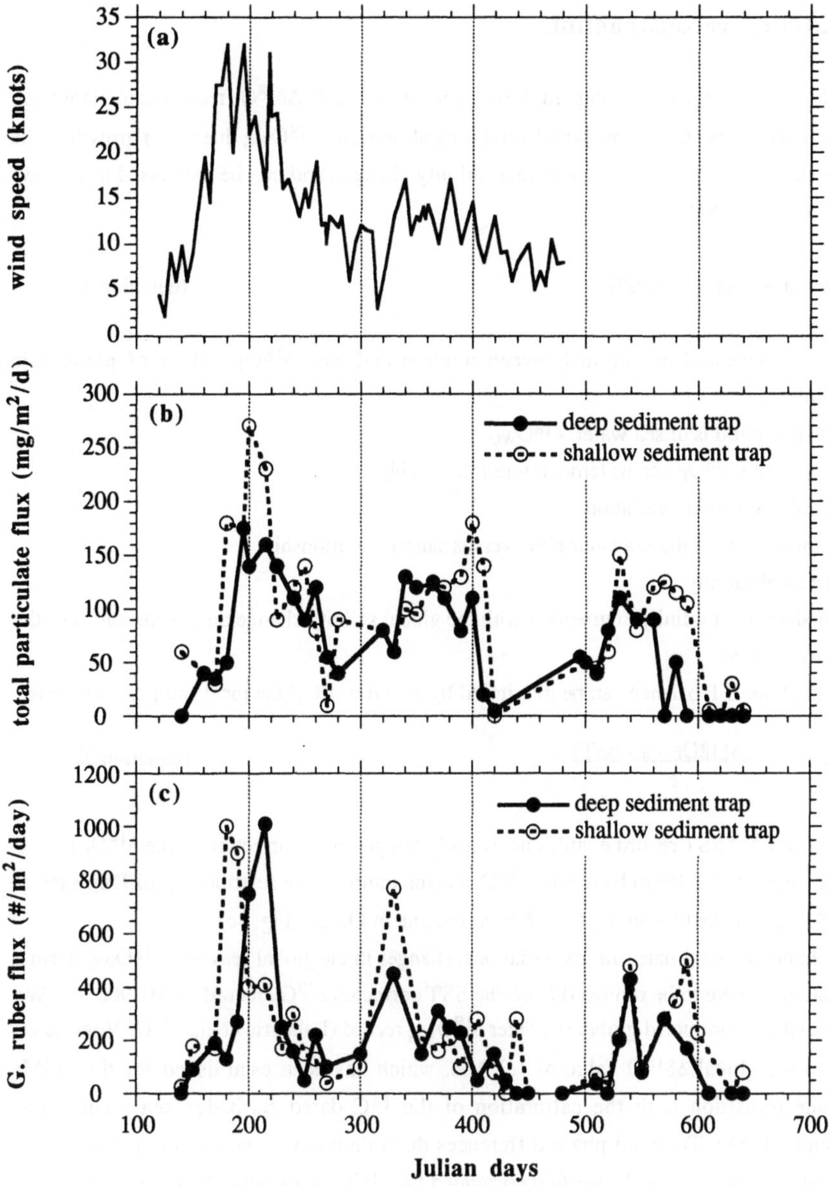

Figure 5: Wind speed measurements (a), total particulate flux (b) (Nair et al., 1989), and foraminiferal flux (*G. ruber* >150μm) (c) (Curry et al., 1992) from a sediment trap study in the central Arabian Sea (1986-1987). The shallow trap is at 1000 m water depth and the deeper trap at 2900 m water depth.

3. Salinity reconstruction

The glacial to interglacial change in the oxygen isotope ratio $\Delta\delta^{18}O_F$ recorded in planktonic foraminifera are controlled by global variations of seawater $\delta^{18}O_{WG}$ due to the growth of the ice sheets, by local SST and sea surface salinity changes and can be expressed as follows (after Broecker, 1989):

$$\Delta\delta^{18}O_F = a + b\,\Delta T + c\,(S-S^*) \qquad \text{(equation 1)}$$

where

$\Delta\delta^{18}O_F$ = measured deviation between modern and past $\delta^{18}O_F$ values of planktonic foraminifera,

a = global variations of sea water $\delta^{18}O_{WG}$

b = slope of the $\delta^{18}O_F$ versus temperature relationship

ΔT = local temperature variation

c = regional slope of the seawater $\delta^{18}O$ versus salinity relationship

S = past local salinity

S* = modern local salinity augmented with the global salinity change due to variations of the global ice volume.

Changes of the E-P balance can be monitored by an index ΔS of the local salinity variations:

$$\Delta S = S - S^* = \frac{(\Delta\delta 18O_F - a - b\Delta T)}{c} \qquad \text{(equation 2)}$$

The U^{k}_{37} SST record enables us to apply temperature correction to the $\delta^{18}O_F$ record using a slope of -0.2 (term b) for the $\delta^{18}O_F$ versus temperature relationship of Duplessy et al. (1981) established for living *G. ruber* in the Indian Ocean (Fig. 3c).

In order to estimate salinity variations, changes in the global seawater $\delta^{18}O_{WG}$ (term a in equation 3) have to be removed from the SST corrected $\delta^{18}O_F$ record of MD900963. We normalized a recontructed global seawater $\delta^{18}O_{WG}$ record (Labeyrie et al., 1987; Vogelsang, 1990) to the global $\Delta\delta^{18}O$ value of 1.2 ‰, which has been established for the LGM/ Holocene transition with the calibration of the ^{14}C dated Barbados sea level curve (Fairbanks, 1989). To avoid phase differences due to a point-by-point subtraction of two independent isotopic records, we first correlated the global seawater $\delta^{18}O_{WG}$ reconstruction precisely to the timescale of the SST corrected isotope record of Core MD900963 (using the LINAGE software developed by D. Paillard & L. D. Labeyrie, CFR Gif-sur-Yvette) before substracting it (Fig. 3c). Moreover, the raw data were first smoothed using a seven point least square procedure (Savitzky & Golay, 1964).

Sea surface salinity and $\delta^{18}O_W$ of seawater are directly linked to the E-P balance (Wüst et al., 1954) and a linear correlation was found between these parameters for most of the global ocean (Craig & Gordon, 1965). In order to reconstruct this relationship (term c in equation 2), we used 52 seawater $\delta^{18}O_W$ and salinity measurements between 20°S and 21°N from the Arabian Sea (J.C. Duplessy, pers. comm.). Since for the Bay of Bengal no seawater $\delta^{18}O_W$ measurements are available, a $\delta^{18}O_W$/salinity relationship was established by using 22 $\delta^{18}O_F$ *G. ruber* core top measurements (Duplessy, 1982). Duplessy et al. (1981) studied the ^{18}O fractionation between foraminiferal shell carbonate and seawater as a function of temperature. We used their temperature equation established for living *G. ruber* in the Indian Ocean to convert the $\delta^{18}O_F$ core top measurements from the Bay of Bengal into seawater $\delta^{18}O_W$. Since this equation is established on living foraminifers, we took into account a $\delta^{18}O$ enrichment of 0.85 ‰ due to gametogenic calcification below the euphotic zone (Duplessy et al., 1981).

$$\delta^{18}O_W = \delta^{18}O_F + 0.20\ T\ (°C) - 2.55 - 0.85 \qquad \text{(equation 3)}$$

This procedure enables to represent the $\delta^{18}O_W$/salinity relationship of the two basins on the same graph (Fig. 6). The $\delta^{18}O_W$ versus salinity slope (0.37) of the composite data set is probably the best estimate for the variations at Site MD900963. The downcore ΔS changes as expressed in equation 2 are presented in Figure 7. To eliminate high frequency variations due to the subtraction of two independant isotope records, we applied a seven point least square filter to the ΔS curve (Savitzky & Golay, 1964). Taking into account the standard reproducibility (± 1 sigma) on the SST (±0.3 °C, Müller et al., 1993), on $\delta^{18}O_{WG}$ (±0.07 ‰, Labeyrie et al., 1985) and on $\delta^{18}O_F$ (±0.05 ‰, Labeyrie et al., 1992), it is possible to estimate an average analytical error of about ± 0.3 ‰ for the reconstructed salinity values.

For stage 5.5 (125 to 115 kyr) and the early Holocene (9 to 6 kyr) negative ΔS values of about -1 to -0.5 ‰ are observed. In stages 6.4 and 6.3 (160 and 140 kyr) and stages 5.0 to 3.1 (75 and 25 kyr), ΔS is consistently positive with values ranging between 0.5 and 1 ‰ . Three periods of more than 1 ‰ increase of ΔS are clearly present in stages 6.2 to 6.0 (140 to 125 kyr), in stage 5.2 (90 to 80 kyr) and in stage 2 (20 to 15 kyr).

The precipitation over India and the Indian Ocean is related to the strength of the SW monsoon (Cadet & Reverdin, 1981), and consequently negative ΔS values observed during stage 5.5 and the early Holocene, suggest that the SW monsoon was relatively stronger during these time intervals. Our results are in agreement with those of Clemens et al. (1991) and Anderson & Prell (1993) who found strong SW monsoon events in geological monsoon tracers from the upwelling in the western Arabian Sea for the same time intervals.

Higher ΔS observed during glacial stages 6 and 2 and substage 5.2 on the Core site MD900963 were probably caused by increased E-P. Consequently, it can be suggested that

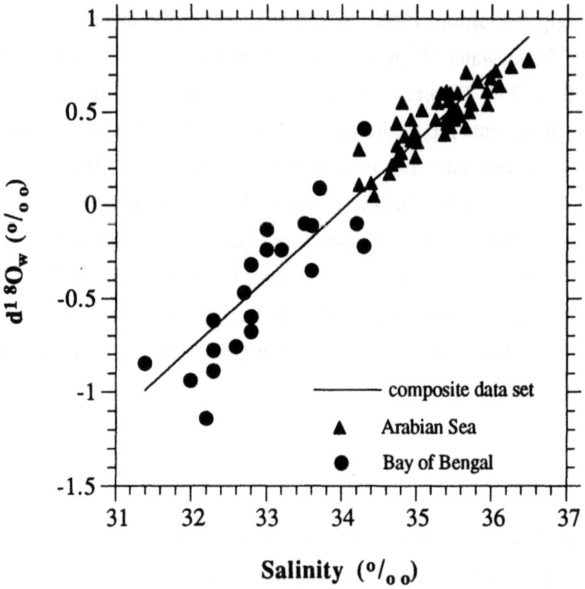

Figure 6: $\delta^{18}O_W$/salinity relationships for the Arabian Sea and the Bay of Bengal. The regression line was calculated for the composite data set.

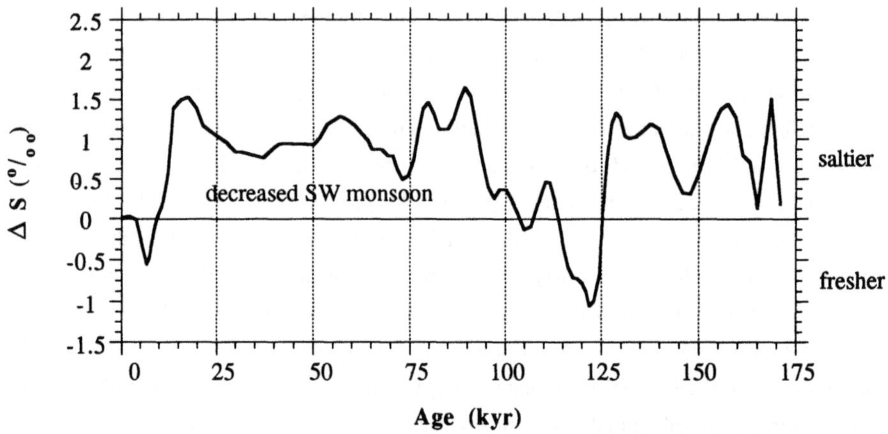

Figure 7: Salinity variations ΔS for Core site MD900963. ΔS changes are calculated for the $\delta^{18}O_W$ versus salinity relationships of the composite data set of the Arabian Sea and the Bay of Bengal. ΔS is directly related to local variations of the E-P balance.

the dry NE monsoon was the dominating seasonal feature during these periods and/or that the humid SW monsoon was weakened (Prell et al., 1980; Duplessy, 1982; Fontugne & Duplessy, 1986; van Campo, 1986; An et al., 1991). Microfossil evidence suggests that during the LGM the decrease in precipitation led to considerably higher salinities in the Bay of Bengal (Cullen, 1981). The generally high values obtained for the ΔS index (with the exception of stage 5.5 and the Holocene) suggest that in this part of the Indian Ocean, the E-P budget was different from the conditions of today during most of the last glacial cycle.

4. The organic carbon record: productivity or preservation

About 99% of the primary production of organic carbon is degraded in the water column and on the seafloor and only a small part is preserved in the sediment. Organic carbon variations in deep-sea sediments are mainly controlled by the carbon flux and the sedimentation rate (Müller & Suess, 1979). Organic carbon records from low-latitude deep-sea sediments suggest that primary productivity was increased during the glacial periods, especially in upwelling regions (Müller & Suess, 1979; Sarnthein et al., 1988; Mix, 1989). Various studies even suggest that higher productivity and enhanced upwelling during the cold periods are directly related to increased wind intensities (Parkin & Shackleton, 1973; Molina-Cruz, 1977; Sarnthein et al., 1981; Pedersen, 1983; Lyle et al., 1992).

The importance of the bottom water oxygen content for the degradation of organic material is still a matter of controversity (Emerson, 1985; Pederson & Calvert, 1990; Betts & Holland, 1991). During the glacial periods, the bottom water circulation was probably reduced and the oxygen content of bottom water decreased, which favours preservation (Emerson, 1985; Emerson & Hedges, 1988). In fact, geochemical evidence like an enrichment of organic carbon and authigenic uranium from a core recovered in the eastern Arabian Sea (10°N, 71°50'E; 2523 m water depth) close to Core MD900963 (5°N/74°E, 2450 m water depth) suggests anoxic conditions during the last glacial cycle (stages 4, 3 and 2) due to changes in the deep water circulation (Sarkar et al., 1993).

Primary production in the ocean is mainly controlled by the availability of nutrients, light and temperature in the euphotic zone. In the central and eastern Arabian Sea, where seasonal variations of the latter physical parameters are small, time series of sediment trap studies indicate that the biological productivity, as reflected by total particulate fluxes of organic and mineral matter, are directly related to the wind intensities (Fig. 5; Nair et al., 1989; Curry et al., 1992; Ittekkot et al., 1992). Increased wind speeds during the NE and the SW monsoon lead to a deepening of the mixed-layer (Rao et al., 1989) and to the injection of nutrient-rich subsurface waters into the euphotic zone (Nair et al., 1989).

For the eastern Arabian Sea, Fontugne & Duplessy (1986) suggest that the variations of organic carbon are related to local productivity and an increase was observed during glacial periods in areas where productivity depends on the NE monsoon strength, like off the coast of North India. In contrast, the productivity of the western Arabian Sea increased during the interglacial periods which is related to a reinforced SW monsoon and stronger upwelling (Fontugne & Duplessy, 1986; Prell & Van Campo, 1986; Anderson & Prell, 1993).

Organic carbon: Organic carbon (C_{org}) measurements for Core MD900963 were carried out at intervals of 10 cm at the Department of Geoscience of the University of Bremen. Measurement details are given elsewhere (Müller et al., 1993). The C_{org} content varies between 0.2 and 0.9% and is strongly correlated with the $\delta^{18}O$ record: highest values are associated with glacial $\delta^{18}O$ maxima and lowest values with interglacial minima (Fig. 8a, b).

Total C_{37} alkenone abundance; The alkenones constitute about 8% of the total organic carbon in living cells of *E. huxleyi*. They can therefore provide a record for the paleoproductivity of Prymnesiophyceae, which represent an important contribution to primary production (Prahl & Muehlhausen, 1989). The correlation between the C_{org} variations and the total C_{37} alkenone concentration (Fig.8c) reflects an increase of oceanic productivity and/or enhanced preservation of organic matter during the glacial stages 6, 4 and substage 5.2. The agreement between these records suggests that the variations of organic carbon are related to marine productivity and that the input of terrestrial carbon is negligible.

Foraminifer fragmentation: The break down of foraminifera tests is a sensitive indicator of carbonate dissolution above the lysocline (Peterson & Prell, 1985). In order to establish a foraminifer fragmentation index for Core MD900963, fragments of planktonic foraminifera were counted in the size fraction between 250 and 355 μm. The result is expressed as the percentage of the sum of whole foraminifers plus fragments to whole foraminifers (Bassinot, 1993). The excellent agreement between the foraminifer fragmentation and the organic carbon content (Fig. 8b, d) is probably related to the enhanced release of CO_2 due to the degradation of organic matter during periods of increased carbon flux (Emerson & Bender, 1981).

Coccolith abundance: Since the availability of nutrients such as nitrate, phosphate and silicate is a limiting factor of productivity in the euphotic zone, the surface water is nutrient depleted. In contrast, these biolimiting elements are enriched in sub-surface water and can be returned to the surface by upwelling or by increases in the depth of the wind-mixed upper layer (Raymont, 1963). The dynamics of the mixed-layer and the nutricline depth are opposed to each other and can be monitored by the abundance of the coccolithophorid species *Florisphaera profunda*. This algae is restricted to the lower euphotic zone in contrast to most of the other coccolithophorids which are abundant in the upper euphotic zone (Okada & Honjo, 1973). Consequently, the abundance of *F. profunda* relative to the amount of other coccolithophorids can be used as a control of the nutricline depth (Molfino &

McIntyre, 1990). A deep nutricline (or shallow mixed-layer) indicates that the upper euphotic zone is nutrient depleted and the production of *F. profunda* is enhanced relative to other coccolithophorids. In contrast, a shallow nutricline (or deep mixed-layer) favours the abundance of other coccolithophorid species.

In order to estimate the variations of the nutricline depth (or mixed-layer depth), the percentage of *F. profunda* coccoliths to the total amount of coccoliths in Core MD900963 is used as a paleoproductivity index (Beaufort et al., in prep.). We assume that a high percentage of *F. profunda* coccoliths / total coccoliths in the sediment of Core MD900963 indicates a deep nutricline (shallow mixed-layer) and a low ratio is characteristic for a shallow nutricline (deep mixed-layer).

Figure 8e illustrates that the ratio *F. profunda* coccoliths to the total amount of coccoliths decreases during the glacial periods 6.2 to 6.0, 5.4, 5.2 and 4 (note that axis is reversed as in Molfino & McIntyre, 1990). During these periods the nutricline was shallower than during the Holocene and the interglacial periods. We suggest that this phenomena is related to the predominance of the NE monsoon which causes the deepening of the mixed-layer and the injection of nutrient-rich subsurface waters into the euphotic zone.

Calculation of carbon and biomarker accumulmation rates: In order to estimate the flux of organic carbon and to take into account the effect of the sedimentation rate, we reconstructed organic carbon mass accumulation rates (Corg MAR). First, the sedimentation rate is calculated on the base of the depth-age model of the $\delta^{18}O$ stratigraphy. The sedimentation rate was simplified into a step-function and smoothed using a seven point least square procedure to avoid spurious peaks due to the small number of time controls (Lyle, 1988). In order to obtain Corg MAR the sedimentation rate was multiplied by the dry bulk density and the Corg content (Fig. 9a-d). The variations of the Corg MAR record also follows the climatic cycles with increased Corg MAR during the glacial periods. Applying the same procedure to the alkenone concentration, we calculated the MAR for C37 alkenones (Fig. 9e). The C37 alkenone MAR increases clearly in stage 6.2, 5.4, 5.4 and 4.

Productivity or preservation? The Corg, the total C37 alkenone variations and the MAR of these organic components can reflect both, changes in productivity and/or changes in preservation. By contrast, the distribution of *F. profunda* coccoliths represents a useful proxy to estimate variations of surface productivity. Consequently, we conclude that the oceanic productivity was increased during the glacial periods and that strengthened preservation due to a oxygen minimum or anoxic conditions is not the dominant cause for high Corg contents.

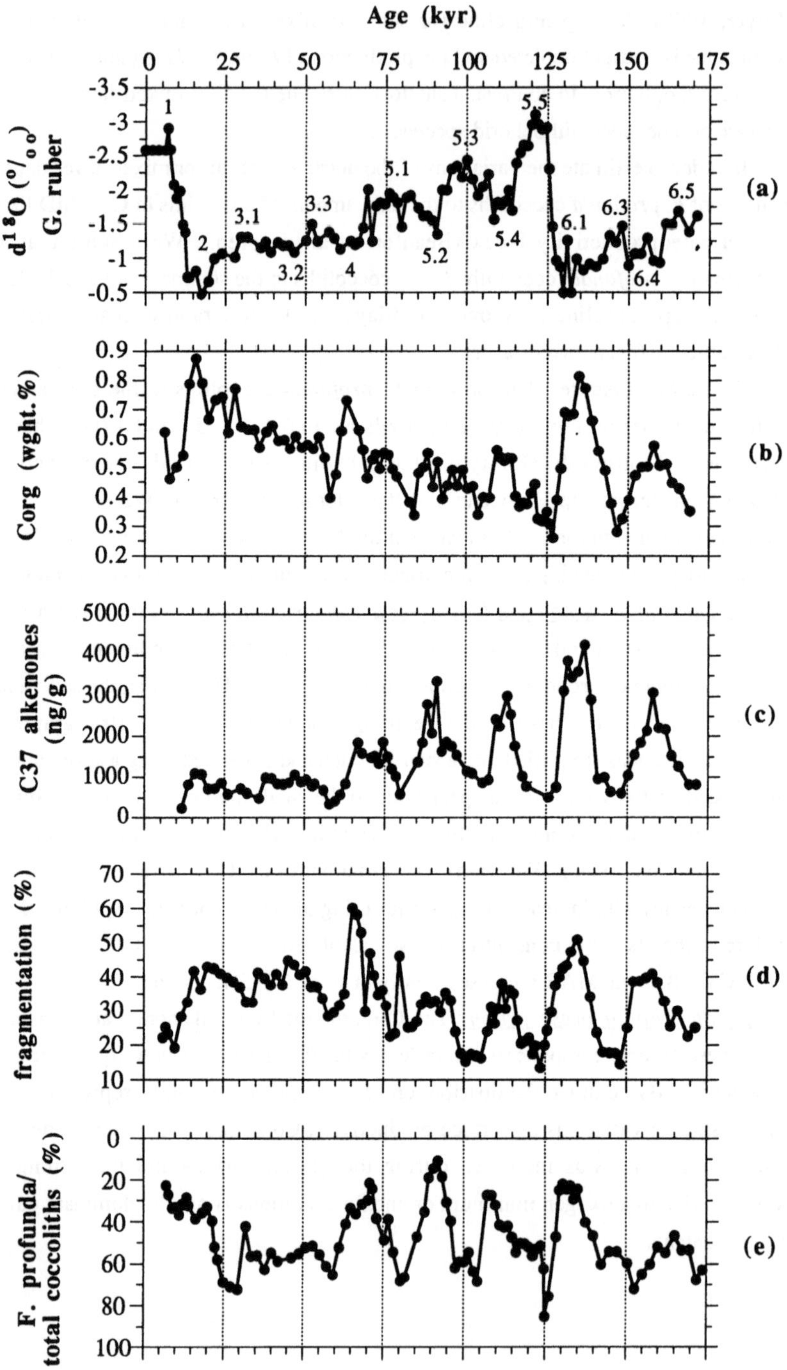

Figure 8: $\delta^{18}O$ stratigraphy (a), organic carbon (b), total concentration of C37 alkenones (c) foraminifera fragmentation (d) and ratio of *F. profunda* /total abundance of coccoliths (e) for Core MD900963.

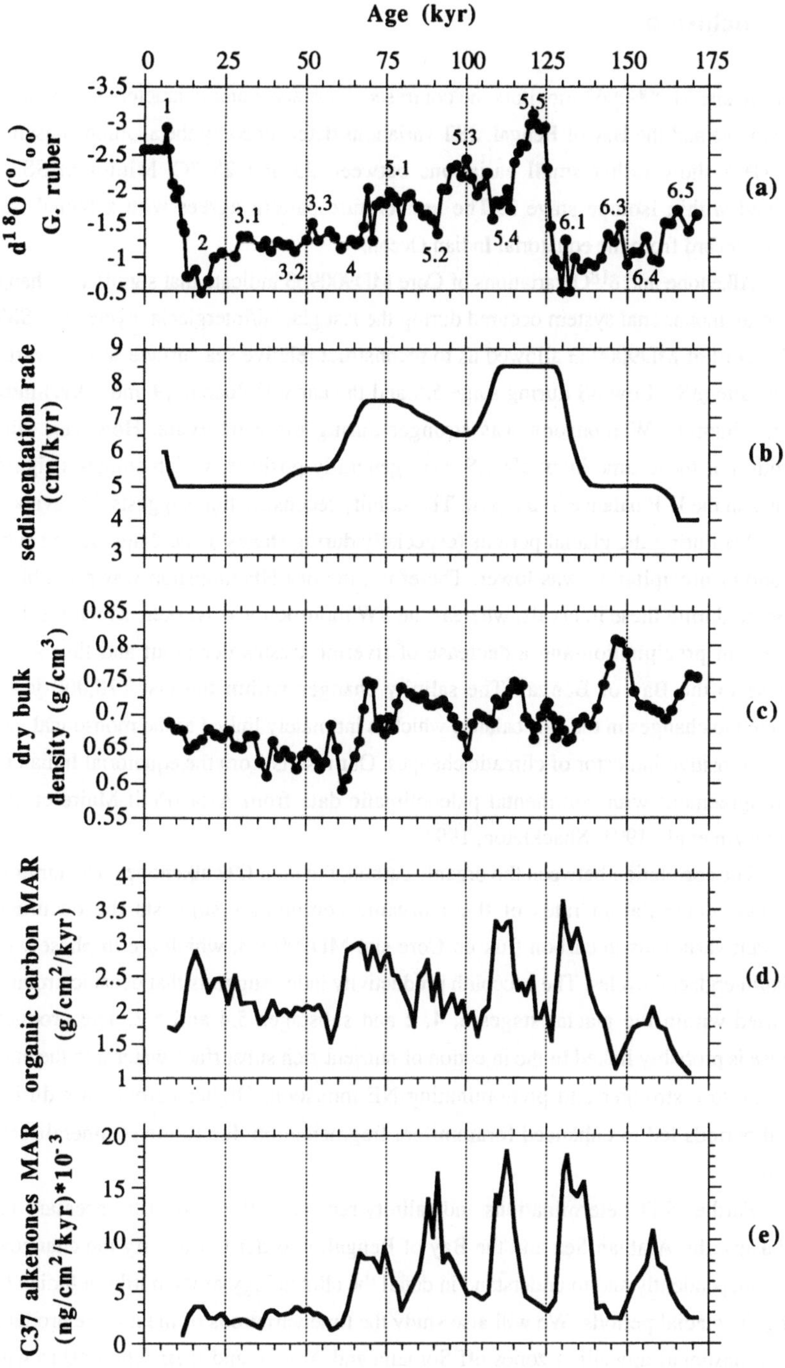

Figure 9: $\delta^{18}O$ stratigraphy (a), sedimentation rate (b), dry bulk density (c), mass accumulation rate of organic carbon (d), mass accumulation rate of C37 alkenones (e) for Core MD900963.

5. Conclusion

The Core site MD900963 monitors the contrasts of surface water characteristics between the Arabian Sea and the Bay of Bengal. SST variations determined by the alkenone unsaturation ratio U^{k}_{37} show rather small variations between 25 and 28 °C. Minimum SST's are observed within isotope stage 3. The temperature pattern agrees with a faunal transfer function record from the equatorial Indian Ocean.

Alkenone and $\delta^{18}O$ variations of Core MD900963 indicate that significant changes of the Indian monsoonal system occured during the last glacial/interglacial cycle. The SST and $\delta^{18}O$ record of MD900963 allowed us to reconstruct relative sea surface salinity variations for this site (ΔS). Low ΔS during stage 5.5 and the early Holocene (9 and 6 kyr) indicates that the humid SW monsoon was stronger during these intervals. However, with the exception of these time intervals, ΔS were generally positive, which suggests significant changes in the E-P balance in the past. The salinity reconstruction suggests that evaporation was higher during the glacial periods (especially during stages 6 and 2 and during substage 5.2) and/or precipitation was lower. Therefore, the dry NE monsoon was probably more important during these intervals, whereas the SW monsoon was weaker, which resulted in a decrease of precipitation and a decrease of riverine freshwater input and thus a salinity increase in the Bay of Bengal. The salinity changes within the last 170,000 years are attributed to changes in the E-P balance which is intimately linked to the monsoonal systems and is a sensitive indicator of climatic changes. Our results from the equatorial Indian Ocean are in agreement with continental paleoclimatic data from Asia (Petit-Maire et al., this volume; An et al., 1991; Shackleton, 1994).

The correlation between the organic carbon, the total C_{37} alkenone concentration and the mass accumulation rates of these organic compounds suggests the occurence of important variations in carbon flux on Core site MD900963, which are in phase with the glacial/interglacial cycles. The coccolith productivity index suggests that oceanic productivity increased within the glacial stages 6, 4, 2 and substages 5.4 and 5.2. The productivity increase is probably linked to the injection of nutrient-rich subsurface water into the euphotic zone due to a stronger and predominating NE monsoon. Higher carbon flux during the glacial periods led to enhanced foraminifera fragmentation due to the remineralization of carbon.

Further SST determinations and salinity reconstructions are still necessary for the two basins, the Arabian Sea and the Bay of Bengal, in order to quantify the changes of E and P independently and to understand in detail the climatology of the northern Indian Ocean during the glacial periods. We will also study the time variations of the marine productivity in the monsoonal upwelling zones off Somalia and Arabia, and their relationship with the salinity index reconstructed in this paper.

Acknowledgements
We thank J.C. Duplessy for providing unpublished $\delta^{18}O$ data. This research is supported by PNEDC, GDR PROMESSE and JE DRED 192.

REFERENCES

An, Z., G.J. Kukla, S.C. Porter and J. Xiao, 1991: Magnetic Susceptibility Evidence of Monsoon Variation on the Loess Plateau of Central China duirng the Last 130,000 Years. Quat. Res., 36, 29-36.

Anderson, D.M. and W.L. Prell, 1993: A 300 kyr record of upwelling off Oman during the late Quaternary: Evidence of the Asian Southwest Monsoon. Paleoceanography, 8, N°. 2, 193-208.

Bassinot, F.C., 1993: Analyse paléocéanographique à haute résolution des carbonates pélagiques des océans Indien et Pacifique en région tropicale. Ph.D. thesis, Université d'Aix-Marseille.

Bassinot, F.C., L.D. Labeyrie, L. Beaufort, L. Meynadier, F. Rostek, J.P. Valet, E. Vincent and Y. Lancelot (in press). High-resolution 900 kyr record in the Maldives Area (tropical Indian Ocean) from a 54 m piston core: monsoon and sea level change signals. Paleoceanography.

Bearman, G., (ed.), 1991: Ocean circulation, Pergamon Press, Oxford, 238 p..

Betts, J.N. and H.D. Holland, 1991: The oxygen content of ocean bottom waters, the burial efficiency of organic carbon, and the regulation of atmospheric oxygen. Palaeogeogr., Palaeoclim., Palaeoecol., 97, 5-18.

Brassell, S. C., G. Eglington, L. T. Marlowe, U. Pflaumann and M. Sarnthein, 1986: Molecular stratigraphy: a new tool for climatic assessment. Nature, v. 320, 129-133.

Broecker, W.S., 1989: The salinity contrast between the Atlantic and Pacific Oceans during glacial time. Paleoceanpgraphy, 4, N°. 2, 207-212.

Cadet, D. and G. Reverdin, 1981: Water vapour transport over the Indian Ocean during summer 1975. Tellus, 33, 5, 476-487.

Chang, A.T.C., L.S. Chin and T.T. Wilheit, 1993: Oceanic Monthly Rainfall Derived from SSM/I. EOS, Transactions, American Geophys. Union, 74, n° 44, Nov. 2, p. 505, 513.

Clemens S., W. Prell, D. Murray, G. Shimmield and G. Weedon, 1991: Forcing mechanisms of the Indian Ocean monsoon. Nature, v. 353, 720-725.

Craig, H. and L.I. Gordon, 1965: Stable isotopes in oceanographic studies and paleotemperatures, (ed. E. Tongiorgi) Spoleto Conference in Nuclear Geology, Spoleto 1965.

Cullen, J.L., 1981: Microfossil evidence for changing salinity patterns in the Bay of Bengal over the last 20,000 years. Palaeogeogr., Palaeoclim., Palaeoecol., 35, 315-356.

Curry, W.B., D.R. Ostermann, M.V.S. Guptha and V. Ittekkot, 1992: Foraminiferal production and monsoonal upwelling in the Arabian Sea: evidence from sediment traps. in : Upwelling systems: evolution since the early Miocene. (eds Summerhayes, C.P., Prell, W.L. & Emeis, K.C.) Geological Society Special Publication No. 64, 93-106.

Duplessy, J.C, P.L. Blanc and A.W.H. Bé, 1981: Oxygen-18 enrichment of planktonic foraminifera due to gametogenic calcification below the euphotic zone. Science, 213, 1247-1250.

Duplessy, J.C. (1982). Glacial to interglacial contrasts in the northern Indian Ocean. Nature, v. 295, 494-497.

Eglinton G., S. A. Bradshaw, A. Rosell, M. Sarnthein, U. Pflaumann and R. Tiedemann, 1992: Molecular record of secular sea surface temperature changes on 100-year timescales for glacial terminations I, II and IV. Nature, 356, 423-426.

Emerson, S. and M. Bender, 1981: Carbon fluxes at the sediment-water interface of the deep-sea. Calcium carbonate preservation. J. Mar. Res., 39, 139-162.

Emerson, S., 1985: Organic carbon preservation in marine sediments. in: The Carbon Cycle and Atmospheric CO2: Natural Variations Archean to Present (eds. E.T Sundquist & W.S. Broecker). Geophys. Monograph, 32, 78-87.

Emerson, S. and J. Hedges, 1988: Processes controlling the organic carbon content of open ocean sediments. Paleoceanography, 3, 621-634.

Fairbanks, R.G., 1989: A 17,000-year glacio-eustatic sea level record: influence of glacial melting rates on the Younger Dryas event and deep-ocean circulation. Nature, 342, 637-642.

Farrington, J.W., A.C. Davis, J. Sulanowski, M.A. McCaffrey, M. McCarthy, C.H. Clifford, P. Dickinson and J. Volkman, 1988: Biogeochemistry of lipids in surface sediments of the Peru Upwelling Area at 15°S. Advances in Geochemistry 1987, Org. Geochem., 13, N° 4-6, 607-617.

Fontugne, M. R. and J.C. Duplessy, 1986: Variations if the monsoon regime during the upper Quaternary: evidence from carbon isotopic record of organic matter in North Indian Ocean sediment cores, Palaogeography, Palaoclimatology, Palaeecology, 56, 69-88.

Imbrie, J., J.D. Hays, D.G. Martinson, A. McIntyre, A.C. Mix, J.J. Morley, N.G. Pisias, W.L. Prell and N.J. Shackleton, 1984: The orbital theory of Pleistocene climate: Support from a revised chronology of the marine $\delta^{18}O$ record, in Milankovitch and Climate. Part I (eds Berger, A., L., Imbrie J., Hays, J., Kukla, G. & Saltzman, B.), D. Reidel, Dordrecht, 269-305.

Ittekkot, V., B. Haake, M. Bartsch, R.R. Nair, R.R. and V. Ramaswamy, 1992: Organic carbon removal in the sea: the continental connection. in Upwelling systems: evolution since the early Miocene. (eds Summerhayes, C.P., Prell, W.L. & Emeis, K.C.) Geological Society Special Publication No. 64, 167-176.

Jasper J. P. and R. B. Gagosian, 1989: Alkenone molecular stratigraphy in an oceanic environment affected by by glacial freshwater events. Paleoceanograpgy, 4, No. 6, 603-614.

Kennedy, J. A. and S. C. Brassell, 1992: Molecular records of twentieth-century El Nino events in laminated sediments from the Santa Barbara basin. Nature, 357, 62-64.

Kleinje, A., D. Kroon and W. Zevenboom, 1988: Phytoplankton and foraminiferal frequences in Northern Indian Ocean and Red Sea surface waters. in G.J.A. Brummer & D. Kroon (eds.): Planktonic foraminifers as tracers of ocean-climate istory. Free University Press, Amsterdam, 271-283.

Labeyrie, L.D.and J.C. Duplessy, 1985: Changes in the oceanic $^{13}C/^{12}C$ ratio duirng the last 140 000 years: high latitude surface water records. Paleogeogr. Paleoclim. Paleoecol., 50, 217-240.

Labeyrie L.D., J.C. Duplessy and P.L. Blanc, 1987: Variations in mode of formation and temperature of oceanic deep waters over the past 125,000 years. Nature, 327, p.477482.

Labeyrie, L.D., J.C. Duplessy, J. Duprat, A. Juillet-Leclerc, J. Moyes, E. Michel, N. Kallel and N.J. Shackleton, 1992: Changes in the vertical structure of the North Atlantic Ocean between glacial and modern times . Quat. Sci. Rev., 121, 401-413.

Lyle, M., 1988: Climatically forced organic carbon burial in equatorial Atlantic and Pacific Oceans. Nature, 335, 529-532.

Lyle, M., F.G. Prahl and M.A. Sparrow, 1992: Upwelling and productivity changes inferred from a temperature record in the central equatorial Pacific. Nature, 355, 812-815.c

Molina-Cruz, A., 1977: The relation of the southern trade winds to upwelling processes during the last 75,000 years. Quat. Res., 8, 324-338.

Molfino, B. and A. McIntyre, 1990: Precessional Forcing of Nutricline Dynamics in the Equatorial Atlantic. Science, 249, 766-769.

Müller, P.J. and E. Suess, 1979: Productivity, sedimentation rate, and sedimentary organic matter in the oceans - I. Organic carbon preservation. Deep-Sea Res., 26A, 1347-1362.

Müller, P.J., R. Schneider and G. Ruhland, 1993: Late Quaternary pCO_2 variations in the Angola Current: Evidence from organic carbon, $\delta^{13}C$ and alkenone temperature. in Carbon cycling in the glacial ocean: constrains on the oceans role in global climate change (eds. R. Zahn, M. Kaminski & T.F. Pederson) NATO ASI Series, 343-366.

Mix, A.C., 1989: Influence of productivity variations on long-term atmospheric CO_2. Nature, 337, 541-544.

Nair, R.R., V. Ittekkot, S.J. Manganini, V. Ramaswamy, B. Haake, A.T. Degens, B.N. Desai and S. Honjo, 1989: Increased particle flux to the deep ocena related to monsoon. Nature, 338, 749-751.

Okada, H. and S. Honjo, 1973: The distribution of oceanic coccolithophorids in the Pacific. Deep-Sea Res., 20, 255-274.

Parkin, D.W. and N.J. Shackleton, 1973: Trade-wind and temperature correlations down a deep-sea core off the Sahara Coast. Nature, 245, 455-457.

Peterson, L.C. and W.L. Prell, 1985: Carbonate dissolution in recent sediments of the eastern equatorial Indian Ocean: Preservation patterns and carbonate loss above the lysocline. Mar. Geol., 64, 259-290.

Pedersen, T.F., 1983: Increased productivity in the eastern equatorial Pacific during the last glacial maximum (19,000 to 14,000 yr B.P.). Geology, 11, 16-19.

Pedersen, T.F. and S.E. Calvert, 1990: Anoxia vs. Productivity: What Controls the Formation of Organic-Carbon-Rich Sediments and Sedimentary Rocks? Amer. Ass. Petr. Geol. Bull., 74, 4, 454-466.

Petit-Maire, N., P. Sanlaville and Z. Yan, 1994: Natural variability of the Asian, Indian and African monsoons over the last 130 ka. in "Global Precipitation and Climate Change" (eds. M. Desbois & F. Désalmand). Nato Advanced Series, this volume.

Prahl, F. G. and S. G. Wakeham, 1987: Calibration of unsaturation patterns in long-chain ketone compositions for palaeotemperature assessment. Nature, 330, 367-369.

Prahl, F.G., L.A. Muehlhausen and M. Lyle, 1988: Further evaluation of long-chain alkenones as indicators of paleoceanograpic conditions. Geochim. Cosmochim. Acta, 52, 2303-2310.

Prahl, F.G. and L.A. Muehlhausen, 1989: Lipid Biomarkers as Geochemical Tools for Paleoceanographic Study. in W.H. Berger, V.S. Smetacek & G. Wefer (eds.): Productivity of the Ocean: Present and Past, 271-289, John Wiley & Sons Limited, Dahlem Konferenzen.

Prell W. L. and W.H. Hutson,1979: Zonal temperature-anomaly maps of Indian Ocean surface waters: modern ice-age patterns. Science, 206, 454-456.

Prell, W.A., W.H. Hutson, D.F. Williams, A.W.H. Bé, K. Geitzenauer and B. Molfino, 1980: Surface circulation of the Indian Ocean during the Last Glacial Maximum, Approximately 18,000 yr B.P., Quaternary Research, 14, 309-336.

Prell, W.L. and E. Van Campo, 1986: Coherent response of Arabian Sea upwelling and pollen transport to late Quaternary monsoonal winds. Nature, 323, 526-528.

Pickard G.L. and W.J. Emery, 1982: Descriptive physical oceanography. An introduction. 4th enlarged Edition, Pergamon Press, Oxford, 249 pp.

Raymont, J.E.G., 1963: Plancton and Productivity in the Oceans. Pergamon Press Book, New York, 660 pp.

Rao, R.R., R.L. Molinari and J.F. Festa, 1989: Evolution of the Climatological Near-Surface Thermal Structure of the Tropical Indian Ocean, 1. Description of Mean Monthly Mixed Layer Depth, and Sea Surface Temperature, Surface Current, and Surface Meteorological Fields. J. Geoph. Res., 94, C8, 10801-10815.

Rostek, F., G. Ruhland, F.C. Bassinot, P.J. Müller, L.D. Labeyrie, Y. Lancelot and E. Bard, 1993: Reconstructing sea surface temperature and salinity using $\delta^{18}O$ and alkenone records. Nature, 364, 319-321.

Sarkar, A., S.K. Bhattacharya and M.M. Sarin, 1993: Geochemical evidence for anoxic deep water in the Arabian Sea during the last glaciation. Geochim. Cosmochim. Acta, 57, 1009-1016.

Sarnthein, M., G. Tetzlaff, B. Koopmannn, K. Wolter and U. Pflaumann, 1981: Glacial and interglacial wind regimes over the eastern subtropical Atlantic and North-West Africa.Nature, 293, 193-196.

Sarnthein, M., K. Winn, J.C. Duplessy and M.R. Fontugne, 1988: Global variations of the surface ocean productivity in low and mid latitudes: Influence on CO2 reservoirs of the deep ocean and atmosphere duirng the last 21,000 years. Paleoceanogr., 3, 361-399.

Savitzky, A. and M.J.E. Golay , 1964: Smoothing and differentation of data by simplified least squares procedures. Analytical Chemistry, 36, N° 8, 1627-1639.

Shackleton, N.J., 1994: Correlations between marine $\delta^{18}O$ and loess sequences. Abstract in "Wind blown sediments in the Quaternary record" The International Union for Quaternary Research, London.

Sikes, E. L., J. W. Farrington and L. D. Keigwin, 1991: Use of alkenone unsaturation ratio Uk37 to determine past sea surface temperatures: core-top SST camlibrations and methodology considerations. Earth and Planetary Science Letters, 104, 34-47.

Sikes, E.L. and J.K. Volkman, 1993: Calibration of long-chain alkenone unsaturation ratios for palaeotemperature estimation in cold polar waters. Geochim. Acta, 57, 1883-1889.

ten Haven, H. L. and D. Kroon, 1991: Late Pleistocene sea surface temperature variations off Oman as revealed by the distribution of long-chain alkenones, in Prell, W. L., Niitsuma, N. et al.: Proceed. of ODP, Sc. Res., (eds W.L. Prell, N. Nitsuma et al.) 117.

Van Campo, E., 1986: Monsoon Fluctuations in Two 20,000-Yr B.P. Oxygen-Isotope/Pollen Records Off Southwest India, Quat. Res., 26, 378-388.

Vogelsang, E., 1990: Paläo-Ozeanographie des Europäischen Nordmeeres anhand stabiler Kohlenstoff- und Sauerstoffisotope. Ph. D. thesis, Universität Kiel, SFB 313: Sedimentation im Europäischen Nordmeer 23, pp. 136.

Volkman J.K, G. Eglinton, E.D.S. Corner and T.E.V. Forsberg, 1980: Long chain alkenes and alkenones in the marine coccolithophorid Emiliania huxleyi, Phytochemistry, 19, 2619-2622.

Wüst, G., W. Brogmus and E.N. Noodt, 1954: Die zonale Verteilung von Salzgehalt, Niederschlag, Verdunstung, Temperatur und Dichte an der Oberfläche der Ozeane. Kieler Meeresforsch., 10, 137-161.

Wyrtki, K., 1971: Oceanogr. Atlas of the Intern. Indian Ocean Expedition. National Science Foundation, Wash. DC, p. 531.

Wyrtki, K., 1973: Physical Oceanography of the Indian Ocean. in: The Biology of the Indian Ocean, Springer-Verlag, pp. 18-36.

ACCUMULATION AT THE SURFACE OF POLAR ICE SHEETS : OBSERVATION AND MODELLING FOR GLOBAL CLIMATE CHANGE

Christophe Genthon

Laboratoire de Glaciologie et Géophysique de l'Environnement du CNRS

Domaine Universitaire BP 96

F-38402 Saint Martin d'Hères Cedex (France)

Jean Jouzel [1]

Laboratoire de Modélisation du Climat et de l'Environnement

CEA DSM, CE l'Orme des Merisiers

F-91191 Gif sur Yvette Cedex (France)

Michel Déqué

Centre National de Recherches Météorologiques, Météo-France

42 Avenue Coriolis

F-31057 Toulouse Cedex (France)

ABSTRACT

Precipitation change over Greenland or Antarctica might have truly global environmental impact because it can affect, directly or through perturbation of the ice flow dynamics, the balance of water between the ocean and the ice sheets, and therefore the global sea level. General Circulation Models (GCMs) of the atmosphere are very popular, and in principle quite appropriate, tools to study the components of a climate change and its consequences, including regional precipitation. They are not altogether successful though, even for reproducing current climate features, and this is particularly true in the polar regions. Models spatial resolution is one important limitation of models performances, particularly at regional scales. We show here how going from rather coarse to very fine resolution, by nowadays standards and considering

[1] Also at Laboratoire de Glaciologie et Géophysique de l'Environnement

NATO ASI Series, Vol. I 26
Global Precipitations and Climate Change
Edited by M. Desbois and F. Désalmand

available computer resources, helps improve the simulation of net annual mean accumulation (precipitation minus evaporation/sublimation) at the surface of Greenland and Antarctica in the Météo-France Arpège GCM. Numerous measurements of accumulation (but few for precipitation) over the two ice sheets exist to check model results, although one may wonder if these observations are all appropriately reliable and if all features of published climatologies are consistently true. On the other hand, ice core analyses bring probably very credible data on accumulation changes over Greenland and Antarctica in the past. Indeed, the agreement is remarkable between accumulation changes over the last glacial-interglacial cycle determined using several fully independent approaches. Such reliable paleoclimate data are essential to check if GCMs used to tentatively predict future climate changes are any good at reproducing observed ones in the past.

KEYWORDS

Precipitation, Antarctic, Greenland, Sea level change, General Circulation Model, Ice core, Paleoclimate.

1 . Introduction

Paleoclimate reconstructions indicate that massive waxing and waning of glaciers and ice sheets, associated with large amplitude (100 m or more) oscillations of the global sea level, were major features of the pleistocene glacial-interglacial cycles. Smaller climate fluctuations in the more recent past are also thought to have contributed to observed trends in sea surface elevation. Likewise, future climate changes may impact on the mass balance of continental ice and therefore on the level of the oceans. IPCC (1990, Chapter 5; 1992, Chapter B) predicts global warming in response to an equivalent doubling of atmospheric CO_2 to be in the range of 1.5 to 4.5°C, and possibly more in the polar latitudes. By the end of next century, the impact of anthropogenic warming on the two major present ice sheets is expected to result in sea level rise anywhere between 0 and +45 cm for Greenland, and -10 to +220 cm for Antarctica (IPCC, 1990, Chapter 9).

As a component of ice sheets mass balance, precipitation bears a substantial part of the forecasting uncertainties. In fact, even the present day net accumulation, that is the current water mass exchange at the interface with the atmosphere (precipitation minus run-off and sublimation), is not well known. As for predicting the contribution of precipitation to mass balance evolution, most attempts to date are local and very much empirically based (which does not necessarily mean that they are crude or irrational): Extrapolation of small fluctuations observed in the recent past; Statistical regression of precipitation versus temperature based on present day spatial distributions or using saturation vapor pressure argument (see section 3); Simulation with energy balance models (IPCC, 1990, Chapter 9).

The General Circulation Models (GCMs) are the most comprehensive and physically based global models available to study atmospheric processes, interactions between climate variables, and mass and energy exchange at the surface. They are also being widely used to tentatively assess the components of a greenhouse affected atmospheric circulation. However, there is still much questioning as to the quality of GCMs produced predictions of climate change. Different GCMs do not generally agree on all aspects of these predictions, and each single GCM is not even consistently good at reproducing all observed features of the current climate (IPCC, 1990, Chapters 4 and 5). On the other hand, as GCMs flaws are being better understood and corrected, and as increasing computer performances allow to reduce many approximations, the prospects of GCMs capabilities is steadily improving.

At present, GCMs tend to be particularly deficient in the polar regions (Simmonds, 1990; Walsh and Crane, 1992; Genthon, 1994). This is partly because these regions have often been considered of lesser importance until recently, and formulations and parameterizations more specifically set up for lower latitudes are not all appropriate for the extreme climate over ice sheets. The role of polar regions in the global climate being now better recognized (Simmonds, 1990), and the global consequences of polar ice sheets response to climate change being identified as potentially large, it becomes clear that improving GCMs performances in the high latitudes is very desirable.

Observation, and first of all present day observation, is necessary to check and constrain any kind of modelling. In the first section of this paper, we remind of the extent and limitations of our knowledge of the present day precipitation or surface mass balance over the two major polar ice sheets. These limitations somewhat hamper progress on GCMs improvement, but do not prevent analyzing the sensitivity of climate over ice sheets in these models. We use in section 2 several versions of a same GCM to explore the impact of resolution. At last, we expose in section 3 why the polar ice sheets provide some of the best available quantitative information on past precipitation fluctuations along with large global climate changes (ice ages). This is a crucial test for the models ability to predict possibly equally large (although opposite) changes in the future.

2 . The present surface mass balance over Greenland and Antarctica

There are not many permanent weather stations in Greenland and Antarctica, and most stations are in the coastal regions where maintenance is easier. In addition, not all stations (and in fact few of them for Antarctica) carry gauge measurements of precipitation because such measurements are particularly difficult, and consequently unreliable, in the harsh environment of the polar ice sheets (Bromwhich, 1988). Precipitation can be very low (a few cm per year), making readings very inaccurate, and frost or wind blown snow can completely obscure the actual meteoric contribution.

Spatial coverage is much larger for the net surface mass balance, or accumulation. Long term mean accumulation can be comparatively easily estimated, either by monitoring stakes burial in snow over years, or by drilling cores or digging pits and dating layers. Water isotopes and snow chemistry can in favorable cases provide annual layers, but they can also be very misleading. Well dated horizons, like the radioactive signatures of atmospheric thermonuclear bomb tests, or ash layers from volcanic eruptions, are generally more reliable when retrievable. For Antarctica, more than 1000 combined such estimates of accumulation are available in data bases (about 1500 are evaluated for the accumulation climatology of Giovinetto and Bentley, 1985, shown on figure 1), largely although not fully comprehensively covering the ice sheet. For Greenland, Ohmura and Reeh (1991, see figure 1) report a network of 251 accumulation data, and precipitation measurements from 35 coastal weather stations.

All data are not equally valid though, and careful and critical selection must be performed when building climatologies (Giovinetto et al., 1989). The large discrepancies between the different accumulation maps produced in the past (Giovinetto and Bull, 1987) clearly illustrate this point and suggest that even the most recently edited maps might be substantially flawed in some areas. Natural variability of accumulation adds up on the technical difficulties of recovering reliable observation. Pettré et al. (1984) and Goodwin et al. (1993) show cases of variability larger than 100% over spatial scales of kilometers to tens of kilometers. Pettré et al. (1984) also evidenced 30% year to year temporal variability at given sites. Many measurements reported in data bases may no adequately fulfill the spatial and time sampling requirements necessary to overcome such difficulties.

Figure 1 displays the most recently published estimates of annual mean accumulation over Antarctica and Greenland. With an accumulation rate of 5 $cm.yr^{-1}$ or less, much of the East Antarctic plateau is more arid than Sahara. West Antarctica gets somewhat more snow, yet it is for a large part dryer than most other places on Earth. Accumulation above 1 $m.yr^{-1}$ is exceptional for Antarctica, but cases have been reported and associated with particular aspects of local meteorology in coastal regions (e.g. Morgan et al., 1991). Because it is not as cold and more subject to ocean moistened air masses than Antarctica, Greenland has a higher nominal surface mass balance. Accumulation over 1 $m.yr^{-1}$ is nonetheless restricted to a small area on the east coast of the southern tip. Accumulation lower than 10 $cm.yr^{-1}$ is also infrequent and mostly confined to the far north of the ice sheet.

The seasonal cycle of precipitation is obviously the feature one expects next, after the annual mean, a model to reproduce. The seasonality has no real impact on sea level that can be felt, so it may appear of limited practical interest to have it correct in a model. On the other hand, it might be a very good test of models dynamics and thermodynamics. For instance, in central Antarctica, precipitation appears larger in winter than in summer (Dolgina and Petrova, 1977; Bromwich, 1988), that is, when atmospheric temperature and therefore mean moisture is lowest. Only a dynamic feature can account for this paradox. Indeed, the frequency of storms produced at the continent edge which are strong enough to travel far inland is much higher in winter. It is consequently though that more moisture transport occurs by perturbations in Antarctica in winter. However, if the direct measurements of the annual mean precipitation are doubtful, those at smaller time scales are even more questionable, and one may wonder how safe it is to use them for models checks.

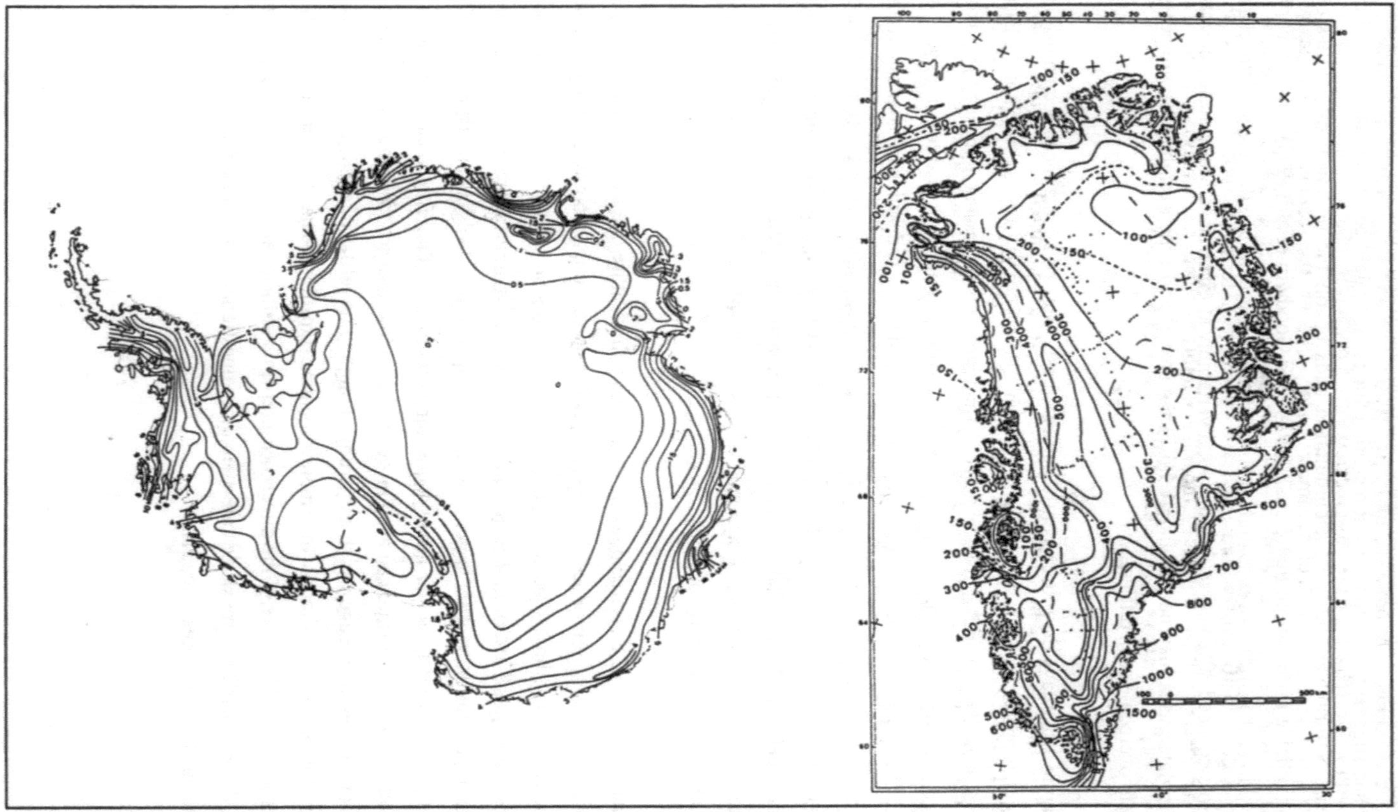

Figure 1: Observed annual mean accumulation at the surface of Antarctica (Giovinetto and Bentley, 1985, in $dm.yr^{-1}$) and Greenland (Ohmura and Reeh, 1991, in $mm.yr^{-1}$).

3. Resolution sensitivity of surface mass balance over Antarctica and Greenland in a GCM

A set of three 10 year simulations of the global climate by the Arpège GCM have been recently completed and archived at Météo-France in Toulouse. The only difference between the three runs is horizontal resolution which ranges from approximately 5.6° for the coarsest (T21) to 2.8° for the intermediate (T42) to 1.5° for the finest (T79) version of the GCM. The experiments where carried out as part of AMIP (Atmospheric Model Intercomparison Project), and they are therefore meant to reproduce the actual climate of the period 1979-1988. They provide a rare opportunity to analyze the modelled climate sensitivity to resolution alone. In particular, the resolution of the T79 model version is close to that nowadays generally available in regional limited area climate models only. Because of the computer resources involved, a 10 year experiment at the resolution of the T79 model is something rather exceptional. T42 is about at the fine end of resolution at which state of the art GCMs are typically used in places where computer resources are not too much of a problem. GCMs with standard resolution close to T21 are not rare.

The Arpège GCM (FCCM, 1991) is a spectral model developed by Météo-France as the climate version of the French weather forecasting model. The spectral truncation is triangular, that is, the leading index of the spherical harmonics retained to integrate the linear dynamics is less or equal to 21 (T21), 42 (T42) or 79 (T79). The associated gaussian grid is almost isotropic in longitude and latitude, so that over the two ice sheets and in cartesian coordinates the grid is much finer in the zonal than in the meridional direction. There are 30 vertical layers, about 10 of which are in the troposphere. The climate of the T42 model over Antarctica is to some extent described and discussed by Genthon (1994). It is not free of flaw, for instance the temperatures tend to be too cold in the coastal regions in winter. Yet it is fairly reasonable compared to some other GCMs (e.g. Tzeng et al., 1993; Hall, 1993; Genthon, 1994).

Figure 2, 3 and 4 display the annual mean accumulation (precipitation minus evaporation/sublimation) over Antarctica and Greenland in the three model runs. It is quite clear that the T21 resolution hardly capture even the largest scale features of accumulation over the two ice sheets. In large areas of Antarctica, accumulation is negative with rates of ablation as high as 50 $cm.yr^{-1}$. Evaporative ablation is known to occur during some years, but only in limited areas and certainly not at such rates and over such long time averages.

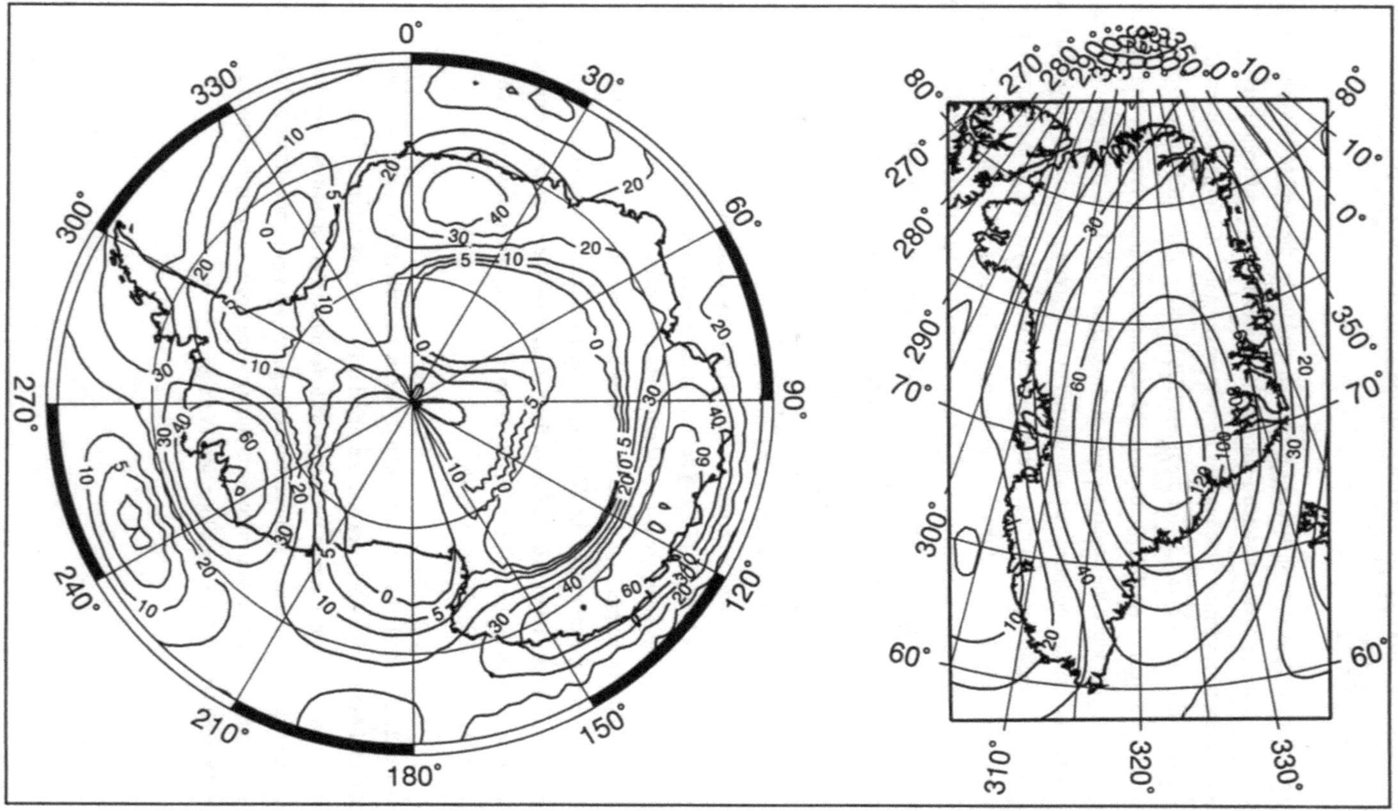

Figure 2: Annual mean accumulation at the surface of Antarctica and Greenland (in $cm.yr^{-1}$), as simulated by the T21 version of the Arpège GCM.

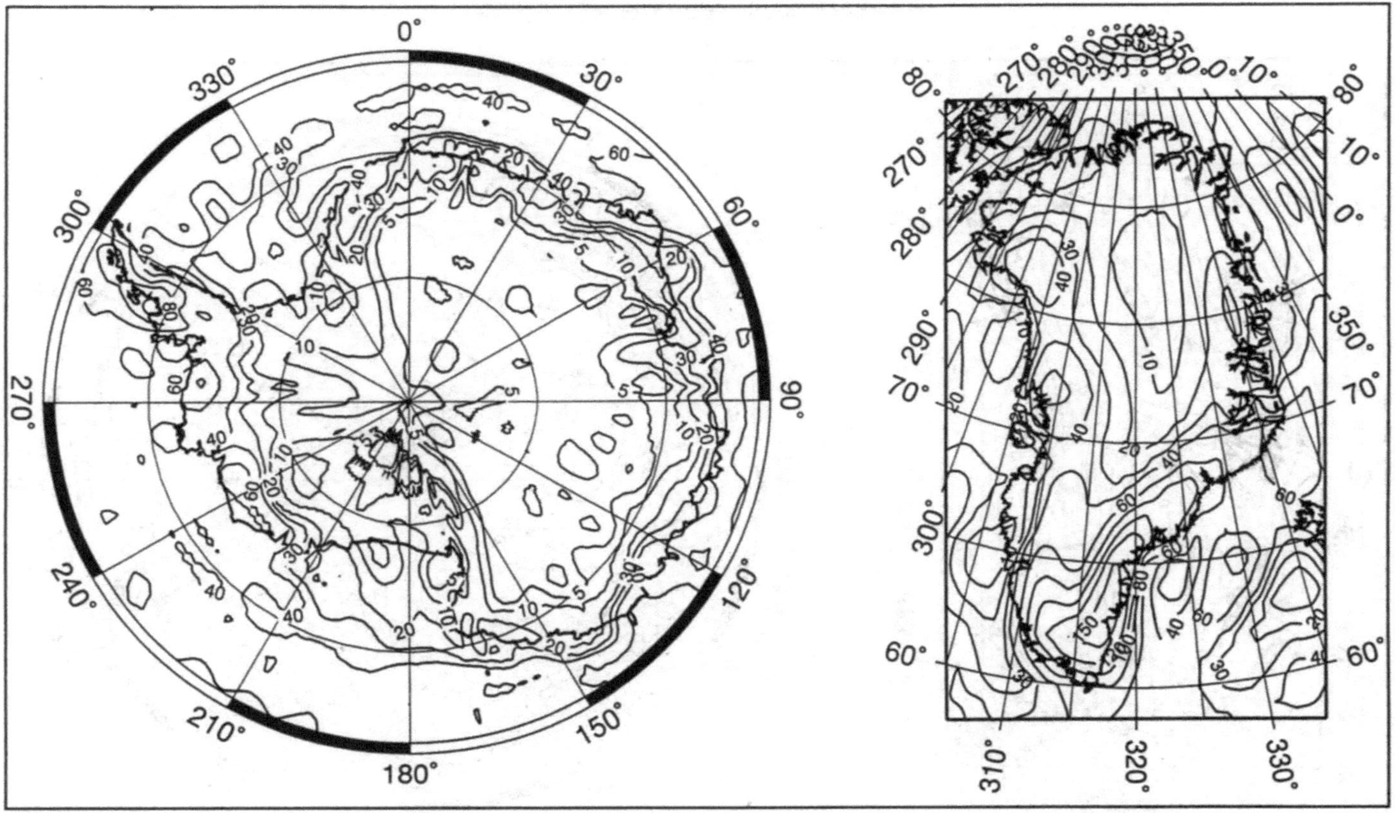

Figure 3: Same as figure 2, for the T42 version.

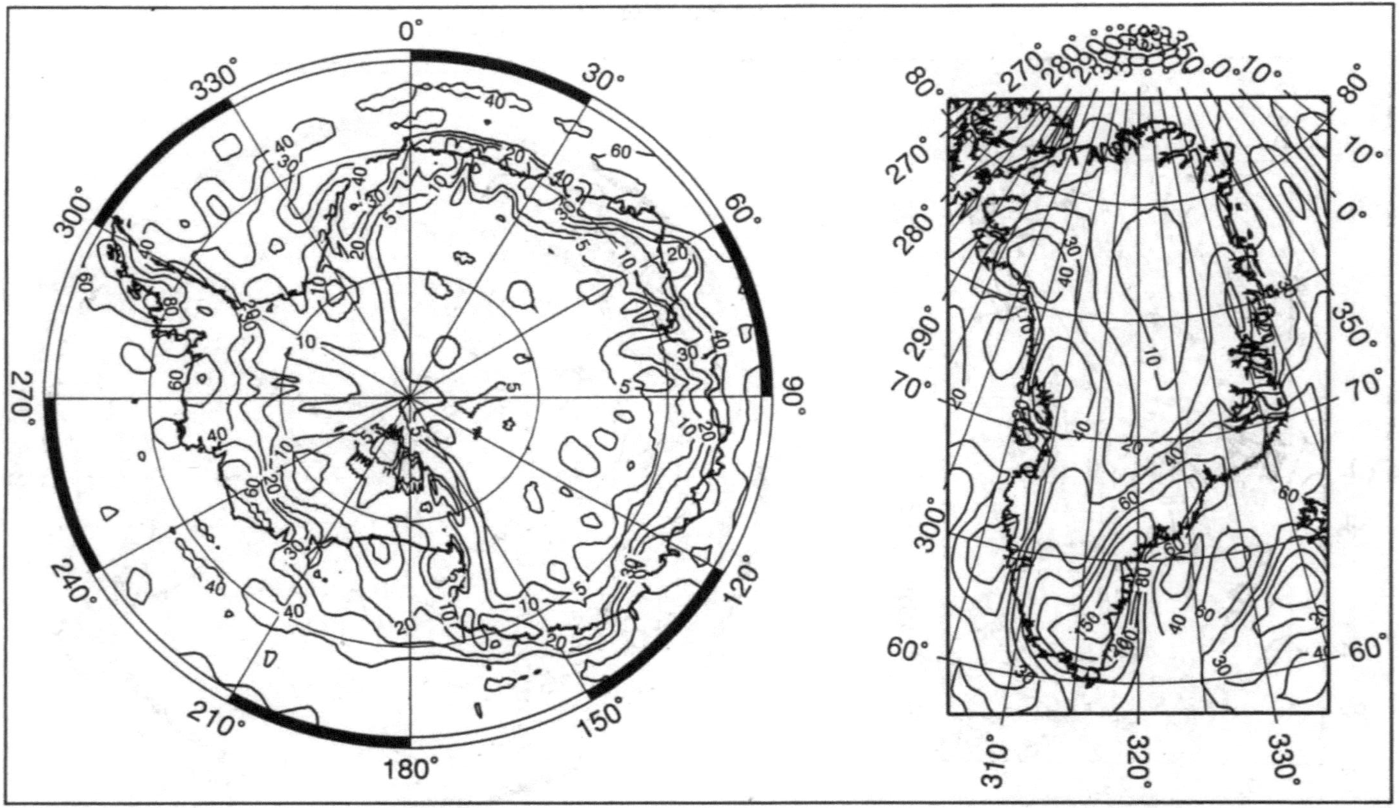

Figure 4: Same as figure 2, for the T79 version.

Doubling resolution (T42) brings a tremendous improvement, both for Antarctica and for Greenland. The East Antarctic plateau and West Antarctic pockets of low accumulation are much better outlined. Altogether though, the model appears too dry inland, with still a lot of ablation, and also dry in the coastal regions. Over Greenland, some realistic features begin to show up, in particular the east-west and north-south contrasts. The tongue of higher accumulation along the west flank of the ice sheet, interpreted by Ohmura and Reeh (1991) as the level of condensation for the air masses raised from sea level along the ice escarpment, is also well reproduced.

At last, T79 brings significant further improvement, indicating that no less than the corresponding resolution should be aimed at to minimize discretization limitations for ice sheets climate modelling with GCMs or other atmospheric models. In Antarctica, negative accumulations are confined to very small and scattered areas. The spotty structure of the accumulation field, already noticeable at T42 resolution, is a characteristic of the spectral advection of moisture. Spectral advection is little diffusive but it generates wavelike distributions and negative values in the wake of strong gradients, as is the case for atmospheric moisture across the coast of Antarctica. A simple scheme to redistribute water is used to eliminate negative mixing ratios, but pockets of artificially low moisture, and therefore low precipitation remain. This, along with a 100% coverage of sea ice around the continent (thus preventing evaporation close to the ice sheet) may explain why the model is on the average too dry.

Over Greenland, the T79 resolution brings in more of the small scale structure than T42, for instance in the southern part of the ice sheet. Interestingly, the region of lowest accumulation in the north, even though it is not a particularly small scale feature, migrates northward and therefore becomes more reasonable. It still extends too far south though. Yet, for both Antarctica and Greenland, many of the maxima and minima at the smallest scale theoretically resolved by the model are not reproduced, and even though these features may be of limited significance for modelling the mass balance of ice sheets, one may wonder how much more of the important information still higher resolution may give access to.

Of course, there is much more than resolution alone into adequately modelling climate over ice sheets, or elsewhere. Fine grids help more accurately resolve small scale aspects of boundary conditions (e.g. topography, of particular importance for ice sheets) and of atmospheric mechanics and thermodynamics (e.g. turbulence, convection). Many essential processes are parameterized instead of being formulated from the fundamentals of physics, either because how these fundamentals act is not well known (e.g. cloud formation) or because computers are not yet powerful enough to cope with the associated volume of numerics. Improving parameterizations is a major topic of GCM research, but improvements can be region dependent. In fact, some of the defects of climate models in the polar regions are apparently due to the fact that they are primarily developed (and hence checked, adjusted and validated) for more temperate latitudes (e.g. snow albedo or cloud parameterizations, see Genthon, 1994).

Parameterizations are also primarily designed and adjusted for best results for modern climate. It remains to check how well they performed when climate is substantially different, before GCMs can be confidently used for prediction of the future consequences of natural or man made perturbations of the environment. This is one motivation for large international programs like AMIP or others in which participating GCMs are run with identical observed or artificially perturbed boundary conditions and the results compared (e.g. Cess et al., 1993). Alternately, paleoclimate observation record natural perturbations of our climate and environment in the past, some of these perturbations being similar in amplitude (but not necessarily in sign or timing) to possible climate changes within the next century. Paleoclimate modelling is therefore adequate to test GCMs ability to reproduce climate changes (e.g. PMIP, the Paleoclimate version of AMIP. See also IPCC, 1990, Chapter 4). Among other sources of paleoclimate information, polar ice core retrieval and analysis brings in the first place data on the atmospheric climate and environment, and therefore provide strong constrains for atmospheric GCMs in polar regions. Precipitation over ice sheets is one crucial information which ice core data can help models better simulate.

4. Glacial-interglacial accumulation changes over polar ice sheets

Deep ice cores from polar ice sheets give access to many climatic and environmental parameters and to their changes in the past. The longest records cover a large part of the last two climatic cycles, i.e. more than 200 ka, both at the Summit site in Central Greenland (Dansgaard et al., 1993; Grootes et al., 1993) and at the Russian station of Vostok in Central East Antarctica (Jouzel et al., 1993). From simple thermodynamic considerations, dependency of saturation vapor pressure on temperature, one can anticipate that precipitation changes are associated with large ice age cooling. It is only recently that such changes have been documented by direct determination of the annual layer thickness, both in the GISP2 (Greenland Ice Sheet Project Two) and GRIP (Greenland Ice Core Project) Greenland cores (Alley et al., 1993; Dahl-Jensen et al., 1993). However, glaciologists have long been interested in this aspect of climate change and in getting indirect estimates of past precipitation rates. One obvious reason is that accumulation is a key parameter for dating deep ice cores. Also, accumulation changes must be taken into account for calculating, from its measured concentration, the annual flux of any compound found in the ice.

The first ice core chronology established accounting for accumulation change was for the 900 m Dome C core in East Antarctica (Lorius et al., 1979). The underlying idea is that the modern precipitation rate over Antarctica appears to be governed by the amount of water in the atmosphere above the inversion layer (Robin, 1977). This amount is itself related to the water vapor saturation pressure and thus to the temperature of formation of precipitation. It is this argument that was used to estimate past precipitation rate noting that the snow isotopic content (deuterium D or oxygen 18 ^{18}O) gives access to the temperature of formation of precipitation in the past (Lorius et al., 1979).

This variable accumulation approach has been refined and applied both to the Dome C (Benoist, 1986; Jouzel et al., 1989) and Vostok cores (Lorius et al., 1985). The precipitation rate is calculated continuously along the cores as the product of its present value by the ratio of the derivatives of the saturation vapor pressure for the time of precipitation and for present conditions. It is the formulation that is directly derived from a simplified unidimensional model, neglecting possible changes in atmospheric circulation and in the local horizontal temperature gradient above the area of precipitation. Keeping this simple unidimensional approach, Ritz (in press) has paid a special attention to determining the processes by which air masses are cooled during their travel above Antarctica. This is because the change in precipitation between glacial and interglacial depends on the relative importance of the different cooling processes (orographic lifting, radiative cooling and subsidence). Ritz (in press) confirmed that on the Antarctic Plateau where radiative cooling and subsidence are predominant, accumulation rate change with time at a given location is governed, in this simple model, by the temperature at the inversion level. For example, this model leads to estimate that during the Last Glacial Maximum (LGM), the accumulation in Central East Antarctica was about half of its present value.

There is little direct justification for the above procedure of calculating paleoprecipitation rate: 1) It assumes that the present correlation between precipitation pattern and saturation vapor pressure was also valid in the past, 2) It implies either that atmospheric circulation patterns have remained stable over time, which seems to be contradicted by terrestrial dust fallout records (Petit et al., 1990) or, at least, that they do not significantly affect this correlation and 3) It assumes that our estimate of the changes in temperature of formation of precipitation from the isotopic record is correct. Moreover, accumulation is too low in this area for a clear annual cycle to be formed which prevents any direct verification of this procedure from the determination of the annual layer thickness.

Surprisingly, support to this very simple 1D model of estimating paleoprecipitation rate came from the measurement of ^{10}Be, a cosmogenic isotope produced by the interaction of cosmic rays with the upper atmosphere. In the Dome C ice, Raisbeck et al. (1981) found a ^{10}Be concentration that was larger in glacial age by a factor of 2 to 3 than in Holocene ice. Among the possible explanations offered was that ^{10}Be deposition flux is constant with time implying that the precipitation rate is inversely proportional to the ^{10}Be concentration. During the glacial period this rate was thus about 50% of that during the Holocene, a value consistent with the simple model described above.

New insight into the ^{10}Be precipitation rate correlation and the vapor pressure procedure came from the Vostok ice core (Yiou et al, 1985; Jouzel et al., 1989). Figure 5 shows the calculated precipitation rate estimated over the last glacial-interglacial cycle from a) the ^{10}Be profile (Raisbeck et al., 1987) assuming a constant deposition flux and b) the paleoaccumulation profile (Jouzel et al., 1987) using the vapor pressure argument. Although there are differences (amounting as much as 50% at some depths), the overall agreement is quite remarkable. It also appears that the annual precipitation in the Byrd West Antarctic site increased some 2.5 times during the transition from the late glacial ice age to Holocene times (Hammer et al., in press), here again in relatively good agreement with the vapor pressure argument. Although results of this type are not yet available for coastal sites, there is little doubt that this about halving of the accumulation during glacial times affected most of the Antarctic ice sheet.

Clearly, the precipitation rates based on ^{10}Be cannot be accurate in detail because there are several causes of variations in ^{10}Be deposition flux (Raisbeck et al, 1981, 1987; Mazaud et

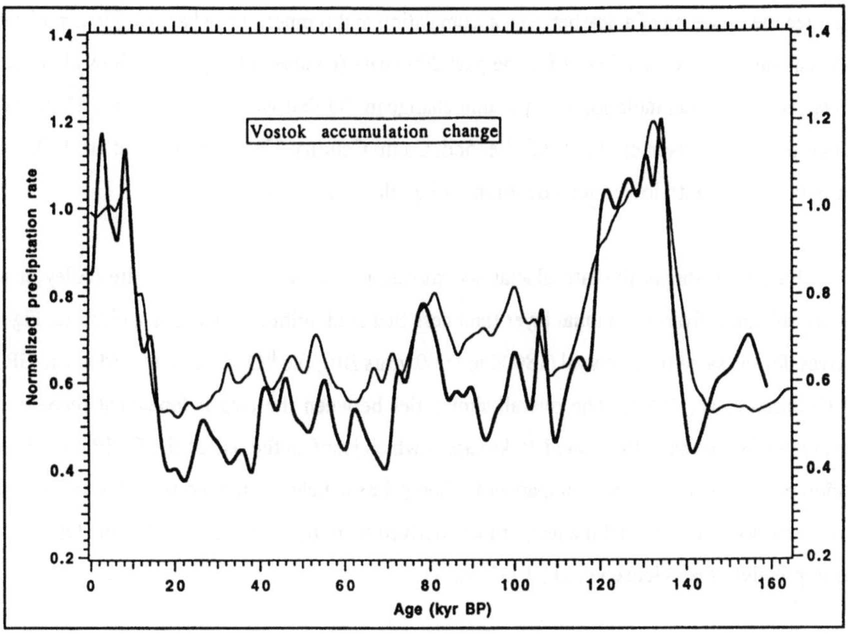

Figure 5: Vostok ice core, normalized precipitation rate with respect to Holocene value. Thick line: from ^{10}Be record; Thin line: from change in the derivative of the saturation vapor pressure.

al., in press). It is obviously the same for those derived through the vapor pressure argument that constitutes a too simplistic approach. However, we have a situation in which two relatively independent procedures for estimating polar paleoprecipitation, neither on a particularly solid theoretical footing give similar and apparently reasonable results. Further, for the Byrd site, accumulation change is supported by the change in the thickness of annual layers (Hammer et al., in press). Taken together these results not only support that Antarctic precipitation were lower during colder periods but also that the accumulation change somewhat follows the temperature change and may indeed be roughly estimated from the derivative of the saturation vapor pressure.

The existence of such a simple temperature / precipitation rate relationship has been attributed to the relative simplicity of the meteorological conditions prevailing over the Antarctic Plateau. Due to more complex weather patterns, it was expected that the situation should not be as apparently simple over Greenland. In fact, prior the results recently derived from the study of the GRIP and GISP2 Central Greenland ice cores discussed below, two earlier results already suggested the existence of a link between accumulation and temperature changes. First, statistical studies at various Greenland sites for the past 200 years (Clausen et al., 1988) show about an 8 per cent change in accumulation rate per mill change in ^{18}O that would translate in a 12 per cent change per °C (Johnsen et al., 1992). Second, Camp Century ^{10}Be data (Beer et al., 1992) are clearly consistent with lower accumulation during glacial periods.

Figure 6a shows the late glacial accumulation history at the GISP2 site (Alley et al., 1993a) estimated from the annual layer thickness that is identified by various indicators. Figure 6b gives, for the same time interval (18000 to 9000 years BP), the ^{18}O climatic record at the GRIP site (Johnsen et al., 1992). The overall similarities between the two independent records are remarkable (the two sites are located 30 km apart which justifies the use of the GRIP record (see Grootes et al., 1993) for such a comparison). Though less detailed at the moment, the comparison between the accumulation and the temperature derived from the isotope record in the GRIP core is as impressive (Dahl-Jensen et al., 1993).

Alley et al. (1993a) note the apparent consistency between the temperature change deduced applying the vapor pressure argument and that derived from the isotope record. For example, both the twofold increase in accumulation at the Younger-Dryas/Preboreal transition and the isotopic record lead to an estimate of about 7°C for the associated temperature change. However a detailed comparison now being performed directly on the GISP2 accumulation and isotope temperature records, suggests that the accumulation/temperature relationship may be time dependent (Kapsner et al., 1993; Alley et al., 1993b). During warm and cold periods of the last deglaciation, accumulation is less sensitive to temperature than expected applying the vapor pressure procedure whereas it is the opposite during transitions between warm and cold states, such as the onset and termination of the Younger-Dryas event. Alley et al. (1993b) interpret this change as resulting from large shifts in storm tracks away from Greenland during major coolings and towards Greenland during major warmings. Another interesting feature of the accumulation change is its extreme rapidity of the doubling associated with the end of the Younger-Dryas possibly occurring in one to three years (Alley et al., 1993a).

These latest findings from ice cores certainly constitute a major and unexpected challenge for climate modelling in the polar regions. Since dynamical aspects of the large scale atmospheric circulation are probably involved, they are bound to particularly challenge GCMs and their ability

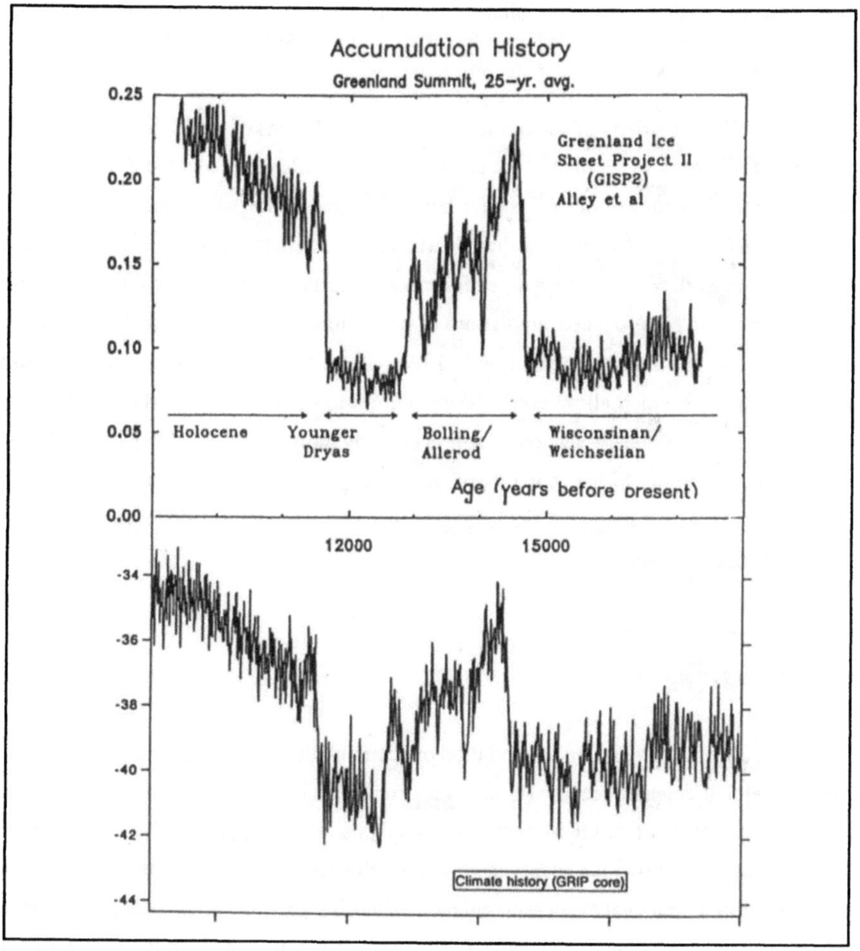

Figure 6: The accumulation history at GISP2 site (from Alley et al., 1993) and the ^{18}O record at GRIP site, a proxy of change in temperature of formation of precipitation (from Johnsen et al., 1992).

to reproduce the components of an atmospheric circulation altered by climate change. However, a lot remains to be understood with regards to the role of the fast responding part of the other components of the climate system (ocean, sea ice, land ice instabilities...) before providing strong constrain to atmospheric modelling. On the other hand, relatively well documented periods of the past like the last full ice age (LGM) already provide adequate boundary conditions and validation information (in particular for precipitation over ice sheets) for snapshot simulations of the atmosphere with GCMs.

Gates (1976) carried out one of the first GCM simulations of the LGM soon after the first global and comprehensive set of relevant paleoclimate data was made available (CLIMAP, 1976, 1981). An increasing number of GCMs have been submitted to the paleoclimate test since (IPCC, 1990, Chapter 4). It is not surprising that GCMs do not consistently agree with observation or with each other, as they do not even agree for modern climate. In particular, only in the past ten years have some GCMs produced simulations of polar climate at a level of quality similar to that attained in the lower latitudes (Simmonds 1990), in parts tanks to the increasing power of available computers which allow ever finer resolution to be used. The surface temperature in some GCMs, in their standard version, is still in error by more than 10°C though (e.g. Walsh and Crane, 1992, Genthon, 1994), but others like Arpège show very encouraging results for precipitation (see also Connolley, 1994, for an other very convincing model result), a variable even more difficult to simulate.

5 . Conclusions

Modelling and predicting precipitation or accumulation over Greenland and Antarctica is certainly of global concern because of the direct impact of ice sheets mass balance on sea level. We have here presented a GCM sensitivity analysis which suggests that the present day accumulation over the two ice sheets is very, though certainly not only, dependent on model spatial resolution. For annual mean accumulation, extensive yet in some cases possibly inaccurate observation and compilations are available to check models results and identify deficiencies. Data sets are much more limited and therefore questionable for precipitation alone. This is unfortunate because the seasonal cycle of precipitation could be a good test of the dynamics which determine the hydrologic cycle over ice sheets.

Even if a model simulates current climate acceptably well, there is no guaranty that it is good at predicting climate change. In this respect, paleoclimate data provide valuable controls for climates significantly different from the current one. We have reviewed evidences that ice cores can yield good estimates of accumulation changes over the past glacial-interglacial cycles. For instance, it is very likely that accumulation at the LGM over the ice sheets was half or less that of nowadays. This is indeed a dramatic change and a strong constrain to models in the polar regions. Thanks to programs like AMIP and PMIP, we expect soon to have a comprehensive view of how state of the art GCMs succeed to reproduce such observation.

It is not unlikely that just going to resolutions even finer than we have tested (and T79 is indeed very fine by nowadays standards for a GCM) would bring more of the observed fine structure of accumulation distribution over the ice sheets. If grids remain equally fine over the entire globe, then only through improvements of computers performances and use can finer resolutions be attained. Massively parallel computing might be part of the answer in a not to distant, yet not immediate future. Alternately, for regional climate modelling, there is a growing interest for using fine grid Limited Area Models (LAMs), necessarily coupled or embedded within coarser GCMs for climate changes experiments (Giorgi and Mearns, 1991), or stretchable grid GCM within which the resolution can be set very fine over a given region only, while the model full continuity is kept. There is certainly a lot to expect from such approaches for ice sheets climate modelling.

Acknowledgments: This research has benefitted from financial support by the Commission of the European Communities (contract EV5V-CT92-0051) and by the Fondation de France. We are greatly indebted to all participants of the Vostok and GRIP projects.

REFERENCES

Alley R.B., Meese D.A., Shuman C.A., Gow A.J., Taylor K.C., Grootes P.M., White J.W.C., Ram M., Waddington E.D., Mayewski P.A. and Zielinski G.A., 1993a : Abrupt increase in Greenland snow accumulation at the end of the Younger-Dryas event, Nature, 362, 527-529.

Alley R.B., Kapsner W.R., Shuman C., Anandakrishnan S., Finkel R.C., Gow A.J., Grootes P.M., Mayewski P.A., Meese D.A., Nishiizumi K. and Waddington E.D., 1993b : Clues to the Younger Dryas from accumulation-rate measurements in the GISP 2 ice core, EOS Supplement, October 1993, 79.

Beer J., Johnsen S.J., Bonani G., Finkel R.C., Langway C.C., Oeschger H., Stauffer B., Suter M. and Woelfi W., 1992 : ^{10}Be peaks as time markers in polar ice cores, *NATO ASI* series "The last deglaciation : Absolute and radiocarbon chronologies". E. Bard and W.S. Broecker Eds., 141-153.

Benoist J.P., 1986 : Analyse spectrale des signaux glaciologiques: Etudes des glaces sédimentaires déposées au Dôme C, Morphologie du lit d'un glacier, PhD thesis, Université Scientifique, Technologique et Médicale de Grenoble, 329 pp.

Bromwhich D.H., 1988 : Snowfall in high southern latitudes, Rev. Geophys., 26, 149-168.

Cess R.D., and 29 co-authors, 1993 : Uncertainties in carbon dioxide forcing in atmospheric general circulation models, Science, 262, 1252-1255.

Clausen H.B., Gundestrup N.S., Johnsen S.J., Bindschdler R. and Zwally J., 1988 : Glaciological investigations in the Crete area, Central Greenland: Search for a new deep drilling site, Annals of Glaciology, 10, 10-15.

CLIMAP Project Members, 1976 : The surface of the ice-age Earth, Science, 191, 1131-1136.

CLIMAP Project Members, 1981 : Seasonal reconstructions of the Earth's surface at the last glacial maximum, Geological Society of America Map and Chart Series, MC-36.

Connolley W.M. and Cattle H., 1994 : The Antarctic climate of the UKMO Unified Model, Antarctic Science, 6, 115-122.

Dahl-Jensen D., Johnsen S.J., Hammer C.U., Clausen H.B., Jouzel J., 1993 : Past accumulation rates derived from observed annual layers in the GRIP ice core from Summit, Central Greenland, NATO ASI Series "Ice in the climate system" (W.R. Peltier Ed.), Vol. I 12, 517-532.

Dansgaard W., Johnsen S.J., Clausen H.B., Dahl-Jensen D., Gundestrup N.S., Hammer C.U., Steffensen J.P., Svienbjörnsdottir A.E., Jouzel J. and Bond G., 1993 : Evidence for general instability of past climate from a 250-kyr ice-core record, Nature, 364, 218-220.

Dolgina I.M. and Petrova L.S., 1977 : Handbook of Antarctic climate (in Russian), Vol.2, Hydrological Institute, St.Petersburg, 493 pp.

(The) French Community Climate Model, specifications, 1991 : By Météo-France, Les Universités and CNRS. Available from Daniel Cariolle, Météo-France, CNRM, Toulouse, France, 44 pp.

Gates W.L., 1976 : The numerical simulation of ice-age climate with a global general circulation model, J. Atmos. Sci., 33, 1844-1873.

Genthon C., 1994 : Antarctic climate modelling with General Circulation Models of the atmosphere, J. Geophys. Res., accepted.

Giorgi F. and Mearns L.O., 1991 : Approaches to the simulation of regional climate change: A review, Rev. Geophys., 29, 191-216.

Giovinetto M.B. and Bentley C.R., 1985 : Surface balance in ice drainage of Antarctica, Antarct. J.U.S., 20, 6-13.

Giovinetto M.B. and Bull C., 1987 : Summary and analysis of surface mass balance compilations for Antarctica 1960-1985, Byrd Polar Research Center Report 1, Columbus, Ohio, 90 pp.

Giovinetto M.B., Bentley C.R. and Bull C., 1989 - Choosing between some incompatible regional surface mass balance data sets in Antarctica, Antarct. J. U.S., 24, 7-13.

Grootes P.M., Stuiver M., White J.W.C., Johnsen S. and Jouzel J., 1993 : Comparison of oxygen isotope records from the GISP2 and GRIP Greenland ice cores, Nature, 366, 552- 554.

Hall A., 1993 : The interaction between climate and ice sheets, with special reference to the boundary layer of the ablation zone. PhD thesis, Dept. of Meteorology, U. of Edimburg, 251 pp.

Hammer C, Clausen H.B. and Langway C.C., 1993 : ECM stratigraphic dating of the Byrd station core, Annals of Glaciology, in press.

Intergovernmental Panel on Climate Change (IPCC), 1990 : Climate change, the IPCC scientific assessment. Houghton J.T., Jenkins G.J. and Ephraums J.J., Eds, Cambridge University Press, Cambridge, 365 pp.

Intergovernmental Panel on Climate Change (IPCC), 1992 : Climate change, the supplementary report. Houghton J.T., Callander B.A. and Varney S.K. Eds, Cambridge University Press, Cambridge, 200 pp.

Johnsen S.J., Clausen H.B., Dansgaard W., Fuhrer K., Gunderstrup N., Hammer C.U., Iversen P., Jouzel J., Stauffer B. and Steffensen J.P., 1992 : Irregular glacial interstadials recorded in a new Greenland ice core. Nature 359, 311-313.

Jouzel J., Lorius C., Petit J.R., Genthon C., Barkov N.I., Kotlyakov V.M. and Petrov V.N., 1987 - Vostok ice core: a continuous isotope temperature record over the last climatic cycle (160 000 years). Nature 329, 403-408.

Jouzel J., Raisbeck G., Benoist J.P., Yiou F., Lorius C., Raynaud D., Petit J.R., Barkov N.I., Korotkevitch Y.S. and Kotlyakov V.M., 1989. A comparison of deep Antarcti ice cores and their implications for climate between 65 000 and 15 000 years ago, Quaternary Research, 31, 135-150.

Jouzel J., Barkov N.I., Barnola J.M., Bender M., Chappellaz J., Genthon C., Kotlyakov V.M., Lipenkov V., Lorius C., Petit J.R., Raynaud D., Raisbeck G., Ritz, C., Sowers T., Stievenard M., Yiou P., Yiou F., 1993 - Extending the Vostok ice core record of paleoclimate to the penultimate glacial period, Nature, 364, 407-412.

Kapsner W.R., Alley R.B., Anandakrishnan S., Shuman C., Grootes P.M., Meese D.A. and Gow A.J., 1993 - Response of snow accumulation to temperature variations in Central Greenland. EOS Supplement, October 1993, 79.

Lorius C., Merlivat L., Jouzel J. and Pourchet M., 1979 : A 30 000-year isotope climatic record from Antarctic ice, Nature, 280, 644-648.

Lorius C., Jouzel J., Ritz C., Merlivat L., Barkov N.I., Korotkevitch Y.S. and Kotlyakov Y.M., 1985 : A 150 000 year climatic record from Antarctic ice, Nature, 316, 591-596.

Mazaud A. , Laj. C. and Bender M., 1993 : A geomagnetic chronology for Antarctic ice accumulation, Geophys. Res. Lett., in press.

Morgan V.I., Goodwin I.D., Etheridge D.M. and Wookey C.W., 1991 : Evidence from Antarctic ice cores for recent increases in snow accumulation, Nature, 354, 58-60.

Ohmura A. and Reeh N., 1991 : New precipitation and accumulation maps for Greenland. J. Glaciol., 37, 140-148.

Petit J.R., Mounier L., Jouzel J., Korotkevitch V.I., Kotlyakov V.I.and Lorius C., 1990 : Paleoclimatological and chronological implications of the Vostok dust record, Nature, 343, 156-158.

Pettré P., Pinglot J.F., Pourchet M. and Reynaud L., 1986 : Accumulation distribution in Ţerre Adélie, Antarctica : Effect of meteorological parameters, J. Glaciol, 32, 486-500.

Raisbeck G.M., Yiou F., Fruneau M.M., Loiseaux J.M., Lieuvin M., Ravel J.C. and Lorius C., 1981 : Cosmogenic ^{10}Be concentrations in antarctic ice during the past 30 000 years, Nature, 292, 825-826.

Raiseck G.M., Yiou F., Bourles D., Lorius C., Jouzel J. and Barkov N.I., 1987 : Evidence for two intervals of enhanced deposition in Antarctic ice during the last glacial period, Nature, 326, 273-277.

Ritz C, 1993 : Chronology of the Vostok ice core by ice flow modelling, J. Glaciol., in press.

Robin G de Q, 1977 : Ice cores and climatic changes, Philosophical Transactions of the Royal Society of London B, 280, 143-168.

Simmonds I., 1990 : Improvements in General Circulation Model performances in simulating Antarctic climate, Antarctic Sci., 2, 287-300.

Tzeng R.I., Bromwich D.H. and Parish T.R., 1993 : Present day Antarctic climatology of the NCAR CCM version 1, J. Climate, 6, 205-226.

Walsh J.E. and Crane R.G., 1992 : A comparison of GCM simulations of Arctic climate. Geophys. Res. Let., 19, 29-32.

Yiou F., Raisbeck G., Bourles G., Lorius C. and Barkov N.I., 1981 : ^{10}Be in ice at Vostok, Antarctica, during the last climatic cycle, Nature, 316, 616-617.

GLOBAL CLIMATE CHANGES AND HUMIDITY VARIATIONS OVER EAST EUROPE AND ASIA BY HISTORICAL DATA

Andrei O. Selivanov
Water Problems Institute
Russian Academy of Sciences
Novobasmannaya St. 10, P.O.Box 524
107078 Moscow (Russia)

ABSTRACT

Historical, or documentary, data serve as an important independent source of information on past climates, including humidity, precipitation, and river runoff. These data, covering time scales of several hundreds to several thousand millennia, fill in the gap between data of direct, instrumental, observations and paleodata. Extensive data bases of such proxies have been compiled from the Chinese, Russian, and other historical sources. However, a variety of human-induced factors (national mentality, wars, etc.) that affected registration of climatic and hydrological events often prevent scholars from application of such data to the quantification of past climate changes.

An approach is proposed to quantify changes in mean annual air temperature, precipitation, and river runoff derived from historical data on extreme events. Use of a fraction of climate extremes of a certain kind among all extremes registered during the time interval (usually several decades) eliminate partly human-induced factors. Application of this technique is exemplified by the records from East European Plain covering the last millennium.

A qualitative analysis of climate changes during the last 2,500-3,000 years over the whole extratropical Asia and East Europe is also presented.' Major phases of thermal changes during this period, Iron Age cooling, Medieval climate optimum, and Little Ice Age, can be traced over the whole Eurasia, but they are usually aged from earlier times in eastern sectors of the continent. The similar pattern of set in characterizes periods of increased precipitation and river runoff.

Negative correlation of temperature and humidification anomalies appears to be the important feature in the arid sectors of Eurasia, whereas in most other regions correlation of these parameters is usually positive, but less intensive. As a first approximation, eight regions can be discerned in the extratropical Asia and East Europe by the differences in their humidity response to the large-scale changes in air temperature during the last millennia.

KEYWORDS

Climate changes, aridity, humidity, historical data, proxy data, Medieval climate optimum, Little Ice Age,East Europe, Central Asia, East Asia.

NATO ASI Series, Vol. I 26
Global Precipitations and Climate Change
Edited by M. Desbois and F. Désalmand

1. Introduction

Possible future greenhouse-induced warming can result in significant changes of precipitation, river runoff, and evaporation values over the continents, i.e. in a shift to humidity or aridity. It is obvious that such changes will determine regional planning issues, especially in presently arid and semi-arid regions. Therefore, prediction of changes in precipitation, evaporation, and river runoff is of urgent need. Nevertheless, pattern of humidity response to the changes in global air temperature is not universal. Problem of correlation between thermal and regional humidity changes remains one of the key issues in climatology.

Many scholars believe in the universal character of positive correlation between global air temperature and humidity over continents. In a simplified way such a mechanism can be explained as following: higher air temperature results in higher evaporation from an ocean surface, and, therefore, in an increase in precipitation over continents. Nevertheless, accurate analysis of instrumental time series demonstrates that in temperate East Europe and Asia a significant positive correlation between air temperature and precipitation over decadal time scale exists only in maritime areas at continental borders, whereas in the inner continental regions relatively arid conditions occur during warm climatic phases (Drozdov, 1985; Klige, 1990). Correlation pattern can be even more complicated for interannual variations.

However, instrumental time series are too short in time and demonstrate relatively low variability in order to form a data base for long-term prediction, especially under the anticipated human-induced intensification of climate changes. On the other side, paleogeographic proxy data in most cases are rather qualitative and can hardly be used in construction of uniform time series of climatic parameters. Problem of temporal resolution makes application of this kind of information even more complicated.

2. Sources of information on climate and weather during the last millennia

Whereas instrumental time series represent changes over 10-100-year time scale, paleodata covers 1000-100,000-year time scales. Secular changes can not be derived from either of these data sources. Therefore, historical and other indirect sources of information on secular climate changes form a necessary bridge between direct (instrumental) and paleogeographical data bases.

Data on climatic and hydrological changes during the last millennia can be distinguished into historical data (documentary data, DD, according to M.J.Ingram et al., 1984) and proxy data (PD) derived from different natural sources (e.g. glaciological, palynological, faunistic analyses). PD can serve as an independent source of information to verify DD in some cases. This is especially true if the same climate-driven natural parameter can be traced both by DD and PD (e.g. water level of an enclosed lake). This paper deals primarily with DD.

Climatic information can be derived from different kinds of DD, e.g. weather registers, chronicles, diaries, account books, state and local documents, farm and military reports, etc. as well as fiction literature can also be used (Battan, 1983; Lamb, 1987). Neuberger (1970) demonstrated an excellent example of climatic information derived from paintings of different ages and countries. Nevertheless, chronicles, including weather registers, indications of fires, droughts, floods, frosts, heats, and so on, remain primary historical sources of climatic information.

It is widely known that China and Russia, the most severe and stable former totalitarian empires in the whole world, are notable for the duration, continuity, and reliability of their chronicles. They cover time periods from 1000-1100 A.D. in Russia and from the 3rd century B.C. in China. In Russia, chronicles were held in cloisters, whereas in China state officials were responsible for their entity and truthfulness. Occasional climatic data from 800-850 A.D. are contained in Japan chronicles. It is remarkable that, in contrast, in India, a country of one of the oldest civilizations in the world, a tradition of strict temporal attribution of events in a written form did not exist. Ancient Greek and, especially, Roman written sources are numerous, but usually quilted and featured, thus hard to be interpreted from the naturalist point of view.

However, historical, or documentary, data as sources of information on secular climatic changes are characterized by certain basic limitations:

1. DD refer primarily to extreme events, whereas ordinary mean climatic conditions are rarely registered.

2. Reliability and significance of an event registered by DD depend to a great extent upon human factors, such as depopulation or overpopulation, wars, famines,superstitions, national mentalities, etc. Extreme events of any kind and level of intensity, if occured frequently, becomes ordinary for people and can be easily ommitted from the chronicles. Flood of a certain elevation has a better chance to be registered in a populated country than in a depopulated one. Moreover, in ancient China chronicles were often being rewritten after the new dynasty has succeded the previous one.

Valuable methodology of reliability analysis on some kinds of DD was presented by Bell and Ogilvie (1978) and Ingram et al. (1981).

3. Climatic interpretation of a certain registered event is not a commonplace. E.g. famine in the fluvial environment can be caused both by an extreme drought and by intensive summer rainfalls.

4. Incompleteness of DD is significant. Chronicles for certain time periods could possibly be lost as a result of a fire, war, etc.

5. Localization of a certain extreme event represents substantial difficulties because of changes in geographical ideas, national mentality, misundestandings in translation from dead languages, etc.

6. Changes in time measurements (calendars). E.g., attribution of an extreme event registered in Russian chronicles to a certain year is complicated by differences in dates for the New Year's day until the early 18th century: civic calendars marked September 1 as a beginning of the next year, whereas Orthodox church marked March 1.

7. Only few kinds of extreme climatic events, those influencing people to the greatest extent, are usually registered as DD. In the temperate zone these are winter frosts and thaws and summer aridity and heavy rains (Lamb, 1977). Meanwhile, hot summer periods, even short in duration, can cause a relatively cool summer season to be registered as "hot".

On fluvial plains, extreme spring floods and droughts can serve as one of the most reliable and universal indicators of climatic humidity. Spring discharges, accounting for over a half of annual runoff on many plain rivers of the temperate zone, affect crop yields to a great extent. Water levels of enclosed lakes, such as the Caspian Sea, the Aral Sea, etc., very changeable under climate variations and affectable for human beings, serve as an excellent integrator of humidity/aridity changes all over the inner Asia (see Selivanov, 1994, for the review). Meanwhile, water-level changes reflect not only climate changes over the lake surface itself but also in its extensive catchment area, that is not necessarily identical to the former. For example, the Caspian Sea water level depends to the great extent upon water inflow from East European Plain, which is by 80 percent is fed by the Volga River. Moreover, the Aral Sea, as well as the Issyk-Kul Lake repeatedly acted as overflow water bodies, to the Caspian Sea via Uzboy valley and to Central Asian lowland via the Chu River respectively. River inflow to the Aral Sea depended also upon substantial changes in the lower cources of the Amudarya River resulted from alluvial sedimentation over a very flat peri-coastal plain. However, in many cases, especially if a lake is relatively small in its water surface and is situated in an active tectonic region, changes of the water level reflect presumably recent tectonic movements. Only tracing of a horizontal shoreline of the certain age along the coast over relatively long distances can prove its climatic origin.

In some regions specific climate indicators exist. In Eastern China, for example, dust falls are caused by strong winds from the Loess Plateau and their increased frequency indicates climatic aridity.

8. Difficulties in comparison of DD and PD on a specific climate/hydrologic indicator exist. E.g., recurrence periods of floods indicated by slack-water deposits of rivers in arid regions (Baker e.a., 1988) do not necessarily correlate well to the recurrences of extreme floods by DD. The first method is based on tracing of floods of a certain elevation, whereas DD give information on unusual fllods with elevations possbly changing from time to time.

3. Regional patterns of air temperature and humidity changes during the Late Holocene

3.1 China and Japan

The earliest but inevitably discontinuous DD on climate changes in China were extracted from inscriptions made on mammalian bones, tortoise tests, and mollusc shells during the Xia and Shang dynasties. These states existed in the middle Yellow River reaches in the 21th-12th millennia B.C. Faunistic structure of animal remains, as well as inscription texts themselves, led Zhu Kezhen (1973) and Hu (1944) to conclusions on the warm and wet tropical climates over the recently temperate plains of East China. According to P.-K.Wang

(1980), who defined previous suggestions in climatic terms, summer monsoon rains were heavy and frequent. Early Xia period (21th-20th centuries B.C.) was marked by the most disastrous floods on the Yellow and Yangtze Rivers (Zhu K., 1973). Greatest historical floods in the Hai He river lower reaches occurred near 1500 B.C. (Zhai, 1986). Many lakes throughout China transgressed near 3600 B.P. (Fang, 1991).

Nevertheless, during the last decade, Chinese scholars tend to revise suggestions of Zhu Kezhen and his followers on the climate conditions during the 2nd millennium B.C. They believe the data to be unsufficient and hardly interpretable (e.g. Hsu, 1992).

During the dynasties of Zhou (1111-221 B.C.) and Qin (221-207 B.C.) climates in Eastern China became cooler but obviously remained warmer than today. Rivers froze up from time to time. An intensive cold "wave" possibly occurred at the turn from the 2nd to the 1st millennium B.C. (Zhu K., 1973). Major glacial advance (Xuedan stage) in Tibet , Qinghai (Zheng B. and Shi, 1985), and East Tian Shan (Zhang Z., 1988) is dated approx. 3,000 years B.P. Climate humidity over the whole East China was possibly similar to the present level or lower during the 1st millennium B.C. Extremely cool and wet climates were typical of Japan during the 1st millennium B.C. and culminated near 500 B.C. (Yoshino and Urushibara, 1978).

Meanwhile, in Central Mongolia, increasing aridity resulted in the middle 1st millennium B.C. in the drift of aeolian sands over nomadic living sites (Vipper e.a., 1989).

High climatic variability became a typical feature in East China during the Han dynasty (206 B.C.-220 A.D.). Late B.C. and early A.D. was possibly marked by slightly warmer climates in China (Zhu K., 1973). Water levels of many lakes rose (Fang, 1991). The Lop Noor Lake in Tarim desert experienced the greatest transgression in the 1st century B.C.-4th century A.D. (Zhao and Xia, 1984). Glacial readvance in Qinghai-Xizang region (Ruoguo stage) dates from 1920+/-110 years B.P. (Zheng B. and Shi, 1985).

Various indications of sharp cooling from the 3rd century A.D., the Three Kingdoms Period, are cited by Zhu Kezhen (1973) and P.-K.Wang (1980). Mountain glaciers in the Qinghai-Xizang region and East Tian Shan readvanced in the early A.D. In Japan, an intensive cooling began earlier, in the 2nd century A.D. (Suzuki and Yamamoto, 1978). Cold climates preserved during the Northern and Southern dynasties (420-581). Rarer floods on the Yellow and Yangtze Rivers, severe droughts in East China and the south-west of the country (Zheng S. e.a., 1977), as well as dominance of lake water-level falls (Fang, 1991), indicates increasing aridity. The same conclusion can be revealed from the more frequent dust falls and higher intensity of desertification processes in the steppe zone in North-East China, Inner Mongolia, and Xingjian (Zhu Z. and Liu, 1982; Zhang D., 1985).

Great dynasties of Sui (581-618) and Tang (618-907) enjoyed steady climates. Damage from river floods decreased (Wang P.-K., 1980), whereas recurrence period of dust falls increased (Zhang D., 1985). However, climates of that time were not so warm as they were traditionally regarded (Yan Z. e.a., 1992). First evidences of a glacial advance in Tibet and East Tian Shan date from that time.

Cool climates prevailed during the Five dynasties (907-960) and Song dynasty (960-1279). Gradual decrease in air temperature accelerated from the 11th century in Japan (Arakawa, 1956), early 12th century in East China (Zhu K., 1973), and 13th century in Qinghai-Xizang Plateau (Thompson e.a., 1989). Winter river freezing became a typical event for the whole East China. In 1111 A.D. the Tai Hu Lake, now in the subtropics, froze up for the first time. Climatic variability and aridity increased during that time: dust falls in East China (Zhang D., 1985), droughts in the middle and upper Yangtze River reaches (Chen, 1987) and in South-East China (Zheng S. e.a., 1977) frequented, desertification in North China (Zhu Z. and Liu, 1982) and Central Mongolia (Vipper e.a., 1989) intensified, lake water levels throughout China fell (Fang, 1991).

Meanwhile, the greatest part of Japan Archipelago experienced wetter climates during that time, whereas in the southwest of the country drier conditions prevailed (Yoshino and Urushibara, 1978).

Yuan dynasty (1271-1368) and early Ming dynasty (1368-1644) in China enjoyed milder climates with rarer droughts and dust falls, but beginning from the middle 15th century intensive cooling and aridification resumed and droughts became more frequent in East China (Zhu Z. and Liu, 1982; Chen, 1987), whereas wet conditions maintained at Honsu Island (Suzuki and Yamamoto, 1978).

The coldest historical period culminated in Japan in the early 15th century (ibid), in East China in the early (Zhu K., 1973) or middle (Wang S., 1990; Yan Z. e.a., 1992), in Tibet in the late 17th century (Wu e.a., 1990), and in Qinghai region near 1700-1720 (Thompson e.a;, 1989). Annual air temperature over East China decreased by 2 degrees C compared to the present.

In the late 18th century annual and winter air temperatures in East China increased to the present level, whereas in Qaidam Basin cold climates maintained until the early 19th century (Thompson e.a., 1989). The latest cool period, marked presumably by winter frosts, occurred in South-East China after 1850 (Zheng S. e.a., 1977). Climatic variability and general aridity increased significantly in the 17th-19th centuries. Extreme number of droughts and dusty falls was registered in the middle and upper Yangtze River reaches. Oppositely, floods became more frequent in the lower reaches of the river (Zhang J. and Crowley, 1989). Driest climates has set in over South-East China (Zheng S. e.a., 1977). Desertification process in Inner Mongolia gained its strength (Zhu Z. and Liu, 1982). Extreme aridity was registered in Beijing in the middle 18th century and humidification increased with rising air temperature during the subsequent decades (Yao, 1982). The same is true for the middle and upper reaches of the Yangtze River.

Nevertheless, the Qinghai Lake experienced a transgression in the late 17th and early 19th centuries (Wang S. e.a., 1991). Last glacial advance in Tibet-Qinghai region and in East Tian Shan dates from the 17th-19th centuries (Zheng B. and Zhang, 1983; Zheng B. and Shi, 1985). In Japan, extreme cooling in the 15th and late 18th centuries also coincided with wet, rainy weather (Suzuki and Yamamoto, 1978).

Global warming during the last century was marked by lower frequency of droughts in the middle and upper Yangtze River reaches (Chen, 1987), medium humidity in Beijing (Yao, 1982), and mean ratio of floods to droughts in the lower Yangtze course (Zhang J. and Crowley, 1989).

3.2 Central Asia

The term "Central Asia" refers here to the former Soviet republics of Tadzhikistan, Kazakhstan, Uzbekistan, Kyrgyzstan, and Turkmenistan. Inhabitated for several millennia, low-lying coasts of the Caspian Sea, the Aral Sea, the Issyk-Kul Lake, very sensible to changes in water balance components, make these water bodies the major source of information on changes of the regional climate. Nevertheless, a number of radiocarbon dates obtained both from coastal terraces and bottom cores leads various authors to contradictable chronological scales of water-level changes. This is especially true for the Aral Sea (cf.: Mayeva and Mayev, 1991; Serebryanny e.a., 1980; Kes', 1991; etc.). DD give an independent sourse of information, but their interpretation is not easy.

Ancient Greek and Roman authors (Eratosphenus, Aristotle, Patroclus, Poliby, Ptolemy, Strabon, etc.) gave a lot of evidences of the Caspian Sea water-level stands and changes in the cources of inflowing rivers. Critical analysis of these data was presented by Berg (1949), Apollov (1951), Bartold (1965), Gumilyov (1966, 1980). Varushchenko et al. (1990) established the reliable curve of the the Caspian Sea water-level changes during the last 2,500 years. Kes' (1991) succeded in ompiling different kinds of data on the Caspian Sea water levels, Berg (1960), Umurzakov and Vinnik (1975) and Shnitnikov (1979, 1986) - on the Issyk-Kul Lake. A recent summary of PD and DD on Central Asia was published in Russian (Selivanov, 1991).

Low regressive levels of the Caspian Sea during the 12th-7th centuries B.C. are indicated both by submerged shorelines at 9-13 m below the mean Caspian water level (MCWL) in 1992 (minus 26.95 m GMSL). Archeological information delivers strong constraints of drier climates in the Aralo-Caspian watershed during that time (Varushchenko e.a., 1987). The Aral Sea water level in the late 2nd-early 1st millennia B.C. fell down by 15-20 m in comparison to its water level in 1900-1950 (53 m above GMSL) (Veinbergs and Stelle, 1980).

In the second half of the 1st millennium B.C. a significant transgression, overdoubling its water area, occurred in the Aral Sea basin (ibid; Mayeva, Mayev, 1991). Excessive water flowed out to to the Caspian Sea via the presently dried Uzboy River in the 8th-5th centuries B.C. (Tolstov and Kes', 1960) or until the 1st century A;D. (Yusupov, 1986). Previously sparsely populated northeastern corner of the Caspian Sea became intensively developed by the nomadic tribes since the middle 1st millennium B.C. (Varushchenko e.a., 1987), desert vegetation was replaced by cereal and wormwood (Abramova and Turmanina, 1988). Glaciers on the southern slope of the Great Caucasus advanced to valleys in the 5th-1st centuries B.C. (Serebyanny and Golodkovskaya, 1985),. the glacial-fed Sevan Lake in the Caucasus experienced an intensive transgression (Sayadyan, 1985).

Controversial evidences of Greek and Roman authors have been summarized by Berg (1949), Apollov (1951), and Gumilyov (1980) as indicating the low Caspian water level. However, Ptolemy, Patroclus, Pliny, and Strabon believed in the more extensive water surface area of the Caspian Sea, especially in its south-western corner. Muravyov (1987) located the water level of that period at 10-15 m above GMSL! Nevertheless, it should be born in mind that ancient authors compiled evidences on the environment of distant areas that had been already gathered by travellers for several senturies and might become out of date. General standards of ancient mapping (i.e. an absence of techniques for longitude determination) restrict us in application of Greek and Roman maps for paleoclimatic reconsrtuctions.

Present author summarizies DD sources on the Caspian Sea water level during the Antiquity (5th century B.C.-3rd or 4th century A.D.) as indicating a prevalence of high water levels, near 3-5.5 m above MCWL. Water -level falls to 4-5 m below MCWL possibly occurred during the 3rd-1st centuries B.C.

Penultimate transgression of the Issyk-Kul Lake was usually dated from the turn B.C./A.D. (Trofimov, 1978; Shnitnikov, 1979; Sevastyanov, 1986). Chinese sources of the Older Han dynasty (206 B.C.-9 A.D.) regard the Issyk-Kul as a drainage lake (Umurzakov and Vinnik, 1975). Therefore, its water level rose to 1620 m above GMSL in comparison with the present elevation of 1607 m. First radiocarbon dates confirm an earlier set in of the transgressive episode, nearly 2750+/-150 years B.P. (Artemyev e.a., 1992). Since early A.D. the lake water level fell down and climatic humidity of the surrounding areas decreased (Sevastyanov, 1986).

During the same period the Aral Sea water level fell down to 40-41.5 m (Veinbergs and Stelle, 1980) or even 23-27 m (Mayeva and Mayev, 1991) above GMSL. Maps dated from the 4th century A.D. showed the Aral Sea as a small -Oxian swamp+ inflowed by a couple of unsignificant rivers (Bartold, 1965). Irrigation societies in the Amudarya River basin declined during that time. Irrigated areas decreased by 10-15 times (Tolstov and Kes', 1960). Dam destruction and flow diversion over the alluvial plain could contribute significantly to the water-level fall. Prevalence of dry climates over Central Asia in the middle 1st millennium A.D. is supported by mountain glacier retreat in Tian Shan and the Caucasus (Aleshinskaya e.a., 1985; Serebryanny and Golodkovskaya, 1985).

The next transgressive episode, up to 53-55 m above GMSL, began in the Aral Sea basin in the late 1st millennium A.D. and culminated in the 12th-13th centuries when the total irrigated area in the sea basin reached maximum values for the whole historical period (Tolstov and Kes', 1960; Andrianov, 1991; Vinogradov and Mamedov, 1991) .

Pollen data from the Aral Sea bottom cores (Mayev e.a., 1983) and Fergana Valley (Serebryanny e.a., 1980), as well as historical data on excessive snowfalls , summer rains, and disastrous floods in the Amudarya and Syrdarya River basins (Masson, 1948) give additional support to the conclusion on higher climatic humidity during that time. Flourished agriculture, horticulture, and viticulture of the Turkic-speaking Khazar and Oguz peoples in the recently desertous North Caspian Plain (Gumilyov, 1965; Pletneva, 1982) indicate humid conditions in the High Medieval times, whereas pollen data reveal a warm dry climate similar to the present (Abramova and Turmanina, 1988).

According to DD, the Caspian Sea water level stood generally at 5-8 m lower than MCWL (Berg, 1949; Apollov, 1951; Gumilyov, 1980; Kvachidze, 1980; Varushchenko e.a., 1987). However, shells from a large number of coastal accumulative features at 1-4 m above MCWL are dated by radiocarbon method at 1500-700 years B.P. (Leontyev and Rychagov, 1982; Varushchenko e.a., 1987). Therefore, mean position of the Caspian Sea water level in the Medieval times was similar to the present. However, in the 6th and 10th centuries water

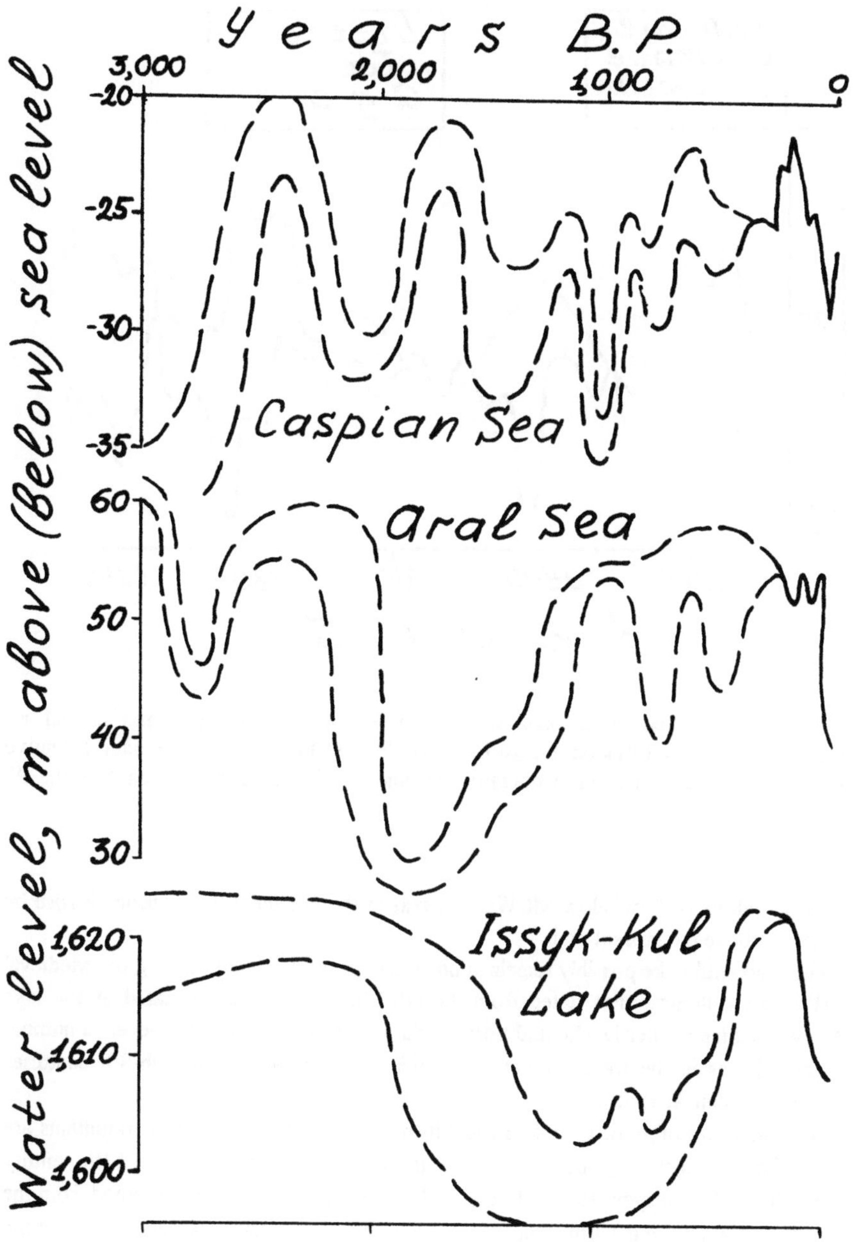

Fig. 1. Water level changes of the largest enclosed lakes in Central Asia during the last 3 000 years by documentary and proxy data. Dotted bars represent a variety of estimates, whereas solid lines show more reliable data.

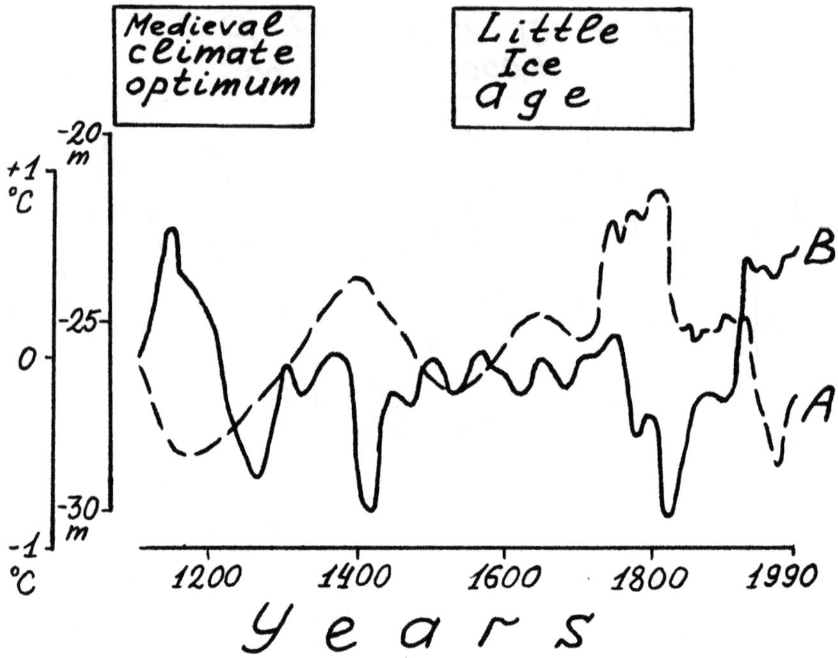

Fig. 2. The Caspian Sea water level during the last millennium (A) and annual air temperatures in the Central Region of East European Plain (B) by documentary data. Negative correlation of water-level and air temperature changes is obvious and significant at the 95 percent level.

level probably fell to 7-10 m below MCWL, whereas in the 9th and 11th centuries it rose by several meters above the present one

The Issyk-Kul Lake possibly remained under a regressive situation during the Medieval times. Many archeological sites dated from the 6th-15th centuries are situated at the lake bottom now. The lake water level could possibly fall down by 12-14 m. However, a number of scholars believes in the transgression and drainage of the lake in the 4th-8th centuries (Umurzakov and Vinnik, 1975).

First evidences of gradual cooling and climatic deterioration in Tadzhik mountains are dated from the 11th century (Mikhailov, 1984), in West Tian Shan - from the 13th century (Mukhametshin and Sartbayev, 1972). In the Aral Sea basin an aridification resumed from the 13th century (Abramova and Turmanina, 1988) and in the Aralo-Caspian watershed - from the 15th century (Varushchenko e.a., 1987).

Cool climates in the 15th-19th centuries was accompanied by humidification all over the Central Asia. Numerous lakes and arboreal vegetation in river valleys in the north-east corner of the Caspian basin abound on the maps compiled in the 15th-17th centuries. Annual and summer air temperature lowered by 3-4 degrees, whereas precipitation doubled

(Abramova and Turmanina, 1988). Inhabitants of the Amudarya River valley suffered from extreme floods in the 18th-19th centuries (Masson, 1948). Mountain glaciers in West Tian Shan advanced in the late 16th-early 19th centuries (Koshoev, 1986; Pomortsev, 1988) and in the late 18th-late 19th centuries in the Caucasus (Serebryanny and Golodkovskaya, 1985).

The Aral Sea rose its water level to 52-53 m above GMSL in 1780s, 1840s, and 1890s (Berg, 1960). The Caspian Sea water level rose in the 16th-19th centuries, at least three times overwhelming the elevations 3 to 5 m above MCWL (Varushchenko e.a., 1987; Karpychev, 1989; Rychagov, 1993). The last rise of the Issyk-Kul Lake water level to its drainage elevation is usually dated from the 18th-early 19th centuries (Umurzakov and Vinnik, 1975; Shnitnikov, 1979). However, our data supports the earlier set in of the transgression, from the middle 15th century (Artemyev e.a., 1992).

General aridification accompanied climatic warming on the Aralo-Caspian watershed since the middle 19th century (Abramova and Turmanina, 1988). In the Aral Sea basin aridification resumed at least two centuries earlier, from the early 18th century (Masson, 1948).

The Caspian Sea water level began to fall since 1810. In 15 years, until 1825, it fell by 3 m (Leontyev and Rychagov, 1982)! The Issyk-Kul Lake water level fell by over 12 m during the last two centuries. The Aral Sea water level maintained a relatively high position until 1950s when it dropped by several meters because of a dramatic increase of water losses from irrigation.

Both DD and PD on the largest lake water levels during the last 3 millennia have been summarized (Fig. 1). It worth to be noted that a close negative correlation of the Caspian Sea water level and air temperature is obvious during the last millennium (Fig. 2), as over longer and shorter time scales (Selivanov, 1994).

3.3 East Europe

A variety of DD on climates of the last millennia in Russia were summarized by Bogolepov (1907, 1911), Buchinsky (1957), Kostin (1975), Drozdov (1980), and others. The most complete inventory was prepared by Borisenkov and Pasetsky (1983, 1988). These authors, as well as Rauner (1981) and his colleagues (Rauner e.a., 1983; Krenke e.a., 1989), were the first to quantify climate changes over the region by comparing DD with climatic extremes during the instrumental period. Different kinds of PD were analyzed by Schostakowitsch (1931, 1934), Shnitnikov (1957 etc.), Klimanov (Klimanov and Yelina, 1980; Klimanov and Nikiforova, 1982; Klimanov e.a., 1985; Klimanov and Serebryannaya, 1986). Klige (1990) presented an independent water-balance estimates based reconstructions of annual river inflow to the Caspian Sea, presumably via the Volga River, during the last 2.5 millennia. An extensive volume in Russian summarized major results of the above studies (Borisenkov, 1989).

Climates of the later B.C.-earlier A.D. in the Central Russia were cool and wet. Aridification from the 2nd-3rd centuries A.D. resulted in frequent droughts, fires, severe winter colds and reduction of winter snows.

Medieval climatic optimum in East Europe lasted from the 9th until the late 14th century, 1100s (Selivanov and Voronov, 1988) or 1200s (Zolotokrylin e.a., 1986; Borisenkov and Pasetsky, 1988; Krenke e.a., 1989) being the warmest period in the Central Russian Plain. According to various estimates, summer temperatures rose by 1.2 (Selivanov and Voronov, 1988) or even by 2 degrees (Borisenkov, 1989), whereas winters were possibly colder than now. Higher seasonal variations can serve as an indicator for drier conditions during that period. This conclusion can be supported by shorter recurrence period of summer droughts and their higher intensity (Drozdov, 1980; Rauner, 1981; Zolotokrylin e.a., 1986). Annual precipitation in Central Russian Plain possibly decreased by 30-40 percent (Selivanov and Voronov, 1988). The Volga River annual flow decreased by 15-20 percent (Klige, 1990).

The14th and 15th centuries became the period of highest climatic variability. Extremely wet and cold seasons were succeded by dry and warm ones. Cold climates possibly culminated in the Central Russian Plain in the 17th century. Winter air temperature decreased by 2.5 degrees during that time. However, coolest summers occurred in 1780s-1820s (Borisenkov, 1989). Runoff of the Volga (Klige, 1990), the Dnieper (Shvetz, 1978), the Don, and some other rivers increased since the 14th century by 20-25 percent. Annual precipitation culminated at 110-120 percent of the present values in the 17th-18th centuries, whereas both summer droughts and heavy rains and extreme floods frequented during that period (Selivanov and Voronov, 1988). For example, heavy rains and river floods in Central Russia in 1672-1674 and 1686 were succeded by severe droughts in 1680, 1687, 1688. Both weather extremes resulted in terrible famines.

Under the general tendency to the rise in air temperature, since the early 19th century climate extremity decreased, as well as did a mean humidification level. Aridification was especially pronounced in lower winter precipitation and river runoff. Lower river runoff served as an important factor of the Caspian Sea water-level fall until 1977. The last decades became the period of humidification. Only during the last 15 years the Volga River runoff increased by 10-15 percent (Golitsyn and Panin, 1989), making up the significant contribution into the Caspian Sea water-level rise.

4. Methodology used in a quantitative climatic interpretation of documentary data

The easiest way to quantify DD is to calculate the number of extreme climatic events registered during successive equal time intervals and to calibrate them by comparing these values for instrumental period to the multi-year mean climate parameters. This promising approach has a number of positive examples (Pfister, 1984; Lamb, 1987; Borisenkov, 1988), whereas PD are more frequently used in such studies. However, in doing so, significant problems arise.

One can consider all extreme events of a particular kind (floods, droughts, severe winter seasons, etc.) either as equal to each other or as unequal. In the last case events (seasons) should be clasified into several groups (e.g. ordinary, extreme, very extreme, catastrophic) or ranks of intensity (0,1,2,....) (see, for example, Easton, 1928). However, this approach cannot increase, in a general case, the uniformity and comparability of data sets, because intensity of a climate extreme can hardly be extracted from DD.

Moreover, the number of registered extremes, or their total rank, usually tend to increase with the age decreased., because older events have a lower chance to be registered or described in details. In addition, secular variations in the number of registered climate extremes depend to a great extent upon human factors, rather than natural ones. Borisenkov and Pasetsky (1983, 1988) calculated the numbers of floods and droughts in the European Russia since the 12th century during 50-year-long time intervals. However, these two time series demonstrate a strong intercorrelation obviously caused by human factors (Selivanov and Voronov, 1988). E.g. the tataro-mongolian invasion into Russia in the early 13th century resulted in the cessation of many chronicles and, therefore, in a sharp decrease in the number of registered climate extremes of each kind. The similar tendency can be followed in Chinese sources during the replace period of Song dynasty by the mongolian Yuan dynasty.

The most popular approach to exclude human factors influencing the numbers of registered extreme events is to compare the numbers of positive and negative extremes of any kind during each time interval (e.g. droughts and heavy rains, great heats and colds, etc.) (Lamb, 1977; Borisenkov and Pasetsky, 1983; Lyakhov, 1984; Zolotokrylin e.a., 1986). However, this approach cannot exclude several time-dependent human factors (see 2).

Analyzing changes in the ratio of a certain kind of climate extreme events to all extremes, one can eliminate a significant part of human-induced variations in the procedure of registration of natural phenomena. Voronov and the present author proposed to calculate the following coefficient for each time interval and each kind of climatic extreme:

$$K = (D1-D2)/D \times (D1+D2)/ \Sigma D,$$

where D1, D2 - numbers of negative and positive extremes of any kind during the time interval (e.g. 30 of 50 years), D - total number of climate extremes of every kinds registered during the time interval, and ΣD - grand total of extremes for the whole period under consideration (e.g. 7 or 10 centuries) (Selivanov and Voronov, 1988).

In other words, we compare fractions of registered climate extremes of every kind in the total set of climate extremes during the time interval and include a weighted coefficient to account for a relative importance of a certain kind of climate extremes.

The important assumption remained is a presumption of equal possibility for negative and positive extremes of a certain kind to be registered.

Uniformity of time series of coefficients K is validated by the Craskell-Wallace parameter:

$$H = 12/N(N+1) \times \sum_{i=1}^{m} n_i (R- (N+1)/2) ,$$

where N - number of elements in a series, n_i - size (number of elements) in the i-sample, $n_1 + n_2 + ... + n_m = N$, R - mean rank for each sample, m = 3.

If the uniformity test demonstrates a positive result, calibration of the time series can be proceeded by calculation of coefficients K for the instrumental period. However, historical criteria can not be applied for the establishment of data sets on extreme climate seasons during the instrumental period. In this case, a climate parameter , A, for any season or a year is regarded as an extreme one if it does not meet the following restrictions of an ordinary season:

$$(\bar{A}+Amin)/2 < A < (\bar{A}+Amax)/2,$$

where $\bar{A}$ is a climatic (multi-year) norm of a parameter, Amax and Amin - highest and lowest values of this parameter during the whole period of instrumental observations (Lyakhov, 1984). Mean error of calculation of annual air temperature by this method ranges from +/-0.2 degrees and +/-20 mm for annual precipitation in the temperate zone.

5. Time series of climate changes over East Europe in 1120-1980

The above methodology was used to generate time series of annual, summer, and winter air temperature, precipitation, and potential river runoff (in mm per a unit surface area) in different regions of East Europe during the last millennium. Four regions differing from each other in their climate variations were distinguished in East European Plain: north-western (NW, at present, Sankt-Petersburg and Novgorog surroundings), central (C, the upper Volga River reaches and surroundings of Moscow), south-western (SW, Kiev and the Dnieper River basin), and western (W, present Baltic states and Belarus). Data on the eastern and northern parts of East European Plain appeared to be insufficient for a quantitative analysis.

Summer, winter, and annual values of K for thermal, precipitation, and runoff extremes were calculated in each region for three decade-long time intervals beginning from 1120-1149. Most of the K-time series (excluding some of the seasonal values of precipitation and runoff) passed the uniformity test.

Positive anomalies of winter precipitation were typical of the 1390-1549, 1630-1719, 1750-1809, negative - of the 19th century. Heavy rainy summer seasons prevailed in the 12th-13th centuries, as well as in 1420-1599, whereas droughts abounded in 1180-1239, 1320-1389, and 1600-1689. It worth to be noted that variations in winter humidity were generally lower than those in summer, whereas thermal changes in winter were significantly higher than in summer.

However, for the period of direct observations (instrumental period) most seasonal values of K did not demonstrate a close correlation with precipitation and, especially, runoff seasonal norms. Therefore, these values of K can not be rigorously used for the reconstruction of changes in seasonal climate norms during the last millennium.

Nevertheless, most annual values of K , as well as seasonal values for air temperature, appear to be closely correlated (at the 95 percent confidence level) to the observed climatic norms of respective parameters. Therefore, changes in annual values of air temperature, precipitation, and runoff can be rigorously calculated from the data on their extremes, i.e. coefficients K (Fig. 3). In NW and C regions highest annual air temperatures, comparable to those in 1970s-1980s, were reached in the 12th century, and lowest ones - in the 15th and early 19th centuries. Thermal variations in the SW and W regions were significantly lower, but the negative culmination in the 15 th century can also be traced. Annual precipitation generally increased in the 12th and the 15th-16th centuries and decreased in the 13th-14th and

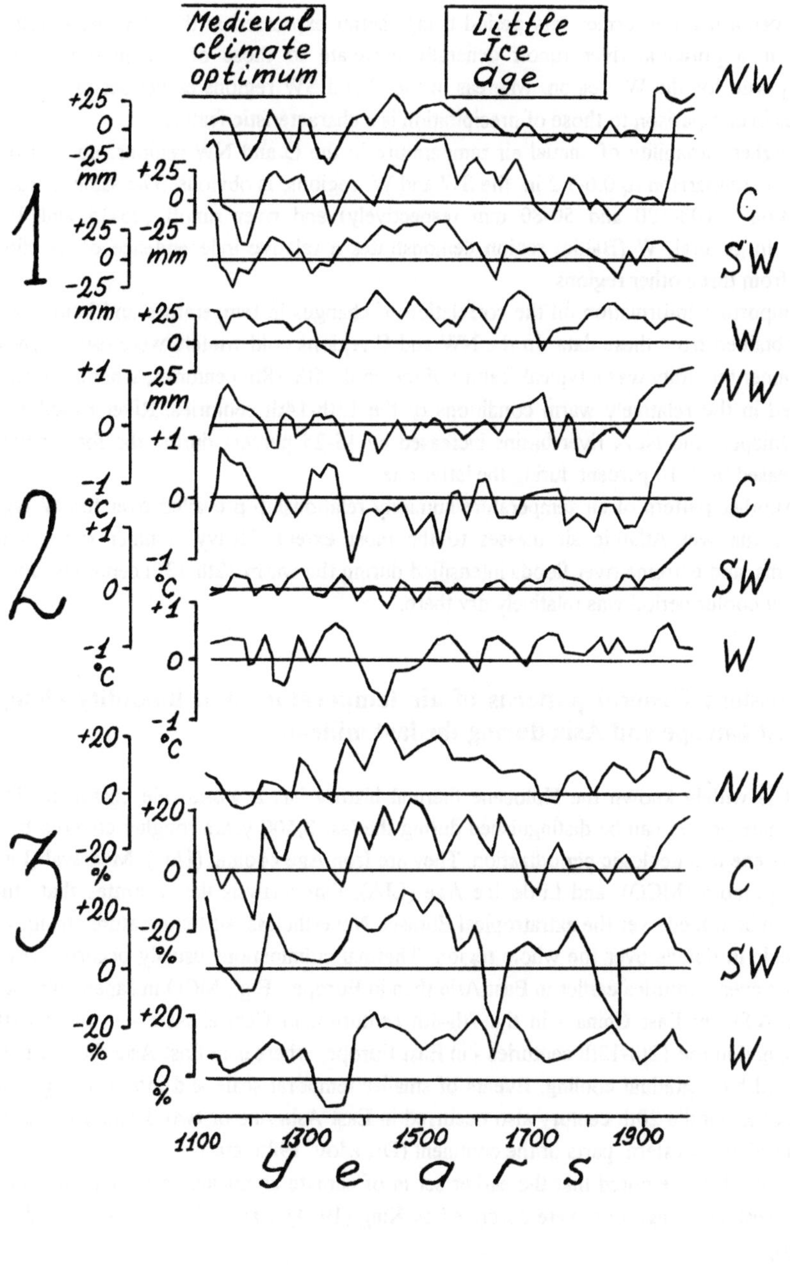

Fig. 3. Changes in climate parameters in East European Plain since 1120 by documentary data. Values of annual precipitation (1), air temperature (2), and possible river runoff (3) are presented in anomalies from the observed climate norms (1890-1988) for different historical regions: NW - north-western, C - central, SW - south-western, W - western.

the 18th centuries, the former dry period being better expressed in the NW and C regions. Changes in a potential river runoff generally correlate to those of precipitation. This is especially true for the W region, whereas in the C and SW regions earlier set in of runoff anomalies in comparison to those of precipitation is a characteristic feature.

Higher variability of annual air temperature in the C and NW regions, up to 1.6-2.0 degrees, in comparison to 0.6-1.2 in the SW and W regions, is obvious. The same is true for precipitation (100-120 and 50-80 mm respectively) and river runoff (25-35 and 10-20 percent). In general, W (Baltic) region demonstrates a relative independence of its climate changes from three other regions.

Important information on the correlation of changes in temperature and humidity can also be obtained from these data. In the NW and C regions cold winters were usually snowy. High summer humidity was a typical feature of the cool 15th-18th centuries, whereas droughts frequented in the relatively warm conditions of the 12th-14th centuries. River runoff in the Volga, Dnieper, and Neva river basins increased by 15-25 percent during the former period and decreased by 5-10 percent during the latter one.

Another pattern of air temperature/humidity relationship prevailed over the W region subject to the wet Atlantic air masses to the most extent. Heavy summer rains, winter snowstorms, and extreme river floods intensified during the warm 12th-13th centuries, and the subsequent cooler period was relatively dry there.

6. Discussion: General patterns of air temperature and humidity changes over East Europe and Asia during the last millennia

It is widely known the Holocene thermal history has a worldwide character. Three major climate phases can be distinguished during the last 2,500 years roughly covered by the Late Holocene in a geologic periodization. They are Iron Age cooling (IAC), Medieval (Little) climatic optimum (MCO), and Little Ice Age (LIA). Our analysis demonstrates that these periods can be traced over the extratropical Eurasia. Nevertheless, air temperature changes did not coincide in details over the whole region. Thermal culminations usually occurred several decades or even centuries earlier in East Asia than in Europe. E.g., MCO in Japan culminated near 500 A.D., in East China - in the 7th-8th centuries, in Central Asia - in the 8th-10th centuries, and in the 12th-13th centuries - in East Europe, whereas in East Asia the last period was marked by a gradual cooling. Events of smaller temporal scale and intensity, e.g. warm periods in the middle 20th century also occurred in East Asia one or two decades earlier that in the central and western parts of the continent (Drozdov, 1985, etc.).

Is worth to be noted that the earlier set in of climate anomalies in the east of Eurasia and their westward migration were described by King (1974), Lamb (1977, 1987) and Wigley e.a. (1986).

Climate humidity represented by any indicator shows larger spatial and temporal variations than air temperature. This phenomenon arises primarily from an integral character of humidification, i.e. from its dependence upon global and regional thermal conditions that

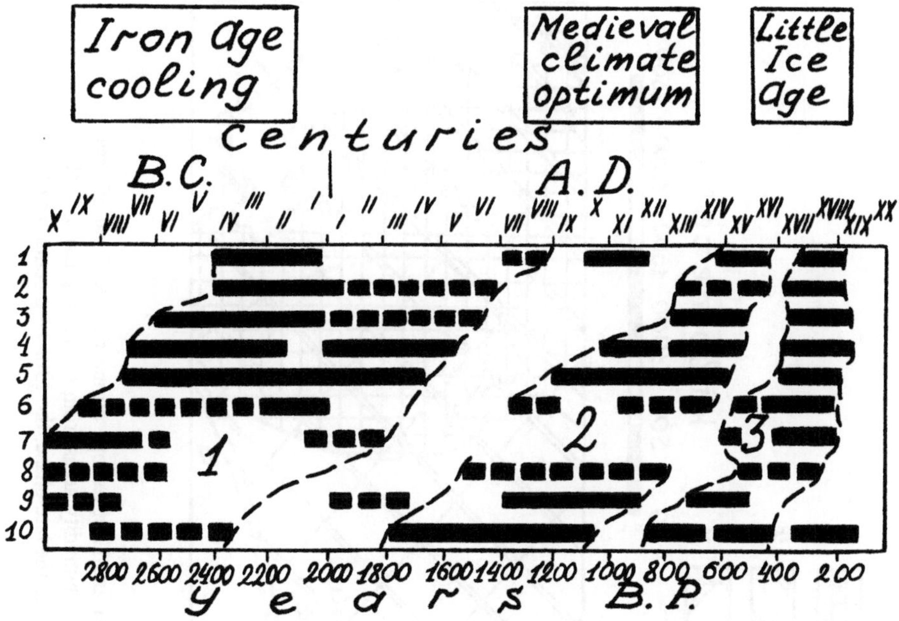

Fig. 4. Periods of high humidity in different sectors of extratropical Eurasia: 1 - England, 2 - East European Plain, central region, 3 - the Caucasus, 4 - the Caspian Sea, 5 - Aralo-Caspian watershed, 6 - West Tian Shan mountains, 7 - Qinghai-Tibet plateau, 8 - Central Mongolia, 9 - plains of East China, 10 - Japan, Honsu Island. Less reliable periods of humidification are shown by dotted bars. Mark that humid periods generally correspond to the Iron Age cooling (1) and the Little Ice Age (3), in most non-arid regions - also to the Medieval climate optimum (2). In western sectors of the continent humid periods set in several centuries later than in eastern ones.

determine a potential evaporation, as well as upon regional and local supplies of water vapour, determining real evaporation.

Analysis of a general character in humidity changes during the last 3 millennia in different regions of temperate and subtropical Eurasia confirms its comprehensive nature (Fig. 4). Generally, periods of higher humidification correspond to both warm and cool periods, IAC, MCO, and LIA. However, the more continental is the climate, the more distinctive is the negative correlation of thermal and humidity conditions in temperate latitudes. During the Medieval climatic optimum increase of humidification, most significant in East China and Central Europe, was not registered at all in the arid Central Asia (Qinghai-Tibet Plateau and, to some extent, Tian Shan mountains), whereas increase of precipitation and river runoff layer during IAC and LIA was very intensive in the last region.

Just as thermal anomalies, humid periods set in earlier in the east sector of the continent. Time lag can be even more than for the thermal anomalies. E.g., a Medieval humidification culminated in the 5th-8th centuries in East China and in the 13th-14th centuries

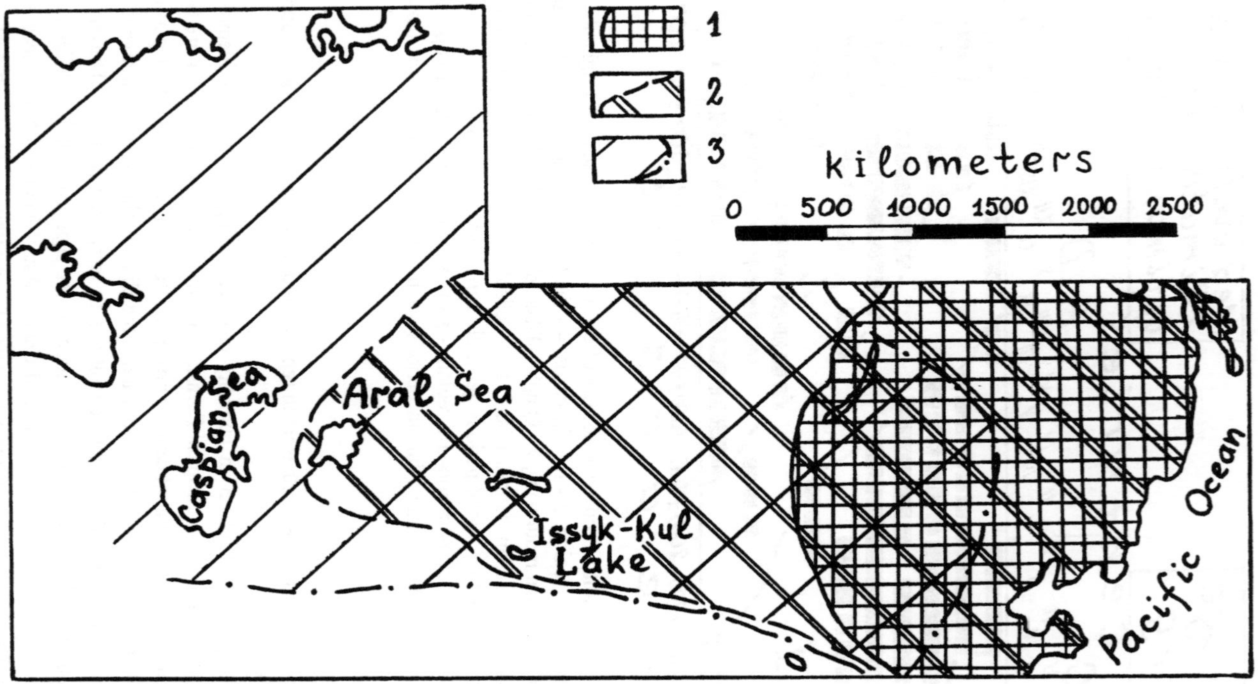

Fig. 5. Areas of increased humidity in extratropical Asia and East Europe during the Medieval climatic optimum (MCO). Note the westward "migration" of high humidity areas from the early phase of MCO, the 7th-8th centuries (1), to the 10th-11th centuries (2), and the late phase, the 13th-14th centuries (3).

in East Europe. As is the case for thermal changes, the time lag is roughly proportional to an amplitude and time scale of the respective climate change.

Nevertheless, time difference in the completion of a humid period is usually smaller, if exists. Periods of high humidification usually come earlier and resume for a longer time in the east sector of the continent than in its central parts. During the MCO area of high humidity broadened from the 7th-8th to the 13th century covering almost the whole extratropical Eurasia excluding its Far-Eastern sector where aridification resumed (Fig. 5).

Having been based on the information summarized in Fig. 4 , one can try to discern regions most similar in their humidity response to the global air temperature changes. Simple four-rank scale (dry, semi-dry, semi-wet, and wet conditions) and a procedure of weighted pair nearest neigbour, or cluster, analysis (Davis, 1986) were used in such a regionalization (Fig. 6). Eight regions discerned cover: (1) plains of East and North-East China, (2) middle and upper reaches of the Yangtze and Yellow River basins, Central and Inner Mongolia, (3) Qinghai-Tibet Plateau, (4) Kyrgyzstan and East Kazakhstan, presumably basins of the Issyk-Kul and Balkhash Lakes, (5) Central Asia, from Fergana valley in the east to the Aralo-Caspian watershed in the west, (6) the Caspian Sea and areas surrounding it from the north and west, including Caucasus, (7) central and southern sectors of East European Plain, (8) western sector of East European Plain (roughly corresponding to the present Baltic states and, partly, Belarus).

The above analysis can result in a sort of preliminary conclusions on future changes in humidity over Eurasia under the anticipated greenhouse-induced global warming. Milder climates and higher river runoff and lake water levels are expected in the monsoon East Asia, whereas arid Central Asia would preserve their aridity, at least for several decades, and aridmost regions will suffer from aridification. Precipitation and river runoff layer in the south of East Europe will possibly decrease by 10-20 percent. In the Baltic states significant changes are of low probability.

7. Conclusions

Historical, or documentary, data (DD) serve as an important independent source of information on past climates, including humidity, precipitation, and river runoff. These data, covering the time scales of several hundreds to several thousands of millennia, fill in the gap between data of direct observations, or instrumental, and paleodata. Extensive data bases of this kind have been compiled from the Chinese, Russian, and other historical sources. However, registration of extreme, abnormal events, possibilities for different climatic interpretation of the same events, their inconcrete spatial and temporal attribution, variety of human-induced factors (national mentality, wars, etc.) prevent direct application of DD to the analysis of past climates.

A proposed quantitative approach makes it possible to achieve valuable results in analysis of air temperature, precipitation and runoff changes over extratropical Asia and East Europe during the last millennia. A fraction of climate extremes of a certain kind among all

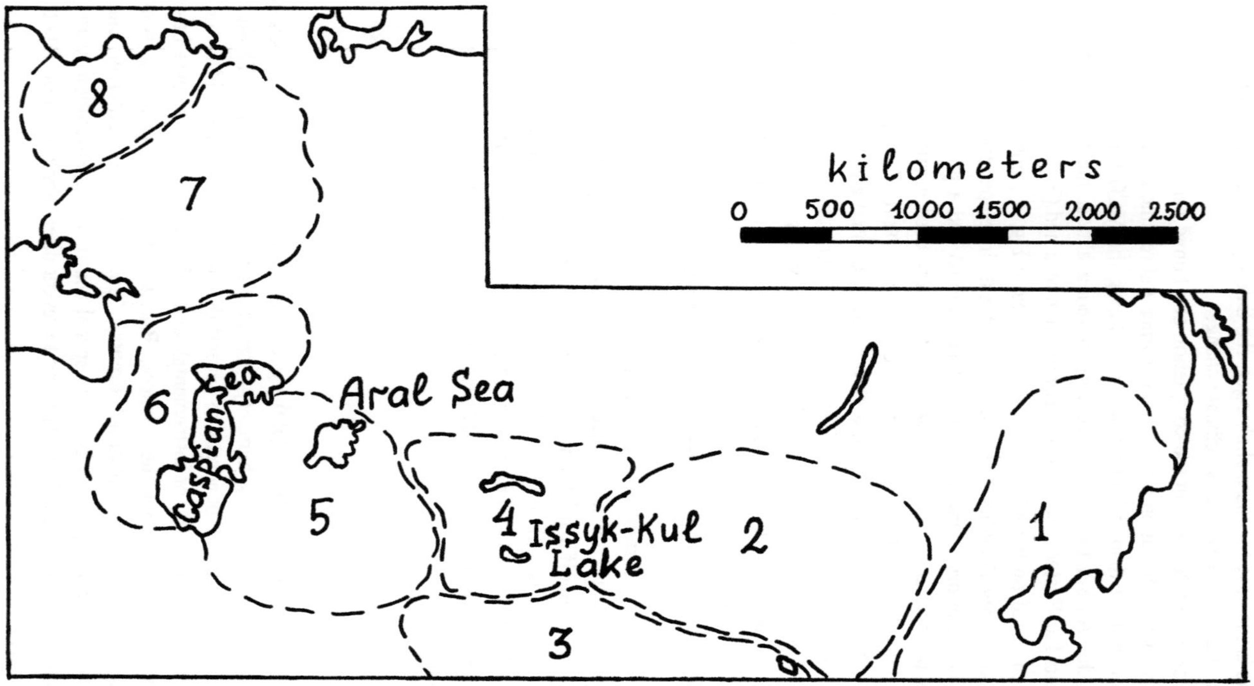

Fig.6. Regions 1-8 of extratropical Asia and East Europe differing from each other in their response to global changes in air temperature during the last millennia.

extremes registered during the time interval eliminate partly human-induced factors. Application of the technique is exemplified by East Europe during the last millennium.

Negative correlation of temperature and humidification anomalies appears to be the important feature in the arid regions of Eurasia, whereas in most other sectors correlation of these parameters is usually positive, but less intensive.

Major phases of thermal changes during the last 2,500-3,000 years, Iron Age cooling, Medieval climatic optimum, and Little Glacial Age, can be traced over the whole Eurasia, but they are usually aged from earlier times in eastern sectors of the continent. The similar time lag, often longer in duration (up to 3-6 centuries), characterize periods of higher precipitation and river runoff. However, time difference in the completion of humid periods is shorter, if exists. Time lag exists also for climatic events of smaller temporal scale and intensity , but it is shorter in duration.

As a first approximation, eight regions can be discerned in the extratropical Asia and East Europe by the differences in their humidity response to the large-scale global changes in air temperature during the last millennia.

- The above approach can possibly be useful in long-term predictive studies.

Acknowledgements: This work was partially funded by Central European University, RSS grant No. 285/93, and by the Russian Foundation for Fundamental Studies, grant No. 93-05-8853.

REFERENCES

Abramova TA and Turmanina VI (1988) Climate changes in the Caspian region during the Late Holocene. In: Velichko AA (ed) Golocenovyye paleoklimaty evropeiskoi territorii SSSR (Holocene palaeoclimates of the European USSR). Akad Nauk SSSR, Moscow, 182-191

Aleshinskaya ZV, Bakov EK, Melnikova AP (1985) Palaeogeography and dynamics of Tian- Shan glaciers during the Holocene. Materialy Glyaciologicheskikh Issledovaniy 52:162-166

Andrianov BV (1991) History of agricultural impact on the environment in the Aral Sea region. Izv Akad Nauk SSSR, ser geogr 4:47-61

Apollov BA (1951) Evidences for the past low stands of the Caspian Sea water level. Voprosy Geogr 24:134-143

Arakawa H (1956) Change of climate as revealed by the blossoming data of the cherry blossoms at Kyoto. J Meteorol 13:599-600

Artemyev VV, Klige RK, Selivanov AO (1992) History of water resources and water-level changes of the Issyk-Kul Lake during the Late Pleistocene and Holocene. Vodnye Resursy 4:71-78

Baker VR, Kochel RC, Patton PC (1988) Flood geomorphology. Wiley, Chichester et al

Bartold VV (1965) Historical sources on the Aral Sea and the Amudarya River lower reaches. In: Sochineniya (Selections) 3:323-414

Battan LJ (1983) Weather in your life. Freeman, San Francisco

Bell WT and Ogilvie AE (1978) Weather compilations as a source of data for the reconstruction of European climate during the medieval period. Clim Change 1:331-348

Berg LS (1949) The Caspian Sea water level during the historical period. In: Berg LS. Ocherki po fizicheskoy geografii (Essays on physical geography). Acad Sci USSR, Moscow, 205-272

Berg LS (1960) The Issyk-Kul Lake. In: Berg LS. Izbrannye trudy (Selected works). Acad Sci USSR, Moscow, 77-135

Bogolepov MA (1907) On the climate fluctuations in European Russia during historical times. Zemlevedeniye (Earth Sci) 14:35-72

Bogolepov MA (1911) Materials on climate changes (1501-1800). Zemlevedeniye (Earth Sci) 18(1):248-257

Borisenkov YeP (ed) (1988) Kolebaniya klimata za poslednee tysyacheletiye (Climate fluctuations during the last millennium). Gidrometeoizdat, Leningrad

Borisenkov YeP and Pasetsky VM (1983) Ekstremalnye prirodnye yavleniya v russkikh letopisyakh (Extreme climatic events in the Russian chronicles). Gidrometeoizdat, Leningrad

Borisenkov YeP and Pasetsky VM (1988) Tysyacheletnyaya letopis neobychainykh yavleniy prirody (Millennial chronicles of extreme natural phenomena) Mysl', Moscow

Chen H (1987) A study on regularity of drought and flood in the Changjiang River basin during last 10 centuries. Dili kexue (Acta Geogr Sinica) 7:197-206

Davis JC (1986) Statistics and data analysis in geology. Wiley, New York et al

Drozdov OA (1980) Zasukhi i dinamika uvlazhneniya (Droughts and humidification changes) Gidrometeoizdat, Leningrad

Drozdov OA (1985) Studies on the correlation between global air temperature and humidification. Trudy Gos Gidrolog. Inst (Ann State Hydrol Inst) 317:3-22

Easton C (1928) Les hivers dans l'Europe occidentale. Leyden

Fang J-Q (1991) Lake evolution during the past 30,000 years in China, and its implications for environmental change. Quatern Res 36:37-60

Golitsyn GS and Panin GN (1989) On water budget and recent water-level changes of the Caspian Sea. Meteorologiya i Gidrologiya (Meteorol and Hydrol) 1:57-65

Gumilyov LN (1965) Les fluctuations du niveau de la mer Caspienne. Cahier monde russe et sov 6:331-366

Gumilyov LN (1980) History of the Caspian Sea water-level fluctuations during the last 2000 years. In: Andrianov BV et al (eds) Kolebaniya uvlazhnennosti Aralo-Kaspiyskogo regiona v golocene (Fluctuations in humidification of the Aralo-Caspian region) Nauka, Moscow, 32-48

Hsu S (1992) Debate on the ancient climate of China. Abstr. 27th Int Geogr Congr, Washington, 265-266

Hu HS (1944) Climate change and a study of climatic conditions of the Yin dynasty. Bull Chin Stud 4:1-84

Ingram MJ, Underhill DJ, Farmer J (1981) The use of documentary sources for the study of past climates. In: Wigley TML (ed) Climate and history. Cambridge Univ Press, Cambridge, 180-213

Karpychev YuA (1989) Changes of the Caspian Sea level in the Holocene according to radiocarbon data. Water Res 16:1-12

Kes' AS (1991) Natural history of the Aral Sea and Aral region. Izv Akad Nauk SSSR, ser geogr 4:36-43

King JM (1974) Weather and the Earth's magnetic field. Nature 247:131-134

Klige RK (1990) Influence of global climatic processes on the hydrosphere regime. In: Paepe R, Fairbridge RW, Jelgersma S (eds) Greenhouse effect, sea level and drought. Kluwer, Dordrecht et al

Klimanov VA and Yelina GA (1980) Palaeoclimates of the northwestern European USSR during the Holocene. Doklady Akad Nauk SSSR 252:419-423

Klimanov VA and Nikiforova LD (1982) Climate changes in the northeast Europe during the last 2000 years. Doklady Akad Nauk SSSR 267:164-168

Klimanov VA, Koff TA, Punning Ya-MK (1985) Climatic conditions in the northwestern Peribaltic region during the last 2000 years. Izv Akad Nauk SSSR 4:93-97

Klimanov VA and Serebryannaya TA (1986) Vegetation and climate changes of Middle Russia highland during the Holocene. Ibid 2:93-102

Koshoev MK (1986) Assessement of dangerous nival and glacial events in the Issyk-Kul/Chu territorial complex: Dr's thesis (geography). Moscow State Univ, Moscow (unpublished)

Kostin SI (1975) Droughts in the Russian plain during the last millennium. Trudy Gos Geofiz Observ (Ann State Geophys Observ) 330:69-79

Krenke AN, Zolotokrylin AN, Popova VV, Chernavskaya MM (1989) Reconstruction of humidity and thermal changes during the historical period by natural indicators. In: Khotinsky NA (ed) Klimaty pozdnelednikovya i golocena (Palaeoclimates of the late glacial and postglacial period). Nauka, Moscow, 34-38

Kvachidze VA (1980) Byandovan archeological site as a source of information on the Caspian Sea water-level changes. In: Andrianov BV et al (eds) Kolebaniya uvlazhnennosti Aralo-Kaspiyskogo regiona v golocene (Fluctuations in humidification of the Aral-Caspian region) Nauka, Moscow, 152-155

Lamb HH (1977) Climate: Present, past and future V 2: Climate history and the future. Methuen, London

Lamb HH (1988) Weather, climate and human affairs. Routledge, London

Leontyev OK and Rychagov GI (1982) On the Holocene history of the Caspian Sea. In: Kaplin PA (ed) Geograficheskiye issledovaniya chetvertichnogo perioda (Geographical studies of the Quaternary) Moscow St Univ, Moscow, 134-146

Lyakhov MYe (1984) Climatic extremes in the center of European USSR in the 13th-20th centuries. Izv Akad Nauk SSSR, seri geogr 6:68-74

Mayev YeG, Mayeva SA, Nikolayev SD, Parunin OB (1983) New data on the Holocene history of the Aral Sea. In: Mayev YeG (ed) Paleogeografiya Kaspiyskogo i Aralskogo morey v kainozoye (Palaeogeography of the Caspian and Aral Seas during the Cenozoic) 2:133-144

Mayeva SA, Mayev YeG (1991) Water-level changes of the Aral Sea during the last millennia. Izvestiya vuzov, ser geodeziya i aeros'emka 1:124-132

Masson MYe (1948) On climate fluctuations in Central Asia in connection with the problem of changes in water regime during the historical period. Trudy Uzbek Geograf Obshchestva 21:3-23

Mikhailov NN (1984) Variability in tree growth during the last millennium in the Central Tadzhikistan mountains. Vestnik Leningrad Univ 18:92-96

Mukhametshin KD, Sartbayev SK (1972) Cyclicity of archa growth in high Tian Shan mountains. Izvestiya Akad Nauk Kirgiz SSR 2:34-41

Muravyov SN (1987) Caspian Sea water level in the ancient Greek sources. Priroda (Nature) 4:74-84

Neuberger H (1970) Climate in art. Weather 25(2):46-61

Pfister C (1984) Klimageschichte der Schweiz: 1525-1860. Paul Haupt, Bern Stuttgart

Pletneva SA (1982) Kochevniki srednevekovya (Medieval nomads). Nauka, Moscow

Pomortsev OA (1988) On the fluctuations of the Issyk-Kul Lake water level and glaciers in its mountain belt. In: Pomortsev OA (ed) Fiziko-geograficheskiye issledovaniya ozera Issyk-Kul i ego beregov (Fisico-geographical investigations of the Issyk-Kul Lake and its shores). Ilim, Frunze, 39-48

Rauner YuL (1981) Dynamics of humidification extremes during the historical period. Izvestiya Academii Nauk SSSR, ser geogr 6:5-22

Rauner YuL, Zolotokrylin AN, Popova VV (1983) Fluctuations in climate humidity at the European USSR territory during the last 4000 years. Ibid 1:50-59

Rychagov GI (1993) The Caspian sea water-level changes during Holocene. In: Water Resources 20:3444-350.

Sayadyan YuV (1982) The Sevan Lake as a natural Holocene climatograph. In: Yanshin AL (ed) Voprosy geologii golocena (Problems of Holocene geology) Yerevan. 61-67

Schostakowitsch WB (1931) Die bedeutung der untersuchung der bodenablaugerungen den seen fur einige freagen der geophysik. Int Verein theor angew limnol 5

Schostakowitsch WB (1934) Lacustrine mud sediments and periodic changes in natural phenomena. Zapiski Gidrol Inst 13:95-140

Selivanov AO (1991) Changes in climate and lake water levels in Middle Asia during the last millennia. In: Koshoev MK (ed) Geomorfologiya i paleogeografiya Issyk-Kul'skoi kotloviny i Vnutrennego Tyan Shanya (Geomorphology and palaeogeography of Issyk-Kul Basin and Inner Tyan Shan). Ilim, Frunze, 20-38

Selivanov AO (1994) Fluctuations of water level in the Caspian Sea over different time scales as an indicator of climatic changes. Palaeogeogr Palaeoclimatol Palaeoecol (in press)

Selivanov AO and Voronov AM (1988) Spatial and temporal regularities of climate and river runoff changes in the European USSR during the preinstrumental period (by historical data). In: Popov IV and Zamyshlyaev VM (eds) Voprosy gidrologii sushi (Problems of land hydrology) Gidrometeoizdat, Leningrad, 243-248

Serebryanny LR, Pshenin GN, Punning YaMK (1980) Tyan Shan glacial periods and the Aral Sea water-level changes. Izv Akad Nauk SSSR, ser geograf 2:52-65

Serebryanny LR and Golodkovskaya NA (1985) Lichenometry and fluctuations of glaciers and climate. In: Velichko AA e.a. (eds) Metody rekonstruktsii paleoklimatov (Methods of palaeoclimatic reconsrtuctions) Nauka, Moscow, 96-103

Sevastyanov DV (ed) Ozero Issyk-Kul i tendentsii ego prirodnogo razvitiya (THe Issyk Kul Lake and its natural trends) Nauka, Leningrad

Shnitnikov AV (1957) Izmenchivost' obshchei uvlazhnennosti materikov Severnogo polushariya (Variability in general humidification of continents in North hemisphere). Zapiski Geograf Obshchestva SSSR, novaya ser 16

Shnitnikov AV (1979) Issyk-Kul: Priroda, okhrana i perspektivy ispolzovaniya (The Issyk Kul Lake: Environment, protection and management perspectives). Ilim, Frunze

Shnitnikov AV (1986) Variability of Tyan Shan glaciation during the Holocene and history of the Aral SEa. Izv Vsesoyuzn Geograf Obshchestva 118:306-312

Svets GI (1978) Mnogovekovaya izmenchivost stoka Dnepra (Secular variability in the Dnieper River runoff). Gidrometeoizdat, Leningrad

Suzuki H and Yamamoto H (1978) Climate and civilization, climate and history. Asakura, Tokyo

Tolstov SP and Kes' AS (eds) (1960) Nizovya Amudaryi, Sarykamysh and Uzboy (The Amudarya River lower reaches, the Sarykamysh Lake and the Uzboy River). Izd Akad Nauk SSSR, Moscow

Trofimov AK (1978) History of the Issyk-Kul Lake during the Holocene. Bull Komissii Izuch Chetvertich Perioda 48:79-86

Umurzakov SU and Vinnik DF (1975) Historical and archeological evidences on the changes in the Issyk-Kul Lake water level. In: Umurzakov SU (ed) Problemy geografii Kirgizii (Geographical problems of Kyrgyzia). Ilim, Frunze, 35-46

Varushchenko SI, Varushchenko AN, Klige RK (1987) Izmeneniya rezhima Kaspiyskogo morya i besstochnykh ozer v paleovremeni (Changes in the regime of the Caspian Sea and enclosed water bodies in palaeotimes). Nauka, Moscow

Veinbergs IG and Stelle VYa (1980) Late Quaternary evolutionary stages of the Aral Sea and their connection with climatic changes. In: Andrianov BV e a (eds) Kolebaniya uvlazhnennosti Aralo-Kaspiysskogo regiona v golocene (Fluctuations in humidification of the Aralo-Caspian region). Nauka, Moscow, 175-181

Vinogradov AV and Mamedov ED (1991) Changes in climate and landscapes of the Amudarya and Syrdarya watershed. In: Andrianov BV (ed) Aral'sky krizis: Istoriko-geograficheskaya perspektiva (Aral crisis: Historical and geographical perspective) Aral Coord Center, Moscow, 66-75

Vipper PB, Dorofeyuk NI, Meteltseva YeP, Sokolovskaya VT (1989) Landscape and climate changes in Central Mongolia during the Holocene. In: Khotinsky NA (ed) Paleoklimaty pozdnelednikovya i golocena (Palaeoclimates of the Late Glacial and Holocene). Nauka, Moscow, 160-167

Wang P-K (1980) On the relationship between winter thunder and the climatic change in China in the past 2200 years. Clim Change 3:37-46

Wang S, Wang Y, Wu R, Li J (1991) Qinghai Lake fluctuation and climatic change since the last glaciation. Chin J Oceanol Limnol 9:170-183

Wigley TML, Jones PD, Kelly PM (1991) Warm world scenarios and the detection of climatic change induced by radiatively active gases. In: Bolin B e a (eds) The greenhouse effect, climatic change and ecosystems. Wiley, New York e a, 271-322

Wu C, Zhang D, Sun C (1990) Reconstructing Middle-Tibet climate during the last 600 years by dendroclimatological method. Acta Meteorol Sinica 4:294-304

Yao C (1982) A statistical approach to historical records of flood and drought. J Appl Meteorol 21:588-594

Yoshino MM and Urushibara K (1978) Palaeoclimate in Japan since the Last Ice Age. Climatol Notes 22:1-24

Yusupov Kh (1986) Drevnosti Uzboya (Antiquities of the Uzboy River). Ylym, Ashkhabad

Zhai Q (1986) Flood legemnds and the highest sea level in Holocene. In: Qin Y and Zhao S (eds) Late Quaternary sea level changes. China Ocean Press, Beijing, 221-225

Zhang D (1985) Meteorological characteristics of dust fall in China since the historic time. In: Liu T (ed) Quaternary geology and environment of China. China Ocean Press, Beijing; Springer Verlag, Berlin e a, 101-106

Zhang J, Crowley ThJ (1989) Historical climate in China and reconstruction of past climate. J Climate 2:833-849

Zhang Z (1988) Glacier variations at the northwestern slope of Nangqiabava mountain since the last glaciation. Binchuang duntu (J Glaciol and Cryopedol) 10:181-188

Zhao S and Xia X (1984) Evolution of the Lop Desert and the Lop Nor. Geogr J 150:311-321

Zheng B and Zhang Z (1983) Fluctuations of glaciers during neoglaciation in Bogda and the Urumqi River heard, Tian Shan. Binchuang duntu (J Glaciol and Cryopedol) 5:133-142

Zheng B and Shi Y (1985) Glacial variation since Late Pleistocene on the Qinghai-Xizang (Tibet) Plateau of China. In: Liu T (ed) Quaternary geology and environment of China. China Ocean Press, Beijing; Springer Verlag, Berlin e a, 44-53

Zheng S, Zhang F, Gong G (1977) Study on the humidity variations in southeastern China during the last 2000 years. Proc. Conf Climate Change and Long-Term Forecasting. Science Press, Beijing, 29-32

Zhu K.Z.(1973) Preliminary study on the climatic fluctuations during the last 5 000 years in China. Kexue tongbao (Sci Sinica) 16:226-256

Zhu Z and Liu S (1982) Deserts in China and combating desertification. Science Press, Beijing

Zolotokrylin AN, Krenke AN, Lyakhov MYe, Popova VV, Chernavskaya MYe (1986) Climate fluctuations in the European USSR during historical past. Izv Akad Nauk SSSR, ser geograf 1:26-36

REGIONAL CLIMATE CHANGE SCENARIO BASED ON PALEOCLIMATIC DATA

Marina Shabalova

Royal Netherlands Meteorological Institute (KNMI),
P.O. Box 201, 3730 AE De Bilt, the Netherlands

ABSTRACT

Based on paleoclimatic reconstructions of three past warm climates, Mid-Holocene, Eemian and Mid-Pliocene, the verification of paleoanalog hypothesis is extended to regional level for both precipitation and temperature and to mean latitudinal precipitation as well. It is found that paleoanalog hypothesis holds also for zonal temperatures in cold epoch, Last Glacial Maximum.

A typical structure of regional warming and precipitation change is revealed based on the paleoclimatic data. It is shown that over land in temperate northern latitudes the current warming climate generates the similar structure. The observed increase in annual precipitation over NH extratropics agrees with the paleoscenario for precipitation.

KEYWORDS

paleoclimatic reconstructions, paleoanalog method, climate change, temperature and precipitation change scenarios.

1 Introduction

The scenario of climate change based on paleoclimatic data (Budyko et al. , 1992) can be considered as an independent approach along with GCM simulations. Since the hydrological cycle is a secondary feature of the general circulation and depends on a sequence of complex processes, GCM calculations of the spatial precipitation distribution are hardly interpretable even for the case of modeling the current climate. Thus, to get some quantitative estimations of large scale precipitation change under the equilibrium global warming condition we suggest using paleoclimatic data.

The proposed approach is an extension of that suggested earlier by Budyko et al. (1978), so-called paleoanalog method. This method was used to get a notion of future anthropogenic climatic change by Budyko and Izrael (1991), Budyko et al. (1992), Zubakov and Borzenkova (1988) and in IPCC (1990). The basic hypothesis is that the change in the large-scale distribution of climate parameters is determined in first approximation by only one parameter - mean global surface air temperature change. The fruitful idea to scale climatic fields of the past to a unified level of global warming may provide for discovering the typical geographical patterns in spatial distributions of climatic variables

NATO ASI Series, Vol. I 26
Global Precipitations and Climate Change
Edited by M. Desbois and F. Désalmand

for different epochs. If this paleoanalog hypothesis verifies, paleoclimatic reconstructions of past warm or cold epochs may be used to develop a unified equilibrium climate change scenario and to determine some characteristic uncertaity for the estimation as well.

So far the paleoanalog hypothesis has been tested empirically only for zonal mean temperatures of various warm climates (Budyko and Izrael, 1991; Hoffert and Covey, 1992). Direct verification of the hypothesis on a regional level or for other elements has not been carried out, although GCM equilibrium simulations indicate a proportionality of the amplitude of some regional climate characteristics with the value of global warming, ΔT (IPCC, 1990, p. 157). In this paper the empirical verification of the paleoanalog hypothesis is extended to mean latitudinal distribution of precipitation and to regional distributions of temperature and precipitation.

The results of this paper relies heavily on the reliability of the Russian paleoreconstructions. Since the literature about this is not easily accessible to non-Russian readers, we briefly summarize the Russian works in paleoreconstructions. In past years, most attention in Russia has been given to the climatic reconstructions for Mid-Holocene (~ 5-6 kyr BP), Eemian (~ 125 kyr BP), Mid-Pliocene (~ 4 Myr BP) and the Last Glacial Maximum (~18-20 kyr BP) (in English: Barash, 1971; Velichko 1988, 1989; Velichko et al. , 1992, 1993; Zubakov and Borzenkova, 1988, 1990; Borzenkova et al. , 1992). A separate list of references refers to the literature in Russian. The reliability of the initial paleoclimatic data has been discussed extensively by CLIMAP (1976), Barash (1988), Velichko (1985), Crowley and North (1991) and Lapenis (1994). The uncertainty of paleoreconstructions of about 1°C for temperature and 50 mm/a for precipitation is usually cited. The generalization of the data in temperature and precipitation maps is done by Borzenkova, 1992. The reconstruction of summer temperature in Last Glacial Maximum (18-20 kyr BP) is based on CLIMAP data (1976) suplemented with data by Burashnikova et al. (1978), Velichko (1973, 1984), Nikolaev et al. (1983) and Safarova et al. (1984) (see list of references in Russian). The maps of reconstructed temperature and precipitation distributions of the three warm eras mentioned above, were recently published also in English (Borzenkova et al. , 1992). The actual Borzenkova's maps show patterns on scale of continents. Although the original data are highly unevenly distributed over the globe with relatively few sea points, the combination of volume and intrisic uncertainty of the data indicate that for the Northern Hemisphere patterns of that scale can be resolved succesfully, at least for the land area. Taking different epochs together results in a reduction of the uncertain part of the reconstructions. Presented regional estimations relate to the areas with sufficent paleoclimatic data available.

2 Data and results

The present work used the digitized paleoclimatic reconstructions of winter and summer temperatures and annual precipitation over land of Borzenkova, 1992. These are gridded data over the whole globe of the difference in temperature δT and precipitation δP relative to present (reference period 1951-1980) with a resolution 10 degrees latitude and 20 degrees longitude. The difference in mean global air temperature with respect to its present value, ΔT, was obtained by averaging the points. Similarly, the mean zonal temperature and precipitation differences dT and dP in the Northern Hemisphere were calculated for 10 degrees belts.

Modern data are obtained from World Climate Disk (1992). According to this database, last decade 1980–1990 was extremely warm, with a global mean temperature of 0.23°C over the norm (reference period 1951–1980). Winter warming in Northern Hemisphere (NH) was about 0.3°C and winter warming over NH land was about 0.55°C over the norm. The regional anomalies of temperature and precipitation for last decade were calculated in the same resolution as the paleodata.

Data presented in Figure 1 (top) show the consistency among the scaled temperature fields from four very different paleoclimatic states. Given here are the latitudinal distributions of mean annual temperature anomalies reconstructed for the Mid–Holocene (global warming ~1°C), Eemian (~1.8°C) and Mid–Pliocene (~3.7°C) from (Borzenkova, 1992), normalized by their mean global air temperature changes ΔT. Also given in the figure are the data of summer temperature anomalies from a cold epoch, the Last Glacial Maximum (LGM, cooling ~ 4.7°C). For cold period, its negative zonal anomalies were normalized by the negative value of global summer warming. The least-squares curve (solid), suggested by Hoffert and Covey (1992) for the three warm eras, fits the data of cold LGM also well. Latitudinal distributions are broadly the same though the range of global temperature change ΔT is rather wide — from -4.7°C to 3.7°C. These results indicate that even in a case of cold climate the paleoanalog hypothesis is roughly verified for mean latitudinal temperatures.

Figure 1 (bottom) shows that for precipitation over land the mean latitudinal anomalies decreased by the global precipitation change ΔP is similar for the three warm epochs. The global precipitation change is estimated from the mean global temperature change: $\Delta P = \alpha \cdot \Delta T$. Parameter α can be evaluated using GCM simulations with doubled atmospheric CO_2 concentration (IPCC, 1990), which indicate that global precipitation increase by about 2–4% is associated with a global warming of 1°C. In Figure 1 (bottom), Holocene data were left without transformation, Eemian data were reduced by 40mm, and Mid–Pliocene data — by 100mm. It is evident from Figure that the latitudinal dependencies are similar with the largest divergence of paleoclimatic precipitation data at low latitudes. Since the absolute precipitation change in tropics is the largest, the relative uncertaity is satisfactory. Thereby, the paleoanalog hypothesis holds for mean latitudinal precipitation also.

To extend the approach to the regional level, we first calculated the average temperature and precipitation anomaly (simple arithmetic mean from the three scaled values at each grid point) and the departures of each anomaly from this mean. Then we calculated the deviation field (the average with the square of the departures at each point). We consider the value of deviation as a measure of the uncertainty of the mean value.

The mean values and deviations were averaged over large areas with homogeneous climate type in Figure 2. Numerical results are given in Table 1 for temperature and in Table 2 for precipitation. For a number of regions in Figure 2 the uncertainty in the scaled precipitation anomalies is less than 50%. We suppose that these regions exhibit the typical geographical patterns of equilibrium precipitation change. The two intracontinental regions with relative uncertainty larger than 100% are the regions with smallest precipitation signal, close to the limit of paleodata resolution. The deviations here are small in absolute sense.

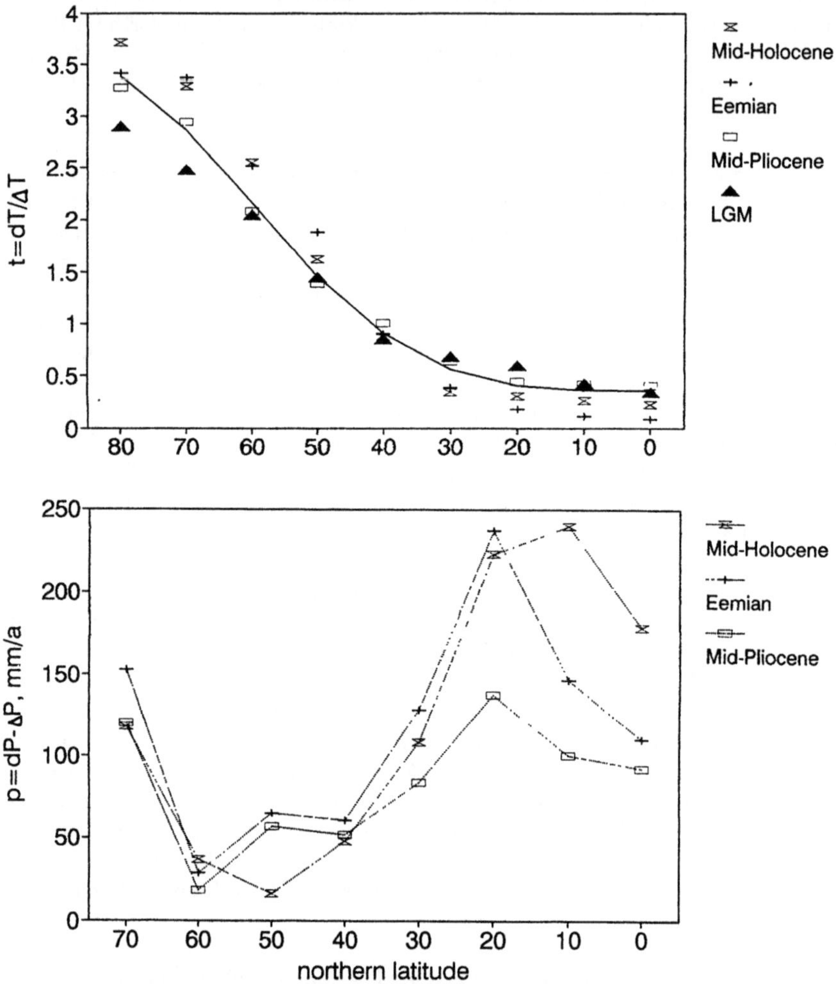

Figure 1: Latitudinal distributions of zonal temperature (top) and precipitation over land (bottom) anomalies, scaled to a global warming of 1°C.

Figure 3 shows maps of summer and winter temperature anomalies based on paleodata for the three warm climates. A marked inhomogeneity occurs in high latitudes in winter, with warming amplified over northern Canada and Asia. A pronounced relatively cold tongue extends over the North Atlantic ocean near Scandinavia. It should be noted, however, that the largest uncertainty of paleodata-derived scenario of winter temperature change occurs in this area. It means that North Atlantic (NA) region is

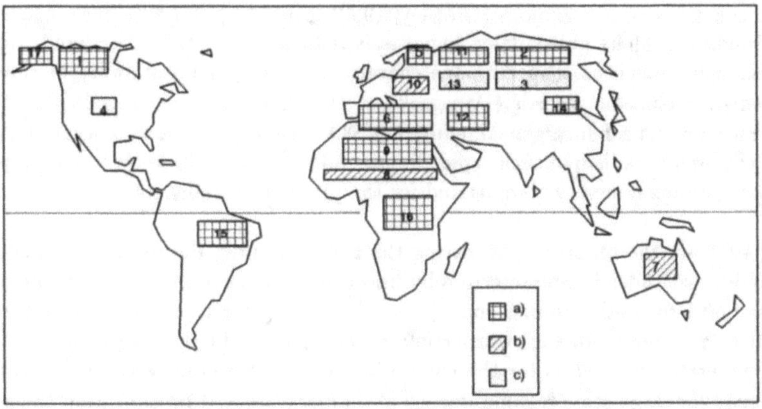

Figure 2: Zones location and numeration. Shading is used to mark regions with the different uncertainty σ of precipitation change estimate $p = \delta P - \Delta P$.
a) $\sigma/p \leq 50\%$; *b*) $50\% < \sigma/p \leq 100\%$; *c*) $\sigma/p > 100\%$.

Table 1: Regional temperature change with global warming of 1°C based on paleoclimatic data from three warm epochs. Current data (winter of 1980-1990 period) are scaled with the Northern Hemisphere warming over land (0.55°C).

Region	$\delta T/\Delta T \pm \sigma$ paleo winter	$\delta T/\Delta T \pm \sigma$ paleo summer	$\delta T/\Delta T$ 1980s winter
Alaska	3.2±0.3	2.1±0.9	3.8
North Canada	2.8±0.5	2.0±0.6	3.1
North Siberia	4.5±0.9	2.2±0.8	3.4
Central Siberia	2.3±0.6	1.1±0.5	2.7
Central Russia	2.7±0.9	1.8±0.5	2.6
Middle Asia	1.1±0.4	0.5±0.4	1.4
Australia	0.3±0.2	0.3±0.2	0.9
Central part of USA	1.2±0.4	0.9±0.2	0.7
Central Europe	1.5±0.8	1.2±0.4	0.7
Mediterranean region	0.7±0.5	0.4±0.3	0.0
Sahel	0.3±0.2	-0.1±0.3	-0.2
Part of Scandinavia	1.8±1.5	2.0±0.4	-0.3
Sahara	0.4±0.3	-0.3±0.3	-0.3

sensitive to the difference in analogs (Crowley, 1990), caused by the difference in sea–ice boundary conditions. The paleoanalog hypothesis fails here. On the other hand, North Atlantic region is known as a specific one characterizing by high instability of climate evident from paleoclimatic records (Dansgaard et al. , 1993; Alley et al. , 1993). Even at the culmination of Last Interglacial, climate of NA region was able to reorganize itself very rapidly. The mechanism of such rapid climate fluctuations is not known. We suppose that this specific region is very complicated for the prediction of any sort.

Apart from the territories bordering the Norh Atlantic, the most of the land in high and middle latitudes is characterized by uncertainty in the temperature reconstruction less than 50% for both summer and winter. Revealed pattern presents a structure of regional warming common for each warm climate. The maps in Figure 3 agree satisfactory with those generated by GCMs for the equilibrium climate with enhanced CO_2 (IPCC, 1990). Thereby for large-scale regional means of temperature and precipitation the paleoanalog hypothesis is roughly verified, apart from some specific areas. For a number of regions the mean temperature and precipitation change can be estimated given only one number, ΔT.

The spatial distribution of winter temperature anomalies over land for the last decade (1980–1990), which may have been the warmest decade of the last few centuries, corresponds to the equilibrium paleoscenario surprisingly well. Using World Climate Disk (1990) we estimated the warming of the 1980s for the same regions as indicated in Figure 2. It is evident form Fig.3 that in winter the warming of 1980's was concentrated over Northern Asia and Alaska (see also IPCC update, 1992), or the location of maxima is the same as in the paleoscenario.

Scaled to the value of NH winter warming over land, 0.55°C, regional temperature anomalies of the 1980's are shown in Table 1. It is evident from Table 1 that the regions with winter temperature anomaly larger than 2°C are the same as in the paleoscenario. (For all of these regions the uncertainty of paleoscenario is less than 50%.) In other words, the current warming climate generates the similar regional structures in winter temperature anomalies over land as those observed in the past warm epochs. It means that found from equilibrium paleodata regional structure holds in transient situation. This result is in line with transient response GCM simulations. As reported by IPCC (1990; p.156), the patterns of equilibrium and transient climate change derived from high resolution models are similar. Also, the empirical study by Wigley et al. (1980) showed that the patterns of change based even on the individual warm years from instrumental record agree well with those found in the present work.

It should be noted that in all regions with scaled winter temperature anomaly larger than 2°C mean annual precipitation of decade 1980–1990 exceeded the norm (see also Vinnikov et al. , 1990, where several regional century-scale trends in NH extratropics are found). This tendency is in line with paleodata (Table 2). On the other hand, considerable decrease of precipitation in Sahara region seems completely contrary to the paleodata, which is, perhaps, a manifestation of transient nature of current climatic change. The possible effect of slow feedback between soil properties and precipitation in North Africa region is discussed by Lapenis and Shabalova (1994).

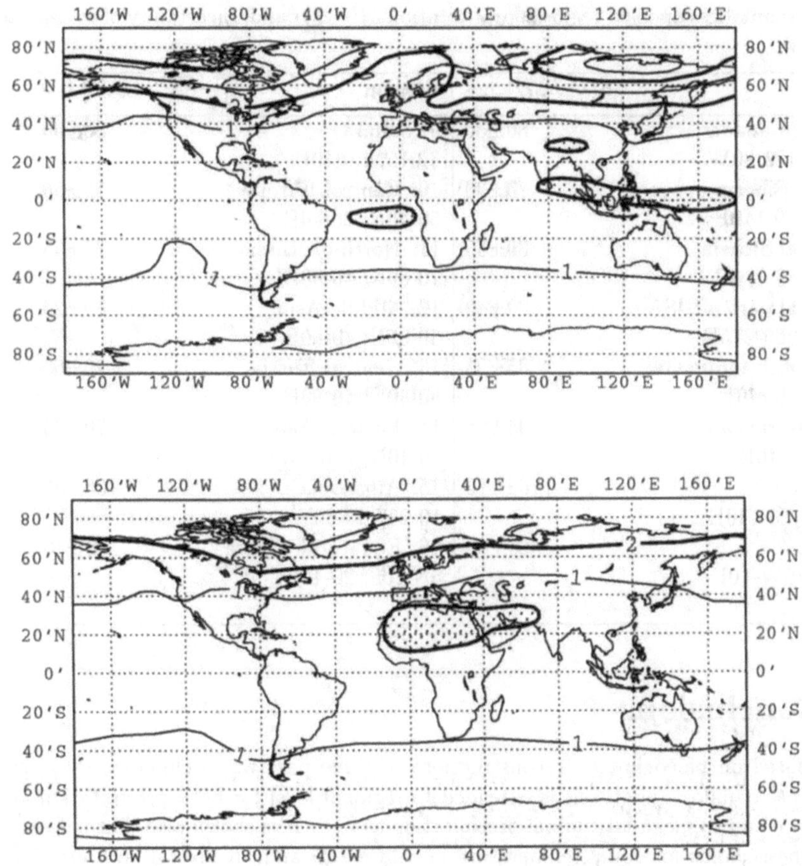

Figure 3: Temperature change scenario based on paleoclimatic data. Global warming of 1°C. Cooling area is stippled. Winter (top) and summer (bottom).

Table 2: Anomalies of scaled regional precipitation $\delta P - \Delta P$ and uncertainty of estimation σ (mm/a).

Region	$(\delta P - \Delta P) \pm \sigma$	Region	$(\delta P - \Delta P) \pm \sigma$
1. North Canada 70-50N; 120-80W	80±30	9. Sahara 30-15N; 0-40E	180±60
2. North Siberia 70-60N; 80-140E	70±30	10. Central Europe 50-40N; 20-40E	55±55
3. Central Siberia 50-60N; 80-140E	30±50	11. North of Russia 70-60N; 40-80E	75±35
4. Central part of USA 35-45N; 260-275E	20±40	12. Middle Asia 50-35N; 40-80E	80±20
5. Part of Scandinavia 70-60N; 20-40E	85±25	13. Central Russia 60-50N; 40-80E	20±50
6. Mediterranean 40-30N; 0-40E	95±25	14. Eastern Asia 50-40N; 120-140E	70±35
7. Australia 20-30S; 120-140E	170±110	15. Amazonia 10-20S;40-60W	165±55
8. Sahel 20-15N;10W-40E	65±60	16. Equatorial Africa 5N-10S; 20-40E	120±40

3 Conclusions

Based on paleoclimatic reconstructions of three past warm climates, the verification of paleoanalog hypothesis is extended to regional level for both precipitation and temperature and to mean latitudinal precipitation as well. It is shown also that paleoanalog hypothesis holds for zonal temperature in cold epoch and, in several land areas, for temperature in winter season of recent decade.

A typical structure of regional warming and precipitation change is revealed from the paleoclimatic data. It is shown that over land in temperate northern latitudes the current warming climate generates the similar structure of winter temperature anomalies. The observed increase in annual precipitation over NH extratropics agrees with the paleoscenario.

The largest disagreement between equilibrium paleoscenarios and current climatic change relates to the North Atlantic region for temperature and North Africa for precipitation. We suppose that this disagreement is caused by the transient nature of current climate as compared with the long-time averaged paleodata.

Acknowledgments: We thank Dr.I.Borzenkova for providing the paleoclimatic data. The support of Dr.K.Selyakov at the early stage of this work is greatfully acknowledged.

REFERENCES

Alley, R.B. et al., 1993: Abrupt increase in Greenland snow accumulation at the end of the Younger Dryas event. Nature, 362, 527-531.

Barash, M.S., 1971: The vertical and horizontal distribution of planktonic foraminifera in Quaternary sediments of the Atlantic Ocean. In: Funnel, B.M. and W.R. Riedel, (eds.), The Micropaleontology of Oceans, Cambridge Univ. Press, p.433.

Barash, M.S., 1988: The Quaternary Paleoceanography of the Atlantic Ocean, Nauka, Moscow, p. 272. (in Russian)

Borzenkova, I.I., 1992: The Changing Climate during the Cenozoic, Hydrometeoizdat, St.Petersburg, p.247. (in Russian)

Borzenkova, I.I., V.A. Zubakov and A.G. Lapenis, 1992: Reconstructions of global climate of the past warm epochs. Soviet Meteorology and Hydrology, 8, 19–26.

Budyko, M.I., K.Ya. Vinnikov, O.A. Drozdov and N.A. Efimova, 1978: Future climate change, Izvestija Akademii Nauk, Ser. geogr., 6, 5–20. (in Russian)

Budyko, M.I. and Y.A. Izrael, 1991: Anthropogenic climate Change, Univ. of Arizona Press, Tucson, p.277.

Budyko, M.I., I.I. Borzenkova, G.V. Mezhulin and K.I. Selyakov, 1992: Future climate changes. Izvestija Akademii Nauk, Ser. geogr., 4, 8–20. (in Russian)

CLIMAP Project Members, 1976: The surface of the ice-age Earth. Science, 191, 1131-1137.

Crowley, T.J., 1990: Are There Any Satisfactory Geologic Analog for a Future Greenhouse Warming? J. of Climate, 3, 1282–1297.

Crowley, T.J. and G.R. North, 1991: Paleoclimatology, Oxford Univ. Press, Clarendon Press, Oxford, p.360.

Dansgaard, W. et al., 1993: Evidence for general instability of past climate from a 250 kyr ice record. Nature, 364, 218–220.

Hoffert, M.I. and C. Covey, 1992: Deriving Global Climate Sensitivity from Palaeoclimate reconstructions. Nature, 360, 573–579.

Intergovernmental Panel on Climate Change (IPCC), 1990: Climate Change. The IPCC Scientific Assessment. Houghton, J.T., G.J. Jenkins and J.J. Ephraums (eds.), Cambridge University Press, p.365.

Intergovernmental Panel on Climate Change (IPCC), 1992: Climate Change 1992. The Supplementary Report to the IPCC Scientific Assessment. Houghton, J.T., B.A. Callander and S.K. Varney, (eds.), Cambridge University Press, p.200.

Lapenis, A.G., 1994: On the accuracy of Russian paleoclimate reconstructions. J. of Climate, submitted.

Lapenis, A.G. and M.V. Shabalova, 1994: Global climate changes and moisture conditions in the intracontinental arid zones. Cl. Change, in press.

Velichko, A.A., 1985: Empirical Paleoclimatology (Methods and Uncertainty). In: Methods of Reconstructions of Paleoclimates, Nauka, Moscow, 7–35. (in Russian)

Velichko, A.A., 1988: Late Pleistocene spatial paleoclimatic reconstructions. In: Late Quaternary Environements of the Soviet Union, Univ. Minnesota Press , Minneapolis, 261–279.

Velichko, A.A., 1989: Evolutionary Analysis of the Contemporary Landscape of the Earth and Prognosis. Quaternary International, 1, 1–17.

Velichko, A.A. et al., 1992: Atlas of the Paleoclimates and Paleoenvironment of the Northern Hemisphere (Late Pleistocene and Holocene). Budapest.

Velichko, A.A. et al., 1993: Greenhouse warming and the Eurasian biota: are there any lesson from the past? Global and Planetary Change, 7, 51–67.

Vinnikov, K.Ya., P.Ya. Groisman and K.M. Lugina, 1990: Empirical Data on Contemporary Global Climate Changes (Temperature and Precipitation). J. of Climate, 3, 662–680.

Wigley, T.M.L., P.D. Jones and P.M. Kelly, 1980: Scenarios for a warm, high-CO_2 world. Nature, 283, 17–22.

World Climate Disc — Global Climatic Change Data, 1992: Chadwyck–Healey Ltd, Software and database intergration.

Zubakov, V.A. and I.I. Borzenkova, 1988: Pliocene Paleoclimates: Past Climates As Possible Analogs of Mid-twenty-first Century Climate. Palaeogeography, Palaeoclimatology, Palaeoecology, 65, 35–53.

Zubakov, V.A. and I.I. Borzenkova, 1990: Global Climate during the Cenozoic. Elsevier Publ. Co., Amsterdam, New-York, Oxford, Tokyo, p.456.

REFERENCES IN RUSSIAN

Abramova, T.A., 1985: Paleogeographical conditions of the Aral -Caspian region during the Late Holocene (according to palynological data). In: Relief and Climate, Moscow, 91–127.

Barash, M.C., 1985: Reconstruction of the Quaternary temperatures using planktonic foraminifers data. In: Methods of Reconstructions of Paleoclimates, Nauka, Moscow, 134-142.

Barash, M.S., N.S. Oskina and Y.I. Ivanova, 1980: The upwellings of African coastal waters in the Late Pleistocene studied by the data on planktonic foraminifers. Oceanology, 20, N1.

Barash, M.S., N.S. Oskina and N.S. Blum, 1983: Quaternary biostratigraphy and surface paleotemperatures by means of planktonic foraminifers, Sites 516 and 518. Init. Rep. DSDP 72, 849–862.

Barash, M.S., V.M. Kuptsov and N.S. Oskina, 1987: The Atlantic: the new data about chronology of Late Pleistocene and Holocene. Bull. of Quat. Study, 56, 3–10.

Bazarov, D.B., 1986: The Cenozoic of the Near-Baical Territory. Nauka, Moscow, p.186.

Blum, N.S., 1982: Paleotemperature reconstructions in different regions of World Ocean for Pleistocene, based on planktonic foraminifers data. Aftoreferat dissertatsii, Moscow, p.24.

Borzenkova, I.I. and V.A. Zubakov, 1984: Climatic Optimum of the Holocene as a model of global climate of the early 21st century. Meteorologiya i Gidrologiya, 8, 69–77.

Borzenkova, I.I., 1987: The features of moisture conditions over the land in the Northern Hemisphere in the geological past. Meteorologiya i Gidrologiya, 10, 15–28.

Borzenkova, I.I., 1990: Moisture conditions in the arid regions of the Northern Hemisphere during the Late Glacial-Holocene. Izv. AN SSSR, Ser. geogr., 2, 12–17.

Borzenkova, I.I., 1990a: Fluctuations in the moisture conditions of the Sahara and the adjacent areas during the past 20 kyr. Izv. AN SSSR, Ser. geogr., 3, 36–41.

Budyko, M.I., I.I. Borzenkova, G.V. Mezhulin and K.I. Seljakov, 1992: Future Climate Changes. Izv. AN SSSR, Ser. geogr., 4, 47–66.

Burashnikova, G.A., M.A. Muratova and I.A. Suetova, 1978: Paleotemperatures over the USRR territory at the maximum of last glaciation. Dokl. AN SSSR, 244, 723-727.

Burashnikova, G.A., M.A. Muratova and I.A. Suetova, 1982: Climatic model of the USSR territory during Holocene Optimum. In: The Development of Nature over the USSR Territory in the Late Pleistocene and Holocene, Nauka, Moscow, 245–259.

Grichuk, V.P., 1969: Reconstruction of some climatic elements for Nothern Hemisphere in Mid-Holocene. In: Holocene, Nauka, Moscow, 41–57.

Khotinskyi, N.A. and S.S. Savina, 1984: Paleoclimatic maps of the USSR territory in Boreal, Atlantic and Subboreal periods of the Holocene. Izv. AN SSSR, Ser. geogr., 4, 18–23.

Klimanov, B.A., 1982: Climate of the East Europe during Holocene climatic Optimum (pollen data). In: The Development of Nature over the USSR Territory in the Late Pleistocene and Holocene, Nauka, Moscow, 251–258.

Klimanov, B.A. and G.A. Elina, 1984: Climate change over the north-east of Russian Plain in Holocene. Dokl. AN SSSR, 274, 1164–1167.

Klimanov, B.A. and T.A. Serebrjannaja, 1986: Climate and vegetation change over the Russian Plain in Holocene. Izv. AN SSSR, Ser. geogr., 2, 93–102.

Kotljakov, V.M. et al., 1988: Isotope analysis of the Komsomolskaja ice core from Antarctica. The Glaciological Matherials, 63, 97–102.

Mamedov, E.D., 1982: Pluvials and arids in the Late Pleistocene and Holocene history of deserts of the USSR and adjacent areas. In: The Development of Nature over the USSR Territory in the Late Pleistocene and Holocene, Nauka, Moscow, 94–108.

Mamedov, E.D., and T.N. Trofimov, 1986: Holocene pluvial lakes of the Caspian Sea area. Bull. Com. for studying the Quaternary, 55, 102–118.

Murzaeva, V.E., V.I. Konopleva, E.D. Devyatkin, and L.R. Serebryany, 1984:

Pluvial situations of the Late Pleistocene and Holocene in the arid zone of Asia and Africa. Izv. AN SSSR, Ser. geogr., 4, 15–22.

Nikolaev, S.D., 1986: The change of temperature of the oceans in Cenozoic (based on oxigen isotope data). Aftoreferat dissertatsii, Moscow, p.47.

Nikolaev, S.D., B.I. Nikolaev and N.C. Blum, 1983: Climate of the Pacific in Pleistocene (mean annual temperature fields 18 and 125 kyr ago). In: The nature of the ocean, Moscow, 257–261.

Safarova, C.A., V.N. Nikolaev and N.C. Blum, 1984: Paleogeograpfy of Pacific 18000 years ago. Oceanology, 24, 643–648.

Varushchenko, S.I., A.N. Varushchenko and R.K. Klige, 1989: Fluctuations of the Caspian Sea and Closed Lake Basins during the Geologic Time. Nauka, Moscow, p.238.

Velichko, A.A., 1973: Natural Proccess in Pleistocene. Nauka, Moscow, p.186.

Velichko, A.A., 1985: Empirical Paleoclimatology (Methods and Uncertainty). In: Methods of Reconstructions of Paleoclimates, Nauka, Moscow, 7–25.

Velichko, A.A. et al., 1982: Paleoclimate reconstructions over the Europe for the Optimum of Mikulino Interglacial. Izv. AN SSSR, Ser. geogr., 1, 15–28.

Velichko, A.A., V.P. Grichuk, E.E. Gurtovaja and E.M. Zelikson, 1983: Paleoclimate of the USSR territory in the Optimum of the last (Mikulino) interglacial. Izv. AN SSSR, Ser. geogr., 6, 30–36.

Velichko, A.A., M.S. Barash, V.P. Grichuk, E.E. Gurtovaja and E.M. Zelikson, 1984: Climate of Northern Hemisphere in the last Mikulino Interglacial epoch. Izv. AN SSSR, Ser. geogr., 1, 5–10.

Velichko, A.A., L.R. Serebriannyi and E.E. Gurtovaja, (eds.), 1985: Methods of Reconstructions of Paleoclimates. Nauka, Moscow.

Velichko, A.A. and A.L. Chapaluga, (eds.), 1987: Climates of the Earth in the Geological Past. Nauka, Moscow, p.225.

Vipper, P., L.A. Dorofjuk, E. Meteltseva and V. Sokolovskaya, 1981: Paleogeography of the Holocene and the Upper-Pleistocene of Central Mongolia. Izv. Akad. Nauk Est. SSR, Ser. biol., 30, 74–79.

Yasomanov, N.A., 1985: Ancient Climates of the Earth. Gidrometeoizdat, Leningrad, p.292.

RECENT PRECIPITATION HISTORY AND EXISTING CLIMATOLOGIES

PRECIPITATION CHANGES OVER THE NORTHERN HEMISPHERIC EXTRATROPICS DURING THE LAST HUNDRED YEARS

Pavel Ya. Groisman [1] and David R. Easterling [2]

[1] State Hydrological Institute, 199053 St. Petersburg, (Russia)

[2] National Climatic Data Center, 23 Second LineFederal Building Asheville, NC 28801 (USA)

ABSTRACT

The first indications that precipitation over land areas of northern extratropics (north of 35°N) has been increasing were published in the late 1980s. In this study of annual precipitation and snowfall changes over Canada, Russia, and the USA all possible precautions and adjustments were made to remove the large-scale inhomogeneities from the analyzed data. Area-averaging by the Thiessen method and scale correction (Groisman et al., 1991a) were used to develop regional, unbiased century-scale annual precipitation time series for the contiguous United States, southern Canada, and the former Soviet Union, together with forty-year time series of snowfall and total precipitation over most of North America. Trends analyses confirm previous results about precipitation increase of northern extratropics and quantify them for North America.

KEYWORDS

Precipitation changes, Snowfall, Russia, North America, Inhomogeneities of precipitation measurements.

1. Introduction

During the last century an anthropogenic rise in carbon dioxide and other greenhouse gas concentrations in the atmosphere has occurred (IPCC, 1990). The inevitable continuation of this rise in the nearest decades will result in some undetermined amount of global warming

NATO ASI Series, Vol. I 26
Global Precipitations and Climate Change
Edited by M. Desbois and F. Désalmand

(Budyko and Izrael 1987; IPCC 1990). The global average surface air temperature has increased to about half a degree during the last hundred years (IPCC 1990). Related with these warming, changes in precipitation (if they can be found) become of special interest; particularly if they can be extrapolated into the future. Keeping this goal in mind, we studied the precipitation and snowfall over three large countries in the Northern Hemisphere extratropics.

To find the possible relationships between precipitation and global temperature, it is necessary to filter from the time and space scales the "weather noise." The simplest of such filters are provided by averaging procedures. Therefore a year was selected as a time scale for precipitation and snowfall data, and the station data were spatially averaged over large regions up to 15 degree latitudinal zones. In doing so, we not only diminished the noise, but revealed the large-scale features of precipitation changes on the decadal and century time-scales.

Contrary to the situation for some other climatological elements, precipitation is still being measured unsatisfactorily (Table 1). During the last 100 years, the weather services have tried to solve this problem and to improve the instrumentation and the technique of precipitation measurements. So, changes in measured precipitation could describe the real precipitation change, but it could also depict the above-mentioned efforts and nothing else.

Table 1: Main components of the systematic error in precipitation measurement (Sevruk 1982)

Component of error	Magnitude
Loss due to windfield deformation above the gauge orifice	Rain: 2 - 10% Snow: 10 - 50%
Losses from wetting on internal walls of the collector and in the container when it is emptied	2-10%
Loss due to evaporation from the container	0-4%
Splash-out and splash-in	1-2%
Blowing and drifting snow	?

Unfortunately, in many countries there is a large amount of time-dependent bias when we look at inhomogeneities in precipitation records. Those which are connected with the changes of rain gauges and techniques of precipitation measurements by the various weather services are schematically depicted in Fig. 1 for several countries from the northern extratropics. It shows that there is no chance to obtain a reliable estimate of precipitation

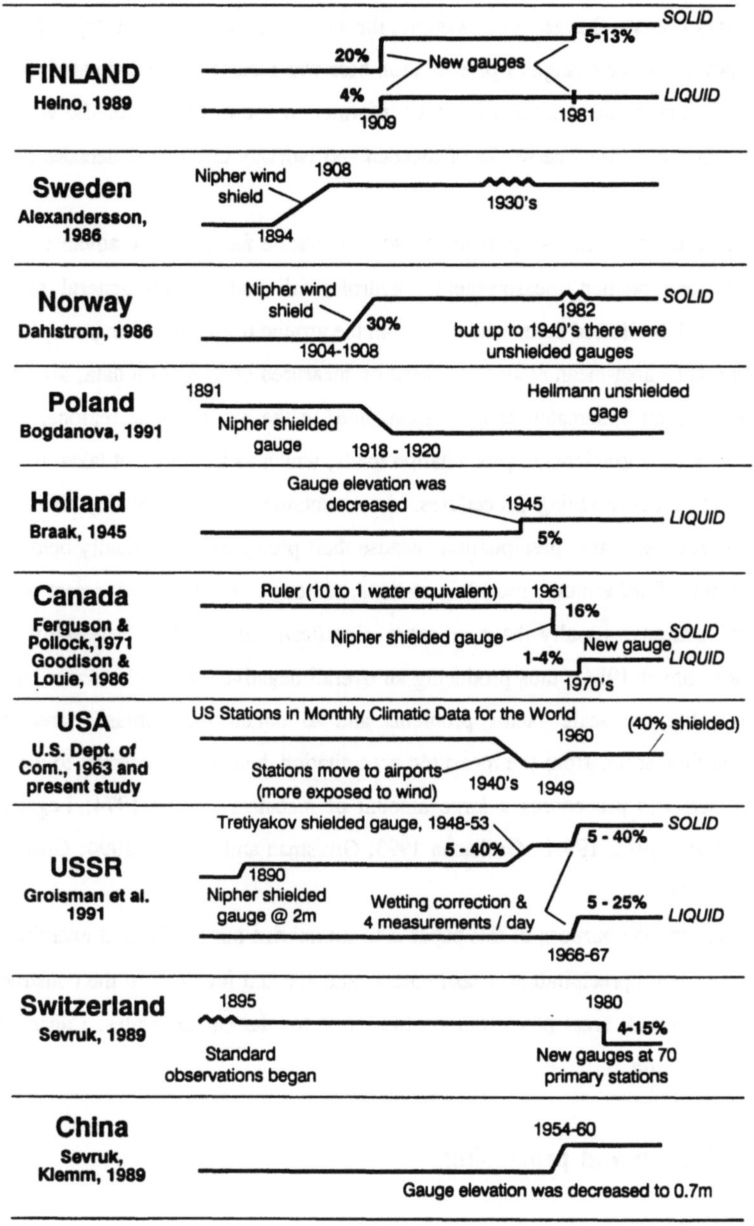

Fig. 1: Some important precipitation discontinuities over the past 100 years at primary observing stations within various mid- and high-latitude Northern Hemisphere countries (Groisman 1991, Karl et al. 1993).

changes during the last century without taking into account the instrumental inhomogeneity of the measurements because all these inhomogeneities provide great nonclimatic changes in measured precipitation totals which could mislead climatologists conducting research on contemporary climate change. For example, the analysis of the gauge history in Scandinavia (Fig. 1) shows that the change of precipitation here was seriously overestimated in Vinnikov et al. (1990) due to the absence of proper information about the date of the wind shields' installation in the national networks of Sweden and Norway in the first decades of the 20th century.

Additional problems arise from the so-called scale factors. The advanced nondirect methods of precipitation measurements, hydrological studies, and general atmospheric circulation models all need their validation from the ground truth area mean precipitation time series. But these area-mean estimates, based on measured precipitation data, are biased due to the precipitation undercatch at the stations and, in the mountainous regions, due to the presence of vertical gradients in precipitation totals, which usually are not taken into account in the existing area-averaging procedures. In mountainous regions, valley stations cannot provide the true mean area precipitation because their precipitation is usually below the true regional mean. Furthermore, mountain stations, although usually a minor part of the total number of stations, usually have a more significant undercatch (especially of solid precipitation, Shver 1984), thus producing an overall negative bias in the water balance.

To solve the scale factor problems and to obtain the unbiased area-averaged precipitation time series from the measured precipitation data, it is necessary to include into the special analysis procedures a large amount of metadata (WWB 1974; Legates 1987; Groisman et al. 1991a, 1991b; Groisman 1993; Groisman and Legates 1994; Groisman and Easterling 1994).

Therefore, the purpose of this paper is to summarize our results that examine various problems with point precipitation measurements and account for them in the construction of unbiased (less biased) areal precipitation time series for the United States, Canada and the former Soviet Union.

2. Data selection and processing

The detail description of the data selection and processing is given in Groisman et al. (1991a) for Russia and in Groisman and Easterling (1994) for the USA and Canada.

In our studies of precipitation over the U.S. and Canada, we used several networks, each selected with a special purpose in mind. To study snowfall we used a network, which consists of 1,107 Canadian (this subset of Canadian stations was selected using the requirement that each station has data available for at least fifteen years out of the last two decades) and 223 U.S. primary stations (including Alaska). The period 1951-1990 was used since all U.S. First-order stations had been relocated to airports with relatively open sites, and it was only after World War II that the networks for northern Canada and Alaska became dense enough to ensure adequate spatial coverage. Therefore, a reliable analysis of changes in snowfall and total precipitation over the U.S. and Canada can be performed only for the last forty years.

To study century-scale changes of annual precipitation over the contiguous U.S. and southern Canada we used a network, which consists of the same Canadian stations, 45 cooperative Alaskan stations, and a subset of the U.S. Historical Climatology Network (Karl et al. 1990) considered to have the most homogeneous data. Although these data are free of the large-scale inhomogeneities shown in Fig. 1, they may still contain undetected inhomogeneities. A special network of Canadian stations that were operating in the 1920s was selected to study Canadian precipitation prior to World War II. The area-averaged precipitation time series based on this network were combined with the corresponding time series developed from the main network of 1107 stations. The precipitation data for the 593 contiguous U.S. stations and the snowfall data were used without further adjustments. However, the Canadian rainfall data were corrected for the gauge change that occurred in the mid-1970s, by increasing all rainfall data before 1975 by factor 1.025 to approximate the results of Goodison and Louie (1986). Area-averaging of annual precipitation over the regions encompassing both Canada and the U.S. was done separately for each country. Only snow-stick measurements were used to represent the Canadian solid precipitation. Each area-mean time series was adjusted to its unbiased precipitation value (using the method described by Groisman et al. (1991a) and relies upon the WWB (1974) estimates of long-term annual precipitation. Then the time series from the two parts of the region were combined.

The network of 622 stations quite evenly spatially distributed over the former Soviet Union was used (Groisman et al. 1991a, 1991b). The station's data were adjusted to the instrumental inhomogeneities as described by Groisman et al. (1991a).

In the following Section we present area-averaged precipitation time series over North America and former Soviet Union. Area-averaging by the Thiessen (1911) method and scale correction (Groisman et al. 1991a) were used to develop regional, unbiased century-scale annual precipitation time series for the contiguous United States, southern Canada, and Russia, together with forty-year time series of snowfall and total precipitation over all of the USA and Canada. To avoid the effect of missing data on the results, we averaged the anomalies (in mm) from the reference period 1971-1990 (1951-1975 for Russia; 1921-1940 while processing the data of old Canadian network that was operating in 1920s) and then reconstructed the area-averaged precipitation, adding the corresponding area-averaged value of the mean precipitation for this reference period. These time series provide our first approximation of contemporary changes of precipitation over the U.S., Canada, and the former USSR. The first experiments using individual gauge corrections (Groisman 1991b, Groisman and Legates 1994) show that this approximation is not changed substantially on a continental scale after the inter-annual variability in gauge errors is taken into account.

3. Current changes in total precipitation and snowfall

3.1 Russia

After we have deleted the effects of non-climatic instrumentation changes on the data, the area mean precipitation time series for the entire USSR (P) shows a strong relationship with the surface air temperature for the Northern Hemisphere (T). Both these time series are depicted in Fig. 2. Here the corrected mean annual precipitation time series, based on the

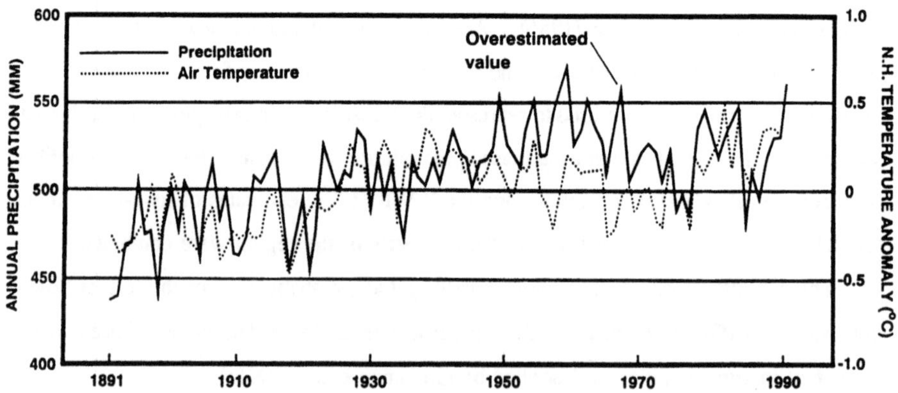

Fig. 2: Unbiased annual area-averaged precipitation over the USSR (37°N - 70°N, 25°E - 140°E) and mean annual surface air temperature of the Northern Hemisphere (Groisman et al. 1991a, Groisman 1991).

data of more than 500 stations, was taken from Groisman et al. (1991a) and hemispheric temperature data from Vinnikov et al. (1990). The correlation between these time series is 0.62 and regression of P to T is equal to 79 mm/°C (or 15%/°C).

There now exists a more-or-less common agreement that the global surface temperature will be increased by some amount due to the greenhouse effect (IPCC 1990, 1992). Hence, there is a high probability that the precipitation over the CIS could increase also. However, when we decomposed the country's precipitation time series regionally, we found that this increase during the 20th century has occurred mostly in the northern regions of Russia (to the north of 55°N). In the main agricultural parts of Russia, Ukraine, and Kazakhstan, there were not found any significant precipitation trends during the century.

3.2 North America

Annual snowfall

Figure 3 shows the zonally averaged snowfall over Alaska, northern Canada (north of 55°N), and over the zones 45-55°N and 35-45°N for the last four decades. The basic statistics of these time series are found in Table 2.

The accuracy of area-averaging of the data in northern Canada is high enough to confirm the strong positive linear trend in annual snowfall revealed in the zone (19% over four decades). This trend, together with the trend in rainfall data (not shown) confirm the systematic increase in annual totals of precipitation during the entire period of instrumental observations.

Snowfall in the zone 45-55°N is closely connected to temperature changes on continental and hemispheric scales. Therefore, the correlation between this time series and

Table 2: Characteristics of zonally-averaged annual snowfall for North America from 1950 through 1990 (from Groisman and Easterling 1994).

Region	Mean and standard deviation (in cm)		Linear trend and standard error (in % per decade)
	Mean	S. D.	
Canada, 55-70°N	195	17	5.1 ± 0.9
North America, 45-55°N	260	27	-0.7 ± 1.4
North America, 35-45°N	140	18	1.8 ± 1.7

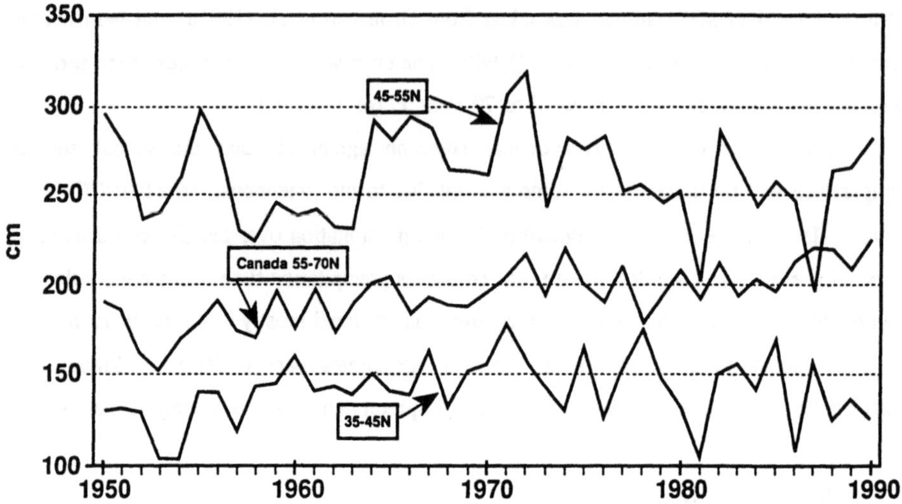

Fig. 3: Annual snowfall over North America during last 40 years (from Groisman and Easterling 1994).

the annual mean surface air temperature over the Northern Hemisphere is about -0.7 (Groisman et al. 1993). Using this relationship and taking into account the intrinsic coherence of annual snowfall totals in southern Canada (that encompassed the main part of the 45-55°N zone) with regional maximum temperature (Groisman 1992) we could foresee a future decrease in snowfall for this zone if the global warming projected by IPCC (1990, 1992) actually occurs.

Annual total precipitation

Figure 4 shows the unbiased annual precipitation totals for the last four decades area-averaged over the same zones as in Figure 3 plus southern part of the U.S. (zone 28-35°N to the east of 105°W). The statistics in Table 3 indicate that annual precipitation has increased during these last decades over the entire continent, and is especially significant over the South-Eastern U.S. (14% per 40 years) and Northern Canada (17% per 40 years).

Table 4 and Fig. 5 provide the same information for the Century-scale precipitation changes over southern Canada and the contiguous U.S. Table 4 confirms the main conclusions derived from Table 3, but shows that the statistical significance of precipitation trends based only on the 41 year record may be misleading when is extrapolated to the entire period of instrumental observations. Therefore, on the Century-scale the linear trend of the U.S. precipitation is not statistically significant.

In order to analyze the spatial field and to identify the independent components of the precipitation pattern, we applied the spatial averaging to the regions depicted in Figure 6a.

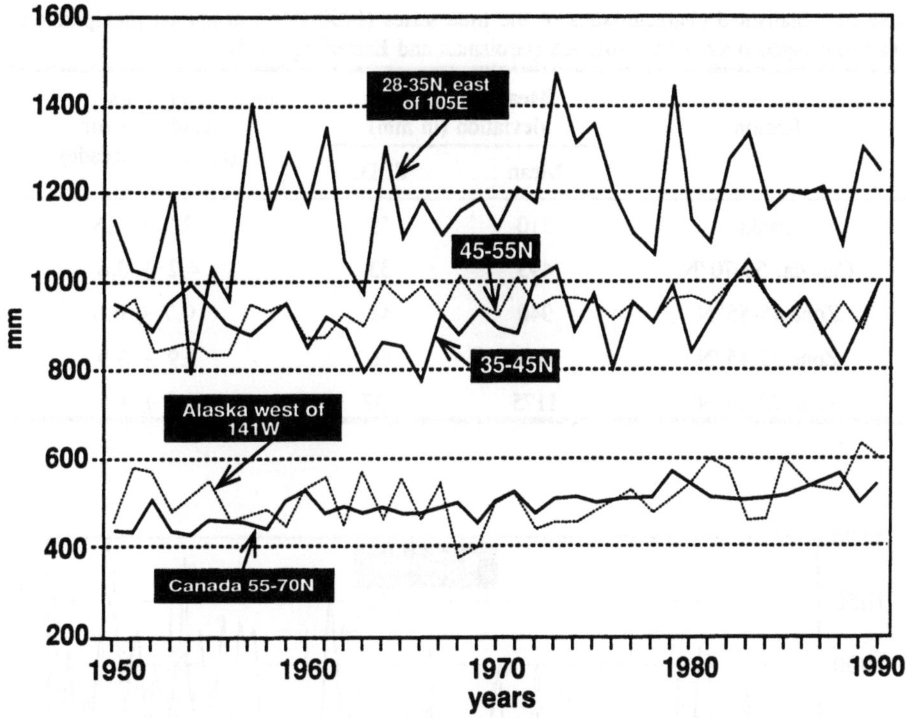

Fig. 4: Annual precipitation over North America during the last 40 years (Groisman and Easterling 1994).

These regions were defined using the principal component analysis of streamflow made by Lins (1985) and extended to Southern Canada. Time series of adjusted annual precipitation area-averaged over these five regions are shown in Figure 6b, with their main statistical features in Table 5.

An increase of more than 10% per century in annual precipitation over Southern Canada (Table 5) is reflected by a corresponding precipitation increase over the North-Eastern and North-Central regions. Linear trends account for up to 20% of variance of annual precipitation over each of these two regions. However, the 5-6% per century trends documented for North-Western and South-Central regions are not statistically significant at 0.05 level, and there have been no systematic changes over the South-Western region during the past 100 years. Therefore, we can conclude that during last century a large-scale increase has occurred over North America as a whole, but the major source of this increase was due to changes in Eastern Canada and the adjacent regions of the U.S. (north-eastern quarter of the country).

Table 3: Statistical characteristics of the time series (1950-1990) of annual precipitation zonally-averaged over North America (Groisman and Easterling 1994).

Region	Mean and standard deviation (in mm)		Linear trend ± its standard error (in % per decade)
	Mean	S. D.	
Alaska	510	59	2.7 ± 1.5
Canada, 55-70°N	495	33	4.2 ± 0.6
Zone 45-55°N	940	41	0.7 ± 0.6
Zone 35-45°N	905	66	1.9 ± 0.9
Zone 28-35°N	1175	137	3.4 ± 1.5

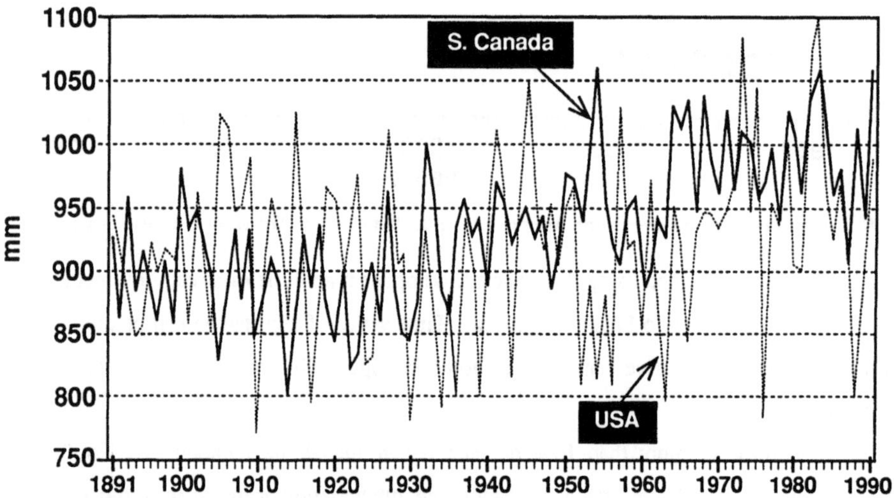

Fig. 5: Changes in areally-averaged annual precipitation over the last century for the continental United States and Canada south of 55°N (from Groisman and Easterling 1994).

Table 4: Characteristics of area-averaged annual precipitation over Canada south of 55°N and the continental United States over the last century (from Groisman and Easterling, 1994).

Region	Mean and standard deviation (in mm)		Linear trend and standard error (in % per century)
	Mean	S. D.	
Southern Canada	935	58	13.7 ± 1.7
Continental United States	915	71	3.9 ± 2.7
Both regions	925	55	7.6 ± 1.9

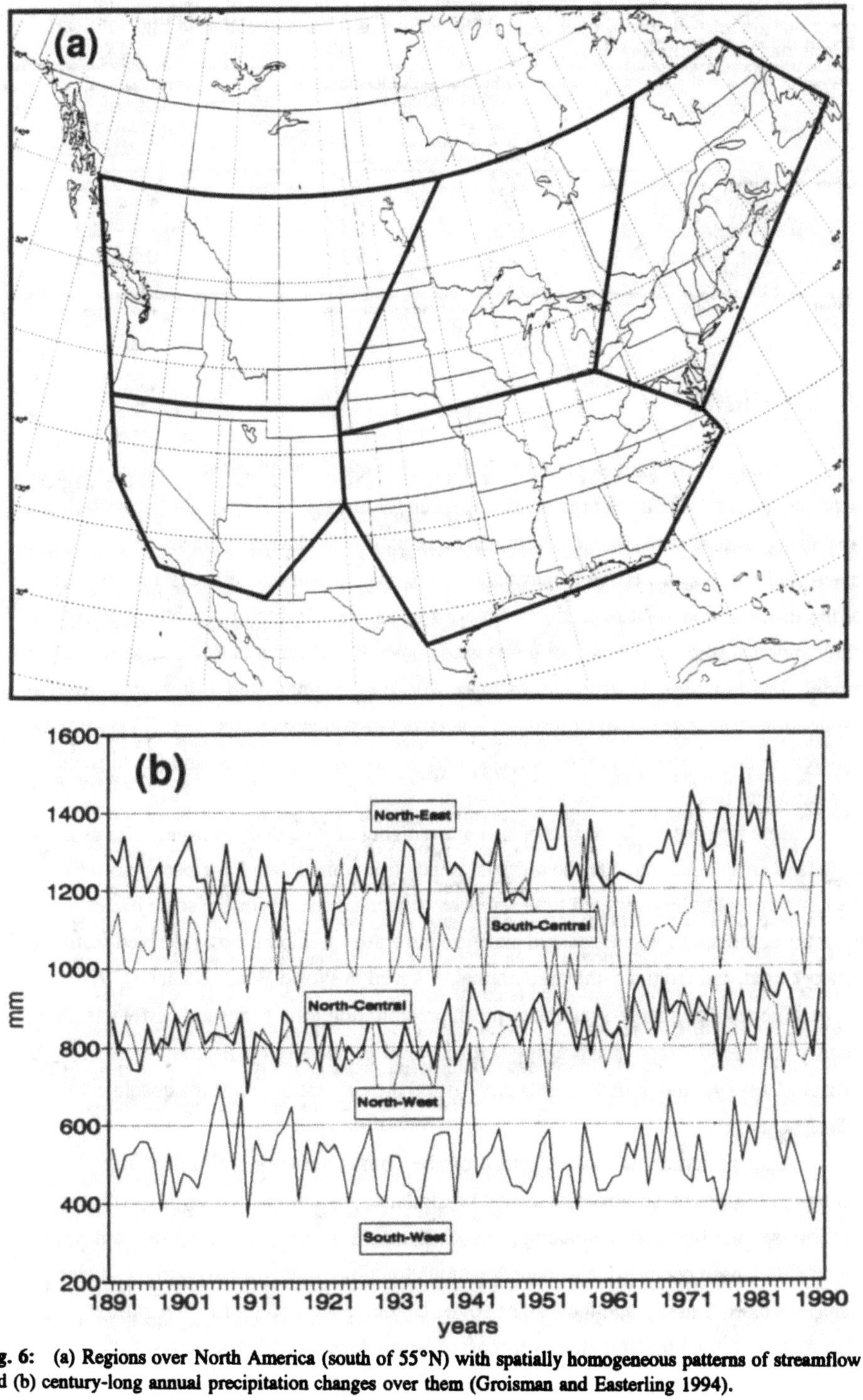

Fig. 6: (a) Regions over North America (south of 55°N) with spatially homogeneous patterns of streamflow and (b) century-long annual precipitation changes over them (Groisman and Easterling 1994).

Table 5: Statistical characteristics of the century-long time series of unbiased annual precipitation area-averaged over regions shown in Fig. 11 and depicted in Fig. 12 (Groisman and Easterling 1994).

Region	Mean and standard deviation (in mm)		Linear trend ± its standard error (in %/100yrs)
	Mean	S. D.	
North-West	825	70	5.5 ± 2.9
North-Central	845	65	12.1 ± 2.4
North-East	1265	90	11.3 ± 2.2
South-West	515	90	0.5 ± 6.1
South-Central	1115	110	5.4 ± 3.4

4. Conclusions

Standard rain gauges installed at the Russian, U.S. and Canadian meteorological networks undercatch "true" precipitation, especially in its solid form. The biases induced by this undercatch have changed during last century due to changes in instrumentation and practice of measurements. This introduces large-scale inhomogeneities in the time series of the precipitation field over North America and Eurasia which must be considered in climatological studies. The use of unadjusted conventional precipitation climatologies for model development, instrument calibrations (e.g., radar), and other hydrological applications can lead to biased results. The errors to which we refer are: (1) systematic undercatch that is present in most widely-used point-source data sets; and (2) subsequent area estimates that do not account for systematic sampling errors.

Here we present the results of spatial averaging of precipitation time series over the large land areas obtained after the artificial trends from initial data have been removed. In the last step, spatially averaged time series have been corrected for the scale biases using the ratio of the area-mean measured and "ground truth" precipitation as derived from the network used, and from the unbiased normals mapped in World Water Balance (1974).

The resulting large-scale, area-mean precipitation time series confirm the sign of century-long tendencies of the changes discovered previously by Bradley et al. (1987), Diaz et al. (1989), and Vinnikov et al. (1990) and provide estimates of the absolute values of these changes.

After adjustments and selection of the "best" network, reliable "first guess" estimates of northern extratropical annual precipitation were obtained. Area-averaging by the Thiessen method and scale correction (Groisman et al. 1991a) were used to develop regional, unbiased century-scale annual precipitation time series for the contiguous United States, southern Canada, and the former Soviet Union, together with forty-year time series of snowfall and total precipitation over most of North America. Trends analyses show that:

(1) during the last hundred years, annual precipitation has increased in Southern Canada (south of 55°N) and in the former Soviet Union by 12-14%, and in the contiguous U.S. by 4-5%;
(2) up to a 20% increase has occurred in annual snowfall and rainfall during last four decades in Canada north of 55°N;
3) in the zone 45-55°N of North America, century-scale total precipitation has increased while a redistribution between solid and liquid form has occurred in conjunction with variations in the zonal maximum temperature;
(4) the main domain of the century-scale precipitation increase over North America to the south of 55°N is Eastern Canada and the adjacent northern regions of the U.S.

These precipitation increases over the USSR territory an North America found by Groisman et al. (1991a) and by Groisman and Easterling (1994) contribute to the global increase in precipitation during the past century in the high latitudes of the Northern Hemisphere that were first reported by Bradley et al. (1987) (cf., Bradley and Groisman 1989, Groisman 1991a).

Acknowledgments : This work was completed while the first author was a senior research associate (U.S. National Research Council) at the National Climatic Data Center/NOAA, Asheville, NC, USA

REFERENCES

Bradley, R.S., H.F. Diaz, J.K. Eischeid, P.D. Jones, P.M. Kelly and C.M. Goodess, 1987 : Precipitation fluctuations over Northern Hemisphere land areas since the mid-19th century. Science, 237, 171-175.

Bradley, R.S., and P. Ya. Groisman, 1989 : Continental scale precipitation variations in the 20th century. In Precipitation measurement, Ed. B.Sevruk,- WMO/IAH/ETH Workshop on Precipitation Measurement. St. Moritz, 3-7December 1989. p.499-503.

Budyko, M.I. and Yu.A.Izrael (Eds), 1987 : Anthropogenic climatic changes. Leningrad, Gidrometeoizdat. 406 pp.

Diaz, H.F., R.S. Bradley, and J.K. Eischeid, 1989 : Precipitation fluctuation over global land areas since the late 1800s. J. Geophys. Res., 94, 1195-1240.

Goodison, B.E. and P.Y.T. Louie, 1986 : Canadian methods for precipitation measurement and correction. WMO/TD-No. 104. Instruments and observing methods. Report No.25 Papers presented at the Workshop on the Correction of Precipitation Measurements, Zurich, Switzerland, 1-3 April, 1985, 141-145.

Groisman, P. Ya., 1991 : Unbiased estimates of precipitation change in the Northern Hemisphere extratropics. AMS Proc., Fifth Conference on Climate Variations, Denver, CO, 42-45.

Groisman, P. Ya., V.V. Koknaeva, T.A. Belokrylova, and T.R. Karl, 1991a : Overcoming biases of precipitation measurement : A history of the USSR experience. Bull. Amer.Meteorol. Soc., 72, 1725-1733.

Groisman, P. Ya., E.M. Mason, and S. DelGreco, 1991b : Metadata for construction of homogeneous long-term precipitation time series. AMS Proc., Seventh Conference on Applied Climatology, Salt Lake City, UT, 119-122.

Groisman, P. Ya., 1992 : Studying the North American precipitation changes during the last 100 years. Proc. of the 5th International Meeting on Statistical Climatology, Toronto, 22-26 June 1992, 75-79.

Groisman, P. Ya., R.W. Knight, T.R. Karl, and R.R.Heim Jr., 1993 : Inferences of the North American snowfall and snow cover with recent global temperature changes Proc. of "Snow Watch '92" Conference, Niagara-on the-Lake, Canada, March 30 - April 1, 1992, 44-51.

Groisman, P. Ya., 1993 : Towards unbiased estimates of North American precipitation, AMS Proc., Eighth Symposium on Meteorological Observations and Instrumentation, Anaheim, CA, J43-J47.

Groisman, P. Ya., and D.R. Easterling, 1994 : Variability and trends of precipitation and snowfall over the United States and Canada, J. Climate, 7, 184-205.

Groisman, P. Ya., and D.R. Legates, 1994 : The accuracy of United States precipitation data, Bull. Amer. Meteorol. Soc., 75, 215-227.

Intergovernmental Panel on Climate Change (IPCC), 1990 : Climate Change. The IPCC Scientific Assessment. J.T. Houghton et al., Eds., Cambridge University Press, New-York, 362 pp.

Intergovernmental Panel on Climate Change (IPCC), 1992 : Climate Change 1992. The Supplementary Report to the Scientific Assessment. J.T. Houghton et al., Eds., Cambridge University Press, New York, 200 pp.

Karl, T.R., C.N. Williams Jr., F.T. Quinlan, and T.A. Boden, 1990 : United States Historical Climatology Network (HCN) serial temperature and precipitation data. NDP-019/R1, Carbon Dioxide Information Analysis Center, Oak Ridge National Laboratory, Oak Ridge, Tennessee, 83 pp. plus appendices.

Karl, T.R., R.G. Quayle, and P.Ya. Groisman, 1993 : Detecting climate variations and change: New challenges for observing and data management systems. J. Climate, 6, 1481-1494.

Legates, D.R., 1987 : A climatology of global precipitation. Publ. Climatol., 40 (1), 84 pp.

Lins, H.F., 1985 : Streamflow variability in the United States: 1931-1978. J. Climate Appl. Meteor., 24, 463-471.

Sevruk, B., 1982 : Methods of correction for systematic error in point precipitation measurement for operational use. Operational Hydrology Report No. 21, World Meteorological Organization, Geneva, 91 pp.

Shver, Ts.A., 1984 : Regularities in distribution of precipitation amount over continents. Gidrometeoizdat, 285 pp. (in Russian).

Thiessen, A.H., 1911 : Precipitation averages for large areas. Mon.Wea. Rev., 39,1082-1084.

Vinnikov, K.Ya., P.Ya. Groisman, and K.M. Lugina, 1990 : Empirical data on contemporary global climate changes (temperature and precipitation). J. Climate, 3, 662-677.

World Water Balance and Water Resources of the Earth (WWB), 1974 : Gidrometeoizdat, Leningrad, 638 pp. (in Russian ; published in English by UNESCO Press, 1978).

THE RECENT WEST AFRICAN RAINFALL VARIABILITY THROUGH EMPIRICAL AND MODELLING INVESTIGATIONS

Serge Janicot

METEO-FRANCE, Laboratoire de Météorologie Dynamique
Ecole Polytechnique
91128 Palaiseau Cédex (France)

ABSTRACT
Interannual statistical connections are investigated in July-August between West African rainfall, atmospheric tropical circulation and equatorial Atlantic and Pacific SST. Analyses performed with rainfall data (1958-1989) and GCM outputs (1970-1988) point out the role of Hadley cells and tropical East-West divergent circulations in reference to two different types of Sahelian drought. A SST warming in the Gulf of Guinea implies a southward and enhanced Hadley circulation along with reduced rainfall in the Central Sahel and increased rainfall in the Guinean region. A SST warming in the eastern equatorial Pacific leads to a stronger East-West divergent circulation over the tropical Atlantic and lower rainfall over all of West Africa. During some years, these two SST forcings interact to modulate West African rainfall variability. In particular the impact of the ENSO event of 1987 on West African monsoon dynamics is different from the statistical signal produced by other ENSO events because of warmer than usual SST in the tropical Atlantic.

KEYWORDS
West African monsoon, sea surface temperature, El Nino and Southern Oscillation, general circulation model, divergent circulations.

1. Introduction

Connections between sea surface temperatures (SST) and the Sahelian rainfall variability have been detailed for fifteen years. Lamb (1978a,b) linked Sahelian rainfall deficits during northern summer to cooler northern (warmer southern) SST in the tropical Atlantic, leading to a reduced northward excursion of the inter-tropical convergence zone (ITCZ). This hypothesis was confirmed by Hastenrath (1984, 1990), but it has also been regularly questioned (Lough 1986, Lamb and Peppler 1991, Janicot 1992, Fontaine and

NATO ASI Series, Vol. I 26
Global Precipitations and Climate Change
Edited by M. Desbois and F. Désalmand

Bigot 1992) since some dry years like 1983 were not concomitant with this SST pattern. At the same time, it was shown that SST in distant basins could exert a forcing on Sahelian rainfall, in particular the northern extratropical Atlantic and the eastern equatorial Pacific (Folland et al. 1986, Palmer 1986, Fontaine and Bigot 1992). General circulation model (GCM) simulations confirmed the empirical results and provided a partial understanding of the dynamical mechanisms involved in the effect of different SST forcings on West African rainfall (Palmer 1986, Semazzi et al. 1988) : warmer equatorial and southern tropical Atlantic, and cooler northern Atlantic cause the weakening of the Hadley cell south of the ITCZ and lead to lower (greater) rainfall north (south) of 10°N; warmer SST in the east equatorial Pacific forces a westerly wind anomaly in the equatorial upper troposphere over the east Pacific, extending over the Atlantic and Africa, and a reduced westerly component of the monsoon, consistant with a decrease of Sahel rainfall. However the respective weight of tropical Atlantic and Pacific surface temperatures in West African monsoon dynamics has not yet been established. The purpose of this paper is to analyse a part of this question for the July-August months over the period 1958-1989 with observations and over the period 1970-1988 with GCM integrations.

2. The data

Tanaka et al. (1975) has previously shown that different types of meridional rainfall profile could exist for Sahelian droughts. This was confirmed by Nicholson (1981) who pointed out two main rainfall patterns linked to rainfall deficiencies in the Sahel, one consisting in a global deficit over all of West Africa , and the other one in negative (positive) rainfall departure north (south) of 10°N. With these two rainfall patterns, most of the West African rainfall anomaly patterns can be derived. Recently, Nicholson and Palao (1993), using rotated principal component analyses, pointed out more clearly coherent rainfall regions covering the "West Coast", the "Sahel", and the "Guinean Coast".

Monthly rainfall fields over West Africa (1958-1989) have been extracted from a dataset built with 110 synoptic stations selected for their quality and their spatial distribution (Moron 1992, 1993). Regionalisations were performed by rotated principal component analyses, which confirm the results of Nicholson and Palao (1993). Monthly rainfall indexes representing coherent regions were constructed. Two standardized indexes, describing the interannual rainfall variability of the corresponding regions, have been retained : a "Central Sahel" rainfall index (CSI) with the stations of Ségou, Dori, Tillaberi, Tahoua, Kano and Maine Soroa, corresponds to the "Sahel" region of Nicholson and Palao, and a "Guinean" rainfall index (GI) with the stations of Sassandra, Abidjan, Lomé, Cotonou and Lagos, corresponds to the "Guinean Coast" region of Nicholson and Palao (Moron 1992, 1993).

Analysed monthly mean SST have been obtained from U.K. Met. Office, Bracknell. (Bottomley et al. 1990) for the 1958-1989 period. These data are analysed on a 5° latitude/longitude grid and have been corrected for presumed observational bias.

3. Equatorial SST and observed West African monsoon variability

Two standardised indexes have been defined to document the SST variability in the equatorial Atlantic (EATL) and in the east equatorial Pacific (EPAC). The EATL index is the July-August averaged SST over a limited domain (5°N-5°S/20°W-5°W). It is strongly correlated with the main mode of variability of the tropical Atlantic (Houghton and Tourre 1992, Fontaine and Bigot 1992), and thus describes a large part of the SST variability in the Gulf of Guinea over which the monsoon flows before reaching West Africa. The EPAC index is the July-August averaged SST over a larger domain (5°N-5°S/180°W-80°W). It is a good indicator of the amplitude of ENSO and anti-ENSO events (Deser and Wallace 1990). These two SST indexes are uncorrelated (see Fig.1) and can be considered as two independent factors (in a statistical sense). They gather also two different tropical dynamics : a zonal mode in Pacific atmosphere variability and a meridional mode in Atlantic atmosphere variability.

The summers of the period 1958-1989 (mean July-August rainfall) have been classified into four types, depending on the CSI and GI indexes signs. Types +/- (rainfall excess north of 10°N and rainfall deficit south of 10°N) include the summers 1958-59-60-61-64-66-67-70-77-78-80-88, types -/+ the summers 1968-73-79-84-87, types -/- the summers 1969-71-72-76-81-82-83-85-86, and types +/+ the summers 1962-63-65-74-75-89.

Fig.1 shows the relationships between the indexes EATL, EPAC, CSI, and the four West African rainfall types. "Significant" signals are emphasized because we want to focus on cases where both the SST forcings and the rainfall indexes are strong enough to overcome the period character. Thus "significant" rainfall patterns and SST anomalies are retained both when each of the two rainfall indexes are greater than 1/4 in absolute value and when the standardized SST indexes are greater than 2/3 in absolute value. These selected thresholds are a compromise between a sufficiently high level of anomaly amplitude and the necessity to retain enough cases out of the 32 available. Then 15 "significant" cases are kept (see Tab.1); they are highlighted in Fig.1.

The diagram related to the connection CSI/EATL (Fig.1) differentiates the rainfall types +/- (cooler temperatures in the Gulf of Guinea) and types -/+ (higher temperatures in the Gulf of Guinea), in association with an abnormal latitudinal location of the ITCZ (Lough

1986, Janicot 1992). The rainfall types -/- and +/+ are not discriminated by the EATL index and are located in the middle of the diagram.

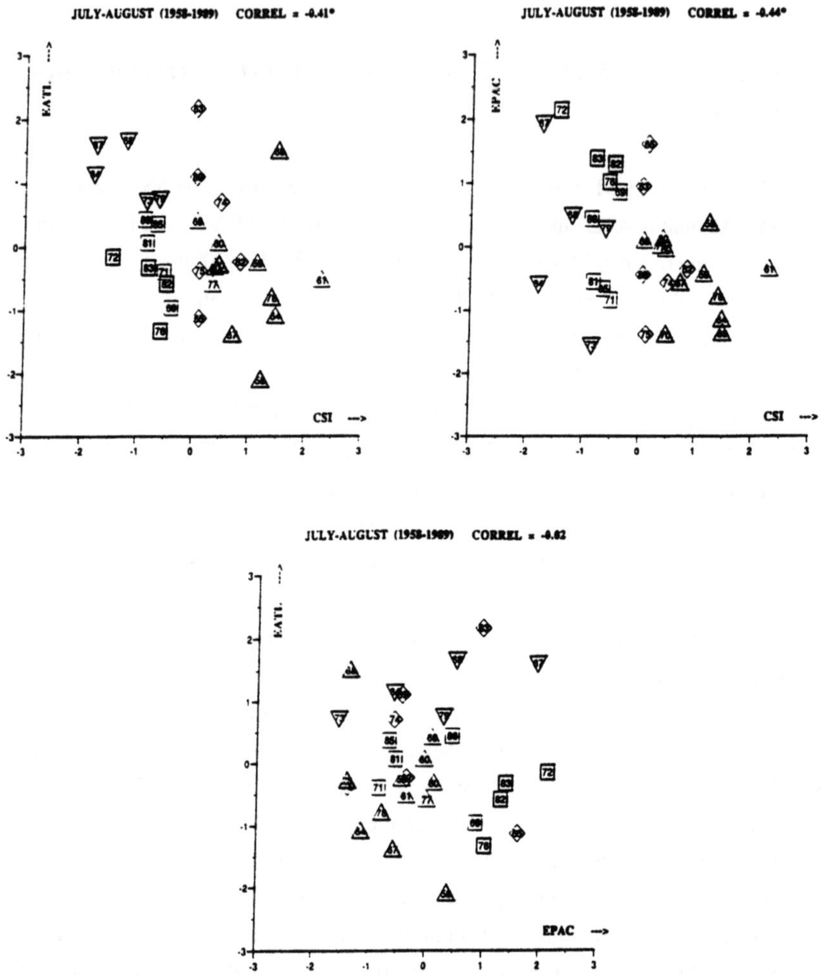

Fig.1 : Scatter plots of July-August (1958-1989) indexes; top left : CSI (abscissa) and EATL (ordinate); top right : CSI (abscissa) and EPAC (ordinate); bottom : EPAC (abscissa) and EATL (ordinate). The rainfall types are indicated by square for (-/-), lozenge for (+/+), upward triangle for (+/-), and downward triangle for (-/+). Heavy figures designate "significant" summers (see text for details).

The diagram related to the connection CSI/EPAC (Fig.1) differentiates the rainfall types -/- (with higher SST in the east equatorial Pacific) and rainfall types +/- (with lower SST). A warm event in July-August (called hereafter ENSO occurrence) can be associated with a global deficit of rainfall over West Africa, whereas a cold event (called hereafter

LNSO -La Nina Southern Oscillation- occurrence) is concomitant with stronger (weaker) rainfall north (south) of 10°N, as with negative EATL indexes. It is interesting to note the non-symmetric effect of opposed EPAC indexes on West African rainfall field, the rainfall type +/+ being not clearly discriminated. The rainfall types -/+ are not differentiated by the EPAC index, except for the summer of 1987.

The connection EATL/EPAC synthesizes the previous results (Fig.1). Since no correlation exists between these two SST indexes (-0.02), different SST forcings (positive -negative- forcing when the SST index is positive -negative-) can coexist and their interaction must be taken into account in the Sahelian and West African rainfall variability. The previous results enable us to suppose that a "significant" positive (negative) EATL index leads to a "expected" "significant" rainfall type -/+ (type +/-) occurrence and that a "significant" positive (negative) EPAC index leads to a "expected" "significant" rainfall type -/- (type +/-). The global rainfall deficit in 1972 was associated with an ENSO occurrence and no SST forcing in the Gulf of Guinea, as for 1982 and 1983. The rainfall types -/+ are concomitant with a positive SST forcing in the Gulf of Guinea and mostly no SST forcing in the east equatorial Pacific, except in 1973 when the two SST forcings were of reversed signs, and in 1987 when the two positive SST forcings worked together with a rainfall deficit in Central Sahel. It may be suggested for these two last summers that the Gulf of Guinea forcing dominated since Guinean rainfall was important (rainfall types -/+ meaning a southern location of the ITCZ). Rainfall types +/- are linked to colder waters in the Gulf of Guinea except in 1988, when the Atlantic forcing was positive (unfavorable to heavy rainfall in the Central Sahel), and synchroneous with a LNSO occurrence (rather favorable to rainfall types +/-); it seems here that the more remote SST forcing was the strongest. In the same way, during summer 1976 (rainfall type -/-), the ENSO pattern forcing was greater than the forcing in the Gulf of Guinea (cooler SST). On the contrary, during summer 1965, in spite of the occurrence of an ENSO SST pattern, Sahelian rainfall responded to cooler SST in the Gulf of Guinea.

Tab.1 shows the correspondence for the 15 selected "significant" cases between observed and "expected" rainfall types, and occurrences of SST forcings. All of the 15 "expected" rainfall type occurrences are observed, which confirms our hypotheses. However, when the two "significant" SST forcings occur, either the strongest forcing, or the nearest one clearly does not impose itself. However, 15 cases is too small a sample to offer any final conclusion. The interaction between the SST forcings cannot always appear clearly. This is the reason why GCM runs have been performed to pursue this investigation.

JULY-AUGUST	EATL	EPAC	EXPECTED RAINF.TYPES	OBSERVED RAINF.TYPES
1958	-	0	+/-	+/-
1964	-	-	+/-;+/-	+/-
1967	-	0	+/-	+/-
1968	+	0	-/+	-/+
1970	0	-	+/-	+/-
1972	0	+	-/-	-/-
1973	+	-	-/+;+/-	-/+
1976	-	+	+/-;-/-	-/-
1978	-	-	+/-;+/-	+/-
1979	+	0	-/+	-/+
1982	0	+	-/-	-/-
1983	0	+	-/-	-/-
1984	+	0	-/+	-/+
1987	+	+	-/+;-/-	-/+
1988	+	-	-/+;+/-	+/-

Tab.1 : Signs of the SST forcing anomalies in the Gulf of Guinea (EATL) and the east equatorial Pacific (EPAC), observed and "expected" rainfall types, for the 15 selected "significant" July-August between 1958 and 1989; (-) : negative values; (0) : non "significant" values; (+) : positive values. For rainfall types, +/- for example indicates "significant" positive (negative) rainfall anomaly north (south) of 10°N. "expected" rainfall types are defined as -/+ (+/-) if there is a "significant" positive (negative) EATL index, and as -/- (+/+) if there is a "significant" positive (negative) EPAC index; these rainfall types are associated when "significant" EATL and EPAC indexes both occur.

4. Equatorial SST and simulated tropical atmospheric dynamics

The GCM of the Laboratoire de Météorologie Dynamique (LMD) is a model defined on a grid constant in longitude and in sine of latitude. It has 64 points in longitude and 50 points in latitude. The vertical coordinate is sigma, the ratio of local pressure to surface pressure, and 11 levels are unevenly spaced, four in the boundary layer, four in the troposphere and three in the stratosphere. The model was described in Sadourny and Laval (1984), and its cycle 5 has been used for this experiment (Polcher et al. 1991).

Seven integrations have been performed (with different initial conditions) with the "COLA/CAC AMIP SST and Sea-Ice Dataset" from 1970 to 1988 for the AMIP project (Atmospheric Model Intercomparison Project). This dataset was assembled from data especially prepared for this project with all available in-situ, climatological and satellite observations.

The GCM results enable us to take into account some dynamical variables like vertical velocity and divergent circulations which are not available with observations over a long period. Fig.2 shows the mean July-August (1970-1988) meridional cross-section of simulated vertical motions averaged around the Greenwich meridian (15°W-15°E). Hadley-type cells are pointed out and the convection inside the ITCZ is located between 10°N and 15°N, with a maximum of 40 $hPa.day^{-1}$. Correlations have been computed between the GCM-grid SST indexes EATL, EPAC, and the simulated vertical velocity in the meridional

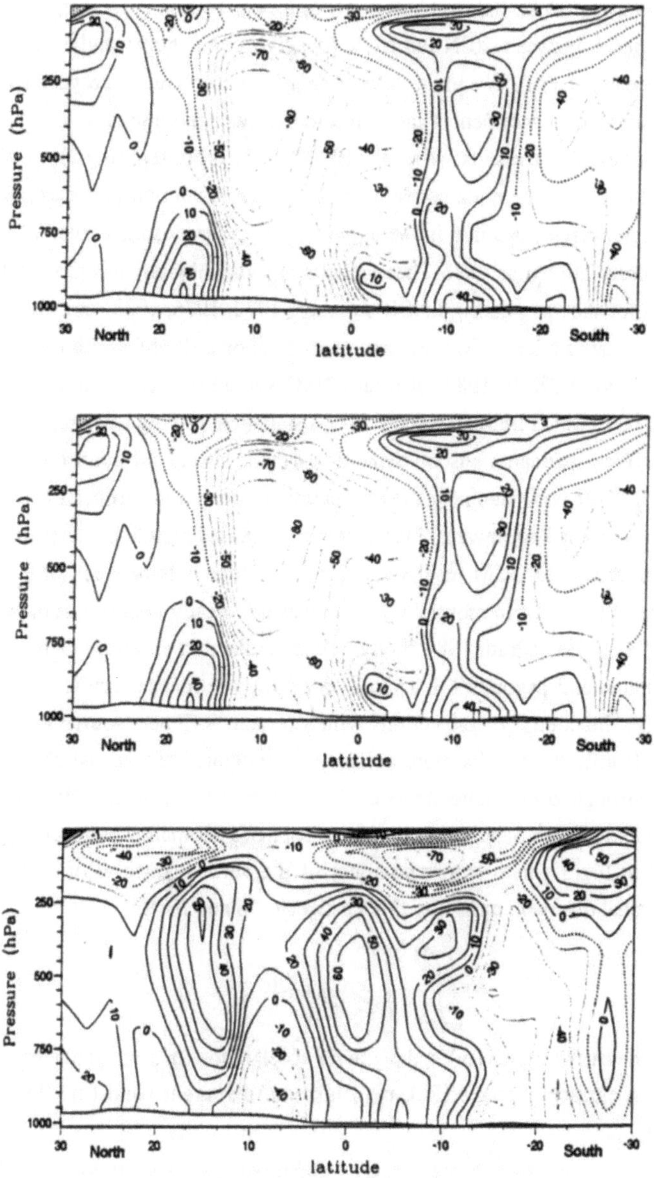

Fig.2 : top : meridional cross-section of the 7-member mean ensemble July-August (1970-1988) simulated vertical velocity (dp.dt^{-1}) averaged from 15°W to 15°E. Units : hPa.day^{-1}; middle : meridional cross-section of correlations (multiplied by 100) betwen GCM-grid EATL and the 7-member mean ensemble simulated vertical velocity averaged from 15°W and 15°E for July-August (1970-1988); bottom : idem with GCM-grid EPAC.

plane. When the EATL index is positive (Fig.2), convection is weaker in the lower half of the troposphere between 20°N and 15°N (positive correlations), and is greater in the whole troposphere between 10°N and the equator (negative correlations), especially over 500 hPa. This indicates an enhancement of the convection inside a more southward ITCZ and confirms the previous empirical results. When the EPAC index is positive (Fig.2), we also observed a southward retreat of the main convection area in the lower half of the troposphere, but we can see decreasing convection in the upper half of the troposphere over all of West Africa (positive correlations) as well. These results are also obvious in individual "significant" summers (not shown) : during ENSO-summers 1972, 1976, 1982, and 1983, the simulated Hadley-type circulation is weaker than average above 500 hPa with a normal latitudinal location of the ITCZ. In 1987 (also an ENSO summer), there is an enhanced and more southern ITCZ, which implies that the EATL statistical forcing was greater than the EPAC one on the West African monsoon dynamics. This is also in accordance with our empirical results. In 1988 (LNSO summer), the convection is stronger in particular above 700 hPa and the ITCZ is extended northward, consistant with increased rainfall in Sahelian areas.

Empirical and GCM results concerning connections between SST in the Gulf of Guinea and West African monsoon dynamics were already pointed out in previous papers (Lamb 1978a,b, Hastenrath 1984, Lough 1986, Folland et al. 1986, Palmer 1986, Janicot 1992, Fontaine and Bigot 1992). Thus it does not seem necessary to pursue further this analysis. On the contrary, connections with a remote SST heat source located in the east equatorial Pacific have to be more precisely investigated. The zonal connections can be described through zonal components of the divergent flow, or East-West circulations, to identify links between Sahel rainfall and atmospheric anomalies in distant regions. This is illustrated in Fig.3 which shows contours of the vertical integral of July-August mean simulated zonal divergent mass flux (MF) between 5°S and 20°N.

$$MF = -a/g. \int_{-\pi/36}^{\pi/9} \int_{1}^{\sigma} p_s.u_{div}.d\sigma.d\lambda$$

where a is the earth's radius (m), g the gravity acceleration ($m.s^{-2}$), sigma the model vertical level, p_s the surface pressure (Pa), u_{div} is zonal divergent wind ($m.s^{-1}$) and lambda is latitude (rd).

These contours can be interpreted as streamlines of the zonal divergent flow and their contribution to the vertical flow (Rowell et al. 1991). The flow is turning clockwise (counter-) around positive (negative) extrema. East-West circulations (or Walker-type cells in an equatorial plane) are clearly depicted. The ascending branch located at 150°E and the subsiding one over the eastern Pacific point out the Pacific Walker cell. An East-West Atlantic circulation is depicted by the ascending branch over Central America and the subsiding one located at 30°W-0°. West African rainfall may be influenced by the displacement and/or the intensity variations of this Atlantic zonal circulation as suggested in

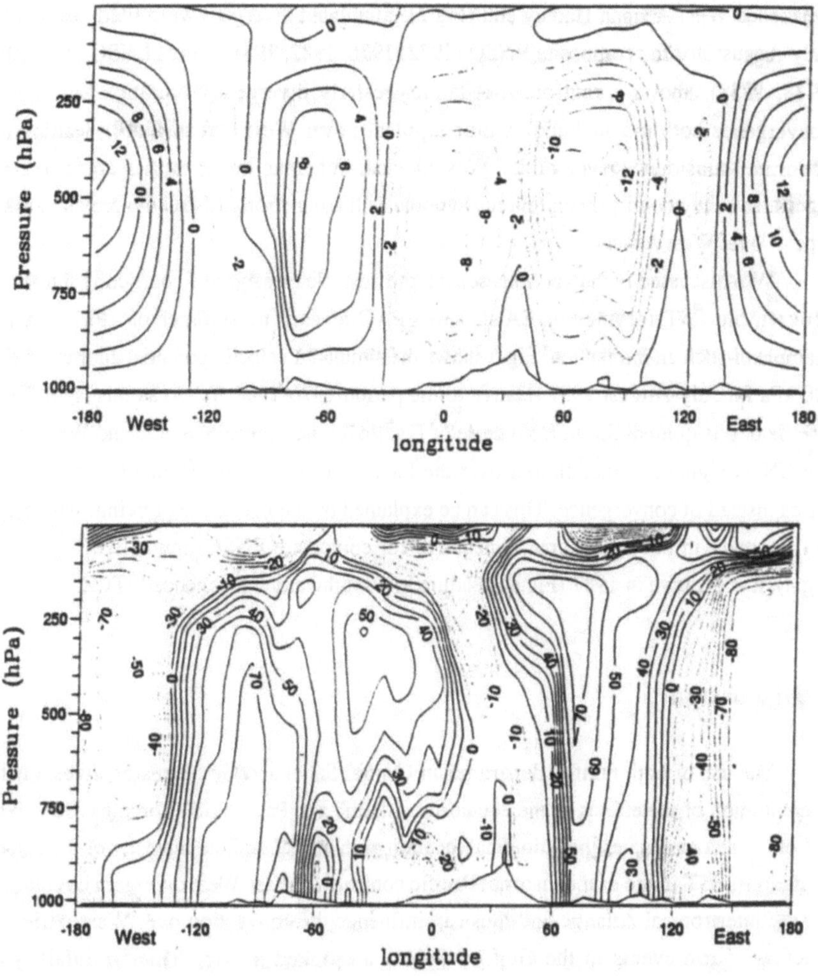

Fig.3 : top : zonal cross-section of the 7-member mean ensemble simulated July-August (1970-1988) vertical integral of the zonal divergent component of mass flux averaged from 20°N and 5°S (see details in the text). Contour interval is 10^9 $kg.s^{-1}$; bottom : zonal cross-section of correlations (multiplied by 100) between the GCM-grid EPAC index and the 7-member mean ensemble simulated July-August vertical integral of the zonal divergent component of mass flux averaged from 20°N and 5°S for the period 1970-1988.

Janicot (1992). Fig.3 shows also the zonal cross-section of correlations between the GCM-grid EPAC index and the vertical integral of simulated zonal divergent mass flux between 5°S and 20°N in July-August (1970-1988). During an ENSO occurrence, the positive correlations over the Atlantic and West Africa indicate an intensification and an eastward extension of the Atlantic zonal circulation, leading to abnormal ascending motions in the whole troposphere over the east Pacific, and a tendency to stronger and larger subsiding motions above 500 hPa over the tropical Atlantic and West Africa, concomitant with a weakened West African Hadley cell (Fig.2). Simulated divergent wind field at 200 hPa in July-August for the composite ENSO (1972, 1976, 1982, 1983) minus LNSO (1970, 1973, 1978, 1988), shows a reinforced Atlantic westerly divergent circulation and a global convergence both in zonal and meridional planes over West Africa (Fig.4), leading to an abnormal subsidence over all of West Africa. Whereas these results are not greatly significant, they confirm the previous hypothesis linking strong ENSO dynamics to rainfall type -/- over West Africa.

We discussed in the previous section the possible existence of conflicting interactions between the SST forcings of EATL and EPAC areas. The summer of 1987 is a good example of such an interaction. Fig.5 shows the simulated velocity potential anomaly field at 200 hPa for July-August 1987 relative to the period 1970-1988. Whereas this is an ENSO-type field, it is quite different from those of Fig.4 over the tropical Atlantic and West Africa. The ENSO signal is rather similar over the Pacific ocean but diffluence occurs over West Africa instead of convergence. This can be explained by the EATL area forcing: whereas the SST anomaly in the EATL area is negative for composite ENSO minus LNSO summers (Fig.4), it is positive in 1987 (Fig.5), leading to a southward and enhanced ITCZ.

5. Discussion

The use of both rainfall departures in Central Sahel and Guinean area gives a better differentiation of the effects of the equatorial Atlantic and Pacific SST forcings. SST in the Gulf of Guinea modulates the latitudinal location of the ITCZ and the intensity of the Hadley circulations. SST in the east equatorial Pacific controls the East-West divergent circulations over the intertropical Atlantic and thus can influence the convection over West Africa. In particular, warm events in the Gulf of Guinea are linked to "dry Sahel" rainfall types -/+,whereas ENSO events, linked to a global decrease of convection over West Africa, lead to "dry Sahel" rainfall types -/-. On the other hand, all "significant" rainfall types -/-, that is 1972,1976, 1982, 1983, are concomitant with ENSO occurrences. The role of the ENSO SST pattern in the occurrence of global rainfall deficiencies over all of West Africa is explained by the enhancement of the Atlantic zonal divergent circulation, leading to a weakening of the Hadley-type cells. However these two equatorial SST forcings can interact. The strong ENSO of 1987 did not have the expected consequence on West African rainfall anomaly pattern since the warm event in the Gulf of Guinea had a stronger effect and was associated with a rainfall type -/+. However, in 1988 the LNSO area forced enhanced Central Sahel rainfall in spite of a warm event in the Gulf of Guinea.

Four points must be highlighted : (1) This study has focused on the 15 most important equatorial SST anomalies which include very strong ENSO events occurring during the period 1958-1989: their effects on the tropical atmosphere and West African

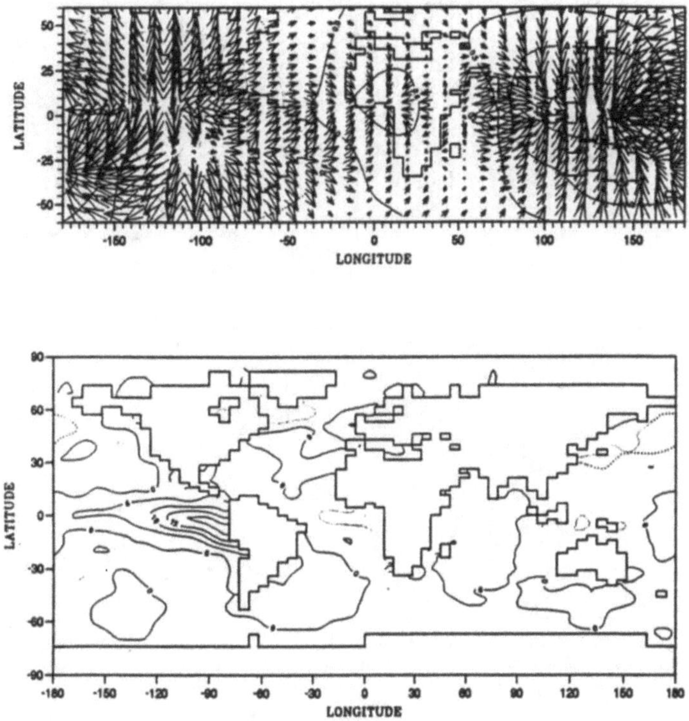

Fig.4 : top : composite simulated field of the 7-member mean ensemble July-August velocity potential and divergent wind at 200 hPa; mean ENSO summers (1972, 1976, 1982, 1983) minus mean anti-ENSO summers (1970, 1973, 1978, 1988) over the domain 60°N-60°S. Units : $10^6.m^2.s^{-1}$ for velocity potential and 15° for 1 $m.s^{-1}$ for divergent wind; for a better clarity, only one divergent wind grid point out of two is displayed; bottom : composite of global July-August analysed SST field for mean ENSO summers (1972, 1976, 1982, 1983) minus mean anti-ENSO summers (1970, 1973, 1978, 1988). Units : 0.1°C.

rainfall anomaly patterns have been all the more marked. (2) Folland et al. (1986) pointed out by principal component analysis an Atlantic mode closely correlated to Sahel rainfall. However, the greatest loadings of this mode are located out of the tropics and the correlation with Sahel rainfall mainly belongs to the low frequency scale. The period used by Folland et al. (1986) (extending through the 20th century) could also lead to different conclusions than the period 1958-1989 which contains a majority of deficient Sahel rainfall years when compared to the long-term mean (Fontaine et al. 1994). (3) The last forty years have been characterized by a strong decreasing rainfall trend in the Sahel, associated with a southward latitudinal trend of the ITCZ. More precisely, there has been a shift from rainfall types +/- towards rainfall types -/+ (Fontaine et al. 1993), linked to reversed SST anomaly patterns in the tropical Atlantic (Lough 1986). However we have seen that the EPAC index discriminates rainfall types +/- from types -/-. Thus a part of the rainfall trend in West Africa

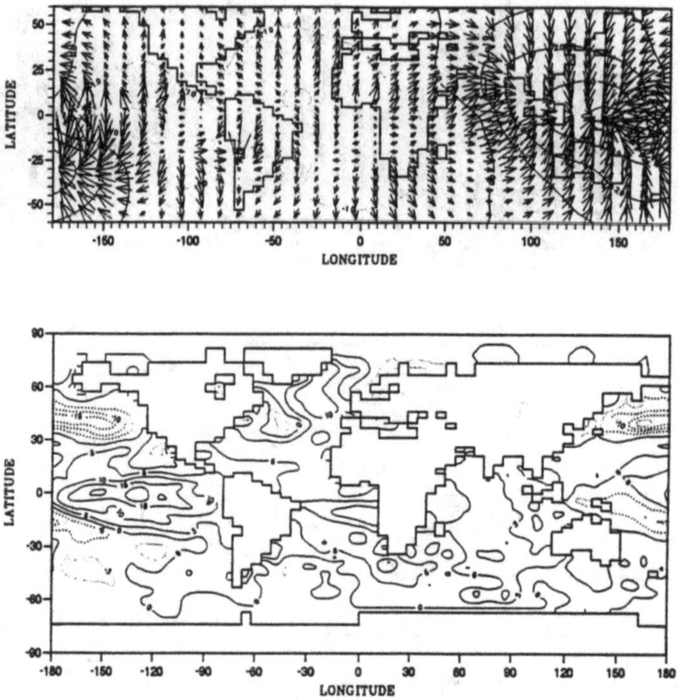

Fig.5 : top : simulated anomaly field of the 7-member mean ensemble July-August velocity potential and divergent wind at 200 hPa in 1987, relative to the period 1970-1988 over the domain 60°N-60°S. Units : $10^6.m^2.s^{-1}$ for velocity potential and 15° for 1 $m.s^{-1}$ for divergent wind; for a better clarity, only one divergent wind grid point out of two is displayed; bottom : global July-August analysed SST anomaly field for 1987 relative to the period 1970-1988. Units : 0.1°C.

can also be linked to SST variability in the east equatorial Pacific. (4) The EPAC forcing seems mostly effective in relation to the upper troposphere circulation. This was already known over the Pacific and West Atlantic (Palmer 1986). It has been shown here for West Africa. However, the GCM results show that equatorial Atlantic SST also has a great effect on the divergent circulation and the vertical velocity.

One question has not be answered : if an ENSO event forces the West African rainfall type -/-, why is a LNSO event associated with a rainfall type +/- and not a rainfall type +/+ ? To solve this question, it may be necessary to take into account extra-tropical atmospheric dynamics, even though these areas show rather low frequency connections with Sahel rainfall (Folland et al. 1986). On the other hand, the reduced Sahelian rainfall production period 1958-1989 may imply that no "significant" rainfall type +/+ is available.

6. Conclusion

We focused here on two particuliar SST areas, one in the Gulf of Guinea, the other one in the east equatorial Pacific. Forcings of the monsoon dynamics of West Africa by these two SST areas can interact together. However it is possible to differentiate their weight. The Gulf of Guinea SST modulates the strength and the meridional location of the Hadley cells on a regional scale, whereas the equatorial east Pacific SST modify the East-West divergent circulation on a planetary scale. Different West African rainfall anomaly patterns can be produced in response to the combination of these two forcings. These results are peculiar to the period 1958-1989, marked by large Şahelian rainfall deficits and strong ENSO events. Extended periods and other SST key-areas should be investigated in particular in extra-tropics to broaden this work.

Acknowledgments: The author is very thankful to B. Fontaine and V. Moron for providing the rainfall data and their participation to this work, and to the U.K. Meteorological Office for providing the SST dataset.

REFERENCES

Bottomley, M., Folland C.K., Hsiung J., Newell R.E. and Parker D.E., 1990: "Global ocean surface temperature Atlas GOSTA", Joint Met. Office / Massachusetts Institute of Technology Atlas. Printed by HMSO, London, 20pp, 313 plates.

Deser, C. and J.M. Wallace, 1990: "Large-scale atmospheric circulation features of warm and cold episodes in the tropical Pacific". J. Climate, 3, 1254-1281.

Folland, C.K., T.N. Palmer and D.E. Parker, 1986: "Sahel rainfall and worldwide sea temperature 1901-1985". Nature, 320, 602-607.

Fontaine, B. and S. Bigot, 1992: "West African rainfall deficits and sea surface temperatures". Int. J. Climatology, 13, 271-285.

Fontaine, B., S. Janicot and V. Moron, 1994: "Rainfall anomaly patterns and wind field signals over West Africa in August (1958-1989)". Submitted to J. Climate

Hastenrath, S., 1984: "Interannual variability and annual cycle: Mechanisms of circulation and climate in the tropical Atlantic sector". Mon. Wea. Rev., 106, 1280-1287.

Hastenrath, S., 1990: "Decadal scale changes of the circulation in the tropical Atlantic sector associated with Sahel drought". Int. J. Climatol., 10, 459-472.

Houghton, R.W. and Tourre Y.M., 1992: "Characteristics of low-frequency sea surface temperature fluctuations in the tropical Atlantic". J. Climate, 5, 765-771.

Janicot, S., 1992 "Spatiotemporal variability of West African rainfall; Part II: Associated surface and airmass characteristics". J. Climate, 5, 499-511.

Lamb, P.J., 1978a: "Case studies of tropical Atlantic surface circulation patterns during recent sub-Saharan weather anomalies: 1967 and 1968". Mon. Wea. Rev., 106, 482-491.

Lamb, P.J., 1978b: "Large scale tropical surface circulation patterns associated with Subsaharan weather anomalies". Tellus, 30, 240-251.

Lamb, P.J. and R.A. Peppler, 1991: "West Africa". Chapter 7 in Teleconnexions: Linkages between ENSO Worldwide Climate Anomalies and Societal Impacts (M.H. Glantz, R.W. Katz, and N. Nicholls, Eds), Cambridge University Press, 121-189.

Lough, J.M., 1986: "Tropical Atlantic sea surface temperature and rainfall variations in Subsaharan Africa". Mon. Wea. Rev., 114, 561-570.

Moron, V., 1992: "Variabilité spatio-temporelle des précipitations en Afrique sahélienne et guinéenne: 1933-1990". La Météorologie, 43/44, 24-30.

Moron, V., 1993: "Guinean and Saharan rainfall anomaly indices at annual and monthly time scales (1933-1990)". Int. J. Climatology, 14, 325-341.

Nicholson, S.E., 1981: "Rainfall and atmospheric circulation patterns during drought periods and wetter years in West Africa".Mon. Wea. Rev., 109, 2191-2208.

Nicholson, S.E. and I.M. Palao, 1993: "A re-evaluation of rainfall variability in the Sahel. Part I : Characteristics of rainfall fluctuations". Int. J. Climatology, 13, 371-390.

Palmer, T.N., 1986: "Influence of the Atlantic, Pacific, and Indian oceans on Sahel rainfall". Nature, 322, 251-253.

Polcher, J., A. Harzallah, S. Bony, S.N. Chalita, E. Cohen-Solal, M. Forichon, F. Hourdin, H. Le Treut, P. Levan, Z.X. LI, and P. Rogel, 1991: "Le cycle 5 du modèle de circulation générale du LMD". Note interne N°170 du LMD, CNRS, Paris, 41pp.

Rowell, D.P., C.K. Folland, K. Maskell, J.A. Owen, and M.N. Ward, 1991: "Causes and predictability of Sahel rainfall variability". CRTN12, Hadley Centre, Met. Office, 23pp.

Sadourny, R. and K. Laval, 1984: "January and July performance of the LMD general circulation model". New perspectives in climate modelling. A. Berger, Ed. Elsevier, 173-198.

Semazzi, F.H., V. Metha and Y.C. Sud, 1988: "An investigation of the relationship between Sub-Saharan rainfall and global sea surface temperatures". Atmos.-Ocean, 26, 118-138.

Tanaka, M., Weare B.C., Navato A.R. and R.E. Newell, 1975: "Recent African rainfall patterns". Nature, 255, 201-203.

EXTREME PRECIPITATION IN THE NORTHERN HEMISPHERE

Khamil Khairullin, Elena Kitaigorodskaya and Georges Shchukin
Main Geophysical Observatory
Karbkysheva Ulitsa 7
196194 St. Petersburg, Russia

ABSTRACT
Atmospheric precipitation is an important natural factor determining the life conditions on the Earth. By the same time, heavy precipitation can cause a huge material damage leading to such menacing natural events as floods, landslides, mud flows, avalanches, water erosion, etc... These dangerous natural disasters can be of various duration - from a couple of minutes (snow avalanches) to a number of months (floods).
In this work, a first attempt to carry out a climatological generalization of the data on maximal precipitation values over Europe, Asia and North America is made. Based on existing publications and on materials of primary observations, a specialized data base of the primary information has been created. Statistical homogeneity of the series has been evaluated, the maximal precipitation's temporal structure has been investigated. Record amounts of precipitation for various durations have been obtained.
In the future, all the generalized climatological information can be displayed in the form of monthly maps, which will be used for applied purposes and for forecasting.

KEYWORDS
Precipitation, map of precipitation, mesoclimate, urban climate, thunderstorm.

1. Introduction

Intensive precipitation causes disastrous events like floods, landslides, avalanches, water erosion etc. To safeguard life and property, it is necessary to know precipitation intensity in different time scales as well as its probability characteristics. Maps of daily maximum values of liquid and solid precipitation amounts were drawn up for territory of former USSR for return periods of two years and one hundred years. Main Geophysical Observatory (MGO) in St. Petersburg has a databank enabling the drawing of such maps

NATO ASI Series, Vol. I 26
Global Precipitations and Climate Change
Edited by M. Desbois and F. Désalmand

for continental areas of the Northern Hemisphere. A.N. Lebedev (Lebedev, 1964-1983) carried out global climatic studies including precipitation daily maximal amounts. In addition to daily and annual maximal values "record" totals of precipitation in the Northern Hemisphere for different time intervals have been determined. In the time most recent interval anthropogenic factors may be having a growing influence on meteorological characteristics, such as precipitation, thunderstorms, hail etc.

2. Extremal precipitation of different probability

For practical purposes information on maximal precipitation of different probability for various time intervals (hour, day, month, year) is essential (Kitaigorodskaya and Shver, 1990; Lebedev, 1964-1983). Most important is daily maximal value with probability of 1%, because this value is incorporated in numerical formulae applied in construction industries and hydrology, as well as for designing drainage and canalization systems. In regions with small climatic variability these characteristics can be determined using 30-year data sets, in Europe series of 40 to 50 years duration are necessary, while in complex physical-geographical conditions, especially in mountainous regions, only 100-years series may be sufficient. In regions with cold climate it is necessary to consider separately the warm and the cold seasons, i.e. liquid and solid precipitation.

Table 1. Annual daily maximal amount of precipitation (mm) (1) and the corresponding percentage of precipitation (2)for Europe.

Latitude		Longitude							
		5	0	5	10	15	20	25	30
50	1			130	130	160	160		
	2			12	12	16	23		
45	1		160	170	170	190	130	170	200
	2		13	14	11	24	19	22	31
40	1	180	260				250		
	2	19	34				31		

In Table 1 generalized data on annual daily maximal amount of precipitation with probability of 1% is given for Europe. Such information is available for all continents. It is interesting to note, that for arid regions the proportion of daily maximal precipitation amount in the annual precipitaion sum can be as big as 30-40%. Corresponding maps can be drawn up. In Fig.1 and 2 annual daily maximal values of precipitation exceeding 100 mm are represented for North America and Europe (with probability of 1%). Carribbean is marked out by maximal amounts of 400 to 500 mm, which is connected with hurricanes. In mountainous regions there are registered daily maximums up to 300 mm (Kitaigodskaya, 1987)

In Table 2 absolute maximal precipitation amounts are given for different time intervals. The "record" for one minute is about 40 mm, for 15 minutes - 200 mm. The highest values of precipitation for long time intervals (week, month, year) were registered in the middle of 19-th century in Cherrapunji.

3. Extremal precipitation and urban mesoclimate

It is necessary to underline, that the meaning of the word "mesoclimate" is not the same as "mesometeorology", which applies to synoptical processes. "Mesoclimate" means mostly "local climate" (Khairullin, 1986).

It can be assumed, that a water reservoir changes micro- and mesoclimate over the area comparable to its own area. In winter the lower non-frozen water areas cause fog formation, and this process can be intensified on the coast by 1000%. At the distances of 500 m from the coast this increase under conditions of cold climate can amount to 200-300%. In this case the number of days with fog increases only slightly, but the duration of foggy condition can rise sharply. Due to the high humidity ice-frost deposition grows.

Cities and nearby water reservoirs form breeze mesocirculation, which is a distinctive feature of the mesoclimate. Mesocirculation has annual and diurnal variations, causing mesoclimatic changes as well. In urban conditions they are more distinctive in winter in the morning hours, and over the water reservoir - in summer in the afternoon (Khairullin, 1986).

Local climatic effects can be quite considerable and influence human activities and health. A correct forecast of possible micro- and mesoclimatic changes, drawing up of schematic maps taking into account the physiographical features make it possible, to mitigate the negative aspects of anthropogenic activities by designing the urban objects, hydroelectric power plants, settlements, etc., thus optimizing living conditions and preventing creation of discomfort zones.

Table 2. Extremal sums of precipitation

Duration	Amount, mm	Place	Date
1 min	38	Baro, Guadalupe	26.11.1970
8 min	126	Fuessen, Bayern	25.05.1920
15 min	198	Plam Point, Jamaica	12.05.1916
20 min	206	Kurta-de-Ardges, Romania	07.07.1947
42 mm	305	State of Montana, USA	22.06.1947
1 h	254	Catskill Mts., USA	26.07.1819
3 h	381	Port-Elizabeth, SAR	01.09.1965
4 h	584	Saint-Citts, West Indies	12.01.1880
12h	1340	Reunion Island	28-29.02.1964
18 h	1041	Paishich, Taiwan	11.09.1963
24 h	1825	Reunion Island	15-16.03.1952
24 h	1652	Eastern China	17.10.1967
24 h	1248	Taiwan Island	10-11.09.1963
24 h	1036	Cherrapunji, India	14.06.1876
24 h	806	Yugoslavia	01.02.1926
24 h	660	Czechoslovakia	03.01.1931
24 h	520	Southern France	24.02.1964
2 days	2253	Shin Lao, Taiwan	17-18.10.1967
4 days	3721	Cherrapunji, India	12-15.09.1974
8 days	3847	USA	01-08.01.1979
15 days	4798	Cherrapunji, India	24.06-08.07.1931
1 month	9300	Cherrapunji, India	July 1861
2 months	12767	Cherrapunji, India	June-July 1861
3 months	16369	Cherrapunji, India	May-July 1861
6 months	22454	Cherrapunji, India	Apr.-Sept. 1861
1 year	26900	Cherrapunji, India	Aug. 1860-July 1861
2 years	40768	Cherrapunji, India	1860-1861

A technique has been developed at the MGO for prognostic quantitative estimation of climate change due to hydroelectrical power plants construction, taking into account the size of the plant, season in the year and physiographical features. At the stage of designing the plant maps can be produced, characterizing possible changes of micro- and mesoclimate under the influence of water reservoir and lower water layers during warm and cold seasons (Khairullin, 1986; Khairullin and Yakovlev, 1990).

Experience accumulated by studying the micro- and mesoclimate near large human structures allows one, to judge on scales of possible effects and on consequences of the power plant construction for the nearby cities.

Antropogenic effects on climate show up mostly in micro- and mesoscales. Urban development and construction of water reservoirs, fields irrigation, quarry working, deforestation can lead to sub substantial changes of basic meteorological elements. These changes can extend over areas from tens of square meters to hudreds of square kilometers and to the heights of 1-1.5 km. K. Khairullin has proposed a table of climatic scales and a relation to evaluate their vertical and horizontal dimensions :

$$\log h = 0.6 \ln 1,$$

where h is height in meters, l is horizontal length in kilometers.

There exist continuous heat and water exchange cycles between all climate scales, mainly connected with turbulence. The direct effect of underlying surface on climate starts with smaller scales (although large from the viewpoint of cartographers). Along with microclimate can be considered, and the climate of soil surface can be determined as microclimate (Khairullin and Yakovlev, 1990 and 1992).

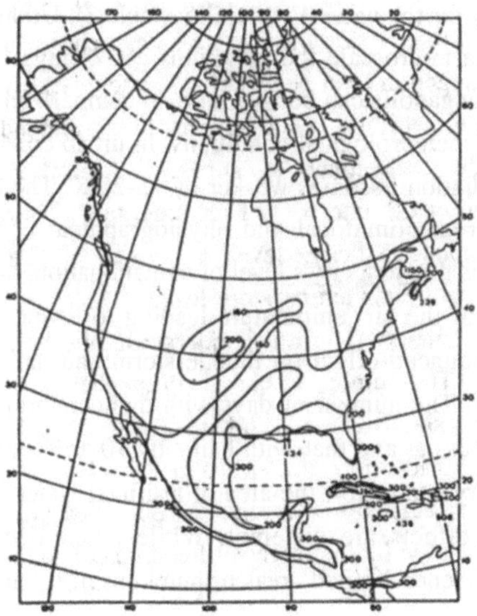

Fig.1 : Annual diurnal sum maximal precipitation over 100 mm. North America.

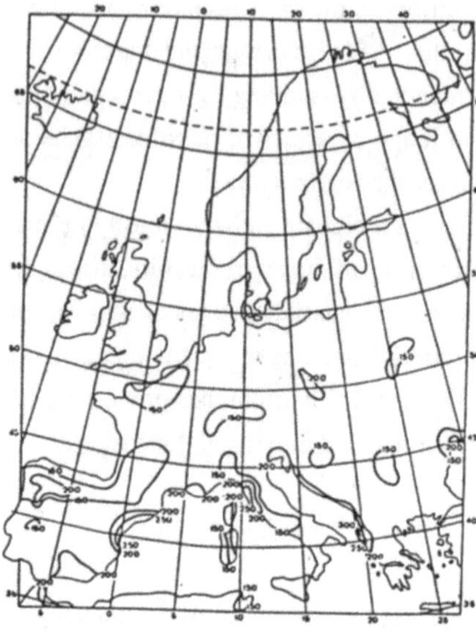

Fig. 2 : Annual diurnal sum maximal precipitation over 100 mm. Europe

By artificial changes of the components of heat, radiation and water balances meteorological conditions in micro- and mesoclimate can change as well, and these changes may exceed by far the naturel variability (table 3). Urban buildings reduce winde speed by 30%, while in microscale the variations can change from -80% to +160%. In winter the center of a megapolis can be "warmer" in comparizon to the suburb regions by 10-15 C under specific meteorological conditions. In urban conditions radiation becomes "redder", and direct radiation becomes weaker by 15-20 K. The number of foggy days in cities situated in different climatocal and physiographical conditions can increase or decrease. The increase is due to a city's level of contamination, and the decrease is caused by a considerable rise of the air temperature leading to lower relative humidity. Urban conditions have a pronounced effect on thunderstorm and hail activity, particularly on extremal characteristics. The number of days with thunderstorms increases by 7-10% on European territory of Russia, and that with hail - by 10-15%. Urbanized regions occupy about 1% oif the country's terrotiry, the area of artificial water reservoirs is of the same order. However, their effects are different. While cities affect extensive areas with pollution, reservoirs influence limited areas of hundreds of square meters or some square kilometers.

Table 3. Maximum deviations of basic meteorological parameters in a city

Parameter	Deviations (differences, ratios, percents)		
	a	b	c
Air temperature c (difference)	10-15	6-9	3-5
Surface temperature c (difference)	6-8	10-15	30-40
Global radiation (ratio)	0.8-0.9	0.6-0.8	0.1
Direct radiation (ratio)	0.6-0.7	0.4-0.6	0.0
Wind speed (ratio)	0.6-0.7	0.3-0.7	0.1-1.4
Relative humidity (difference)	20-25	15-25	25-30
Precipitation (ratio)	15-10	10-5	-
Thunderstorm (%)	10-15	-	-
Hail (%)	15-20	-	-
Fogs (%)	-50 - +50	-75 - +75	-100 - +100

a - city-suburb (mesoclimate);
b - districts in the city (topoclimate);
c - built-up area microclimate.

Anthropogenic impacts on precipitation are of great interest. H. Landsberg was among the first investigators, who made a generalized picture of the effect of urbanization on precipitation. S. Changnon showed, that during recent decades the number of days with showers increased. One of the authors has studied the impact of the cities in former USSR on thunderstorms, hail and precipitation (Gaivoronski et al., 1991; Khairullin and Yakovlev, 1990 and 1992). This impact depends considerably on physiographical features, but comparing the data with the background values allowed it, to reveal the role of cities. It was shown, that the mean values of the above characteristics increase by 5 to 10%. Greater increases have been observed in maximal values. The number of days with hail and showers in a city can increase by 15-25% (Table 4).

Table 4. Large cities precipitation amount's ratio to the background values

Region	Months					Maximal precipitation amount
	IV	V	VI	VII	IV-VII	
Northern Kazachstan and Ural	1.08	1.03	1.01	1.07	1.05	1.25
European part of Russia	1.04	1.02	1.02	1.02	1.03	1.15

Above data was derived from 70-years data sets collected at approximately 100 stations. Similar investigations were carried out for USA, where mean amount of precipitation increases by about 2 to 3% in cities. By not considering the cities located on coasts and near large bodies of water this increase will be 5%. Maximal amount of precipitation increase is 15 to 20% in arid regions.

This information obtained from a large number of stations over a long time period is confirmed by radar measurements. The distribution of average precipitation is more complicated. A tendency is detected to precipitation increase on the leeward side of large cities. This increase is less frequent in the city itself, since considerable time is necessary for such "heat islands" to have an effects on cloudiness transported to the suburbs. Nevertheless, there seems to exist about 5% increase of precipitation in suburbs due to city's effect (Khairullin, 1986).

Most of the long-series of meteorological stations are situated in large cities. Therefore, by considering the problem of global climate change it is necessary to take into account the effects connected with urbanization. In many cases these effects exceed natural climatic changes. This is especially true regarding temperature, precipitation and some other meteorological events (Gaivoronsky et al., 1991; Khairullin, 1993).

4. Conclusions

1. Extremal precipitation is an important indicator of the climate variability.
2. Generalized quantitative chracteristics of extremal precipitation over the continents territory in Northern Hemisphere have made it possible to produce the maps necessary for practical purposes.
3. Investigation of the urbanization effects on precipitation shows the main impact being the increase of extremal precipitation amounts.

REFERENCES

Gaivoronsky Yu.F., Stasenki V.N., Tschchan Zui Binh, Yurtchak B.S., 1991 : Radiolocationny issledovania taifunov v SRV. Tesisi dokladov 5 Meshdunarodnogo symposiuma po tropitcheskoi meteorologii. Obninsk, 79.

Kitaigorodskaya E.M., and T.A. Shver, 1990 : Osobennosti raspredellenia ekstremalnich sutotchnich summ osadkov na kontinentach Trdy GGO, vyp.532, 177-185.

Kitaigorodskaya E.M., 1987 : Sutotchny maksimum osdakov v Kanade i SSHA. Trudy CGO, vyp.512, 113-119.

Khairullin K.Sh., 1986 : Nenilina Yu.M. Mesoklimat bolshich gorodov SSSR. Trudy CGO, vyp.501.

Khairullin K.Sh., and B.A. Yakovlev, 1990 : Vliyanie urbanizazii na grosu, grad i osadki. Trudy CGO, vyp.527.

Khairullin K.Sh., Yakovlev B.A., 1992 : Godovoy stok basseina Atlantitcheskogo okeana i kometnaya aktivnost. Tesisi soveshchaniya, Kaliningrad.

Khairullin K.Sh., 1993 : Antropogenic activities and the problems of micro- and the problems of micro- and mesoclimate of cities. Technical conference on tropical urban climates. WCACP-24. WMO/TD No.538. Dhakka, Bangladesh.

Lebedev A.N. i dr., Seria po klimatografii semnogo shara : Asia, Afrika, Australia, Amerika, Sapadnaya Europa, SSSR. Leningrad, Hydrometeoizdat, 1964-1983.

POSSIBLE IMPACTS OF CLIMATE CHANGE UPON THE WATER BALANCE IN CENTRAL EUROPE

M. Lapin
Slovak Hydrometeorological Institute, Bratislava

Introduction

Predicted climate change connected with the increase in the atmospheric greenhouse effect will probably cause changes of atmospheric circulation and, from this, resulting changes of precipitation, cloudiness, snowcover, air humidity and other climatic elements all over the world. These changes can seriously influence the natural environment and socio-economic activities in many countries. The United Nations Conference on Environment and Development (UNCED) expressed the risks of possible climate change and the necessary measures for reduction of those risks in the Framework Convention on Climate Change and in Convention Agenda 21. In this paper some model calculations and scientific assessments of the possible impacts of the temperature rise and the precipitation regime change upon water balance, relative air humidity and snow cover variability in Central Europe are presented. Calculation of water balance elements was carried out according to the Zubenok - Budyko complex method. At assessment of climate warming impacts upon changes of relative air humidity and snow cover characteristics was used an empirical correlation method. Two climatological stations (Hurbanovo 115 m a.s.l. and Liptovský Hrádok 640 m a.s.l.) were used as a basic source for all calculations, and seven other ones in Slovakia for the correlation method applications. This paper represents a brief review of results obtained in completing some of the Czecho-Slovak National Climate Programme tasks in 1991 and 1992.

Methods and data

The Slovak Hydrometeorological Institute (SHMI) provides meteorological observations for climatological purposes in network of more than 100 stations. After the data quality control only 12 of them are usable for time series analysis. Two stations were selected for calculation of water balance elements change. Station Hurbanovo represents the lowlands in southern Slovakia and station Liptovský Hrádok the valleys in northern Slovakia near the High Tatras.

The potential evaporation (E_o) was calculated according to the Zubenok (1968) nomograms using monthly air temperature and relative air humidity averages. For the actual evaporation (E) calculation, the complex Budyko method (1974) was adopted using E_o and monthly precipitation totals, as well as some parameters of the precipitation regime and optimal soil moisture.

NATO ASI Series, Vol. I 26
Global Precipitations and Climate Change
Edited by M. Desbois and F. Désalmand

In the first step, the statistical values of monthly data for the period 1901 - 1980 were calculated and we considered them to be normals. In the second step, the scenarios of temperature rise, as well as the precipitation and relative air humidity changes, for the mean year 2025 were applied (Table 1.). The differences between the observed normals and the scenarios for potential and actual evaporation are shown in Table 2.

In the third step, an original monthly data modification was carried out according to three scenarios for E_o and five scenarios for precipitation (LAPIN 1992). The selected results are in Table 3 and in Figures 3 and 4. The main reason for monthly data modification was to obtain a data time series representative of that expected in year 2025 when the climate has been changed according to the above scenarios.

For the period 1951 - 1991 we prepared means and totals of several climatological elements, both for warm and cold half-years in each year. Then a correlation analysis of relative air humidity and snow cover characteristics with air temperature and precipitation was done for nine stations in Slovakia. Results are shown in the Table 4.

Results

The scenarios of possible climate change in Slovakia (Table 1) follow from the analysis of warmer periods in the recent past and from GCM assessments (LAPIN, ŠIPOCZ, 1991). We suppose a stronger warming effect in winter than in summer, a rise of precipitation (P) in winter, a decrease in spring and autumn, and small changes in summer. The relative air humidity will probably decrease on average all the year round. The calculated results of the basic water balance elements after the 1-2 °C warming show a significant rise of the potential evaporation (E_o), but only small changes of actual evaporation (E) (decrease in lowlands and increase in mountainous areas) (Table 2 shows results for Hurbanovo only).

A very important result is the significant decrease of soil moisture as well as a decrease in the balance runoff for the whole territory of Slovakia by approximately 20 %. These results follow from the calculation using long-term mean input data. Therefore we carried out, in the next step of calculation, the elaboration according to the modified individual monthly input data of E_o and P observed in the period 1901 - 1990. There has to be stressed that also linear trends of basic climatological elements in the period 1901-1990 are similar to those predicted in next decades (Figure 1 and 2). In the next step of calculation we used three scenarios of additional changes and trends in time series of individual monthly data in the period 1901-1990:

a) change of all individual monthly data by 10 or 20 %

$Y_i = c * X_i$, $c = 1.1$ (0.9, 1.2 or 0.8)
X_i are original and Y_i modified individual monthly data

b) additional linear trend in time series of data by 20 %

$Y_i = (1 + 0.2\ i/n) * X_i$, $n = 90$, $i = 1,2...90$, for E and P

$Y_i = (1 - 0.2\ i/n) * X_i$, $n = 90$, $i = 1,2...90$, for P

X_i are original and Y_i modified individual monthly data

c) rise of the individual monthly data variability by 25 %

$Y_i = X_i + 1{,}25\ (X_i - X_o)$, X_o is 90 years' mean

Selected final results are shown in this paper only (Figures 3 and 4, Table 3). According to the (b) scenario, the additional trend of precipitation P was considered negative by 20 % in spring and autumn months, positive by 20 % in winter months and the P data were not changed in summer months. The trend of E_o was considered positive by 20 % in all months. The mean annual usable soil moisture of upper 1 m layer (W) and the balance runoff (Q) are the final results of the water balance calculations. We can see that comparing to the W and Q values calculated for the original data in the period 1901-1990, the resulted W and Q are considerable lower after the modification of data. If E_o and P change according to the (b) scenario, we can expect the changes in annual regime of W and Q too (Table 3). In mountainous areas of Slovakia a surplus of precipitation is observed at present in average in all the year round comparing to the evapotranspiration and water balance demands. After 1-2 °C warming the territory with precipitation surplus will cover the high mountainous area only and in warm half year the deficiency of water for runoff will appear in the great majority of Slovakia.

The results of correlation analysis are shown in the Table 4. Significantly close correlation was found between mean relative air humidity (U) and precipitation totals (P) in warm half-year on the whole territory of Slovakia. If P totals decrease by 100 mm in average, U will probably decrease by 2 % in lowlands and by 1 % in other regions. Acceptable correlation was found also between U and mean air temperature (T) in warm half-year. The rise of T by 1 °C will cause decrease of U by about 2 % in all the Slovakia. In cold half-year no significant correlation was found there between U and P or U and T, respectively.

Snow characteristics are in a close correlation with mean air temperature of cold half-year. The average number of days with snow cover (N) will decrease probably by about 15 days if T increases by 1 °C. That is about 30 % in lowlands but only 10 % of the present average of N in mountains. More important will be the decrease of the daily depths of snow cover totals (S) if the T increases. The 1 °C rise of T in cold half-year will probably cause the decrease of S by 100 to 300 cm in lowlands (approx. by 50 %) and by 400 to 1400 cm in the middle altitudes (approx. by 35 to 15 %). The correlation between U and snow characteristics N and S is not significant in cold half-year. At Hurbanovo were found similar results for the period 1901-1991.

Fig.1. Linear trends of annual sums of the selected water balance elements Eo, P and E at Hurbanovo in 1901-1992

Eo and E are calculated according to Zubenok-Budyko method

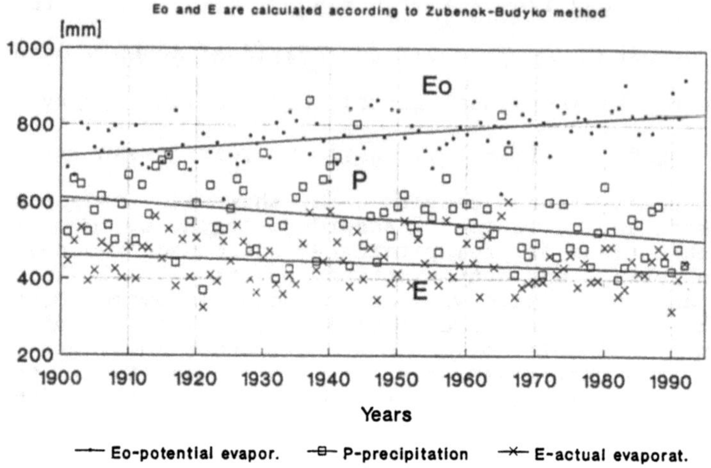

Hurbanovo, 115 m a.s.l., Slovakia

Linear trends of calculated annual means of W and annual sums of Q at Hurbanovo in the period 1901-1992

W and Q are calculated according to Zubenok-Budyko method

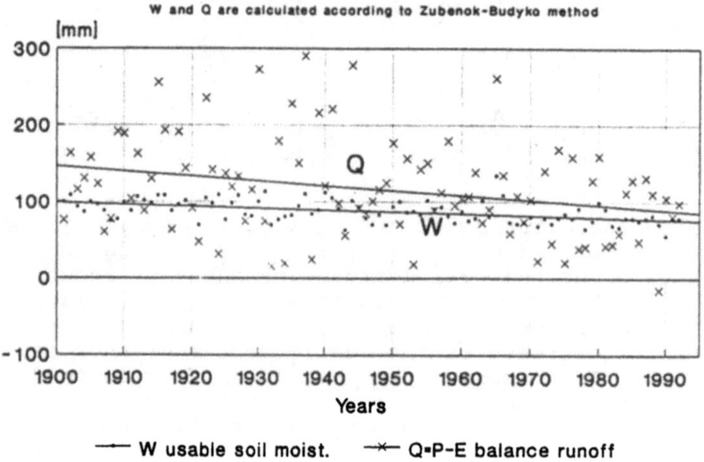

Hurbanovo, 115 m a.s.l, Slovakia

Fig.2. Linear trends of annual sums of the selected water balance elements Eo, P and E at Liptovsky Hradok in 1901-1992

Eo and E are calculated according to Zubenok-Budyko method

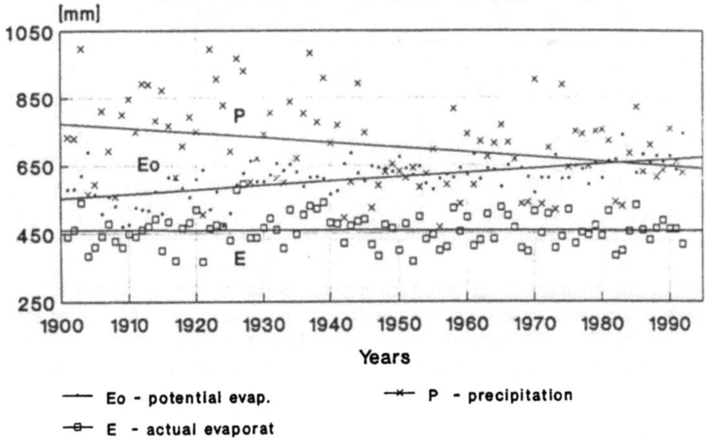

Liptovsky Hradok, 640 m a.s.l., Slovakia

Linear trends of annual means of W and annual sums of Q at Liptovsky Hradok in the period 1901-1992

W and Q are calculated according to Zubenok-Budyko method

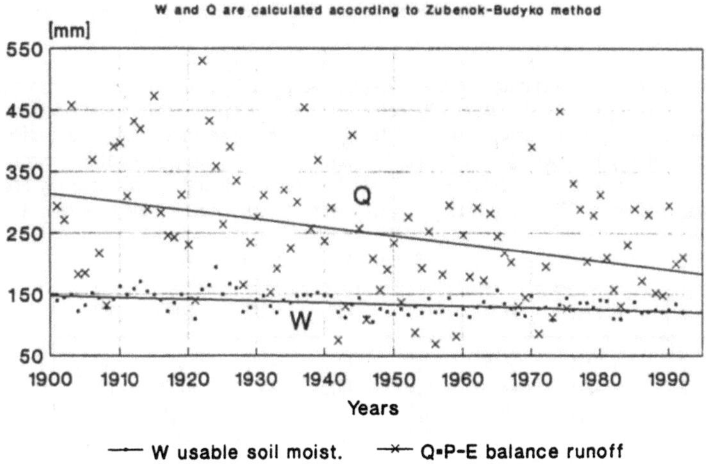

Liptovsky Hradok, 640 m a.s.l., Slovakia

Table 1. Scenarios of the average deviations δ of the 2025 climate characteristics in Slovakia compared to long term means 1901-1980. (T - air temperature in °C, P - precipitation in mm, U - relative air humidity in %).

	I	II	III	IV	V	VI	VII	VIII	IX	X	XI	XII
	MONTHS											
δT_1	1,5	1,4	1,2	1,0	0,8	0,6	0,5	0,6	0,8	1,0	1,2	1,4
δT_2	3,0	2,9	2,5	2,0	1,5	1,1	1,0	1,1	1,5	2,0	2,5	2,9
δP_1	0	0	0	0	0	0	0	0	0	0	0	0
δP_2	6	4	-5	-9	-5	5	2	0	-5	-5	0	3
δP_3	0	0	-5	-10	-10	0	-10	-5	-5	-10	0	0
δU_1	0	0	0	0	0	0	0	0	0	0	0	0
δU_2	-2	-2	-4	-4	-3	-2	-2	-2	-2	-2	-2	-2

Table 2 Deviations δ (in mm) of the potential evaporation monthly sums E_0 and the actual evaporation monthly sums E calculated according to combination of the scenarios in Table 1, from sums calculated for the period 1901-1980. The calculations were made from the long term means of air temperature T, relative air humidity U and precipitation P, (e.g. for E(223) the scenarios T_2, U_2, P_3 were used; see text)

Station Hurbanovo

	I	II	III	IV	V	VI	VII	VIII	IX	X	XI	XII	Year
	MONTHS												
$\delta E_0(11)$	4	5	2	3	2	1	0	2	3	5	6	3	36
$\delta E_0(12)$	5	7	7	8	8	6	3	5	6	9	7	5	76
$\delta E_0(21)$	5	6	6	5	4	3	2	3	5	8	7	4	58
$\delta E_0(22)$	6	9	11	11	10	8	4	7	8	11	10	6	101
$\delta E(111)$	3	3	1	0	-2	-1	-1	1	1	2	5	2	14
$\delta E(222)$	7	7	4	-2	-11	-7	-1	2	0	2	6	6	13
$\delta E(221)$	7	7	4	0	-5	-5	-3	1	1	4	7	6	24
$\delta E(223)$	7	7	3	-4	-14	-12	-8	-5	-3	0	5	6	-18

Table 3 The selected deviations between 90-year means (Y) calculated according to the scenarios a, b, c, (see text) and means (X) calculated according to the original data in the period 1901-1990, (for E(a0) scenario a was used for E_0 and unchanged totals P in the period 1901-1990, P - precipitation, E - actual evaporation, W - usable soil moisture, Q - balance runoff Q=P-E, all values are in % of averages for period 1901-1990 for stations Hurbanovo and Liptovský Hrádok).
WHY - warm half-year, CHY - cold half-year, WIN - winter, SPR - spring, SUM - summer, AUT - autumn

Hurbanovo

Y X	Year	WHY	CHY	WIN	SPR	SUM	AUT
	SEASONS						
P(a) - P	-2,6	-4,5	-0,5	10,3	-10,3	0	-10,3
P(b) - P	-2,5	-4,5	-0,5	10,3	-10,3	0	-10,3
E(a0) - E	1,6	0,6	6,0	X	4,4	-1,5	1,4
E(0a) - E	-2,7	-2,8	-2,4	X	-2,5	-2,1	-4,3
W(aa) - W	8,0	-9,6	-5,9	X	-5,9	-10,7	-13,1
Q(aa) - Q	-11,9	-22,2	-1,8	10,5	-62,0	13,0	-17,1
Q(cc) - Q	8,4	15,6	2,5	1,1	16,0	8,7	4,3

Table 3 continued

Liptovský Hrádok S E A S O N S

Y X	Year	WHY	CHY	WIN	SPR	SUM	AUT
P(a) - P	-3,0	-4,3	-1,1	10,3	-9,8	0	-10,4
P(a0) - P	3,9	3,3	5,7	X	5,6	-1,8	2,7
E(0a) - E	-1,7	-1,8	-2,9	X	-2,1	-1,8	-2,7
W(aa) - W	-6,7	-8,0	-4,9	X	-5,2	-8,8	-8,4
Q(aa) - Q	-11,6	-40,4	-3,1	11,0	-90,1	-6,2	-18,6
Q(cc) - Q	4,0	14,0	1,5	1,0	18,1	6,2	3,1

Table 4. Correlation coefficients (r) and linear terms (l) of the linear regression equations for the relation of the variables: U - relative air humidity in %, T - air temperature in °C, P - precipitation in 100 mm, N - number of days with snowcover in 10 days, S - daily snowcover depths totals in 100 cm. In the columns 1 and 2 the values for warm half-years and in other ones for cold half-years of the period 1951-1991 are presented.

C O R R E L A T I O N S

Stations (altitude)	r l	1 (UT)	2 (UP)	3 (UT)	4 (UP)	5 (UN)	6 (US)	7 (NT)	8 (ST)
Hurbanovo	r	-0,4	0,7	-0,3	0,4	0,4	0,4	-0,8	-0,7
(115)	l	-2,0	2,5	-0,6	1,2	0,4	0,3	-1,3	-1,8
Bratislava	r	-0,4	0,7	-0,2	0,5	0,2	0,2	-0,7	-0,6
Koliba (286)	l	-1,8	2,3	-	1,5	-	-	-1,6	-3,2
Piešťany	r	-0,3	0,6	-0,2	0,6	0,4	0,2	-0,7	-0,6
(165)	l	-1,2	2,0	-	2,1	0,6	-	-1,0	-1,3
Košice,	r	-0,5	0,7	0,0	0,6	0,1	0,0	-0,6	-0,5
letisko (230)	l	-2,3	2,5	-	1,7	-	-	-1,1	-1,9
Sliač,	r	-0,4	0,6	0,1	0,3	0,2	0,1	-0,7	-0,6
letisko (313)	l	-1,9	1,8	-	0,7	-	-	-1,5	-3,9
Liptovský	r	-0,4	0,6	0,0	0,4	0,4	0,1	-0,4	-0,5
Hrádok (640)	l	-1,2	1,4	-	0,9	0,3	-	-1,0	-4,3
Poprad,	r	-0,3	0,6	-0,3	0,4	0,4	0,2	-0,6	-0,6
letisko (695)	l	-1,1	1,6	-0,6	1,7	0,4	-	-1,1	-4,2
Telgárt	r	-0,4	0,5	0,1	0,4	0,2	0,1	-0,6	-0,6
(901)	l	-1,5	1,2	-	1,0	-	-	-1,4	-13,2
Štrbské	r	-0,5	0,4	-0,2	0,3	0,0	0,1	-0,1	-0,4
Pleso (1360)	l	-1,8	1,1	-	1,4	-	-	-	-13,9

Fig.3. Linear trends of annual balance runoffs (annual sums) Q at Hurbanovo in the period 1901-90 (see text)

Q and Q(b,b) are calculated from monthly values P and E

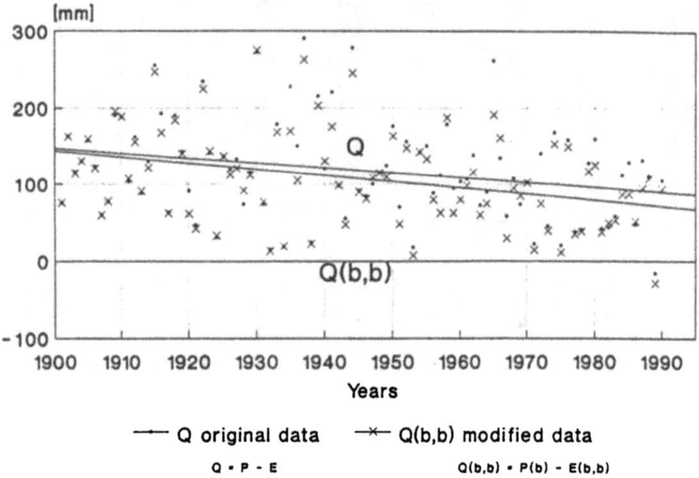

Linear trends of calculated usable soil moisture W (annual averages) at Hurbanovo in the period 1901-90 (see text)

W and W(b,b) are calcul. according to Zubenok-Budyko method

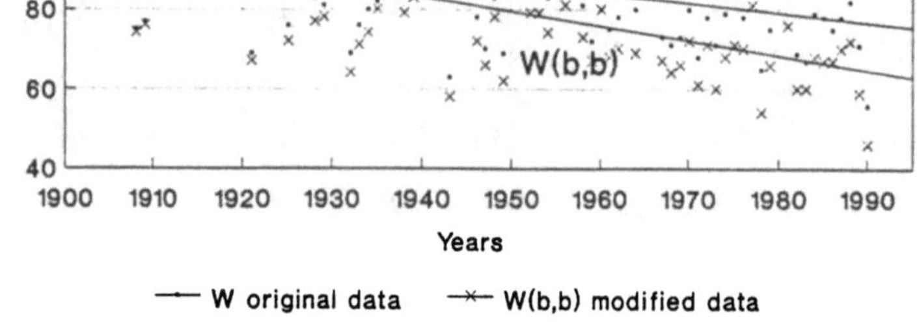

Fig.4. Linear trends of balance runoffs (annual sums) Q at Liptovsky Hradok in the period 1901-90 (see text)

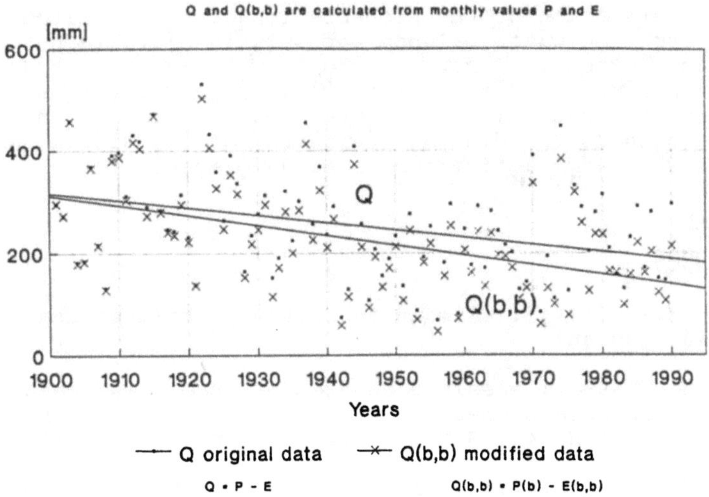

Linear trends of calculated soil moisture W (annual averages) at Liptovsky Hradok in period 1901-90 (see text)

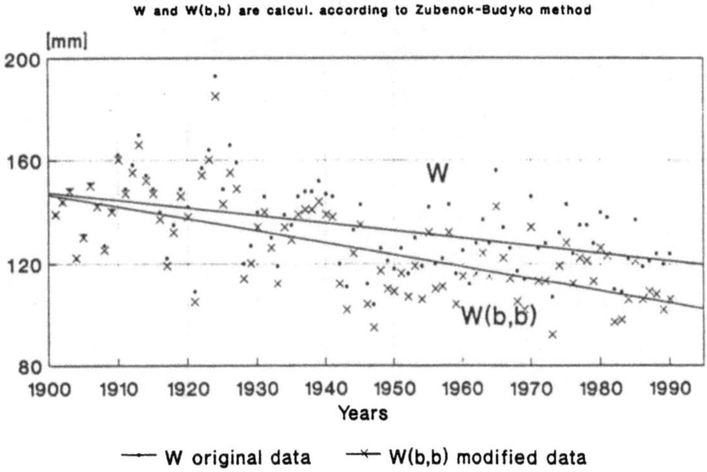

Conclusions

Results presented in this paper represent only a small part of those generated. We consider they are sufficient for a first estimation of the probable impact of estimated climate change upon the water balance not only in Slovakia but also at comparable sites in Central Europe. The significant decreases of soil moisture and runoff as well as snowcover characteristics are stressed. In the future a more detailed analysis by various methods within the framework of the Slovak National Climate Programme activities is being prepared.

References

BUDYKO, M.I. (1974): Climate and Life. Gidrometeoizdat, Leningrad (in Russian)

LAPIN, M. (1992): Possible changes of the water balance elements in Slovakia connected with supposed climate change: In: Práce a štúdie No 44, SHMÚ, Bratislava, p. 17 - 29 (in Slovak)

LAPIN, M. et. al.(1992): Impacts of possible climate change upon various socio-economic branches in ČSFR. Final reports of the National climate programme project, Bratislava, 14 reports (in Czech and Slovak)

LAPIN, M., ŠIPÖCZ, M. (1991): Possible impacts of expected climate changes on the water balance in lowlands of Slovakia. Meteorologické zprávy, 44, Prague, No 3, 79 - 83 p.(in Slovak)

ZUBENOK, L.I. (1968): Estimation of areal evaporation in the individual years. Trudy GGO, vyp.233, Leningrad (in Russian)

METHODOLOGIES OF LARGE SCALE ESTIMATION OF PRECIPITATION FROM CONVENTIONAL DATA

TERRESTRIAL PRECIPITATION ANALYSIS: OPERATIONAL METHOD AND REQUIRED DENSITY OF POINT MEASUREMENTS

B. Rudolf, H. Hauschild, W. Rueth and U. Schneider
Global Precipitation Climatology Centre (GPCC)
Deutscher Wetterdienst
Postfach 10 04 65
D-63004 Offenbach a.M. (FRG)

ABSTRACT
The Global Precipitation Climatology Centre (GPCC) has established a semi-automatic data processing system to obtain monthly precipitation analyses. This includes components for the collection, storage and quality-control of all available gauge-measured data. The calculation of areal mean precipitation on a 2.5° grid by using an objective analysis method and the merging of these analyses with results from other sources to get complete global gridded data sets are a part of the routine system as well.

First investigations of the spatial sampling error of the areal mean precipitation totals are performed by using Monte-Carlo-techniques, from which a station density required to get an error of less than 10% can be obtained. Up to now the GPCC has prepared analyses of 24 months (1987, 1988). Some intercomparisons of large areal means over the continents are presented for different precipitation data sets (GPCC analyses, Legates climatology, satellite data, accumulated daily forecast results of the ECMWF model).

1 Introduction

The GPCC, a central element of the Global Precipitation Climatology Project (GPCP), was initiated by the WMO and is carrying out the tasks as follows: 1) Collection and storage of all available conventionally measured precipitation data world-wide on a monthly basis received via the World Weather Watch Global Telecommunication System and acquisition of data from other sources on a bilateral basis. 2) Quality-control of the conventionally measured data. 3) Calculation of areal precipitation totals on a 2.5° grid based on the conventionally measured data by using an objective analysis method. 4) Correction of the analyses with regard to the systematic gauge measuring errors depending on the instruments used. 5) Estimating the accuracy and error of the gridded data. 6) Intercomparison studies for the gridded data from different sources. 7) Merging the results from the different sources to obtain complete global gridded precipitation data sets. 8) Publication of the results and dissemination of the gridded data sets to the World Data Centres for Meteorology.

NATO ASI Series, Vol. I 26
Global Precipitations and Climate Change
Edited by M. Desbois and F. Désalmand

The first step was the development of a semi-automatic data processing system (Rudolf, 1993) to enable the GPCC to derive the products on a routine basis, which was realized in the research phase 1989 - 1991. Since the beginning of 1992 the GPCC routine system is operating on a trial basis.

2 The GPCC Quality-Control System For Gauge-Measured Precipitation Data

Since almost every large dataset contains, on a more or less frequent basis, erroneous data (e.g. due to typing or coding errors), the routine operating system at the GPCC is including procedures for an extensive quality-control of the gauge-measured precipitation data received from all over the world. According to the different input data types to the routine analysis system (synoptic precipitation reports, monthly precipitation data from various sources), there are different quality-control procedures. However, the quality-control procedures for both the synoptic, as well as the monthly precipitation data, consist of two basic parts: 1) an automatic check of the full datasets, including the correction of obviously erroneous data by computer programmes, supplemented by 2) a manual revision of the remaining questionable data, which have been identified and flagged by the programmes.

In order to improve the database of monthly precipitation totals on the 2.5° grid, a computer programme (named RRRTRW) has been developed to calculate monthly totals from the short-term (6-24 hourly) precipitation depths contained in SYNOP reports. Details of the calculation of the monthly precipitation totals from SYNOP reports by the programme RRRTRW are described in Schneider et al. (1992). To support the manual quality-control of the questionable SYNOP reports and the reports corrected by RRRTRW, an interactive programme has been developed and installed on the graphics-workstation, which can show all information relevant to the station being checked on the screen.

The quality of the monthly precipitation totals derived from SYNOP reports can be estimated by comparing them to the monthly precipitation depths from CLIMAT reports, which have been extensively quality-controlled at the DWD/SWA and the GPCC. These comparisons for the period January 1987-August 1988 reveal that the monthly precipitation data derived at the CAC (Climate Analysis Center/NOAA, Washington), which undergo an automatic-only quality-control, show a generally smaller relative RMS difference than do the GPCC results quality-controlled only by the programme RRRTRW (see Figure 1). However, the GPCC results subsequent to the manual revision are in much better agreement with the CLIMAT reports than are the CAC results.

Investigations of the inaccuracies remaining in the quality-controlled GPCC results (cf. Figure 2) indicate that the main source of uncertainty is the incomplete coverage of the monthly period by SYNOP reports. As a consequence, it is planned to improve the handling of missing information, for example by evaluating a parameterization of precipitation from synoptic weather information, as is practiced at the CAC, or by looking at synoptic precipitation at neighbouring stations.

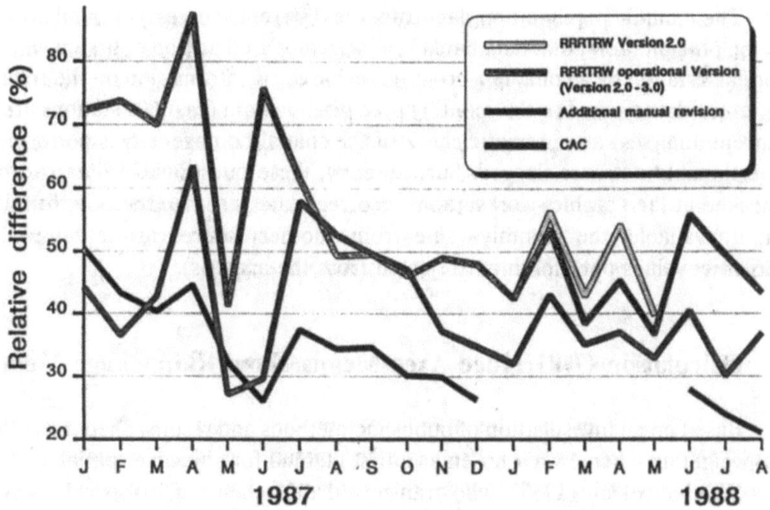

Fig. 1: Relative RMS difference between monthly precipitation totals derived from SYNOP reports at the CAC and at the GPCC (automatic-only, as well as automatic plus manual quality-control) and CLIMAT precipitation data.

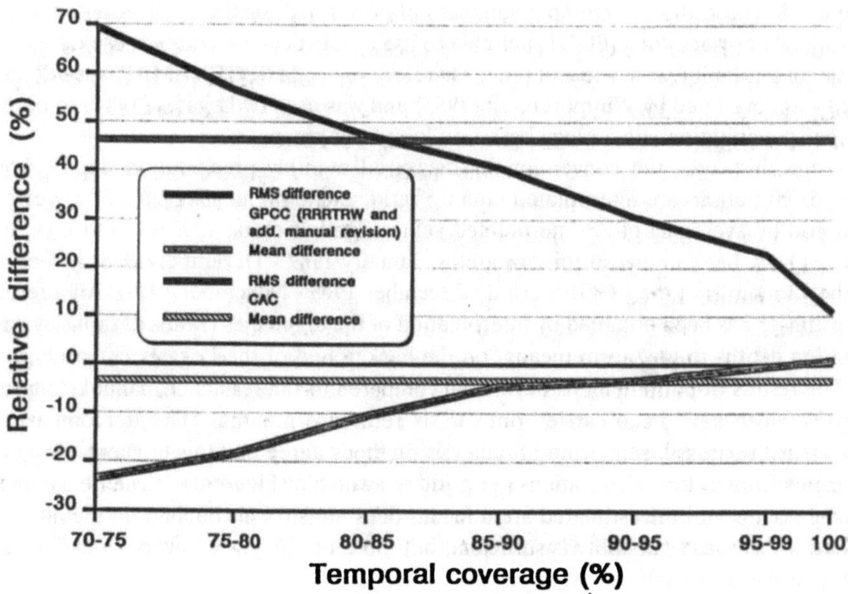

Fig. 2: Relative RMS and mean difference between monthly precipitation totals derived from SYNOP reports at the CAC and at the GPCC (automatic plus manual quality-control) and CLIMAT precipitation data as a function of the temporal coverage of the month averaged for 1987 (average differences for the CAC results, because no information on the temporal coverage has been available).

The monthly precipitation data from the different sources, which all are being stored in the precipitation point data bank, are intercompared and an optimum value is selected for each station by a computer programme based on information on the reliability of the different data types. For the monthly precipitation data (ca. 6600 stations are used in the current analyses) an automatic check of the spatial homogeneity is performed, whereby questionable data are flagged. Subsequently, these questionable data can be manually checked at the graphics-workstation. Incorrect data is set to the code for missing values or, if available, the monthly value from another source can be selected (about 50 incorrect values per month are rejected from the analysis).

3 Calculation Of Gridded Area-Means From Rain-Gauge Measurements

Based on an investigation of published methods and results, Shepard's (1968) empirical weighting scheme was chosen as a first method for objective spatial analysis. Following Willmott et al. (1985), who transferred the scheme to spherical coordinates, this method avoids some deficiencies which are typical for a pure distance weighting, 1) by using only the 4 to 10 nearest stations, 2) by clustering near neighbouring measurements (directional weighting), and 3) by the extrapolation of estimated gradients of the precipitation field to yield extrema being not covered by measurements. Bussieres and Hogg (1989) found by intercomparison studies of the empirical approaches of Shepard (1968), Barnes (1973) and Cressman (1959) and a Gandin-based statistical method that Shepard's scheme was the best of the empirical methods and is nearly as good as the statistical one. However, this intercomparison was not performed for the application on a global scale. At present, the GPCC is not able to use a statistical method because of insufficient data availability in space and time. Therefore the code SPHEREMAP was adopted which was developed by Willmott et al. (1985) and was used by Legates (1987) to derive a global precipitation climatology based on long-term means.

Using this code, the conventionally measured monthly precipitation depths from about 6,700 stations are interpolated on a 0.5° grid, and areal means on a 2.5° are calculated grid by averaging of 25 interpolated values (Rudolf et al., 1992). Up to now, the analyses have been prepared for 24 months, January 1987 - December 1988, as well as for the 12 months of the FGGE period (December 1978 - November 1979). An alternative estimate has been obtained by interpolation of the anomalies (ratios of monthly precipitation depths to long-term means) on the background of the Legates climatology.

The results of both methods have been compared to the results of manual isohyetal analyses which have been carried out for six reference months. The intercomparison shows that the areal means from all analysis methods agree quite well where the measurements from at least five stations per grid are available (Figure 3). It can be assumed that the accuracy of the estimated areal means does not strongly depend on the analysis method if the data availability is sufficient, but more on the given density, location and quality of the data itself.

A detailed study of the results indicated that the extrapolation of gradients can cause unrealistic extremes over areas not covered by measurements, if the absolute values are

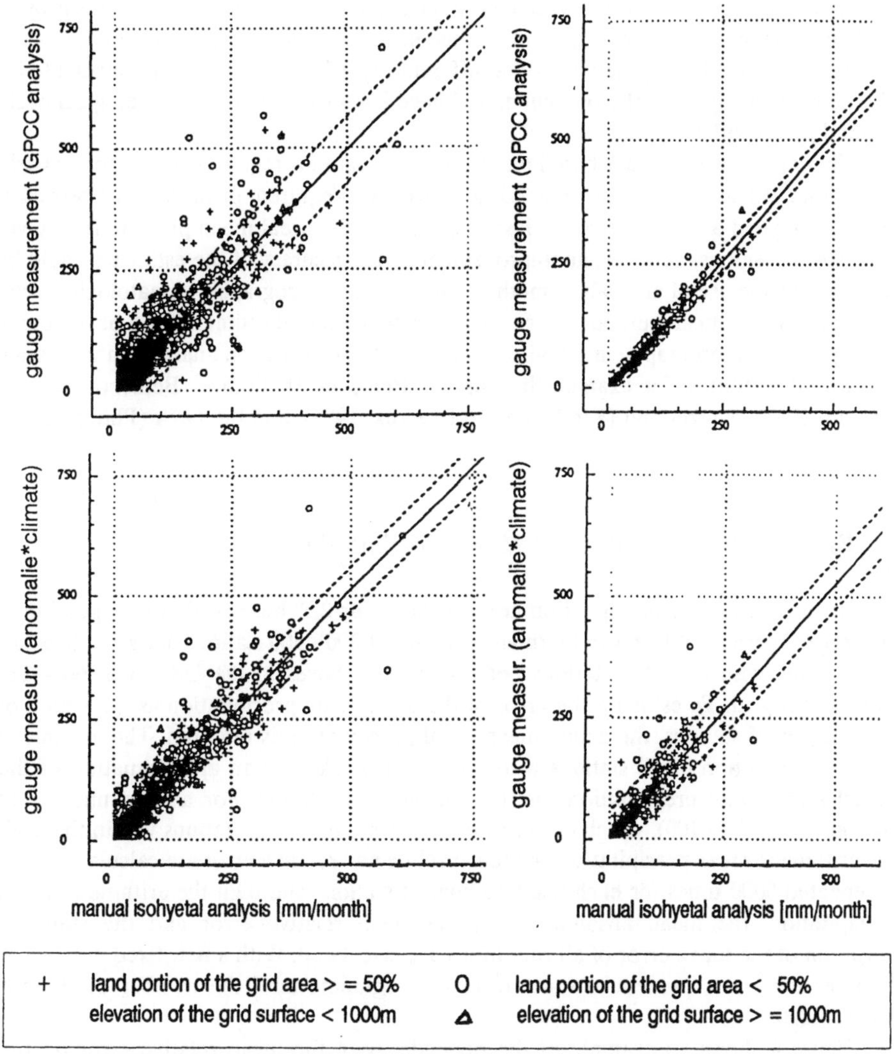

Fig. 3: Scatter diagram of precipitation totals on a 2.5° grid for October 1987 derived from gauge measurements using
upper: the objective analysis method for monthly precipitation depths,
lower: the objective method for precipitation anomalies on the climatological background
versus the results from a manual isohyetal analysis,
left: for all grids with 0 - 4 stations,
right: for all grids with 5 or more stations.

interpolated. This deficiency is much less pronounced if the analysis of anomalies is overlaid on the climatological background. In this second way, however, significant errors can be produced by inhomogeneities of the different sets of stations, which are available for the anomaly analysis on the one hand, and have been used for the climatological background on the other hand.

First error investigations based on all available data for regions with dense networks show, that at least 10 stations on a 2.5° grid area are required, or more if the terrain is not flat, to yield a relative error of less than 10% of the areal mean precipitation. Over data sparse regions, however, an improvement of the accuracy of the estimates could be expected by the use of an analysis method which includes orographic-climatological relationships or statistical methods - perhaps with satellite-based estimates as a background. At present, such orographic methods are applied for precipitation analysis on mesoscale by several institutes, for example in Austria (Behr, 1993), France (Benichou and Le Breton, 1987), Switzerland (Kirchhofer and Sevruk, 1992), and the USA (Phillips et al., 1992).

4 Error Estimation Of Areal Mean Precipitation

Error statistics for precipitation are calculated for 2.5° boxes with a very good data coverage. There are 4 boxes in Germany with about 300 - 500 stations per grid, 22 boxes in Australia with 80 - 240 stations per grid and 20 boxes in the USA with 60 - 114 stations per grid. To estimate the error of the areal mean precipitation as a function of the number of stations, the actual mean areal precipitation is required. The arithmetic mean precipitation of all stations within the grid is taken as an approximation of the actual mean areal precipitation for the following calculation. For a fixed number of stations (from 2 to 100), which are selected randomly out of all stations within the grid, the arithmetic mean precipitation and the absolute and relative error are calculated. This is repeated 5000 times for each fixed number of stations and then the arithmetic mean precipitation, the mean absolute error, the mean relative error and the standard deviation of the mean error of all repititions are calculated. With a new fixed number of stations, the entire procedure starts from the beginning until the number of stations exceeds 100.

Figure 4 shows an example of the dependency of the mean relative error on the number of stations selected from the dense raingauge networks in Australia, Germany and the USA. Generally more stations are needed to get a fixed relative error, if the percentage standard deviation (of the actual mean) is increasing. The lines in Figure 4 in three cases (USA and Australia) are showing no constant slope, if the selected number of stations exceeds 20% of all stations in the grid, which is caused by the limited number of all stations in these 2.5° grids. In Australia there are some months and grids with very low mean monthly precipitation amounts (< 1 mm) and a relative large standard deviation (about 3 mm), so the percentage standard deviation reaches values of more than 300 %. In these cases a lot of stations are needed and with the maximum number of selected stations (100), the mean relative error is still larger than 10 %. In order to eliminate these cases which still exhibit, instead of the large relative error, a very small mean absolute error, a limit of 1 mm is set for the mean absolute error.

Table 1 shows the number of stations necessary to get a mean relative error and the mean relative error + 2 * standard deviation of less than or equal 10 %. In Australia, the fluctuation of the number of stations needed to calculate an error less than 10% is larger than in Germany. This depends on the large values of the relative standard deviation in arid 2.5° boxes and in grids, which are only in part influenced by the ITCZ.

	mean relative error <10 %	mean rel. error + 2* STD <10%
Germany	3 - 10	20 - 50
Australia	2 - 70	2 - 100
USA	3 - >60	10 - >60

Tab. 1: Number of stations for which the mean relative error or the mean relative error plus double standard deviation is <10% or the mean absolute error <1 mm.

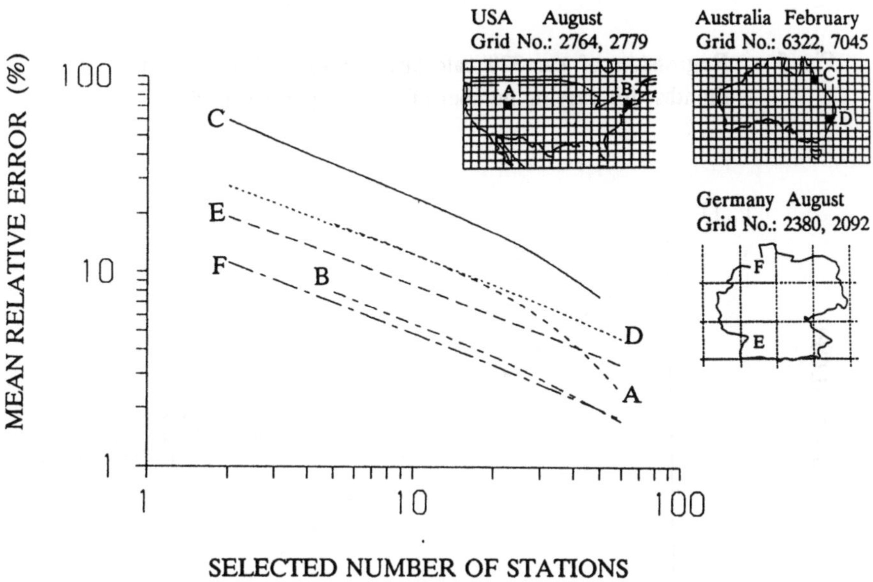

Fig. 4: Mean relative error (%) as a function of the selected number of stations per grid in Germany and Australia for August 1987.

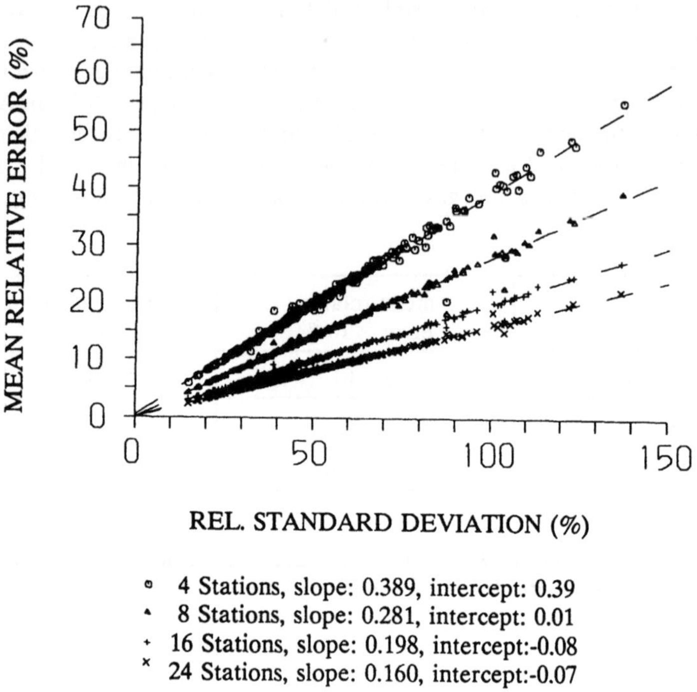

Fig. 5: Regression of ca. 320 calculated cases (different grids and months) for varying number of stations per 2.5° grid.

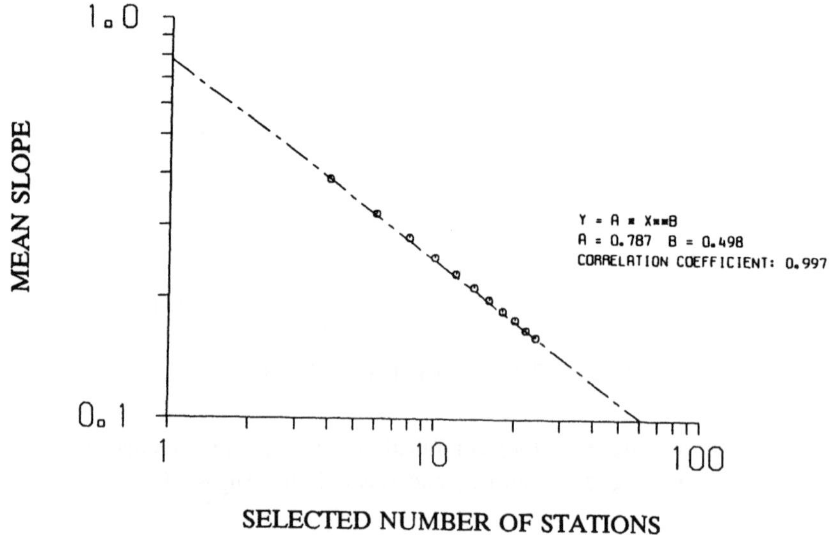

Fig. 6: Regression of the slope calculated in Fig. 5.

Jenne and Joseph (1985) have done a similar investigation in a 240 x 240 km and a 410 x 410 km box in Texas, USA. Although the method of calculation is not completely the same - they take the long time average precipitation as the actual mean areal precipitation and out of all stations in the grid 5, 10 and 18 stations are seleceted once for all calculations - the results are showing the same tendency.

Presently, the mean error for 320 cases (different grids and months) has been estimated. From these cases, it can be deduced that the mean relative error depends on the percentage standard deviation and the number of stations in a 2.5° grid (see Figure 5). For each fixed number of stations per grid, the slope and the intercept can be calculated by simple regression. The slopes depend on the number of stations per grid and are fitted by a multiplicative regression model with the equation:

$$\ln y = \ln a + b \ln x \tag{1}$$

with a = 0,865 and b = - 0,555

The combination of the regression equations in Figures 5 and 6 leads to the following equation:

$$RE = c + a * \frac{RS}{NS^{b}} \tag{2}$$

RE = mean relative error
RS = percentage standard deviation of the precipitation in a 2.5° grid
NS = number of stations in a 2.5° grid
a = 0.865, b = 0.555, c = 0.37 (c is the mean intercept of y - axis in Figure 5)
a,b,c are constants which are calculated by regression

The equation 2 characterizes the curves in Figure 4 with percentage standard deviation values. So if the number of stations per grid and their spatial distribution is sufficient to calculate a representative standard deviation, then it is possible to estimate the mean relative error for the grid with equation 2.

So far, the latitude - dependent size of 2.5° grid areas has not been taken into consideration. The results gained for grids over Australia, Germany and the USA (except Alaska) in the latitude range of 10° to 50° do not show significant alterations. However, for a global use the actual station density should be used instead of the number of stations per grid.

5 Intercomparison Of Results From Different Data Sources

Different precipitation data sets are available at the GPCC, namely precipitation estimates derived from satellite data, as there are IR-estimates based on the GPI method (Arkin, 1979, Janowiak, 1992) and SSMI-estimates calculated at NASA (Wilheit et al.,

1991), ECMWF T106-model 0-24h and 12-36h forecasts (Arpe, 1991), objective analyses of gauge measurements (GPCC, 1992 and 1993) and the Legates (1987) climatology (not corrected for the systematic gauge measuring error). Large areal means of these data sets are compared with one another over the continents in a 60°-wide "tropical" latitude belt around the equator (for spring and autumn seasons between the latitudes 30° N to 30° S large areal means have been calculated; for summer and winter seasons the 60° latitude belt was shifted by 10° to the North, respectively to the South, see Figure 7) for the years 1987 and 1988. The IR-estimates tend to overestimate the precipitation in almost all months for 1987 and 1988 over land areas (see Figure 7) and also show the largest month to month variability in the large-areal means. Between the IR estimates and the GPCC-analyses on average a difference of about 40% of the monthly large areal precipitation amount is found over the two-year period.

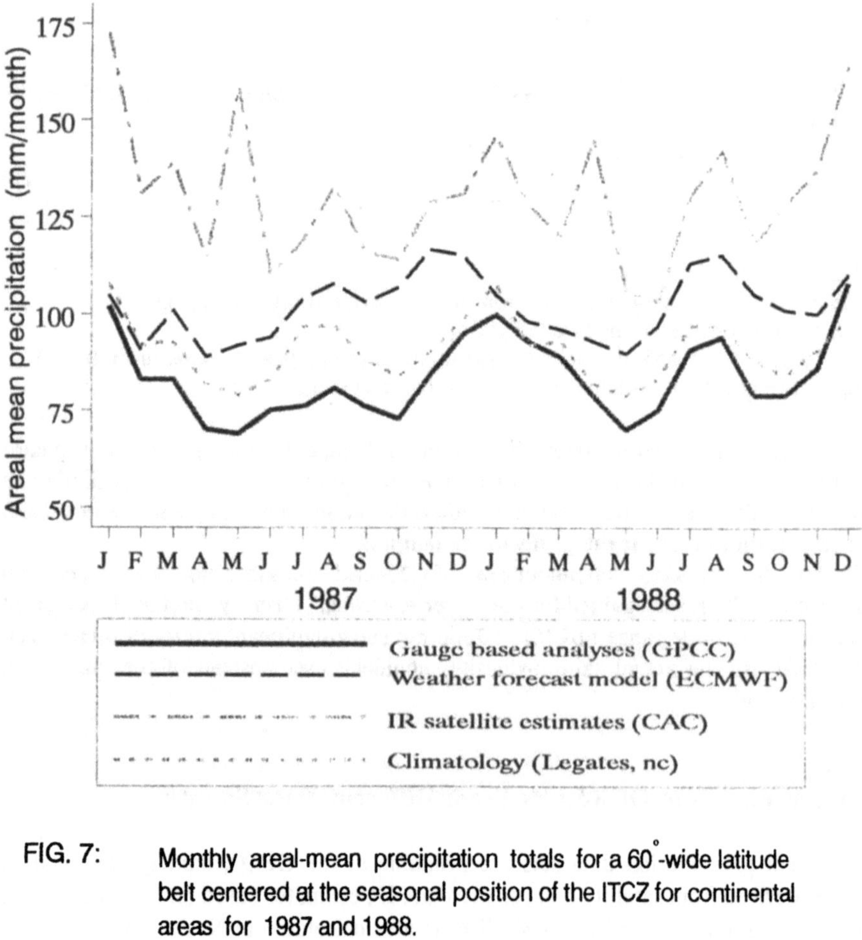

FIG. 7: Monthly areal-mean precipitation totals for a 60°-wide latitude belt centered at the seasonal position of the ITCZ for continental areas for 1987 and 1988.

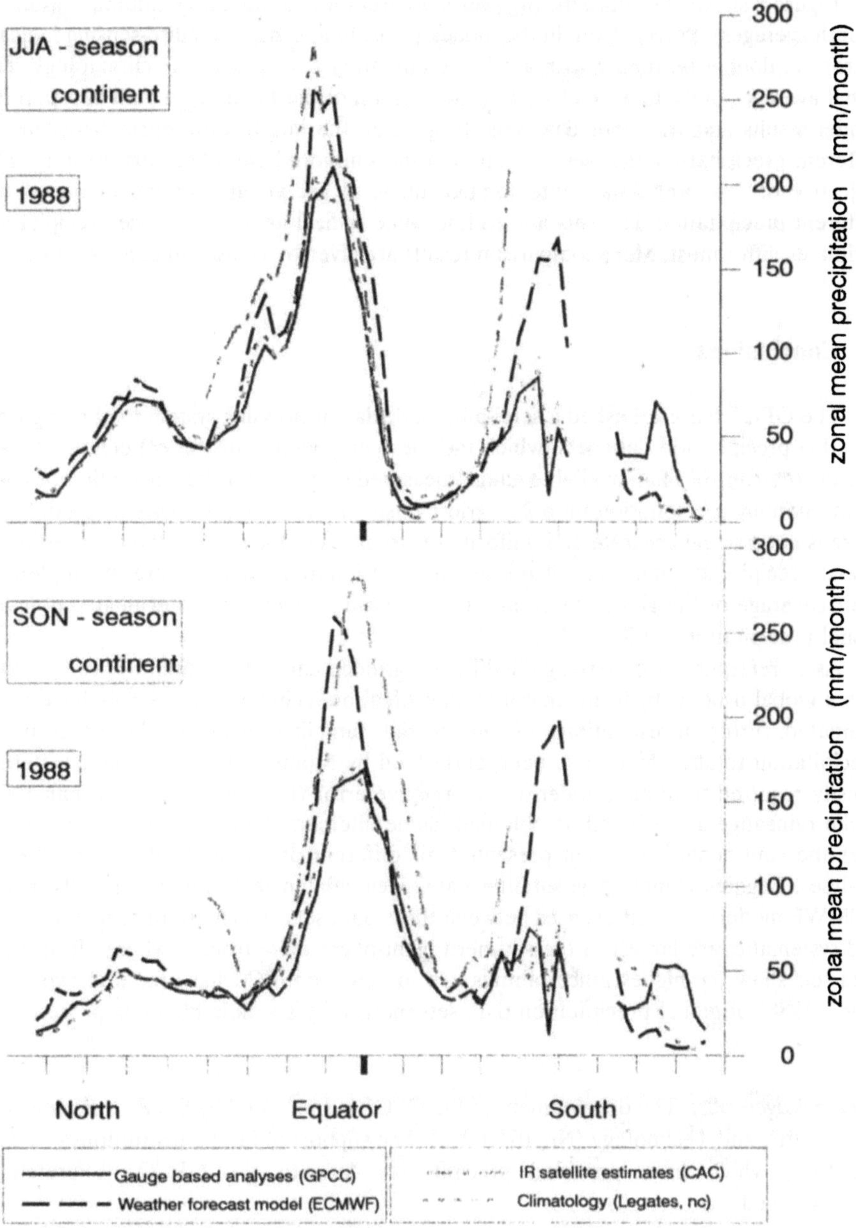

FIG. 8: Zonal averages for summer and autumn season 1988 over land areas

Figure 8 shows the zonal means over land areas for summer and autumn seasons (3 month averages). Precipitation in the model and estimates from satellite-sensing indicate a tropical double maximum with a relative minimum also seen in the climatology. The zonal averages indicate a small shift in the location of the ITCZ maximum between the model results and the other data sets. In general, the largest differences between the different precipitation analyses are found in the southern hemisphere extratropics. The objective of this preliminary intercomparison is to get an idea of the errors of the different precipitation data sets and to find some critical areas, where the precipitation estimates differ most. More comparison results are given by Hauschild et al. to this issue.

6 Conclusions

The GPCC has established a semi-automatic data processing system to obtain global monthly precipitation data sets, which includes components for the collection, storage and quality-control of all available gauge-measured data and for the calculation of areal mean monthly precipitation on a 2.5° grid by using an objective analysis method. Since there is not a single accurate and uniform way to measure the space/time distribution of global precipitation and most of the existing precipitation data sets are incomplete in their coverage of the globe, different data sets have to combined to evaluate complete global precipitation fields.

As a prerequisite for merging the different gridded data sets of monthly precipitation into a global product by using an optimum estimation technique their errors have to be estimated. From investigations of the spatial sampling error of the areal mean precipitation totals, which have been performed by Monte-Carlo-techniques, a station density required to remain under a fixed relative error (for example of less than 10%) in the raingauge analysis can be obtained. Some intercomparisons of large areal means over the continents have been presented for different precipitation data sets (GPCC analyses, Legates climatology, satellite data, accumulated daily forecast results of the ECMWF model). The differences between these data sets, which are in part substantial and systematic, are largest in the southern hemisphere extra-tropics, where all of these data sets show the highest uncertainties. Up to now the GPCC has prepared two years (1987, 1988) of global precipitation data sets merged by a simple blending procedure.

Acknowledgements: The development of the GPCC is sponsored by the Federal Ministry of Research and Technology (No. 07KF005). The support of all the institutions of many countries, which have provided precipitation data to the GPCC, is gratefully acknowledged.

REFERENCES

Arkin, P.A. (1979): The relationship between the fractional coverage of the high cloud and rainfall accumulation during GATE over the B-scale array. Monthly Weather Review, **107**, 1382-1387.

Arpe, K. (1991): The hydrological cycle in the ECMWF short range forecats. Dynamics of Atmospheres and Oceans, **16**, 33-60.

Barnes, S.L. (1973): Mesoscale objective map analysis using weighted time-seriesobservations. NOAA Technical Memorandum, ERL NSSL-62 U.S. Department of Commerce, 60 p.

Benichou, P. and O. Le Breton (1987): Pise en compte de la topographie pour la cartographie des champs pluviométrique statistique. La Météorologie, Oct. 1987 (see also the contribution of Benichou to this issue).

Behr, O. (1993): Methods for national and regional precipitation distribution assessment. Proc. International Workshop on Analysis Methods of Precipitation on a Global Scale, Koblenz, 14-17 Sept. 1992, WCRP-81, WMO/TD-No. 558, A/77-A/86.

Bussiéres, N. and W.D. Hogg (1989): The objective analysis of daily rainfall by distance weighting schemes on a meso-scale grid. Canadian Meteorological and Oceanographic Society, Atmosphere-Ocean, **27** (3), 521-541.

Cressman, G.P. (1959): An operational objective analysis system. Monthly Weather Review, **87**, 367-374.

GPCC (1992): Monthly precipitation estimates based on gauge measurements on the continents for the year 1987 (preliminary results) and future requirements. Rep.-No. DWD/K7/WZN-1992/08-1, Offenbach,August 1992 - to order from GPCC.

GPCC (1993): Global area-mean monthly precipitation totals for the year 1988 (preliminary estimates, derived from rain-gauge measurements, satellite observations and numerical weather prediction results). Rep.-No. DWD/K7/WZN-1993/07-1, Offenbach, July 1993 - to order from GPCC.

Janowiak, J. (1992): Tropical rainfall: A comparison of the satellite-derived rainfall estimates with model precipitation forecasts, climatologies and observations. Monthly Weather Review, **120**, 448-462.

Jenne, R. and D. Joseph (1985): Sensitivity experiments with different 1 to 500 km scale networks (NCAR). World Meteorological Organisation, WMO/TD-No. 115.

Kirchhofer, W. and B. Sevruk (1992): Mittlere jährliche korrigierte Niederschlagshöhen 1951-1980. In: Hydrologischer Atlas der Schweiz, Kartenblatt 2.2, Bern.

Legates, D.R. (1987): A climatology of global precipitation. Publ. in Climatology **40** (1), Newark, Delaware, 85 pp.

Phillips, D.L., J. Dolph and D. Marks (1992): A comparison of geostatistical procedures for spatial analysis of precipitation in mountainous terrain. Agricultural and Forest Meteoology, **58**, 119-141.

Rudolf, B. (1993): Management and analysis of precipitation data on a routine basis. Proceedings WMO/IAHS Int. Symp. on Precipitation and Evaporation, Vol. 1, 69-76, Slovak Hydromet. Inst., Bratislava, Sept. 1993.

Rudolf, B., H. Hauschild, M. Reiss and U. Schneider (1992): Beiträge zum Weltzentrum für Niederschlagsklimatologie - Contributions to the Global Precipitation Climatology Centre. Meteorologische Zeitschrift N.F. **1**, 7-84.

Schneider, U., D. Henning, H. Hauschild, M. Reiss and B. Rudolf (1992): Zur Berechnung monatlicher Niederschlagshöhen aus synoptischen Meldungen. Meteorologische Zeitschrift N.F. **1**, 22-31.

Shepard, D. (1968): A two-dimensional interpolation function for irregularly spaced data. Proc. 23rd ACM Nat. Conf., Brandon/Systems Press, Princeton, NJ, 517-524.

Wilheit, T.T., A.T.C. Chang and L.S. Chiu (1991): Retrieval of monthly rainfall indices from microwave radiometric measurements using probability distribution functions. Journal of Atmospheric and Oceanic Technology, **8**, 118-136.

Willmott, C.J., C.M. Rowe and W.D. Philpot (1985): Small-scale climate maps: A sensitivity analysis of some common assumptions associated with grid-point interpolation and contouring. The American Cartographer **12** (1), 5-16.

WMO/ICSU (1990): The Global Precipitation Climatology Project-Implememtation and Data Management Plan. World Meteorological Organisation, WMO/TD-No. 367, Geneva, 45 p., 6 app. (see also the Contribution of Arkin to this issue).

CARTOGRAPHY OF STATISTICAL PLUVIOMETRIC FIELDS WITH AN AUTOMATIC ALLOWANCE FOR REGIONAL TOPOGRAPHY

Patrick Bénichou
METEO FRANCE
Service Central d'Exploitation de la Météorologie
42, Avenue Gustave Coriolis
31057 Toulouse Cedex (France)

ABSTRACT

The AURELHY method (Analysis Using RELief for HYdrometeorology) takes into account the possible statistical relationships between the pluviometric (or thermometric) field to analyse and the regional topography. At any given point, assessment of rainfall is made by adding two terms, the first being the result of a linear regression with the elements of local topography, the second being obtained by interpolation by kriging of residuals.

KEYWORDS

Cartography, precipitations, relief, topography, kriging, climatology.

1. Introduction

In cartography more and more pressing needs for pluviometric parameters have led us to elaborate a new method, for analyzing pluviometric fields, that takes into account the possible statistical relationships between the field to analyze and regional topography. This method has been successfully tested on statistical fields such as mean monthly rainfalls. It was christened AURELHY (Analysis Using the RELief for HYdrometeorology). We have used it here to interpret the annual variations of the statistical relationships obtained by AURELHY over a mountainous area, i.e. the Massif Central.

NATO ASI Series, Vol. I 26
Global Precipitations and Climate Change
Edited by M. Desbois and F. Désalmand

2. Presentation of the Aurelhy method

2.1 Scope of the method

A rainfall cartography, whatever it may be, cannot be carried out in a credible way without allowing for parameters likely to be significantly linked. Hence the idea of setting up an analysis system involving, through linear regression, the local topography coded in vector form. Once the regression has been carried out, it only remains to analyze "mathematically" (kriging) the regression residuals whose variance is hoped to be much lower than that of the basic field. The "mathematical" portion of interpolation is then greatly completed with a "physical" portion by making allowance for topography. All this can be formalized (Benichou, 1986) by expanding the value P observed at site Si whose geographical coordinates are (x_i, y_i) and topographical environment is T_i:

$$P(S_i) = P(x_i, y_i, T_i) = \underset{\text{(regression)}}{f(T_i)} + \underset{\text{(residual)}}{\varepsilon(x_i, y_i)} \qquad (1)$$

Assessment of rain at any given point M(x,y,T) is made as follows:

$$P^*(M) = P(x,y,T) = f(T) + \varepsilon^*(x,y) \qquad (2)$$

(application (interpolation of regression) of residuals)

2.2 Coding the topography by P.C.A.

From a file acquired from the "Institut Géographique National", providing an estimated elevation at any point of a kilometric grid, we carried out a smoothing of France topography by calculating at each of the 27,000 points of a 5 km-mesh grid, an average elevation Z. The "landscape" surrounding a grid point G is determined by considering a square matrix (11*11) of mean elevations surrounding the point G. The site G is thus characterized by 121 values.

It is obvious that 121 values are redundant when characterizing a site. Hence the necessity to condense this information. The Principal Components Analysis was thus applied to "relative landscapes", obtained from the basic landscapes by taking out the mean elevation of the central point from each of the 121 elevations. The first 15 components suffice to account for 90% of the field variance, so that the relative landscape in each point G is given with a satisfactory accuracy through the linear combination of 15 eigenvectors $V_1,...,V_{15}$, each V_j being weighted by a principal component C_j. The relative landscape in G is:

$$\Sigma\ C_j(G) * V_j$$

In view of this, each eigenvector V_j appears as a typical relative landscape. Each landscape in France is a certain combination of these typical landscapes. The first five eigenvectors or typical landscapes are visualized on figure 1. They are characterized by:

V_1: global embankment effect (39% of variance)
V_2: west/east slope (24% of variance)
V_3: north/south slope (9% of variance)
V_4: north/south saddle (3% of variance)
V_5: north-east/south-west saddle (3% of variance)

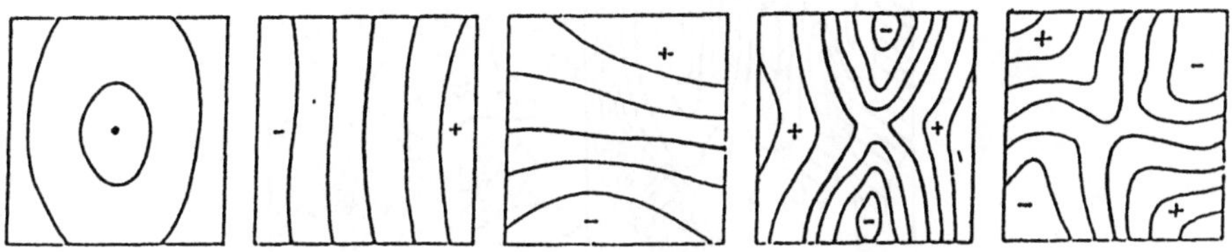

Fig.1: First five eigen vectors or typical landscapes

The other eigenvectors reveal finer spatial structures but the essential of a relative landscape is accounted for by giving its first five principal components. To sum up, each site G is now coded by a vector with 16 predictors:

- the average altitude Z of central point,
- the first components C_1,...,C_{15} of the relative landscape surrounding the point G.

The spatial distribution of Z, C_1, C_2, C_3 on the Massif Central is shown on figure 2.

Finally, the coding shown above are used as predictors to a regression. These scalar variables present the advantage to be independent. Once the topography coding has been carried out, at the 27,000 grid-points as well as at the 10,000 rainfall reporting stations S_i, it is suggested to quantify the fraction accounted for by the landscape in a precipitation field, given that the more this fraction will be important the more the fraction to interpolate will be reduced.

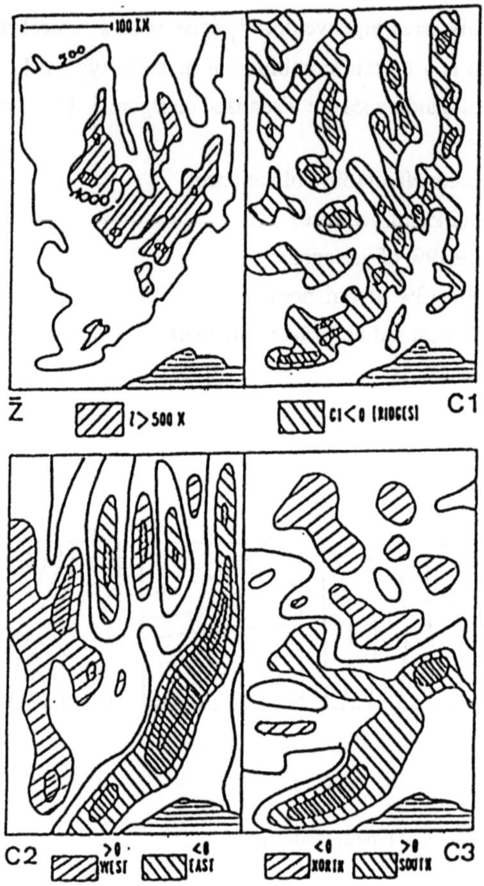

Fig. 2 : Spatial distribution of Z, C_1, C_2, C_3 over the Massif Central.

2.3 Regression against topography and kriging of the residuals

The studied field must be vast enough for the sample of measuring points to be sufficient, but it must not be too vast either, for the regression equation to make sense, which brought us to divide France into ten regions, each of them representing a geographical entity: Brittany, Pyrénées... and in particular the Massif Central (250 km * 370 km): in this field, 400 measuring points show pluviometric values. From this sample of 400 values, we select, among the 16 potential predictors, the significant ones (through an ascending progressive selection method and according to a Fisher test on the multiple correlation coefficient). As an example, for the January normals over the Massif Central, 6 predictors have thus been selected, leading to about 60% of the variance being explained, and to the equation:

$$1000 * f(T) = 2.5\ C_3 - .7\ C_1 + 24.5\ Z - 1.5\ C_9 + .4\ C_2 - .1\ C_{11} + 722000 \quad (3)$$

This equation is then applied at any grid point G in the studied field. However, it is necessary to ensure that both samples of predictors concerning the sites on the one hand and the grid points on the other hand are extracted from the same population. After taking this precaution, we can map f(T). The relief having explained 60% of the variance, the residual $\varepsilon_{x,y}$) represents only 40% of it and it is this residual that we are going to assess through a kriging method.

The kriging method is a geostatistical method which amounts to a probabilistic approach to interpolation. The theory of kriging is explained in (Delfiner,1973). It implies some hypothesis on the spatial structure of the field to analyze (existence of a variogram). The method looks for an unbiased estimate P* at any point G:

$$P^*(G) = \Sigma\ \lambda_i * P(S_i)$$

The λi are computed to minimize the variance of the estimation error:

$$\sigma_\varepsilon{}^2 = E(\ (P-P^*)^2\ )$$

It is reasonable to think that the residual $[\varepsilon(x,y) = P(x,y,T) - f(T)]$ is more suitable for applying the kriging method. The kriging program makes an automatic structural identification out of the residual field. The assessment $\varepsilon^*(x,y)$ at the grid points yields a chart of this field.

2.4 Reconstitution of the rainfall field. Validation of the AURELHY method.

At each point G of the grid covering the Massif Central we know the value f(T) given by the regression equation (3), together with the assessed residual r(x,y). We infer

$$P^*(x,y,T) = f(T) + \ \varepsilon(x,y)$$

and the final chart of January normals over the domain (see figure 3).

Now let us assess the quality of the results thus obtained and let us compare it with that obtained through more classical methods such as kriging directly applied to the basic values. This comparison has been carried out: it emerges that in areas in which network is dense and homogeneous, the error statistics for the two methods are comparable and this can be easily understood: in the case of direct kriging, the values alone are sufficient to ensure the dissemination of the "topography" information and, at any rate, with an ideally laid out network, kriging provides in theory quite a satisfactory cartography, on condition that the spatial structure of the field is suitable for it (existence of the variogram). In areas where measuring points are sparse, the results yielded by the comparison are altogether different. A number of posts have been taken away from an area in order to simulate an area lacking in, but not without, measuring points. The comparison of retrieved values may be summarized as follows: direct kriging yields conservatively a flat field in the area which gives a mean absolute error of 25 mm, an error which exceeds the standard deviation of the estimation error provided by the program, the experimental variogram being certainly underestimated. AURELHY method takes down the error from 25 to 11 mm for an average value of the field of 80 mm.

Lastly, experiments concerning strength of the method give the following results. Regressions carried out on a more and more reduced sample (100 to 50% of the total population) have proven to be almost identical, suggesting the presence of the strong stability of a good portion of the field ("physical" portion) under less favourable conditions.

The use of AURELHY method in automatic cartography seems therefore fully justified. It can still be said that using kriging only over the field of regression residuals can make it possible to go round certain obstacles that are important for kriging in the recognition of spatial structure, like anisotropy or even the non-existence of the variogram.

The statistical justification of the method being recognized, an interest can be taken in its climatological spin-offs. As a matter of fact, a statistical relationship between rainfall and topography is implicit to each geographical area and to each studied pluviometric field. It is the annual and interannual variabilities of these relationships that will be now investigated over the mountainous area of the Massif Central.

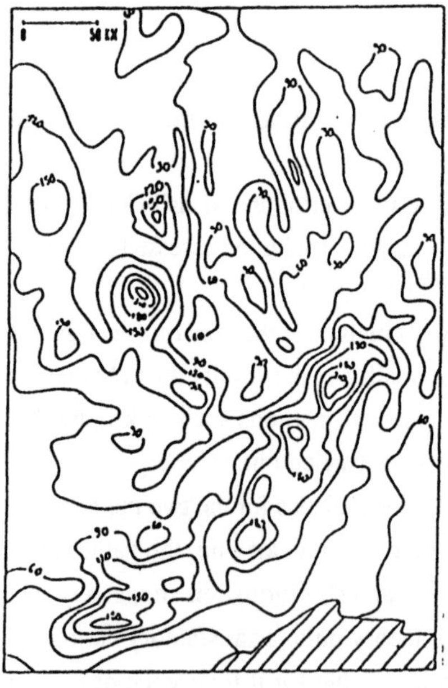

Fig. 3 : January normal precipitation (mm) analyzed by the AURELHY method.

3. Annual evolution of statistical relationships between topopography and two average parameters : rainfall amounts and number of rainy days

In section 2.3 an example of the application of the AURELHY method was shown as applied to normal January rainfalls. This method has been also applied over 12 months over the Massif Central, for three parameters:

- normal monthly rainfalls (R);
- average number of rainy days (N_o);
- average number of days with precipitation greater than 5 mm (N_5).

Three sets of 12 regression equations have been obtained.

3.1 Strength of the relationships

The percentage of variance accounted for by topography does not show any marked annual variation; for the three fields, it lies between 50 and 65%, showing that revealed relationships are greatly significant, in view of the size of the samples (400 to 500 points). The whole of obtained relationships involves, in at least 80% of cases the first Principal Components C_1, C_2, C_3 of the relative landscapes (cf 2.2), as well as the mean altitude Z. This indicates that it is the relief in its main characteristics (embankment, large scale slope effects) that seems to govern the spatial distribution of studied parameters. Therefore now, the annual evolution of regression coefficients associated with Principal Components C_2 and C_3 are investigated in detail.

3.2 Influence of easterly-westerly exposure (C_2)

Let us bear in mind that a landscape with westerly exposure has C_2 positive and conversely. Figure 4a represents the annual variation of the coefficient of C_2 in the regression equations obtained for R, N_o, N_5. As regards rainfall amounts (R), this coefficient is in general positive, which shows that slopes with westerly exposure are more watered than slopes with easterly exposure, owing to oceanic origin of disturbances. On the other hand in autumn, a complete reversal occurs, slopes with easterly exposure being then more watered; as a matter of fact, autumn is a season favourable to the development of Mediterranean lows, bringing a warm and moist south-easter which originates the famous Cévennes precipitations.

With regard to numbers of rainy days (N_o and N_5), the obtained coefficients are positive: western slopes are not only more abundantly watered but also more frequently watered than the eastern slopes; this means that the pluviometric deficit observed on the eastern slopes is not only due to a weakening of oceanic disturbances but sometimes also to precipitation being stopped over the ridge line. The theoretical difference concerning rain frequency between eastern and western slopes varies from 0.5 to 1 day in 10,000 C_2 units (C_2 varies from -60,000 to +40,000 over the Massif Central). The minimum value (0.5day/10000) is observed in autumn: the Cévennes rains fill the gap between rain frequencies over western and eastern slopes respectively, but their period of activity, compared with that of oceanic disturbances, does not seem to last long enough to change the sign of the regression coefficient of C_2.

3.3 Influence of North-South exposure (C_3)

A lanscape with southerly exposure has a positive component C_3 and vice-versa. Component C_3 has proved - over the Massif Central and on an average - to be the best predictor to account for the spatial distribution of the studied pluviometric parameters. Annual variation of coefficient of C_3 in regression equations are represented in figure 4b.

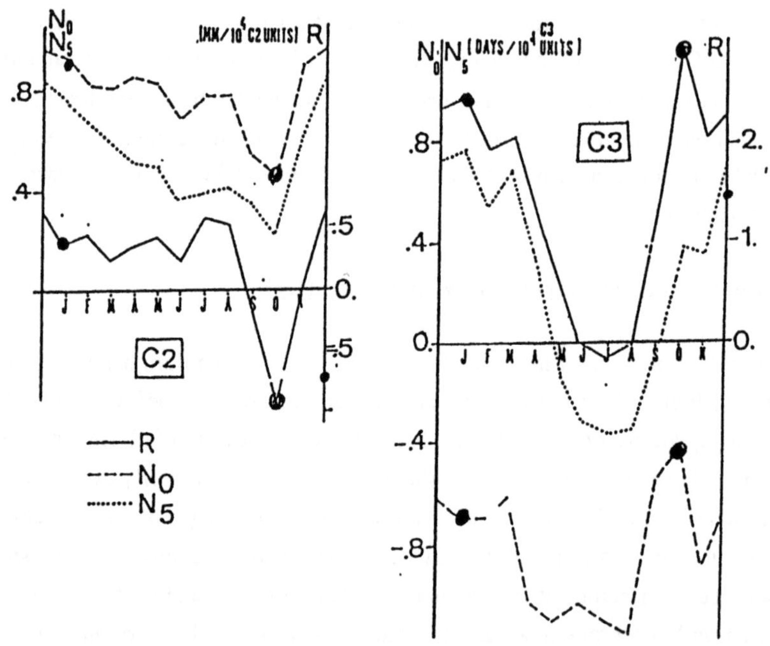

Fig. 4: Annual variation of the coefficient of : a) C_2 (west-east slope) (left) ; b) C_3 (north-south slope) (right)

It is surprising to notice the diametrically opposite influences of northerly and southerly exposures on spatial distribution of amounts (R) and frequencies (N_0) of precipitation. As regards rainfall amounts (R), C_3 coefficient remains positive all year round: reliefs facing South are most sensitive to the amount of precipitation. As a matter of fact, most of the rainfall comes in a general flux with south component (south-west for oceanic disturbances and south-east for Mediterranean disturbances). Effect of C_3 is almost null in summer, at which time rains are mostly convective; it is strong in winter, corresponding to a maximum activity of oceanic disturbances, and becomes maximum in autumn, at which time Mediterranean disturbances develop.

As regards rainfall frequency: in a general way, the coefficient concerning N_0 is undoubtedly negative all the year round, which shows that it rains more often on the north slopes, which is typical of the contrast which exists between oceanic and Mediterranean climates: South slopes are abundantly watered, but rather rarely, while North slopes receive repeated but weak precipitation. With regards to frequency of heavy rains (N_5), the results are more complex. In summer, it is again the northern slopes which see most often heavy rains (differences of 0.35 day/10000 C_3 units). As a matter of fact, it is again the effect due to the estival drought in a Mediterranean climate, the bulk of southern slopes in the Massif Central being part of the Mediterranean area (cf fig2). In the winter, on the other hand, it was seen that rain frequency N0 is more important over northern slopes (differences of about 0.7 days/10000 C_3 units), the reversed effect is observed for heavy rains N_5, the southern slopes being more often abundantly watered (-0.7 day/10000 C_3 units). Therefore it is for light rains (<5mm) that the highest spatial variations of frequency are observed: the north-south contrast can be estimated at 1.4 day/10000 C_3 units. Once again, it happens that light rains water northern slopes and heavy rains water southern slopes.

The statistical relationships revealed on the monthly normals of pluviometric parameters have therefore contributed to a better cartography of these parameters. It has been seen that their climatological interpretation allowed a better knowledge of the meteorological events which govern their spatial distribution. It is now advisable to test on a monthly basis the validity of relationships obtained over average fields. This is the purpose of the following chapter.

4. Interannual variability of statistical relationships between topography and january precipication over the massif central.

The AURELHY method has been applied 30 times to the January rainfalls from 1951 to 1980, providing therefore, in addition to corresponding charts, a set of 30 regression equations and their associated multiple correlation coefficient. Before studying the monthly results individually, it is interesting to seek to know more about the structure of the field of interannual variability (standard deviation) of January monthly rainfalls. For that purpose, we have also applied AURELHY to this field of standard deviation.

4.1 Analysis of standard deviation of January rainfalls (σ).

As we proceeded in 2.3 for January normal precipitation, we obtain the equation:

$$1000 * \sigma = 2.3\ C3 - 0.3\ C_1 + 17.\ Z - 0.3\ C_2 - 1.1\ C_{11} + 39000 \qquad \textbf{(4)}$$

with a percentage of 83% of variance accounted for by topography. It may be remarked that it is virtually the same predictors (C_3, C_1, Z, C_2) which account for the spatial distribution of σ and also that of mean rainfall (R) (cf.2.3). The percentage of variance accounted for by this set of predictors shows that even over a statistical parameter of second order (the variance), topography plays a major role. It can be said that relief not only influences the spatial distribution of pluviometric fields, but also its temporal variability. Figure 5 shows the distribution of σ analyzed by AURELHY. The moderate and quasi permanent feature of oceanic influences is found again, in opposition to the more irregular and violent feature of Mediterranean influences. It is this play of influences that will be seen again along the different months of January that have been studied, through the medium of the different regression coefficients associated with the selected predictors.

4.2 Interannual variability of statistical relationships

In the same way as for monthly normals in more than 2 thirds of cases the selected predictors are altitude or the first 3 principal components of landscapes (cf 2.2). The distribution of regression coefficients associated with C_1, C_2, C_3 is schematized in figure 6. It will be noted that these coefficients vary within a wide bracket about the value determined for the field of normal precipitation (cf. equation (3)). This means therefore that it is not legitimate to use a statistical relationship determined uniquely for an average field in order to map all occurrences of this field. This is true in particular for coefficient of C_2 (east - west slope) which fluctuates widely on both sides of 0, which shows that contrasts between western and eastern slopes can be reversed from year to year. As regards C_1 (embankment) and to a lesser degree C_3 (north-south slope), it seems all the same that the behaviour is somewhat steady, at least in a qualitative way, from year to year. Reagardless it is shown that spatial contrast due to topography can be, on a monthly basis, 3 or 4 times more violent than, if not opposite to, those observed over the average field.

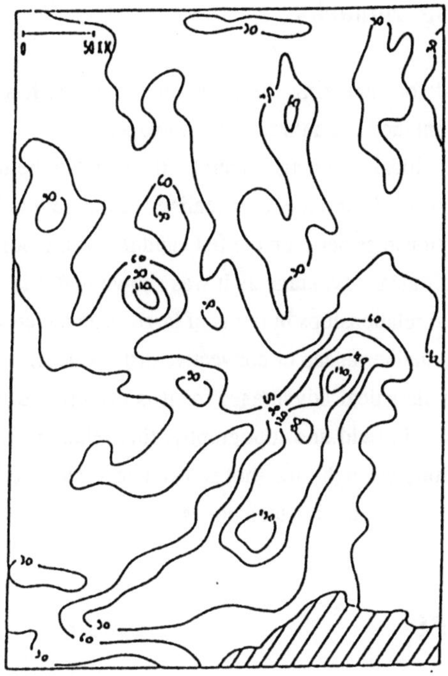

Fig. 5 : January monthly rainfall standard deviation (mm) analyzed by AURELHY.

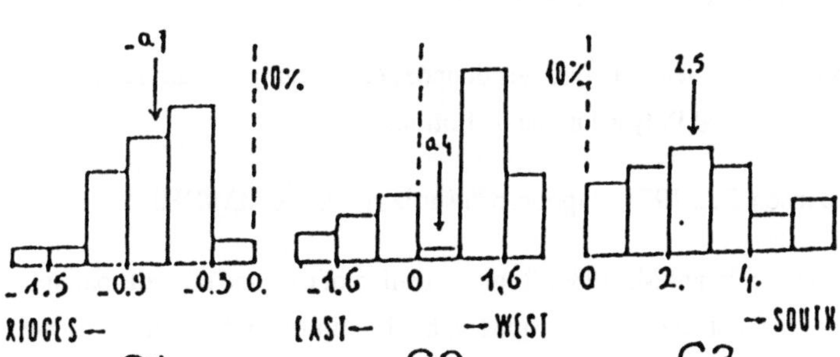

Fig. 6 : Distribution of the regression coefficients of C_1, C_2, C_3 over the period 1951-1980.

5. Concluding remarks

A method for analyzing the pluviometric fields has been discussed (AURELHY method) and this method has a threefold advantage.

A cartographical advantage owing to the fact that the method provides a chart which is both the most probable and the most realistic of the field to be analyzed since it integrates the statistical relationships between the field and regional topography.

A climatological advantage as it was shown that it is relatively easy to interpret the obtained statistical relationships in terms of large scale meteorological influences (combined effect of oceanic, mediterranean or convective disturbances).

Finally, a statistical advantage, for it is established, that if there is a relationship between pluviometric field and topography, this relationship is not universal and can be extremely fluctuating through time (not shown here). This method provides an improvement over other objective interpolation techniques.

REFERENCES

Benichou P., 1986: Cartography of statistical pluviometric fields with automatic allowance for topography. 3rd Int.Conf. on Stat. Climatology. Vienna. Austria. June 1986.

Creutin D., Obled C., Tourasse P., 1980: Analyse spatiale et temporelle des épisodes pluvieux cévenols. La Météorologie VI-20.

Delfiner P., 1973: Analyse objective du géopotentiel et du vent géostrophique par krigeage universel. La Météorologie V-25.

Laborde J.P., 1984: Analyse des données et cartographie automatique en hydrologie. Institut National Polytechnique de Lorraine.

Nordenson T.J., 1971: Rapport n°6 sur les projets OMM/DHI.

Obled C., Slimani M., 1986: Regionalization of extreme rainfalls parameters through kriging and correlation with topography descriptors. Int. Symp. on flood frequency and risk analysis. Baton-Rouge, May 1986.

Peck E.L., Brown M.J., 1962: An approach to the development of isohyetal maps for mountainous areas. Jour. of Geoph.Res. vol 67.

Schulze R.E., 1977: Practical considerations in the application of surface fitting techniques for estimation of areal rainfall. 2nd Conf. on Hydromet. Toronto.

WMO/OMM N°326, 1977: Geilo (Norway) Symp. on orographic influence on the distribution of precipitation.

APPROACHES FOR REGIONALIZATION OF PRECIPITATION CLIMATES IN THE CONTEXT OF GLOBAL CLIMATE CHANGE MONITORING

Henry F. Diaz
NOAA/ERL,
325 Broadway, Boulder, CO, 80303, USA

ABSTRACT

Two precipitation regionalization methodologies are used and evaluated. In one, precipitation in the tropics, consisting of 34 stations, each considered to be representative of a 20-degree square region, was analyzed using the time scale identification techniques of singular spectrum analysis and the multitaper method of spectrum estimation. A second approach consisted of defining a spatial domain based on the distribution of the principal vegetation types dominant in that region in order to develop area-mean indices from the climate stations located within it. The geographical domain of this second approach was the general region of the North American Great Plains.

The set of tropical stations with significant variance in the 3-7 yr ENSO time scale were all located in areas where the ENSO phenomenon significantly modulates the amount of seasonal precipitation over time. However, even in areas of the western tropical Pacific where an ENSO signal in precipitation is shown to be of the same sign, correlation between stations can be relatively low. In the specific example discussed here, the association between Darwin, Australia and the island station of Apia to the east was only r=0.35. In the region of the U.S. Great Plains, however, strong decadal variations were evident in a number of grassland regions, encompassing a rather large area. Additional tests for regions with different vegetation types are needed to evaluate the validity of using dominant vegetation type as a regionalization tool.

KEYWORDS

Tropical rainfall regimes, El Niño-Southern Oscillation (ENSO), regionalization techniques, U.S. Great Plains precipitation.

NATO ASI Series, Vol. I 26
Global Precipitations and Climate Change
Edited by M. Desbois and F. Désalmand

1. Introduction

The purpose of this study is to evaluate some different methods for defining regional station ensembles in order to optimize information on the time behavior of areal averages of precipitation. The overall aim is to study variations in large-scale precipitation in order to understand the nature of possible causal mechanisms of such changes. Interannual to decadal-scale precipitation variability is often spatially distributed in preferred large-scale patterns. Such patterns reflect movements in airmass boundaries, shifts in storm tracks, etc. Here we discuss some approaches to define "homogeneous" climatic regions in order to analyze the time behavior of precipitation averaged over these spatial domains.

Two geographical regions subject to different large-scale climatic influences are considered: the tropics and the U.S. Great Plains. Two quite different approaches are used to define the spatial domain of the target regions and compare the behavior of precipitation variability. In the tropics, as an initial screening, we examined the similarity in the characteristics of the seasonal precipitation cycle. It was felt that this information could be used as an important indicator of spatial differences in precipitation regimes. Our main focus here, however, is to identify those areas where the El Niño/Southern Oscillation (ENSO) phenomenon strongly modulates both the seasonal cycle and interannual variability.

Singular spectrum analysis is used to define the principal time scales of climatic variability in the tropics, and to illustrate its utility for defining regional boundaries. The focus is primarily on interannual to decadal time scales. Additional information is extracted and summarized in tabular form for general information purposes, but will not considered in any detail here. The rationale for approaching regionalization in this manner is that stations over a given area that have similar temporal features of precipitation changes are responding to similar large-scale forcing; hence, the development of regional averages from stations possessing such common temporal characteristics, might be advantageous for monitoring a variety of climate mechanisms (for example, an index of rainfall fluctuations in the African Sahel region has been very useful in a variety of climate studies).

In the second approach we make use of the dominant vegetation types within a general physiographic province, e.g., the U.S. Great Plains, to define the regional boundaries (the old-fashioned Koeppen-style of climate classification). The rationale for this approach is that a particular vegetation ensemble reflects the integration of several aspects of the long-term climate in a region, and therefore, the development of regional averages based on such vegetation zones will be useful for monitoring a variety of climate mechanisms, over a broad range of frequencies. Having defined such regional boundaries, we then can assess the degree of agreement of precipitation variations in the time domain among the different regions.

2. Methodology

2.1 Characteristic time scales

The method of singular spectrum analysis (SSA) has been used to identify the characteristic scales of variation on climatic time scales. SSA, an extension of empirical orthogonal function analysis, considers *M* lagged copies of a time series process, x_i, $1 \leq i \leq N$, sampled at equal time intervals t_s, and calculates the eigenvalues λ_k and eigenvectors ρ_k, $1 \leq k \leq M$ of their covariance matrix **C** (see Ghil and Vautard, 1991; Vautard *et al.*, 1992). The matrix **C** has a Toeplitz structure with constant diagonals corresponding to equal lags. The choice of window width, $t_w = Mt_s$, represents a tradeoff between the amount of information desired to be retained (large M), and statistical significance (small M). Subsets of eigenelements and associated principal components provide for noise reduction, the removal of trend and the identification of oscillatory components (Rasmusson *et al.*, 1990; Penland *et al.*, 1991; Vautard *et al.*, 1992).

The spectral characteristics of the significant principal components were evaluated via both maximum entropy spectral analysis (MESA, Burg 1967) and the multitaper filtering method (Thomson, 1982). The latter allows an estimation of significant spectral peaks and was used for evaluating the spectrum of interannual variability components. Selected SSA components provide for optimal reconstruction of particular processes at precise time scales. As in spatial EOF analysis, the complete sum of all eigenelements returns the original input series. In contrast to other standard filtering techniques, where the basis functions are selected a priori, SSA functions as a data-adaptive filter.

A Geographic Information System (GIS) approach was used to manipulate the vegetation and climate station data. GIS is simply an analysis tool that allows the superposition of different data fields or attributes as a means to explore mutual commonalities. Stations with long records of monthly precipitation within specified grasslands regions were extracted from the database and averaged to form the regional indices.

2.2 Vegetation-based classification

Vegetation types have long been used as a key component in climate classification. We have used a global vegetation type database developed by Komarov (1964) and translated by D. L. Varlyguin, obtained from the National Geophysical Data Center in Boulder, Colorado. A second database of vegetation types for the contiguous United States developed by Kuchler (1967) and later modified by the U.S. Environmental Portection Agency was also used to ascertain how a classification based on another (independent) landcover source might change the results. Precipitation-recording stations located within particular vegetation zones were averaged together to form regional indices. By and large, the spatial distribution of the stations used here is fairly even, so it was felt that there was no need to resort to objective analysis schemes such as optimal averaging, etc to calculate the regional mean indices. The temporal behavior of the annual regional precipitation totals are compared. Results are shown for the Komarov vegetation types and for a couple of the Kuchler grassland zones which partially overlap with the others.

3. Analysis

3.1 Regional classification in the tropics

The global tropics from 20°N to 20°S were divided into 20-degree squares and stations were identified with at least 30 yr of continuous record. A total of 34 stations from the Global Historical Climate Network (GHCN) developed at the National Oceanic and Atmospheric Administration (NOAA), comprising the best individual sets of records were chosen as representative of rainfall variations in these rectangular areas. The stations are listed in Table 1 and their locations illustrated in Figure 1. The results of the SSA decomposition of the annual cycle components of the monthly rainfall are summarized in Table 2a, while a summary of the characteristic time scales of interannual variability for annual rainfall totals is given in Table 2b. We will focus below on classification based on similarities in year-to-year variations in rainfall to illustrate the efficacy of this technique.

It is well-known that the ENSO phenomenon has a strong influence on the precipitation climate of large areas in the tropics (Lau and Sheu, 1988; Nicholls, 1988). To illustrate this feature, Figure 2a highlights some of the key regions where an ENSO signal in December through February precipitation is found (see also Fig. 5 in Lau and Sheu, 1988). The figure shows the location of five-degree boxes (also based on GHCN

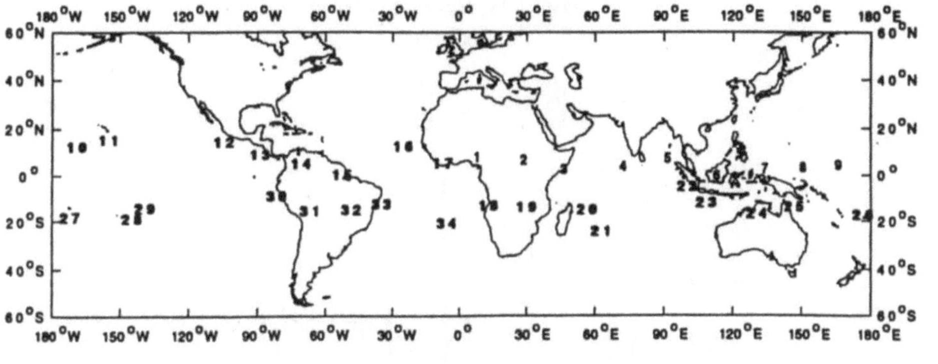

Figure 1. Map illustrating the location of the 34 rainfall stations used in this study.

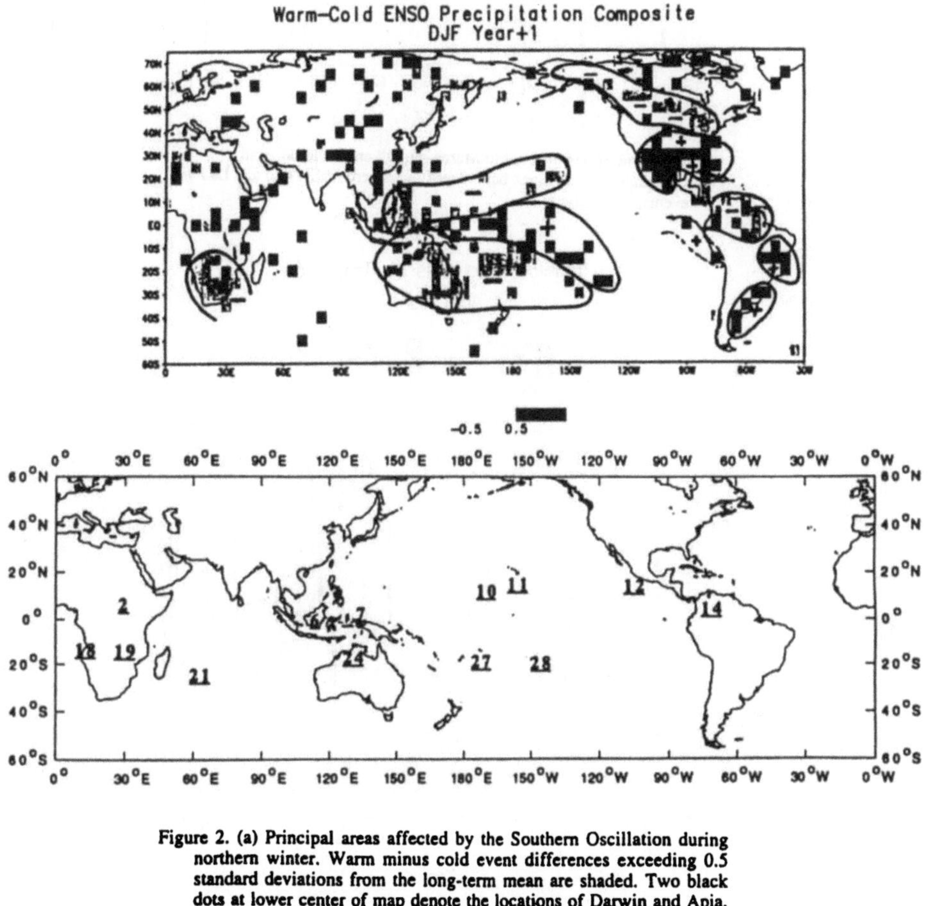

Figure 2. (a) Principal areas affected by the Southern Oscillation during northern winter. Warm minus cold event differences exceeding 0.5 standard deviations from the long-term mean are shaded. Two black dots at lower center of map denote the locations of Darwin and Apia, (b) Stations where interannual variability in the 3-7 yr band comprises more than 15% of the total variance. (see Table 1).

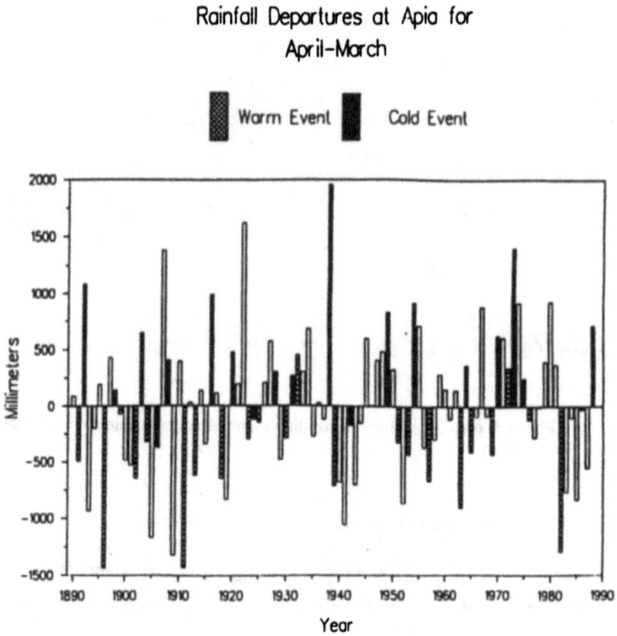

Figure 3. Time series of April to March rainfall anomalies at Apia. Crosshatched and black bars identify year zero of El Niño and La Niña events, respectively.

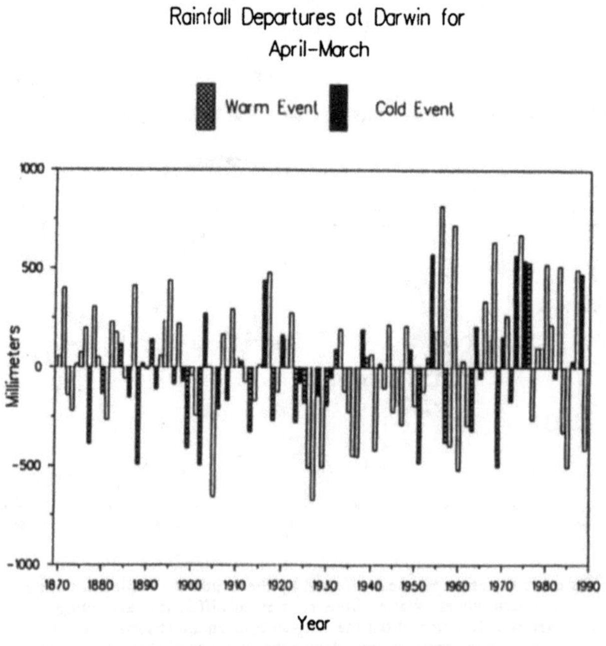

Figure 4. Same as Fig. 3, except for Darwin, Australia.

Table 1. Stations used in the analysis of regional groups based on their characteristic time scales of variability. Elevation in meters above sea level, unknown station elevation is coded as -999. The period of record (P.O.R) and the annual average rainfall (in mm) are also shown.

No.	Station Name	Lat.	Long.	Elev.	P.O.R.	Ann. Avg.
1	Kano, Nigeria	12.05	8.53	476	1905 1988	814
2	Kadugli, Sudan	11.00	29.72	499	1910 1989	725
3	Gaalkacyo, Somalia	6.85	47.27	-999	1933 1981	147
4	Minicoy, India	8.30	73.15	1	1891 1990	1629
5	Port Blair, Andaman, Is.	11.67	92.72	73	1868 1990	2976
6	Miri, Malaysia	4.33	113.98	17	1917 1990	2991
7	Koror, Palau Is.	7.33	134.48	29	1924 1990	3765
8	Truk, Caroline Is.	7.47	151.85	2	1935 1990	3524
9	Kwajalein, Marshall Is.	8.73	167.73	2	1945 1990	2574
10	Johnston Island	16.70	-169.50	5	1952 1984	669
11	Hilo, Hawaii	19.72	-155.07	9	1880 1990	3433
12	Manzanillo, Mexico	19.05	-104.33	3	1921 1981	1018
13	Zacatecoluca, El Salvador	13.50	- 88.90	170	1950 1981	2193
14	Agua Viva, Venezuela	9.60	- 70.60	110	1941 1986	973
15	Cayenne, Fr. Guiana	4.83	- 52.37	9	1847 1989	3317
16	Mindelo, Cape Verde Is.	16.88	- 25.00	62	1921 1976	102
17	Odienne, Ivory Coast	9.50	-7.57	434	1921 1990	1580
18	Luanda, Angola	-8.85	13.23	74	1879 1985	370
19	Mporokoso, Zambia	-9.40	30.10	-999	1919 1988	1263
20	Agalega Is.	-10.43	56.75	2	1951 1990	1694
21	Rodriguez Is.	-19.68	63.42	58	1951 1989	1151
22	Padang , Sumatra	- 0.88	100.35	-999	1879 1989	4306
23	Cilacap, Java	- 7.73	109.02	-999	1952 1989	3357
24	Darwin, Austr.	-12.40	130.87	31	1870 1990	1575
25	Port Moresby, Papua-N.G.	-9.43	147.22	28	1891 1990	1086
26	Rotuma	-12.50	177.05	26	1912 1990	3517
27	Apia	-13.80	-171.78	2	1890 1990	2892
28	Takaroa	-14.48	-145.03	2	1952 1990	1572
29	Atuona	-9.80	-139.03	51	1929 1990	1294
30	Talara, Peru	-4.57	- 81.25	85	1943 1990	41
31	Cobija, Bolivia	-11.00	-66.70	280	1950 1981	1752
32	Porto Nacional, Brazil	-10.52	-48.72	-999	1915 1989	1672
33	Recife, Brazil	-8.05	-34.92	11	1875 1989	1810
34	St. Helena Is.	-15.93	-5.67	435	1892 1990	853

precipitation data) where there are significant differences in December-February precipitation totals between warm (El Niño) versus cold (La Niña) years (see Diaz and Kiladis, 1992 for a list of ENSO event years and for additional references). The northern winter season is examined here, as this is the time of year when the strongest teleconnections are observed, and it also coincides with the mature phase of ENSO events(see Rasmusson and Carpenter, 1982; Rasmusson and Wallace, 1983). The stations with the highest proportion of interannual variance in annual total precipitation in the 3-7 yr ENSO band (arbitrarily chosen at 15% or more of the total series variance, see Table 2b) are: Kadugli, Miri, Koror, Johnston Is., Hilo, Manzanillo, Agua Viva, Luanda, Mpokoroso, Rodriguez, Darwin, Apia and Takaroa. For comparison with the ENSO-sensitive areas illustrated in Figure 2a, the station index given in Table 1 is is plotted at the location of the stations in Figure 2b. The information summarized in Table 2a provides a means to develop regional groupings based the characteristics of the annual precipitation cycle at each station. High relative interannual variance in the ENSO frequency band (Table 2b) provides yet another way.

Two of the stations with the longest precipitation records and located near the center of action of one pole of the Southern Oscillation (Trenberth and Shea, 1987) are Darwin, Australia, and Apia, Samoa Islands. The location of these two stations are shown by a solid dot in Figure 2a. The stations are seen to be located within a broad area that is affected in the same general manner by teh ENSO - drier conditions during El Niño, and the opposite during the opposite extreme. Figure 3 shows a time series of annual departures from the record mean for Apia. A year is defined to run from April to March to account for the tendency of the Southern Oscillation to drift to the positive or negative phase around northern spring (see Bradley *et al.,* 1987). As one would expect, the ENSO phenomenon has a dominant influence on the annual rainfall of the island, with cold ENSO events associated with above average rainfall and viceversa for warm (El Niño) conditions. Note that for much of the 1980s, rainfall was below normal at Apia. This period corresponds to a prolonged interval when sea surface temperature (SST) has been relatively high in the central equatorial Pacific. Similar periods with anomalies of one sign are evident in the early 1940s (also a time of persistent warm conditions in the equatorial Pacific) and in the early 1970s (associated with cold SSTs and above average rainfall).

The time series of rainfall anomalies at Darwin, Australia is shown in Figure 4. Although both of these stations respond in a similar fashion to the extremes of the SO, the correlation between the two time series of April to March anomalies is only 0.35, significant at better than the 1% level, but accounting for just slightly more than 12% of their respective variance. Furthermore, the low frequency behavior of precipitation appears to be somewhat different at Darwin compared to Apia. Note that since the mid-1960s, Darwin, displays a preponderance of *wet* years, averaging about 200 mm above the long-term period mean. Figure 5 illustrates how different the distribution of April-March precipitation totals are in both Darwin and Apia as a function of the Southern

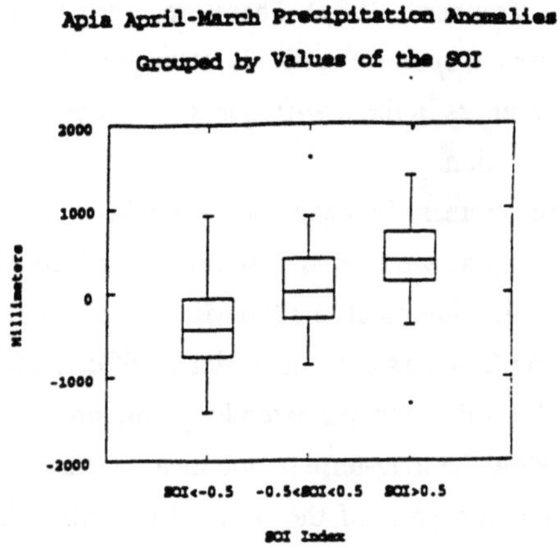

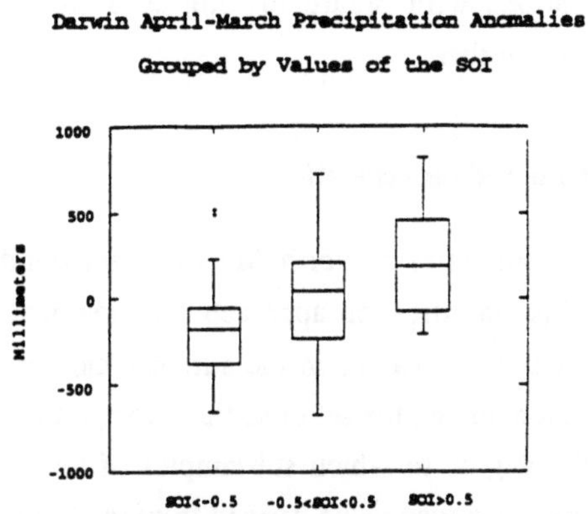

Figure 5. A comparison of the frequency distributions of April to March precipitation at Apia and Darwin, as a function of the mean value in the Tahiti-Darwin Southern Oscillation Index.

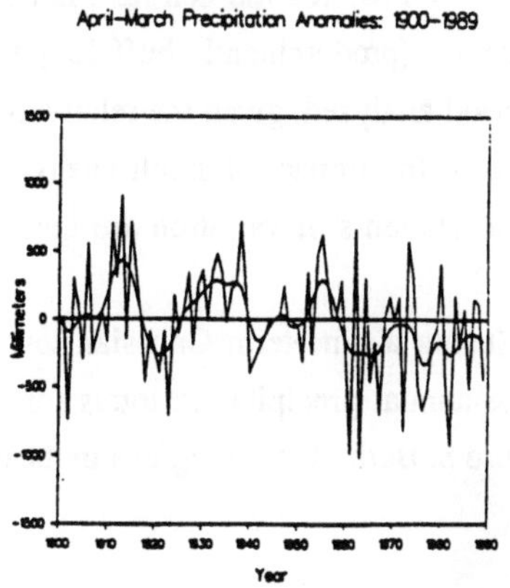

Figure 6. Time Series of mean annual precipitation departures for the four-station ensemble of Miri, Koror, Padang and Cilacap

Oscillation Index (SOI, see Trenberth, 1984). Therefore, while the behaviour of seasonal precipitation in the western tropical Pacific is strongly modulated by the ENSO phenomenon, there are differences in the low-frequency behaviour of precipitation in this region of high annual precipitation.

To check whether the increase in mean annual rainfall at Darwin in the last couple of decades is a real climate signal, we looked at station records from the very wet region immediately north of Australia. Four stations (#6 Miri, #7 Koror, #22 Padang, and #23 Cilacap) were used to compile a western equatorial Pacific rainfall index (Figure 6). Indeed, these stations exhibit a dry interval extending from around 1960 to the present, averaging about 200 mm below the long-term period mean. If these signals are legitimate, they may represent a southward shift of the tropical rainfall belt in the far western equatorial Pacific during the last few decades. The Tahiti minus Darwin Southern Oscillation Index has displayed a tendency since the late 1970s to be in the negative (El Niño) phase. The implications with regard to long-term changes in the Southern Oscillation need to be studied further.

3.2 Regional classification based on vegetation

Figure 7 shows the boundaries of North American grasslands vegetation which were chosen to illustrate this classification approach. The regional averages display a strong annual cycle with maximum values in the summer months and minima in the winter. The average correlation among the set of stations comprising each region gives a quantitative measure of the degree to which subsamples of the full set of observing stations can represent the area-averaged value derived from the full set of station records (Wigley *et al.*, 1984). The mean correlation among the stations in each region for annual total precipitation ranges from 0.31 (N=129) for the northern Great Plains region labelled "Prairie-Bunchgrass," to 0.45 (N= 116) for the central Plains region labelled "Grass Prairies with Subtropical Elements" (predominantly buffalo grass). Since most stations are available during the full period analyzed, given the relatively high spatial correlation among the stations in the regions, the impact of small changes in the regional station ensembles should be minor. Coefficients of variation are quite similar among all four regions, ranging from 20-25%.

Smoothed time series (using a nine-term Gaussian low-pass filter to highlight decadal scale variations) of the annual precipitation totals for the two regions located farthest south are shown in Figure 8. Both of these regions in the central United States

Table 2a. SSA and harmonic analysis of the annual (12-month) and semi-annual (6-month) cycle of mean monthly precipitation. The first two columns refer to the SSA components (M=30), the second two columns refer to a simple harmonic fit to the long-term monthly averages. SSA values are the percent of variance of the monthly means about the record mean, the second set of values are the percent of variance of the monthly means about the long-term annual mean.

Station	12	6	12	6
1. Kano	62.4	16.7	69.0	22.9
2. Kadugli	65.3	6.8	90.1	8.0
3. Gaalkacyo	nil	6.0	4.9	47.5
4. Minicoy	46.2	9.0	82.3	10.5
5. Pt. Blair	62.8	3.8	88.3	2.7
6. Miri	30.1	10.2	65.8	26.1
7. Koror	22.3	10.0	79.3	15.5
8. Truk	16.5	9.7	64.7	24.2
9. Kwajalein	30.3	9.1	82.3	16.3
10. Johnston	12.4	9.7	49.4	36.3
11. Hilo	11.1	8.2	23.4	20.3
12. Manzanillo	46.0	10.0	71.4	20.8
13. Zacatecoluca	71.2	nil	87.6	1.3
14. Agua Viva	11.0	26.1	22.0	61.9
15. Cayenne	46.8	12.7	80.2	12.0
16. Mindelo	24.0	9.2	38.8	27.3
17. Odienne	72.0	9.1	85.2	11.7
18. Luanda	22.1	12.5	38.9	27.2
19. Mporokoso	78.0	nil	92.7	0.7
20. Agalega	32.0	7.8	76.1	20.1
21. Rodriguez	24.8	6.5	80.7	10.9
22. Padang	14.0	10.0	54.6	36.8
23. Cilacap	16.6	12.4	43.3	44.4
24. Darwin	60.3	9.2	84.0	12.6
25. P. Moresby	47.7	7.8	89.4	9.0
26. Rotuma	17.5	9.4	86.4	0.4
27. Apia	37.5	9.3	96.9	2.0
28. Takaroa	25.3	6.6	84.2	8.4
29. Atuona	16.2	5.2	68.7	6.2
30. Talara	14.0	nil	nil	nil
31. Cobija	64.0	3.2	97.6	1.3
32. P. Nacional	68.0	3.2	96.4	0.1
33. Recife	56.0	3.2	96.2	3.6
34. St. Helena	28.0	13.4	68.9	18.9

Table 2b. SSA analysis of interannual variability of precipitation in the global tropics. Values are percent of total interannual variance for the strongest statistically significant spectral peaks within the indicated frequency band. Stations with a strong ENSO component are highlighted in bold face.

Station	2 to 3 yr	3 to 7 yr	> 7 yr
1. Kano	21	11	6
2. **Kadugli**	14	**16**	6
3. Gaalkacyo	22	14	3
4. Minicoy	2	2	4
5. Pt. Blair	14	2	4
6. **Miri**	13	**21**	4
7. **Koror**	10	**20**	2
8. Truk	8	3	3
9. Kwajalein.	14	10	2
10. **Johnston Is.**	8	**39**	2
11. **Hilo**	12	**17**	3
12. **Manzanillo**	14	**27**	4
13. Zacatecoluca	32	nil	4
14. **Agua Viva**	16	**23**	3
15. Cayenne	14	2	4
16. Mindelo	6	2	11
17. Odienne	14	12	5
18. Luanda	4	**15**	2
19. **Mporokoso**	16	**15**	5
20. Agalega	20	12	1
21. Rodriguez	24	**17**	2
22. Padang	16	3	6
23. Cilacap	4	14	8
24. **Darwin**	15	**17**	2
25. Pt. Moresby	9	13	2
26. Rotuma	22	2	3
27. **Apia**	10	**29**	4
28. **Takaroa**	31	**15**	3
29. Atuona	2	10	7
30. Talara	nil	nil	nil
31. Cobija	1	nil	7
32. Porto Nacional	2	9	4
33. Recife	nil	7	21
34. St.Helena	5	6	6

exhibit significant decadal scale variability, with major droughts occuring in the 1930s and 50s and wet spells in other periods, most recently in the early 1980s. These geographical areas represent ecologically coherent regions that are of interest, from the point of view of studying climate/vegetation responses, as well as for purposes of climate system monitoring for evidence of climatic change.

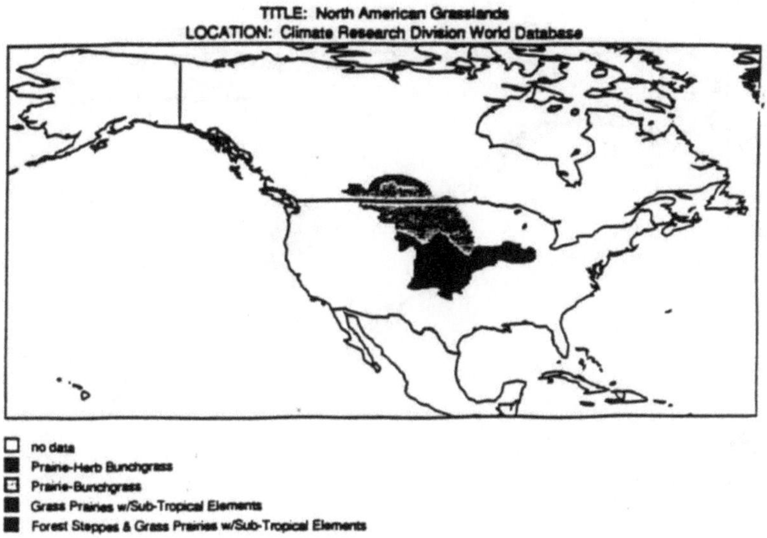

Figure 7. Map illustrating the boundaries of the North American grasslands region based on the Komarov classification.

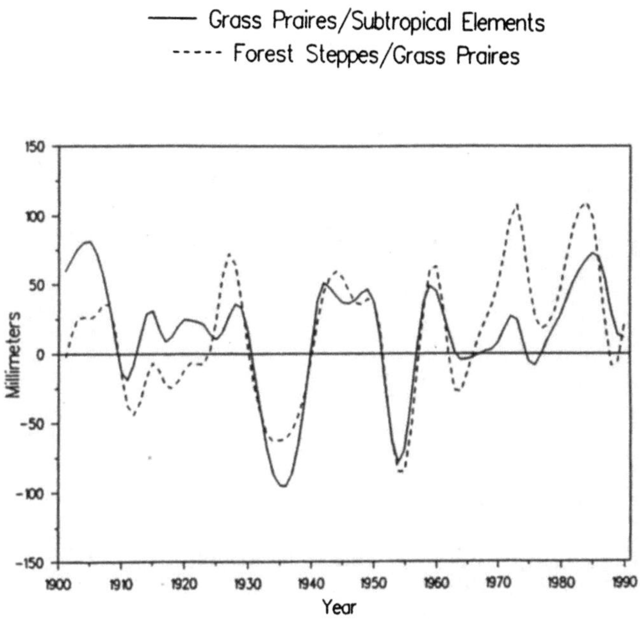

Figure 8. Low-pass filtered time series of annual (January to December) precipitation anomalies in two North American grasslands regions.

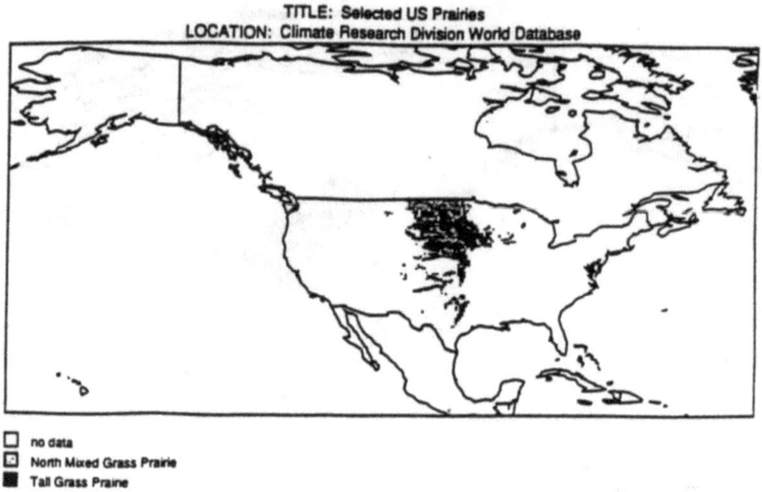

Figure 9. Map illustrating the boundaries of the North American grasslands region based on the Kuchler classification.

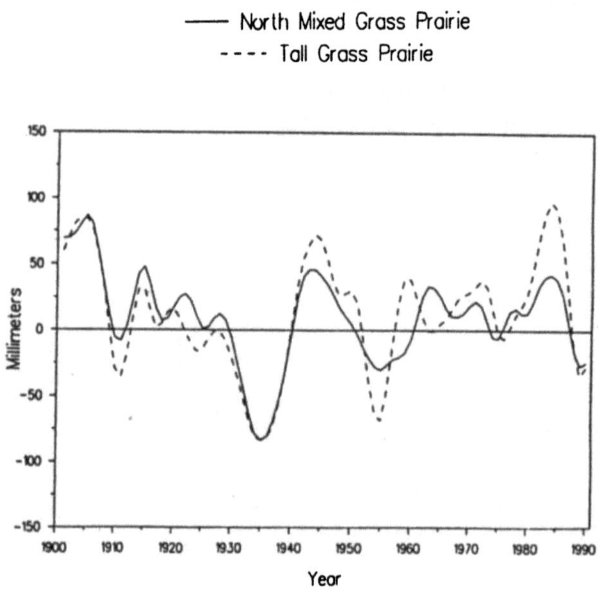

Figure 10. As in Fig. 8, except for the two regions labelled "North Mixed Grass Prairie," and "Tall Grass Prairie."

There are other vegetation type data sets available which yield somewhat different regional boundaries. To illustrate such differences, I have used a vegetation classification data set, based on the work of Kuchler (1967). A map illustrating the boundaries of the two regions considered here is shown in Figure 9 (compare with the map in Figure 7). The smoothed precipitation series for these two of Great Plains grasslands regions are displayed in Figure 10. The geographical domains of these two sets of grassland systems differ in some detail from each other, but their temporal signals overlap considerably. The differences in regional boundaries, notwithstanding, there is good agreement between the respective smoothed series, with the major dry and wet periods since 1900 being common to all the regions.

4. Conclusions

Regionalization approaches depend on the analysis objectives, and it is likely that no one technique will work optimally in all cases. Regional groupings based on characteristic time scales of variability (from the climatological annual cycle to interannual variations) can be useful in studies aimed at understanding the underlying nature of the forcing mechanisms that lead to climatic variability at specified time scales. However, even a classification approach in tropical areas known to be strongly influenced by ENSO extremes, based on quantitative measures of the SO, such as the SOI or SST in the central equatorial Pacific, may lead only to a moderate degree of temporal association. Nevertheless, selecting regions which may behave in an out-of-phase manner with respect to a given ENSO phase, could be useful in studies of the long-term behavior of ENSO teleconnections.

Regional classification based on vegetation types using GIS techniques is found to be useful for defining broad regions of the North American interior with similar low-frequency behavior in annual precipitation. While only rudimentary kinds of classification approaches have been tried in this study, the work thus far indicates that use of GIS tools might be helpful in climate monitoring, as well as in other diagnostic work for other vegetation zones in the U.S., as well as in other parts of the world.

Acknowledgments: I would like to thank Craig Anderson and Jon Eischeid of the Cooperative Institute for Research in Environmental Sciences (CIRES) of the University of Colorado-Boulder for their help in producing the figures.

REFERENCES

Bradley, R. S., Diaz, H. F., Kiladis, G. N. and Eischeid, J. K., 1987 : ENSO signal in continental temperature and precipitation records. Nature, 327, 497-501.

Burg, J. P., 1967 : Maximum entropy spectral analysis. Reprinted in Modern Spectrum Analysis (1978), D. G. Childers (ed.), IEEE Press, NY, pp. 42-48.

Diaz, H. F. and Kiladis, G. N., 1992 : Atmospheric teleconnections associated with the extreme phases of the Southern Oscillation, In Diaz, H. F. and Markgraf, V. (eds.), El Niño, Historical and Paleoclimatic Aspects of the Southern Oscillation. Cambridge University Press, pp. 7-28.

Ghil, M. and Vautard, R., 1991 : Interdecadal oscillations and the warming trend in global temperature time series. Nature, 350, 324-327.

Komarov, 1964: Physical Geographic Atlas of the World. Academy of Sciences of the USSR and Geodetic and Cartographic Institute of the USSR, Moscow, Plate 75.

Kuchler, A. W., 1967 : Potential Natural Vegetation of the Conterminous United States. American Geographical Society, Special Publ. No. 36.

Lau, K.-M. and P. J. Sheu, 1988 : Annual cycle, quasi-biennial oscillation, and Southern Oscillation in global precipitation. J. Geophys. Res., 93 : 10, 975-10, 988.

Nicholls, N., 1988 : El Niño-Southern Oscillation and rainfall variability. J. Climate, 1 : 418-421.

Penland, C., Ghil, M., and Weickmann, K., 1991 : Adaptive filtering and maximum entropy spectra with application to changes in atmospheric angular momentum. J. Geophys. Res., 96 : 22, 659-22, 671.

Rasmusson, E. M. and Carpenter, T. H., 1982 : Variations in tropical sea surface temperature and surface wind fields associated with the Southern Oscillation/El Niño. Mon. Wea. Rev., 111, 517-528.

Rasmusson, E. M. and Wallace, J. M., 1983 : Meteorological aspects of the El Niño/Southern Oscillation. Science, 222, 1195-1202.

Rasmusson, E. M., Wang, X. and Ropelewski, C. F., 1990 : The biennial component of ENSO variability. J. Marine Syst., 1, 71-96.

Thomson, D. J., 1982 : Spectrum estimation and harmonic analysis. Proc. IEEE, 70, 1055-1096.

Trenberth, K. E., 1984: Signal versus noise in the Southern Oscillation. Mon. Wea. Rev., 112, 326-332.

Trenberth, K. E. and Shea, D. J., 1987 : On the evolution of the Southern Oscillation. Mon. Wea. Rev., 115, 3078-3096.

Vautard, R., Yiou, P., and Ghil, M., 1992: Singular-spectrum analysis: A toolkit for short, noisy chaotic signals. Physica D, 38, 95-126.

Wigley, T. M. L., Briffa, K. R. and Jones, P. D., 1984 : On the average value of correlated time series, with applications in dendroclimatology and hydrometeorology. J. Climate Appl. Meteorol., 23, 201-213.

SPATIAL AND TEMPORAL INHOMOGENEITY OF GLOBAL PRECIPITATION DATA

Boris Sevruk
Swiss Federal Institute of Technology
Department of Geography ETH
CH-8092 Zurich (Switzerland)

ABSTRACT
A serious drawback of point precipitation measurement is that it is subject to the systematic errors including especially wind-induced error. This error depends on the particular observational method used, the climate conditions and the gauge site exposure. Since different observational methods are used in different countries and the gauge site exposure in a network varies from exposed to protected, the wind-induced error varies from site to site, causing the precipitation data to be spatially inhomogeneous and non-compatible. Temporal inhomogeneity also can exist owing to changes in observational methods and exposure degrees of a particular gauge site during the observational period. The adjustment for this kind of inhomogeneity using statistical methods does not eliminate the systematic error. Both kinds of inhomogeneities can be eliminated only using the correction procedures. In this way compatible and homogeneous data sets of precipitation for any time period and any size of area can be provided.

KEYWORDS
Precipitation measurement error, correction, inhomogeneity

1. Introduction

Spatial precipitation studies are based primarily on information obtained from a network of point precipitation measurements. A serious drawback of such measurements is that they are biased. These biases are related to the particular observational method used and to environmental conditions affecting the micro-climate and exposure of a given gauge site. The reason for the bias is that the point observation is subject to the systematic error of precipitation measurement. This error has more components causing losses of measured precipitation, the most important of which is considered to be the loss due to wind-field deformation above the orifice of the precipitation gauge. The gauge elevated above ground acts as an obstacle to free airflow and causes aerodynamic blockage, resulting in rising and

NATO ASI Series, Vol. I 26
Global Precipitations and Climate Change
Edited by M. Desbois and F. Désalmand

accelerating airflow over the gauge orifice. Consequently, a part of the small precipitation particles is deflected before it reaches the orifice and instead of falling into the gauge, falls to the ground. Such a loss amounts to 3 to 15 % for rain and up to 100 % for snow. Second in the order of magnitude is the loss due to the wetting of inner walls of the precipitation gauge, which is usually larger than the loss due to splashing or evaporation of water accumulated in the gauge.

The wind-induced error depends on several variables. In addition to the aerodynamic characteristics of a particular type of gauge (Sevruk, 1993) and precipitation structure, there is also the wind speed during the period of precipitation at the level of the precipitation gauge orifice to be considered. The latter depends considerably on the gauge site exposure.

Inhomogeneities of precipitation time series can be caused in a number of ways by the systematic error of precipitation measurement. In particular, changes in the wind-induced losses caused by the replacement of instruments and changes in their elevation heights and the environmental changes at precipitation gauge sites or their surroundings (growing or cutting of woods, urbanization, etc.). In this way the exposure of gauge site can also be affected.

1.1 Problem description

The problem is that from the point of view of systematic measurement error, the adjustment of time series for inhomogeneities represents only half of the job if the spatial analysis of precipitation has to be made as well. What matters here is the compatibility of measurements between the points. Yet compatibility does not exist very frequently due to the spatial variability of systematic error. The latter is affected by the gauge site exposure, wind speed and temperature, latitude and altitude, and by different observational methods used in different countries. In other words, two homogeneous time series, even if each of them was obtained from measurements using the same type of gauge but having different site exposure are not necessarily spatially homogeneous. The systematic error for each of them could be different, as well as the deviation of the measured precipitation amount from "the true precipitation". This is why the adjustment of point precipitation time series for inhomogeneities does not eliminate the spatial inhomogeneities. The adjustment procedures are effective only for reducing the precipitation time series to the level before or after the time of occurrence of the change. Moreover, the reduction is related either to the statistical characteristics of a particular time series or to the time series from the next located stations. Consequently, they eliminate only that part of variability of measured values that is imposed by temporal changes of variables other than those causing the natural variability of precipitation. Thus the systematic error could be reduced or enlarged through the adjustment procedure, but not eliminated; it still exists. Hence the effect of inhomogeneity elimination is relative from the point of view of the "true precipitation", which still remains unknown.

The basic difference between the adjustment of time series for inhomogeneities and the correction of precipitation measurements for systematic error is that while the former eliminates only a partial effect of a change in systematic measurement error due to the change in a particular variable, the latter reduces the measured values of precipitation to the "true precipitation". Consequently, only the correction of point precipitation measurements eliminates all sorts of bias and provides compatible data sets for any time period and any size of area from a local to a global scale.

1.2 Purpose

The present paper is a follow-up to the previous paper of the author on reliability of precipitation measurement as presented during the Workshop on Precipitation Measurement in St. Moritz in 1989 (Sevruk, 1989). One of its aims is to show the progress on the topic of spatial and temporal inhomogeneity of precipitation measurement from the point of view of systematic error as accomplished during the four years since that workshop took place. The highlights include (i) the WMO Solid Precipitation Measurement Intercomparison near completion, (ii) the first, detailed precipitation map of Switzerland corrected for systematic error, (iii) investigations of wind-field around precipitation gauges in a wind tunnel, (iv) applications of computational fluid dynamics to the solution of the problem of wind-induced error, and (v) the analysis of the effects of gauge site exposure. For more details and references see (Sevruk 1993, Sevruk et al 1993, Sevruk and Zahlavova, 1993, Sevruk and Nespor, 1994).

In support of the above, a series of figures will be shown further on to demonstrate the effect of systematic measurement error in general and the exposure degree in particular on temporal and spatial inhomogeneities of precipitation values. In a further paper of the author in this volume, the problems of wind-induced error are discussed.

2. Type of gauge

Figure 1 shows the dependence of wind-induced error on wind speed for snow and selected national types of standard gauges as obtained from the preliminary results of the WMO Solid Precipitation Measurement Intercomparison.

From the comparison of errors for gauges with and without wind shield in Figure 1, it follows that the use of wind shield can reduce the error by half. However, the error is still great, between 20 and 40 % of measured values for wind speed of 5 ms^{-1}, depending on the type of gauge. For instance, the difference in wind-induced error for the Slovak gauge and the Finnish gauge is 60 % and between the two neighbouring countries of Germany and Austria it is 12 %. It is clear now that in such a case the winter precipitation amounts between these counties are not compatible. This is why the spatial inhomogeneities of precipitation values exist between countries using a different type of a gauge.

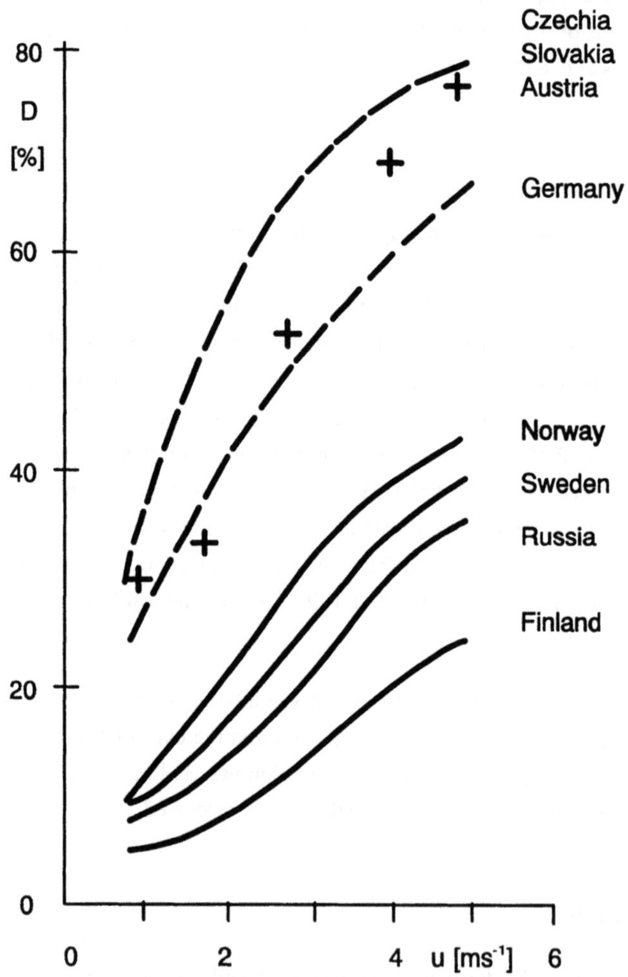

Fig. 1 Wind-induced losses, D, for snow, as a function of wind speed, u. D is expressed by differences between precipitation values as measured by the double-fence international reference gauge (DFIR) and various national standard gauges, with and without wind shield, as indicated in the legend by solid and dashed lines, respectively. Crosses indicate control measurements from other places for the German (Hellmann) gauge.

With the above in mind, it appears that even in the same country and at the same gauge site, the replacement of one type of gauge with another can produce considerable temporal inhomogeneities. This is demonstrated in Figure 2, which shows the double mass curve of annual precipitation at the Geneva Airport.

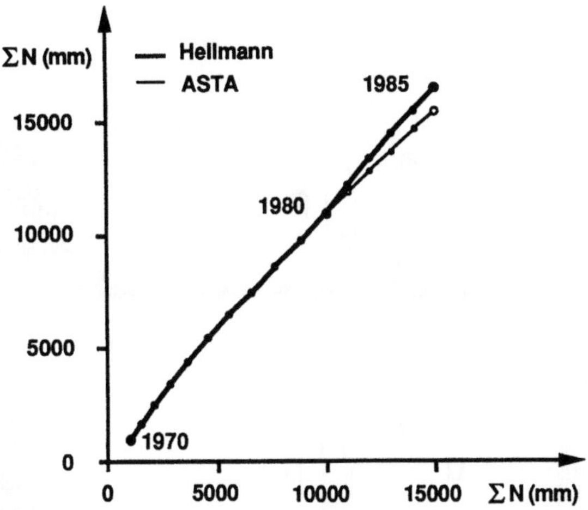

Fig. 2 Double mass curve of annual precipitation series at Geneva Airport. The break in 1980 indicates the replacement of the old, manual Hellmann gauge by the new, heated, tipping-bucket gauge (Sevruk 1989).

In the course of automatization of meteorological observations in Switzerland, the manual gauge of the Hellmann type was replaced at 70 sites by the heated, recording tipping-bucket gauge. Both gauges have more or less the same size, but the latter gauge has a considerably thicker orifice rim to accommodate for the heating device.

Investigations in a wind tunnel indicated that thicker orifice rims cause greater wind acceleration above the gauge orifice and consequently greater wind-induced losses. The respective wind profiles indicate the dependence of the wind speed increase over the gauge orifice on the thickness of orifice rim. (See Sevruk and Nespor, 1994, in this Volume). In any case, the annual precipitation sums for the gauge with a thicker orifice rim in Geneva were up to 15 % smaller than for the manual gauge. Similar differences have also been observed at other gauge sites in Switzerland.

3. Exposure of the gauge site

As shown in Figure 3, the wind speed at the level of the precipitation gauge orifice is greater at an open site than at a protected one. It follows that despite the same wind conditions, changes in a gauge site exposure degree can cause increase or decrease in the measured precipitation values and consequently, temporal as well as spatial inhomogeneities in the time series.

Fig. 3 Dependence of the wind speed at the level of the gauge orifice on the exposure of a gauge site. From the top: Open site, partly open site, partly protected site and protected site. Wind speed is indicated by the length of the arrow. P is precipitation amount as measured in the gauge and alpha is the vertical angle of obstacles. The latter is used to characterize the degree of gauge site exposure, as shown by Sevruk and Zahlavova (1993a).

To illustrate temporal inhomogeneities, the double mass curve in Figure 4 shows the effect of growing woods and the accompanying gradual change of exposure degree on precipitation values for a gauge site originally situated at an open mountainous pass which was later afforested. The differences between the increase in precipitation for the summer and winter seasons reflect the greater effect of exposure on snow measurement than on rain measurements.

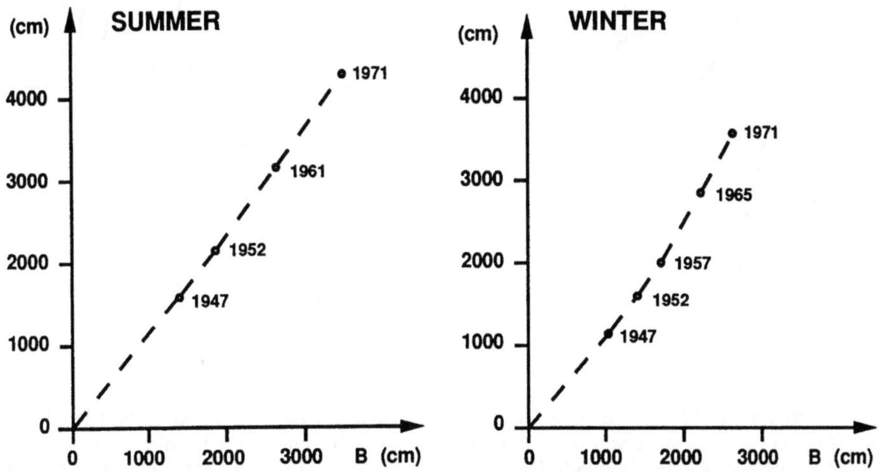

Fig. 4 The effect of growing woods on measured precipitation amounts for the winter (right) and the summer (left) season at a gauge site located in an afforested area, as expressed by the double mass curve of precipitation time series. The breaks are identified by years. According to Sevruk (1973).

Figures 3 and 4 demonstrate that the wind-induced error can account for a considerable amount of spatial variability in precipitation values in a network of gauges having different degrees of exposure. As an example, Figure 5 shows precipitation distribution in a small mountainous basin situated above Lake Geneva in Switzerland in the summer season from June to September.

One set of isohyets in Figure 5 was based on data from elevated storage gauges and the other on the ground-level gauges. The first type of gauge is subject to the wind-induced error but the second is not. As can be seen from the comparison of both sets of isohyets, the spatial distribution of precipitation looks very uniform and appears to be not depending on the altitude if ground-level gauge measurements are considered, but it seems to be very heterogeneous and altitude-dependent for the measurements of elevated storage gauges. Moreover, due to the wind-induced error the isohyets in the last case are also too small.

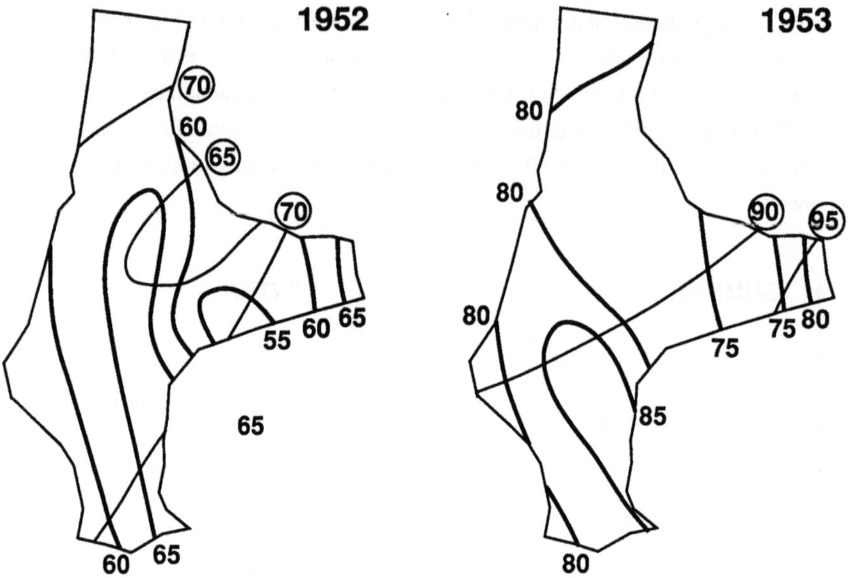

Fig. 5 Comparison of isohyets based on data from elevated (thick line) and ground-level (thin line) storage gauges in a small mountainous basin. The encircled numbers indicate the summer season precipitation amount of ground-level gauges in [cm] from June to September in 1952 and 1953. According to Sevruk (1975).

As mentioned above, the spatial inhomogeneity of precipitation data sets over an area where the same type of precipitation gauge is used as the standard depends on the variability of gauge site exposure. Information on gauge site exposure can be obtained either from the direct measurements of the angle alpha as shown in Figure 3, or station history records (metadata). An assessment of the degree of exposure can be made based on such records, and a classification into four groups as shown in Figure 3, (from the top), is possible as follows: Class 1 - open site, (alpha = 0-5 degree); class 2 - partly open site (alpha = 6-12 degree); class 3 and 4 -partly protected and protected site, respectively (alpha = 13-19 and 20-27 degree). A frequency distribution histogram of exposure classes of 398 Swiss stations for the year 1920, 1950 and 1980 is presented in Figure 6. In the three different years 1920, 1950 and 1980, more than 50 % of gauge sites were partly open (class 2) and the proportion of open stations gradually increased from 10 % in 1920 to 25 % in 1980. As shown in Figure 6, the distribution of exposure classes changes also according to the altitude zone, and the proportion of open stations (class 1) increases with increasing altitude.

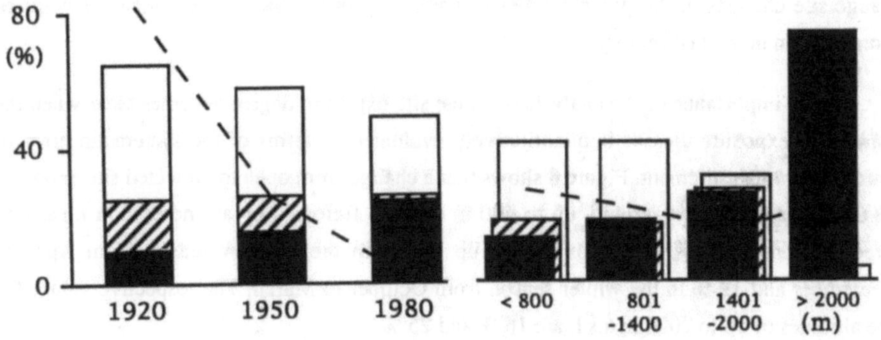

Fig. 6 Frequency distribution histograms of gauge site exposure after 30 and 60 years (left) and for four altitude zones in Switzerland (right). The black column indicates open sites, the empty one partly protected sites and the hatched one protected gauge sites. According to Sevruk and Zahlavova (1992).

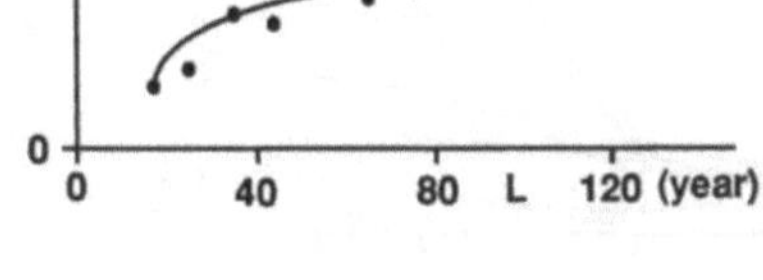

Fig. 7 Dependence of the number, n, of changes in exposure class per gauge site on the length, L, of observational period; average values. According to Sevruk and Zahlavova (1992).

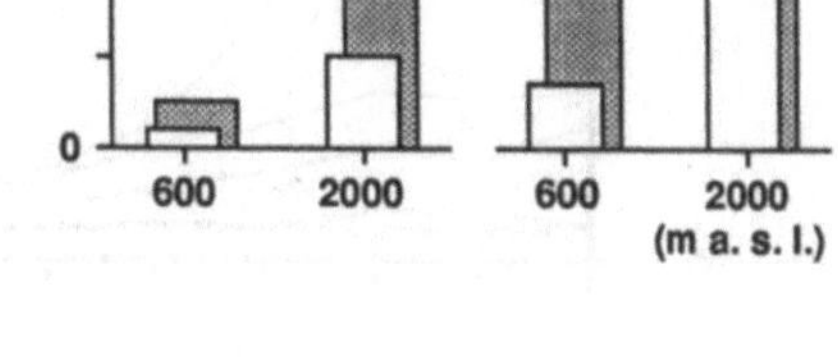

Fig. 8 Comparison of the mean seasonal wind-induced error for open (black) and protected (empty) gauge sites at altitudes of 600 and 2000 m a.s.l. in Switzerland for the Hellmann precipitation gauge. The summer season is from April to September and the winter season from October to March.

Figure 6 indicates a considerable variability of gauge site exposure with time and altitude in Switzerland. This is, of course, the cause of temporal and spatial inhomogeneities of precipitation data. In addition, there is on average one exposure change each 27 years per gauge site, as shown in Figure 7. However, the rate of changes is increasing for the gauge sites having longer observational periods than 100 years. This documents the very frequent gauge site changes at the pioneer times of nationally organized network measurements of precipitation in the last century.

The importance of the analysis of gauge site exposure degree becomes clear when the change in exposure classes is quantitatively evaluated in terms of the systematic error of precipitation measurement. Figure 8 shows that a change from open to protected site exposure in the lowlands in Switzerland, up to 600 m a.s.l., is followed by an increase in measured precipitation amounts of approximately up to 5% in the summer season from April to September and 12 % in the winter season from October to March. The respective values for the altitudes of up to 2000 m a.s.l. are 16 % and 25 %.

4. Season and altitude

The seasonal and altitudinal dependence of systematic error of precipitation measurement is evident from Figure 9, where the annual pattern of this error for five Swiss stations is plotted.

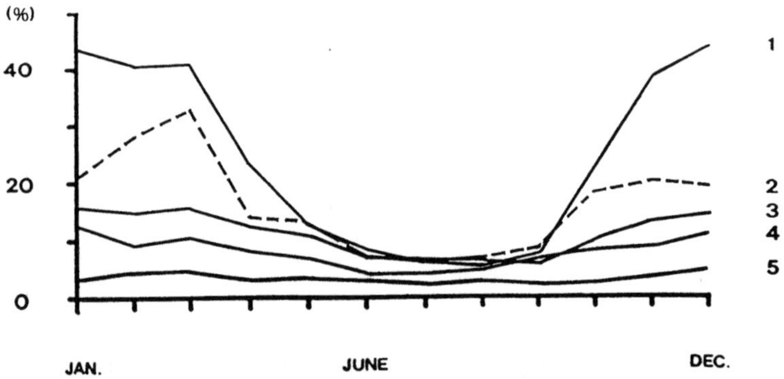

Fig. 9 Monthly mean values of the systematic error of precipitation measurement for five representative Swiss stations: 1 - San Bernardino (1628 m a.s.l.); 2 - Davos (1580 m a.s.l.); 3 - Geneva (430 m a.s.l.); 4 - Zurich (416 m a.s.l.) and 5 - Locarno (336 m a.s.l.).

The Locarno station, situated in a valley in the southern part of Switzerland, shows minor error and a small seasonal pattern. This is due to prevailing heavy rainfalls in the summer and a generally small amount of snow in the winter. In contrast, the error and the seasonal pattern is large in the mountain stations of Davos and San Bernardino situated in altitudes above 1500 m a.s.l. because of frequent snowfalls, even in the summer. The seasonal pattern is generally characterized by large errors in the winter and small ones in the summer. In addition, the San Bernardino station is open and consequently the error is larger than for the partly protected Davos station. However, there are different error values even between the two lowland stations Zurich and Geneva which have almost the same altitude. The reasons are not only the different class of exposure of both stations (Zurich is less exposed than Geneva) but also the generally stronger winds in the western part of Switzerland.

5. Conclusions

The preceding examples show that for studies of spatial distribution of precipitation, the spatial inhomogeneities of precipitation time series as caused by the spatial variability of the systematic error of precipitation measurement should be taken into consideration. In a particular country, where one type of gauge is used as a national standard, the spatial inhomogeneities depend mainly on the spatial variability of exposure, altitude and climatic conditions of gauge sites. In the cold season of the year they are larger than in the warm season. The temporal inhomogeneities depend above all on the gauge site exposure changes. On the regional and global scale, the type of precipitation gauge and gauge parameters such as the shape of gauge and orifice rim thickness as well as the use of wind shield also have to be considered.

The spatial and temporal inhomogeneities of precipitation data sets can be greatly reduced if the correction procedures for the systematic error of precipitation measurements are applied. Such procedures exist already for some types of gauges and should be drawn up as soon as possible for the other ones. As an example of the application of corrections, the Swiss map of corrected precipitation can be used (Kirchhofer and Sevruk 1992).

REFERENCES

Kirchhofer, W. and B. Sevruk, (1992) Mean annual corrected precipitation amounts 1951-1980. In: M. Spreafico, R. Weingartner and Ch. Leibundgut (Eds.) Hydrological Atlas of Switzerland, Part 2.2, Landeshydrologie und -geologie, Bern.

Sevruk, B. (1973) The effect of growing woods on the precipitation measurement in the Baye de Montreux basin (in German). Wetter und Leben, 25(1-2).

Sevruk, B. (1975) Inaccuracy of precipitation measurements - A serious problem of water resources instrumentation. In: Proc. 2nd World Congress on Water Resources, New Delhi, Vol. III, 429-440.

Sevruk, B. (1989) Reliability of precipitation measurement. In: B. Sevruk (Ed.). Precipitation measurement. Swiss Federal Institute of Technology, ETH, Zurich, 13-19.

Sevruk, B., W. Kirchhofer, R. Tihlarik and L. Zahlavova, (1993) Precipitation corrections in Switzerland. In: B. Sevruk and M. Lapin (Eds.) Proc.Precipitation and Evaporation, Vol. 1. Slovak Hydrometeorological Institute, Bratislava and Swiss Federal Institute of Technology, Zurich, 155-156.

Sevruk, B. und V. Nespor (1994) The effect of dimensions and shape of precipitation gauges on the wind-induced error. (See this Volume).

Sevruk, B. and L. Zahlavova, (1992) Statistics for precipitation gauge site exposure. In: F. Zwiers (Ed.) Proc. 5th International Meeting on Statistical Climatology. Canadian Climate Centre, Toronto, 383-386.

Sevruk, B. and L. Zahlavova, (1993) Estimation of past changes in precipitation gauge site exposure. In: B. Sevruk and M. Lapin (Eds.) Proc. Precipitation and Evaporation, Vol. 1. Slovak Hydrometeorological Institute, Bratislava and Swiss Federal Institute of Technology, Zurich, 181-182.

THE EFFECT OF DIMENSIONS AND SHAPE OF PRECIPITATION GAUGES ON THE WIND-INDUCED ERROR

Boris Sevruk and Vladislav Nespor
Swiss Federal Institute of Technology
Department of Geography ETH
CH-8092 Zurich, Switzerland

ABSTRACT

The results of various kinds of studies into the physics of precipitation gauges including field intercomparison measurements, wind tunnel experiments and simulation using computational fluid dynamics are reviewed. The focus is on the effects of precipitation gauge construction parameters on the wind-induced error regarding practical applications such as the development of correction procedures or the selection of suitable design criteria for the sensitive parts of precipitation gauges. In this context, the following gauge parameters are to be noted: the proportions and the shape of both gauge body and orifice rim. In addition, the use of wind shielding devices is one of the most critical factors. The ideal precipitation gauge should have a streamlined body in a shape of a flat plate protected by the wind shield. Yet such gauges are not used at all. A common precipitation gauge consists of an upright cylinder having rather poor aerodynamic properties. Therefore, with regard to a considerable systematic error due to wind, the measured precipitation values have to be corrected.

KEYWORDS

Precipitation measurement, precipitation gauge dimensions, aerodynamics, wind-induced error, simulation, wind tunnel

1. Introduction

As the development of accurate measuring instruments is limited by financial and physical constraints, the measurement of meteorological variables is carried out using less expensive and more common methods, simple instruments and uncomplicated arrangements. As a consequence, the meteorological observations are subject to systematic errors. These errors fill the gap between the "true" and measured values of a particular variable. They are caused when the physical characteristics of the instruments interfere with other variables

NATO ASI Series, Vol. I 26
Global Precipitations and Climate Change
Edited by M. Desbois and F. Désalmand

involved in the measuring process, and depend on variables which include construction parameters of instrument and weather conditions.

In the case of a common precipitation gauge, which mainly consists of an upright cylinder elevated above ground, the main interfering processes are the wind-field deformation above the gauge orifice, wetting of the inner surfaces of the gauge, evaporation of accumulated water in the gauge between the period of precipitation event and the time of observation, and splashing of raindrops or blowing of snow from and into the gauge.

The phenomenon of wind-induced error of precipitation measurement has attracted the attention of many scientists through the two last centuries. It is a very complex, physical matter. It includes the interaction of the body of a precipitation gauge with the wind flow and the solid and liquid particles falling through the air. The trajectories of particles become distorted in a wind through the displacement and acceleration of wind flow over the top of the gauge as caused by the aerodynamic blockage of the gauge body. The lighter particles are carried beyond the gauge opening, resulting in a reduced catch. The extent of reduction depends on the falling velocity of particles, wind speed and the aerodynamic properties (drag of the air) of a particular type of gauge. For the same gauge, the reduction is smaller for big raindrops but several times greater for light snow and it increases with increasing wind speed. This process of wind-induced error is described by a set of differential equations of fluid dynamics (three-phase flow; gas, liquid and solid phase) and is still not fully understood; in particular, the effect of turbulence has not been studied sufficiently.

A practical solution to the problem at hand would be to place the gauge in a pit with the orifice flush with the ground. However, for various reasons (e. g. drifting snow), the gauges are elevated 20 to 200 cm above ground. The wind-induced loss amounts to 3 to 10% for rain and up to 80% for snow, and even more in the mountains. The diversity of error magnitudes is caused by many variables. Nevertheless, under the same environmental and observational conditions, the instrumental, construction variables play the most important role. The better the aerodynamic properties of a gauge, the smaller the wind-induced error. In this context, the following gauge parameters are to be noted: the proportions and the shape of both gauge body and orifice rim. In addition, the use of wind shielding devices is one of the most critical factors.

The purpose of this contribution is to review the results of various kinds of studies into the physics of precipitation gauges, including field intercomparison measurements, wind tunnel experiments and simulation using computational fluid dynamics. Updated information on the topic, providing a good insight on the problem, can be found in Hall et al. (1993a,b) and Sevruk (1993) (see Culling, 1993), as well as in Sevruk et al. (1989, 1991, 1993), Nespor (1993) and Nespor at al. (1993).

2. Construction parameters of gauges

From the point of view of fluid dynamics, the details of the shape and dimensions of a precipitation gauge are important variables responsible for the extent of the aerodynamic blockage effect of a gauge. Figure 1 shows various shapes of gauge body, whereby the gauge 1 has the worst and gauge 6 the best aerodynamic shape. This can be seen from the extent of the deformation of streamlines and incidence angel of falling particles immediately above the gauge orifice. Only a few gauge types used worldwide at present are of an aerodynamically more suitable conical shape and none have the best shape of a flat dish as shown in Figure 1 (gauge 6). The great majority of common precipitation gauges consists of an upright cylinder usually having the height/diameter ratio between 1.4 and 3.0 (gauge 3 in Fig. 1). The height of most gauges varies from 20 to 70 cm (up to 90 cm) and the diameter from less than 10 to 25 cm. This represents a substantial bulk with rather worse aerodynamic properties. A few types of gauge are of combined form; upright cylinder over a funnel which has a slightly better shape than the cylinder alone.

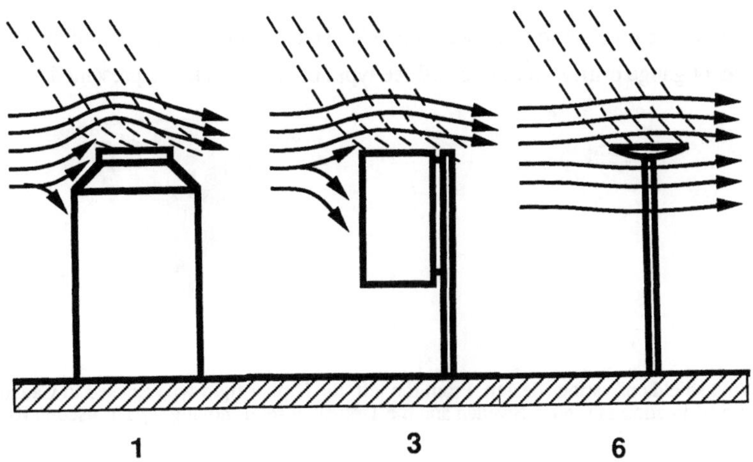

Fig. 1 Shapes of precipitation gauge body and the respective wind fields. The number 1 indicates the shape having the worst aerodynamic properties and the number 6 having the best ones. Arrows show the streamlines and the dashed lines the trajectories of precipitation particles.

The most important construction parameter regarding the wind-induced error seems to be the orifice rim used to reinforce the upper end of the gauge to prevent its deformation. It is a very sensitive part of the gauge immediately affecting the development of the wind field above the gauge orifice. Orifice rims differ in shape and thickness, depending on the type of gauge. Figure 2 shows various shapes and thicknesses of orifice rims used at present.

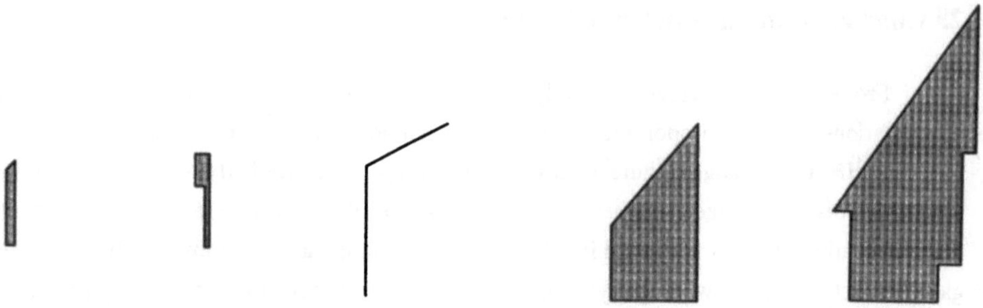

Fig. 2 Shapes and proportions of the precipitation gauge orifice rim. Note the differences in thickness. The thickest orifice rim is used on heated tipping-bucket gauges. It contains the heating device.

A thin and simple sharp rim, as shown in the left in Figure 2, giving the best performance is not used at all. Slight deviations from the parameters of the orifice rim can cause considerable changes in characteristics of the wind field above the gauge orifice, and consequently different precipitation values.

A further parameter to be noted is the wind shield, which is used to reduce wind speed at the level of gauge orifice. In Figure 3, three types of wind shield are presented.

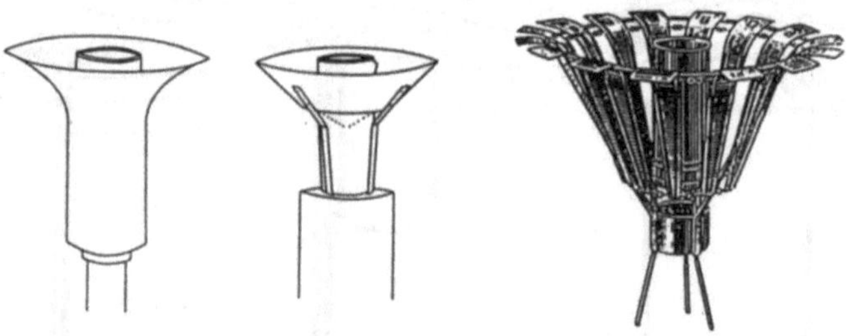

Fig. 3 Wind shields of various design. From the left; the Nipher rigid type as used in Canada and its shortened modification as used in Sweden and the Tretyakov semi-moving type as used in Russia.

It follows from the points set out above that aside from the pit gauge, the ideal model of elevated precipitation gauge should have a body in a shape of a flat plate, allowing for a small drag of air, as for instance, the inverted Frisbee as shown in Figure 1 (gauge 6). However, the pit gauge is not used by meteorological services and only occasionally by special agencies, and in addition it is not suitable for snowfall measurements. Moreover the flat, streamlined gauge has many problems caused by out-splashing, heating during the cold season, and ejection of precipitation which has already entered the gauge by a wind-driven circulating flow inside the gauge, etc. Therefore, with the exception of some experimental and special gauges, the body of almost all types of common gauges consists of a deep upright cylinder as shown in Figure 1 (gauge 6).

The gauge is fixed on a pile or posed on the ground so that the gauge orifice is elevated above ground to be protected against in-splash and drifting snow. The installation height varies from 0.2 to 2.0 m and in the mountains even up to 4-5 m. The higher the gauge, the greater the wind speed and the resulting wind-induced error.

Other sources of systematic errors such as evaporative and wetting losses are less spectacular from the physical point of view and generally less consequential, particularly in the winter and for gauges having a fixed funnel and protected container.

3. Methods

To study the physical laws controlling the phenomenon of wind-induced error, four methods and their combinations can be applied. The first insights into the behaviour of precipitation gauges were acquired already during the period of initial measurement of precipitation in the 18th century, through more or less passive experiment by observation and deduction. The introduction of field intercomparisons of precipitation measurements by different types of installation and gauges in the 19th century represented a significant step forward. Further, in this century, the wind tunnel tests contributed a lot to develop more suitable, aerodynamic shapes of gauges and protective structures against wind such as wind shields and fences. Yet the greatest progress can be expected from the application of computational fluid dynamics used quite recently for the solutions of problems in science and in actual practice.

Each of the above-mentioned methods has its merits, uncertainties and restrictions. This is why a verification of results by independent methods is still necessary and why a combination of more methods will probably yield best results.

3.1 Intercomparison measurements

Thank to passive experiments, the construction of precipitation gauges was improved as early as in the 18th century, particularly regarding the evaporation protection of accumulated water in the gauge. The main purpose of intercomparison measurements of precipitation in the last century was to select the best gauge and the most suitable conditions of installation. Gauges of different types were situated near each other at the same site. Generally speaking, the gauge showing more precipitation was considered to be better. Sometimes the reasons for the better or worse performance of a particular gauge were analyzed taking into account the different construction parameters. Hence, the first ideas on the possible role of gauge parameters and their suitable design relative to the wind-induced error emerged. Because of a variety of construction parameters involved which might have different or even opposite effects on precipitation catch, the correct interpretation of the results of intercomparison measurements from gauges of various shapes, diameters, heights, etc. sometimes proved difficult. For instance, a shorter gauge collector can create, on the one hand, better aerodynamic conditions and at the same time the wetting losses also could be smaller, but on other hand, the reduced depth of collector can provoke out-splash, and in spite of the above-mentioned improvements, reduced gauge catch.

In view of the rather restricted possibilities of theoretical studies, the practical impact of intercomparison measurements can be still substantial. This is one of the reasons why they are used at present with success, for example to derive empirical relationships of various kinds. If they are supported by measurements of other meteorological parameters such as wind speed and intensity of precipitation and temperature, a statistical model can be derived describing the relative effect of a particular gauge parameter on the wind-induced error. And in addition, if the WMO reference measurements are included using the pit gauge for rain and a double fence (DFIR) for snow measurements, correction procedures for the wind-induced losses can be established for a particular type of gauge.

3.2 Wind-tunnel experiments

More theoretical background on the wind-induced error can be obtained from the results of wind-tunnel experiments. In addition to the visualization of stream lines and the trajectories of particles, basic flow characteristics such as wind vectors (velocity and directions), intensity of turbulence and the replacement height over the gauge orifice can be measured under stable conditions. The measurements are made in sensitive positions such as the windward, leeward and the centre of gauge orifice using a pulsed-wire anemometer. In this way, the laws of wind field deformation can be studied. The simulation of particles and the measurement of gauge catch can also be arranged.

Wind-tunnel experiments are more convenient than the field intercomparison measurements. In addition, they are able to not only confirm but also in part explain the results of the field intercomparison measurements.

3.3 Computational fluid dynamics, CFD

The real breakthrough in the study of the physics of precipitation gauges was achieved quite recently by applying the advanced techniques of computational fluid dynamics, CFD.

The whole phenomenon of wind-induced error of precipitation measurement, including also the particles, could be dealt with directly using numerical simulation. As shown by Nespor (1993) this can be made in three steps: (i) computation of the wind-field around the gauge using the numerical simulation of a single-phase flow, (ii) computation of the precipitation particle trajectories in the resulted velocity vector field, and (iii) determination of the wind-induced error from relative areal densities of particles in the level of the gauge orifice. In contrast to the field intercomparison measurements and to the wind tunnel studies, this procedure allows the experimental conditions to be changed immediately and in a very convenient manner.

Unfortunately studies of this kind are still rare. Part of reason might be the lack of attractiveness of this subject to organizations having access to the expensive hardware and software required.

The absence of complex computer simulation studies of precipitation gauges and the inconclusive nature of many of the intercomparison measurements explains why the material as reviewed further in this report originates mainly from wind-tunnel experiments and is not generally based on systematical studies. It is hoped that the gaps will be closed in the next future, resulting in a complete picture of precipitation gauge behaviour in wind during precipitation.

This review is divided into paragraphs dealing separately with some characteristics of the wind field and following precipitation gauge construction parameters: the body of a gauge, including the height/diameter ratio, the orifice diameter and thickness, as well as the shape and wind shield. An empirical model of the wind-induced error is also presented.

4. Results and discussions

4.1 Wind field characteristics

The wind field around the English standard gauge Mk2 and the contour lines of normalized velocity are shown in Figure 4. The arrows indicate the velocity vectors, that is the direction and the speed of airflow, resulting from numerical simulation using the PHOENICS software (IKOSS, GMBH, Stuttgart, Germany). The airflow at the windward side of the gauge is directed upwards and is accelerated. Wind speed increases roughly up to the centre of the gauge orifice where the maximum increase of approximately 30% is obtained. At this point, the direction of airflow changes downwards and wind speed decreases, causing a wind-driven circulating flow inside the gauge and development of eddies in and behind the gauge. The results are based on flow simulation and agree well with results from wind tunnel measurements as shown by Nespor (1993).

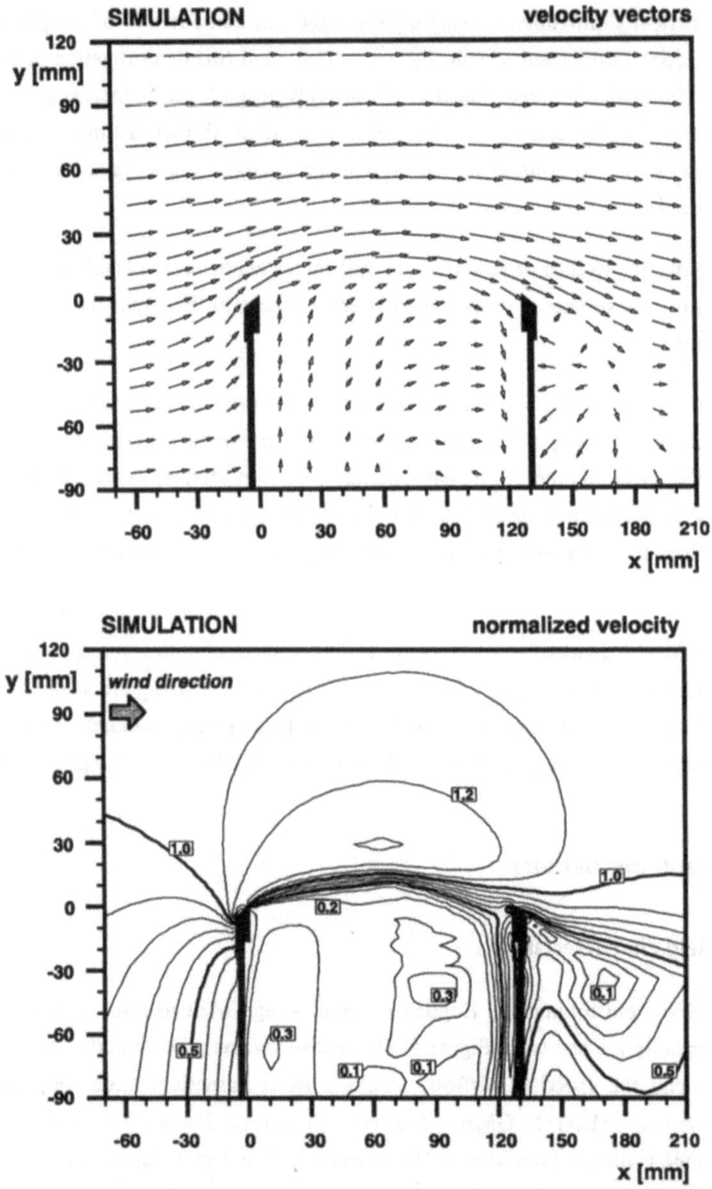

Fig. 4 Wind field (top) and contour lines of normalized velocity (below) based on flow simulation over the orifice of the Mk2 gauge using the PHOENICS program. According to Nespor (1993).

The profiles of normalized wind speed and intensity of turbulence situated windward and leeward of the gauge orifice and at the centre for two different gauges are shown in Figure 5. They provide a more detailed insight into the structure of the wind field above the gauge orifice. It can be seen, for instance, that in addition to the maximum wind speed increase, the height above the gauge at which it occurs differs considerably, not only between the types of gauges, but also between the profiles situated at different positions. The ratio of this height and orifice diameter is called displacement. The wind speed increase and displacement are the basic wind field characteristics used to evaluate the gauge performance. The smaller they are the better.

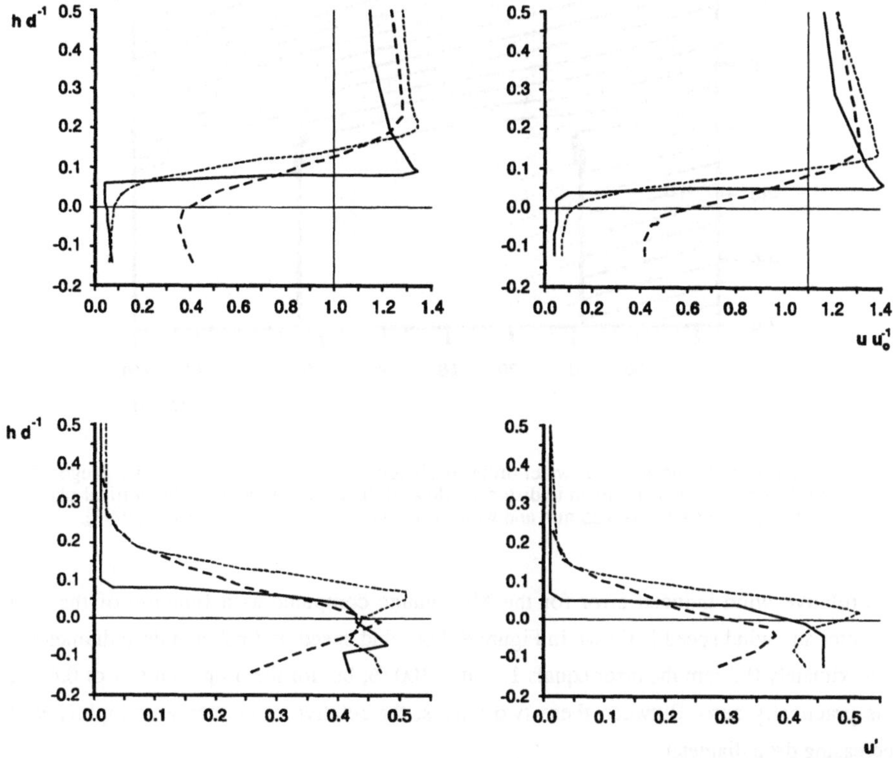

Fig. 5 Comparison of normalized profiles of horizontal wind speed (above) and intensity of turbulence (below) over precipitation gauges with thin (left) and thick orifice rim (right), according to Sevruk et al. (1993). The solid line indicates the windward profile, dense point line the centre and dashed line the leeward profile: d indicates the orifice diameter; h is height above the orifice; u is profile wind speed; u_o is the free wind speed in the wind tunnel and ù is turbulence intensity.

An example of movement of water drops in the numerically simulated wind field over the Mk2 gauge is shown in Figures 6 and 7. The wind speed is 3 m/s and the drop diameter 0.25 mm. Due to the wind field deformation water drops are deviated from their original paths as demonstrated in Figure 6. This results in their areal redistribution as shown in Figure 7 by contour lines of relative areal density of drops in the horizontal plane level with the gauge orifice. This relative areal density tends to be lower in the windward part and over the gauge orifice, and higher in the lee of the gauge.

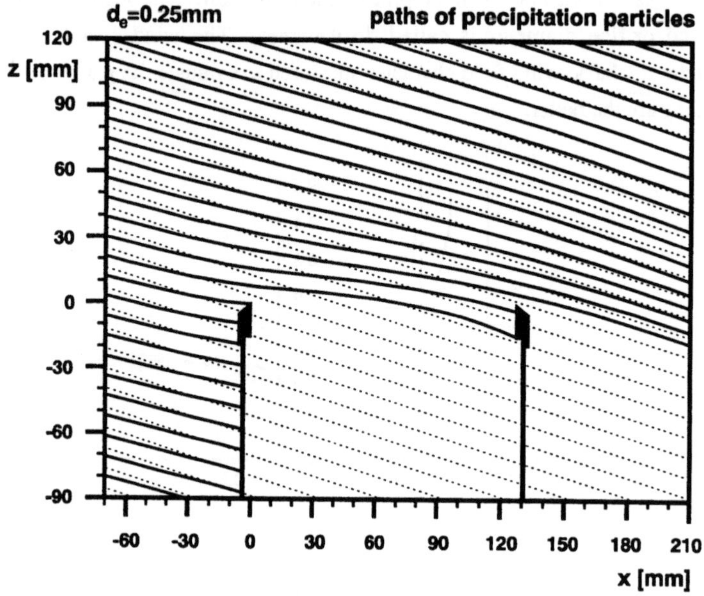

Fig. 6 Comparison of trajectories of water drops as altered by the flow around the Mk2 gauge (solid lines) with the original trajectories in undisturbed flow (dotted lines) plotted in the vertical plane of symmetry. The drop diameter is 0.25 mm and wind speed 3 m/s. According to Nespor (1993).

The relative wind-induced error for the Mk2 gauge computed as a function of the drop diameter and wind speed is shown in Figure 8. For wind speed of 3 m/s and drop diameter of approximately 0.2 mm the error equals 1, that is 100 %, but for the drop diameter of 0.6 mm it is practically zero. Between these two values, the relative error increases rapidly with decreasing drop diameter.

4.2 Height/diameter ratio

As pointed out by Sevruk et al. (1989), the increase of wind speed above the gauge orifice depends on the gauge type, that is on the construction parameters. One of these parameters, affecting the extent of the blocking of a gauge, is the ratio of cylindrical gauge height to the diameter. Hall et al. (1993a) investigated this ratio in the wind tunnel. They found a rapid increase of wind speed above the orifice centre with increasing ratio up to 0.5.

Beyond this ratio value only a modest rate of increase was observed. The maximum increase of wind speed was 25%. Along with wind speed the displacement also increased. Its maximum value was 0.2.

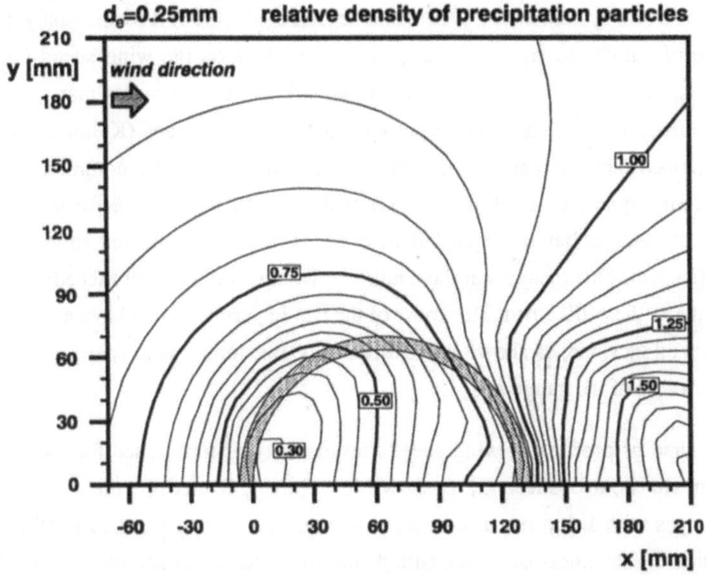

Fig. 7 Contour lines of the relative number of falling drops as compared with the free air stream. Because of the symmetry the gauge rim is indicated by a half circle. The free wind speed is 3 m/s.

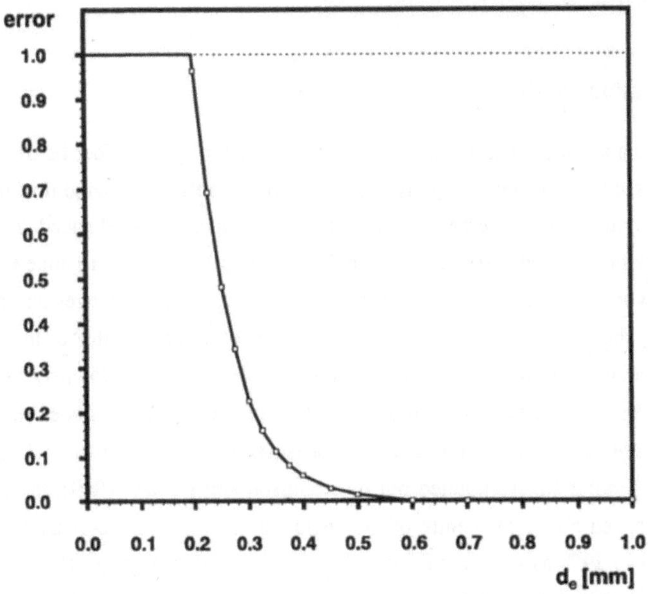

Fig. 8 Wind-induced error for the Mk2 gauge as a function of the drop diameter, d_e. Wind speed is 3 m/s. According to Nespor (1993).

4.3 Orifice area and rim thickness

Almost all types of gauge used worldwide at present have a height/diameter ratio greater than 0.5, and in most cases it varies between 1 and 3 (Sevruk and Klemm, 1989). In the light of the results by Hall et al. (1993a) and Sevruk et al. (1993) it seems firstly, that for a common type of gauge the effect of the gauge orifice area on the wind speed increase is small. There is, however, evidence from intercomparison measurements that small orifice gauges (less than 70 cm^2!) catch more precipitation than large ones (Kalma et al., 1969). Secondly, the maximum increase of wind speed above the centre of a common gauge is at least 25%. According to Sevruk et al. (1989), it was between 30 and 45% for seven types of national, standard gauges having height/diameter ratios from 1.5 to 3.2 and various sizes and shapes of orifice rim. This means that the additional wind speed increase exceeding the value of 25% was due to the orifice rim. Thickness of the rim in particular can have a serious effect on the wind speed increase, as shown in Figures 5. The maximum wind speed increases with increasing thickness of the gauge rim.

The course of profiles of wind speed and intensity of turbulence for gauges with a thick or a thin rim is also different, as shown in Figure 5. Moreover, the displacement is greater for gauges with a thin rim as compared with gauges having a thick rim (Fig. 5). It is clear that beside the thickness of orifice rim, its different shapes, as shown in Figure 2, have to be considered as a further, important parameter affecting the wind field characteristics. The respective studies are still not been conducted.

4.4 Shaping of gauge body

Different shapes of gauges are shown in Figure 1. A sloping shoulder below the gauge orifice, as can be seen for example on gauge 1, tends more to deflect the wind upwards than a common upright cylinder, thus increasing the vertical component of wind speed. Therefore it can cause larger wind-induced losses as compared with an upright cylinder gauge. According to Jones (1969), who evaluated field intercomparison measurements of gauges having a short and large sloping shoulder in the U.S.A., the loss varied between 2 and 6%. In contrast, a gauge which has a shape of an inverted cone, tends to deflect the wind downwards from the orifice, and because of reduced wind speed above the gauge orifice, to reduce the loss. This was conformed either by field intercomparison measurements and wind-tunnel tests or by numerical model simulation. As pointed out by Robinson and Rodda (1969), the maximum increase of wind speed above the centre of the orifice of such a gauge, consisting only of a shallow funnel, was 29% as compared with 37% for a cylindrical gauge (Mk2). The rain catch was larger than that of cylindrical gauges for light showers but smaller for heavy or prolonged rains, thus indicating that splashing was taking place. According to Folland (1988), a simple numerical model indicated more catch for a flat "champagne glass gauge" than for the cylindrical gauge. Conclusive results concerning the better performance of gauges having the more suitable, aerodynamic shape of an inverted Frisbee, as for example the gauge 6 in Figure 1, were obtained from the series of wind-tunnel experiments carried out in Warren Spring Laboratory in the UK by Hall et al. (1993a,b). The maximal increase of wind speed was 12% and the displacement was only 0.07.

4.5 Wind shield

Wind shields help to displace the airflow around the gauge downwards. Moreover, they have a considerably larger diameter than the gauge, and in this way they reduce the height/diameter ratio. Consequently, wind speed increase above the gauge orifice is small and does not exceed 15% above the gauge orifice centre. According to the contour lines of wind speed over wind shields of different design as published by Goodison et al., (1983), Bolshakhov (1974), Tammelin, (1982) and Lindroth (1991), it looks as if the gauge is shifted leeward into the wind field of the wind shield, in the zone of lower wind speeds as demonstrated for instance in Figure 4. However, according to Hall et al. (1993a), who investigated different shapes of gauges and wind shields in the wind-tunnel, the shields do not reduce the displacement. The best results with almost no wind increase and no displacement were obtained from streamlined, i. e. shallow, Frisbee-like gauges equipped with a simple, rigid wind shield. This is a very promising development.

4.6 Empirical model of wind-induced error

The state-of-the-art of precipitation gauges from the aerodynamics point of view is that they have rather poor aerodynamic properties, and this situation will not change in the near future. Therefore, with regard to a considerable systematic error due to wind, the measured precipitation values have to be corrected. Since the correction procedures derived using the computational fluid dynamics are lacking, empirical models have been developed. They are based on intercomparison measurements in the field as mentioned in sub-section 3.1. A model of this type explaining the differences in wind-induced losses between two different types of gauges in terms of wind speed and intensity of rain is presented in Figure 9.

The diagram in Figure 9 incorporates 3 one-parameter curves relating the percentage difference to intensity for three wind speed intervals. The overall tendency of differences is obvious. They increase with increasing wind speed and decreasing intensity. Below a certain threshold value and despite the practically unchanged wind speed a sharp increase of differences exists with decreasing intensity for each wind speed interval. Above the threshold value there is only a slight effect of intensity and all three curves appear to run parallel. The course of curves is in any case similar to that as presented in Figure 8. Here, the intensity of rain substitutes the drop size. This seems to conform at least the possibilities of using the computational fluid dynamics to simulate the wind-induced error.

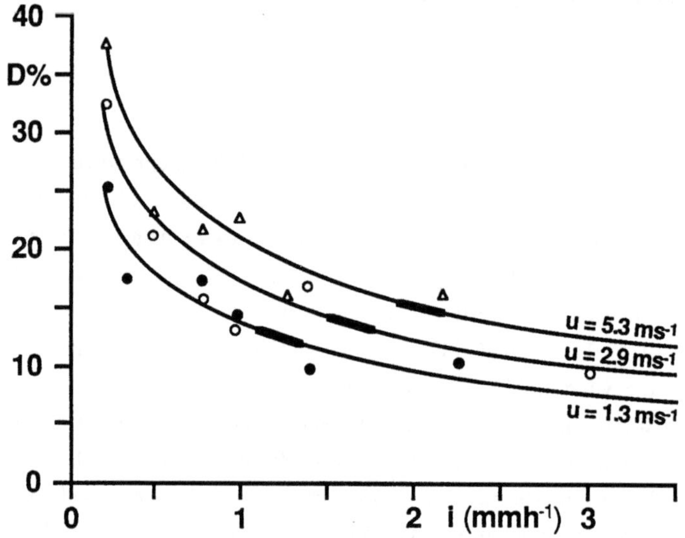

Fig. 9 The dependence of percentage difference, D%, between the precipitation values of different gauges on intensity, i, of precipitation and wind speed, u. Short and heavy lines indicate threshold values of intensity i, below which the difference starts to increase more sharply. According to Sevruk and Tettamanti (1993).

The reason for the considerable differences between the catches of two gauges mentioned above could be the thicker orifice rim of a gauge showing less precipitation (Sevruk 1993).

5. Conclusions

Precipitation measurement is subject to systematic errors. They are caused by the interference of physical characteristics of precipitation gauges with other variables accompanying the measurement process. The most important is the wind-induced error. To study the behaviour of a gauge in the wind more methods can be used. The greatest progress can be expected from the application of computational fluid dynamics. The knowledge of precipitation gauge physics assists the designer of new gauges, and helps to assess the reliability and possible error magnitude of existing types of precipitation gauge and enables to derive the correction procedures for systematic error of precipitation measurement.

REFERENCES

Bolshakhov, A. B., (1974) Experimental investigations of the aerodynamics properties of the wind shields of precipitation gauges. Soviet. Hydrol.: Select. Pap. 4, 2587-262.

Couling, S. (Ed.), (1993) Measurement of airborne pollutants. Butterworth Heinemann, Oxford, 301 p.

Folland, C. K. , (1988) Numerical models of the raingauge exposure problem, field experiments and an improved collector design. Q. J. Roy. Meteorol. Soc., 114, 1485-1516.

Goodison, B. E., V. R. Turner and J. R. Metcalfe, (1983) A Nipher type shield for recording precipitation gauges. In: Proc. 5th Symposium on Meteorological Observations and Instrumentation, Toronto, American Meteorological Soc., Boston, 21-26.

Hall, D. J., J. G. Irwin, B. H. Stone and S. L. Upton, (1993a) Aerodynamics consideration in precipitation collector design. In. S. Couling (Ed.) Measurement of airborne pollutants. 104-148, Butterworth Heinemann, Oxford.

Hall, D. J., S. L. Upton, and G. W. Marsland, (1993b) Improvements in dust gauge design. In. S. Couling (Ed.) Measurement of airborne pollutants. 171-217, Butterworth Heinemann, Oxford.

Jones, D. M. A. (1969) Effect of housing shape on the catch of recording gages. Month. Weath. Rev., 97(8), 604-606.

Kalma, J. D. , J. Lomas, M. Thaller and Y. Shashova. (1969) An accurate small orifice rain gage. Water Resour. Res., 5(1), 300-305.

Lindroth, A. (1991) Reduced loss in precipitation measurements using a new wind shield for raingauges. J. Atmosp. Ocean. Technol., 8(3), 444-451.

Nespor, V. (1993) Numerical simulation and wind tunnel measurements: Aerodynamics of precipitation gauges. In: B. Sevruk and M. Lapin (Eds). Proc. Precipitation and Evaporation, Slovak Hydrometeorological Institute, Bratislava and Swiss Federal Institute of Technology, Zurich, Vol.1, 127-133.

Nespor, V , B. Sevruk, R. Spiess and J.-A. Hertig. (1993) Modelling of wind unnel measurements of precipitation gauges. Atmosp. Envir. (in press).

Sevruk, B. 1994 Spatial and temporal inhomogeneity of global precipitation data. (See this Volume).

Sevruk, B., J.-A. Hertig and R. Spiess. (1989) Wind field deformation above precipitation gauge orifices. Internat. Assoc. Hydrol. Scie., No. 179, 65-70.

Sevruk, B., J.-A. Hertig and R. Spiess. (1991) The effect of a precipitation gauge orifice rim on the wind field deformation as investigated in a wind tunnel. Atmosp. Envir., 25A(7), 1173-1179.

Sevruk, B., J.-A. Hertig and R. Tettamanti. (1993) The effect of orifice rimthickness on the wind speed above precipitation gauges. Atmosp. Envir. (in press).

Sevruk B. and St. Klemm. (1989) Types of standard precipitation gauges. In: B. Sevruk (Ed.) Precipitation measurement. Swiss Federal Institute of Technology, Zurich, 227-232.

Sevruk, B. and R. Tettamanti. (1993) General model of wind-induced error of precipitation measurement. In: Proc. 8th Symposium on Meteorological Observations and Instrumentation. American Meteorological Soc., Boston, 323-324.

Tammelin, B. (1982) The wind-induced error of the Finnish rain gauge (in Finnish). Ilmatietteen Laitoksen Tiedonantojy, Helsinki, No. 36, 58 p.

METHODOLOGIES OF GLOBAL SCALE PRECIPITATION FROM SATELLITE DATA

USE OF MICROWAVE RADIOMETRY TO ESTIMATE PRECIPITATION : PHYSICAL BASIS AND INTERCOMPARISON OF ALGORITHMS

Eberhard Ruprecht and Christian Thomas
Institut für Meereskunde an der Universität Kiel
Düsternbrooker Weg 20
24105 Kiel, Germany

ABSTRACT

The application of microwave radiometry for rainfall estimation is discussed. The first part presents a description of processes by which hydrometeors affect microwave radiation.

In order to show the state-of-art of rainfall estimation with space-borne microwave radiometry five algorithms are intercompared in the second part. Two are based on scattering, one on emission and two are mixed algorithms, which include both emission and scattering.

The algorithms are applied to SSM/I observations over the Atlantic Ocean. The retrieved rainfall rates by the different algorithms partly differ a great deal. The differences depend on climatic regions, demonstrating that the algorithms are probably tuned to certain atmospheric conditions. The question arises whether a globally applicable algorithm is possible at all. A severe problem is the validation of the retrieved precipitation because hardly any direct observations are available. Thus an intercomparison of algorithms is today the only way to understand the behaviour of a scheme in different weather and climate situations and assess its results.

KEYWORDS

Microwave radiometry, hydrometeors, rainfall, satellites, precipitation algorithm

NATO ASI Series, Vol. I 26
Global Precipitations and Climate Change
Edited by M. Desbois and F. Désalmand

1. Introduction

Of all present-day available space-borne remote sensing methods for rainfall estimation, microwave radiometry is the only direct one. That means the signal observed at satellite level is directly related to the amount of rain water in a raining cloud. This fact does not mean that microwave radiometry is a priori superior to other indirect methods as e.g., in infrared or visible spectral range. But the latter methods are limited in their applicability and there is more potential in the microwave methods, which are, however, not yet fully exploited.

In this paper the physical principles are discussed which provide the basis for the rainfall retrieval from microwave observations. The state of art of the rainfall estimation with microwave radiometry is demonstrated with a comparison of the results of five published rainfall retrieval algorithms applied to SSM/I (Special Sensor Microwave/Imager, see Hollinger et al., 1987). The lack of in-situ data makes an assessment of the absolute accuracy of the retrieval results impossible.

2. Basic principles

Hydrometeors i.e., water drops and ice particles, affect the microwave (MW) signal by several physical processes:

- self-emission by hydrometeors,
- scattering and absorption of the atmospheric and surface emission by hydrometeors,
- scattering and absorption of the emitted radiation of water drops by hydrometeors at higher levels.

Over the ocean with small emissivity the first process increases the clear-sky emission. The other processes work in the opposite direction. The total effect depends on the amount of rain water, its vertical distribution, and the amount of cloud water and ice content at higher levels. These competing processes can produce an ambiguity in the rainfall - brightness temperature relation as it is typically found in single channel algorithm. (Brightness temperature T_B is defined as the thermodynamic temperature that a black body would have if it would emit the same amount of MW energy which is received by a radiometer; therefore T_B is just another measure for the radiative energy.)

Absorption (emission) and scattering depend on cloud and rain parameters and on the frequency which is used for the observation (see Ulaby et al., 1981). The following considerations are important (here we have assumed Rayleigh approximation, which is valid for $2\pi\,|m|\,r\,\nu/c \ll 1$ with c = phase speed of light, m = index of refraction, r = droplet radius, ν = frequency):

1. Instrumental parameter (frequency, ν): absorption ~ ν, scattering ~ ν^4:

   Absorption and scattering increase with increasing frequency and for the scattering the increase is much greater. That means, scattering is more important when higher frequencies are used and on the other hand, scattering can be neglected at low frequencies.

2. Cloud and rain parameters

a) number density of particles, n_o: absorption ~ n_o, scattering ~ n_o.

   Both coefficients are directly proportional to the number of droplets or ice particles.

b) radius of particles r: absorption ~ r^3, scattering ~ r^6

   Both coefficients increase strongly with r; the effect of small particles is therefore negligible. For large particles scattering becomes more important than absorption.

c) K parameter ($K = (m^2-1)/(m^2+2)$): absorption ~ Im(-K), scattering ~$|K|^2$

   The index of the refraction, m, and therefore also K depends on frequency and on temperature and phase of the particles. Because the variation of K with ν and T is small, it is neglected.

   The main dependency is on phase of the particles. The imaginary part of K (Im(-K)), which determines the absorption coefficient σ_a, is about 10^{-3} smaller for ice particles compared to water drops. The ratio of $|K|^2$ for both phases is about 5, for water larger than for ice. As a consequence absorption by ice particles is far less than the absorption by water drops, even though both hydrometeors may have similar scattering properties.

For larger drop sizes and higher frequencies Rayleigh approximation is no longer valid and Mie theory shows that absorption and scattering can decrease again. From the above discussion, the following conclusions can be drawn for the application of microwave radiometry to determine rainfall:

- water clouds ($r < 100\ \mu m$) have only a small effect at low frequencies ($\nu < 20$ GHz),
- ice clouds are "almost invisible" for frequencies up to about 40 GHz,
- for high frequencies (85 GHz and higher) scattering is the dominant effect of ice clouds,
- emission is the dominant effect of water clouds and rain for all frequencies but scattering will increase with increasing droplet size.

A further property of the radiation field, which has to be considered, is polarization. Radiation emitted by the sea surface is linearly polarized, in contrast to the radiation of the atmosphere and of water and ice drops (as long as the latter two randomly oriented). The degree of polarization of the received microwave signal is therefore a measure of the contribution of radiation from the sea surface to the total signal. Thus, it is a measure of cloud transmission (rain intensity) and/or the amount of cloud-free ocean viewed by the antenna within the beam width.

These different effects of rain clouds on the microwave signal lead to three fundamentally different rain retrieval algorithms employing channels found on the SSM/I:

- emission type affected however by scattering (19, 37 GHz),
- scattering type including polarization (85 GHz),
- mixed type (19, 37, 85 GHZ).

3. Development of algorithms

Although the physical processes are understood, it is not possible to solve the radiative transfer equation analytically, or to calculate rain intensity directly. In order to develop sufficient algorithms, physical and statistical approaches are accomplished.

A pure physical approach is not carried out. Instead two different methods are applied. First, parameters which are related to rain intensity are theoretically derived or based on physical reasoning. The polarization corrected temperature introduced by Spencer et al. (1989) or the scattering index defined by Petty (1994) are such parameters. Their relation to rain intensity has to be empirically derived on a statistical basis. Second, a method is applied which make use of statistical inversion technique. Brightness temperatures of several channels (frequencies) are simulated assuming a prescribed cloud vertical structure (e.g., Kummerow and Giglio, 1994). See also the paper of Weinman et al. in this volume, they used the method with air-borne observations and retrieve the vertical distribution of hydrometeors from radar measurements.

Most often statistical approaches are applied. The most simple one is to take direct observations of brightness temperatures and associated rain intensities and carry out a multiple regression analysis. Based on physical reasoning the regression can be taken non-linear. There are, however, disadvantages with this procedure. First, direct observations of rain intensities at the times of the satellite overpass are very rare (in particular over the oceans). Second, since brightness temperatures depend not only on rain water but on many other atmospheric and surface parameters, such a regression may only be representative for a certain climatic region or time of the year, i.e., for situations which are similar to those included in the original observation data set.

An alternative to these problems is to apply simulated data: brightness temperatures are calculated given the state of the atmosphere and surface and given a great variety of rain intensities (e.g. Wu and Weinman (1984), Bauer (1992)).This approach, though most often followed, may lead to uncorrect results too. Since atmospheric parameters are well correlated with each other (though the relation is not exactly known), rain water for example cannot be changed independently of the vertical distribution of temperature and humidity and may be correlated to cloud water and ice. Thus, multiple regession analysis may be applied to totally unrealistic situations.

The latter problem can be solved by a combination of the radative transfer model with a numerical cloud model (e.g. Mugnai and Smith (1988), Adler et al. (1991)). This gives an excellent insight into the interaction between cloud and rain particles and microwave radiation.

Fig. 1 (examples from the work of Adler et al.) shows the rain intensity brightness temperature relationship for 19, 37 and 85 GHz. It clearly demonstrates the ambiguity in the rainfall estimation when using a single channel algorithm. The 19 and 37 GHz T_B's increase only for low rainfall rates.

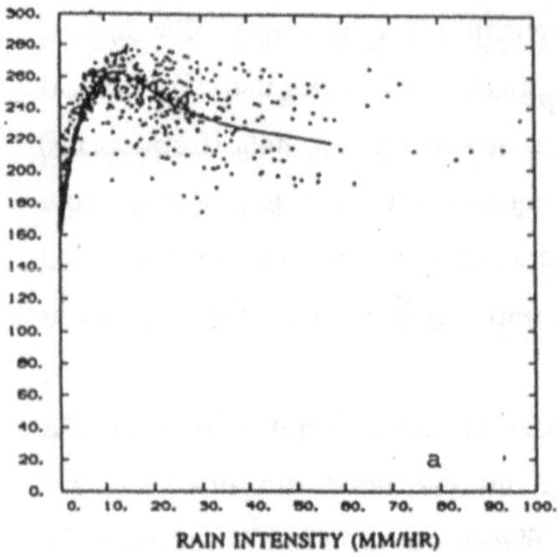

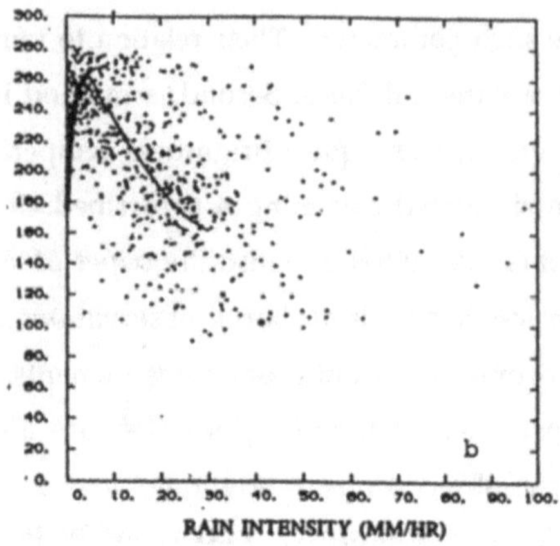

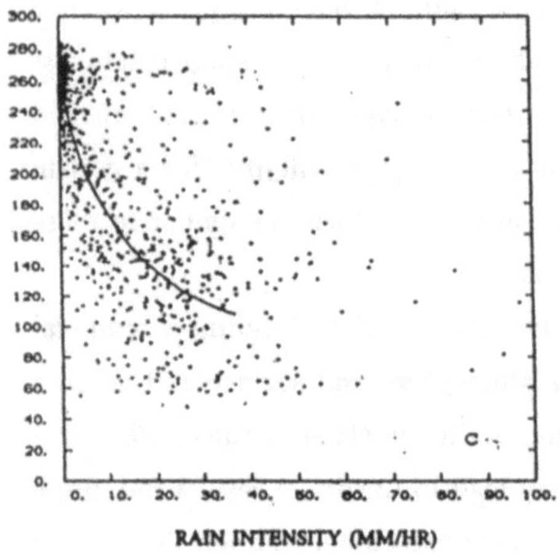

Fig. 1

Brightness temperatures at 19 GHz (a), 37 GHz (b) and 86 GHz (c) versus rain intensities of mature clouds calculated with a cloud model by Adler et al. (1991), each dot represents an average of an 1.5x1.5 km area (horizontal resolution of the cloud model), the curves shown are the best fit to the data points.

Brightness temperature maximum exists at intermediate rain rates. The T_B's ultimately decrease as the rain rates increase because in heavier rain cloud optical thickness is increased, and it contains more large drops which scatter the microwave radiation back to the earth. The reduction of T_B is even more intense at higher frequencies (85 GHz and partly 37 GHz) because of scattering effects caused mainly by ice hydrometeors. The figures also show a large variation of T_B. It demonstrates the effect of the other atmospheric parameters as for example cloud and ice water, temperature and humidity distribution or surface properties. The effect of variable cloud properties can be reduced if the spatial resolution is decreased or results are averaged over larger areas. Then for example a linear decrease of T_B at 85 GHz with increasing rain intensity is found. Thus, Adler et al (1993) have developed an algorithm based on 85 GHz channel observation only. Though this approach is very helpful for the development of an algorithm, rainfall data are necessary to derive regression coefficients and to validate the method.

4. Intercomparison of different algorithms

The many processes of interaction between a rain cloud and the MW radiation field are the basis for different rainfall algorithms. From the above discussion one may come to the conclusion that one single algorithm for all different types of rain clouds may be impossible to derive. In order to learn what different algorithms are able to perform, we have carried out an intercomparison of five rainfall algorithms. All of these were available to us from the literature. All are applicable to SSM/I data and determine rain intensity over the ocean. The five algorithms are cited in Table 1; shown also are the algorithm types, used channels, the methods applied for the algorithm development, and how a threshold rain/no rain is set. The channels, which are given in parentheses, are used for the description of the state of the atmosphere and ocean surface as for Petty and Katsaros or to determine the rain/no rain threshold as for Adler et al. The algorithm of Petty and Katsaros (1990) is based on theoretical consideration how polarisation of the 85 GHz radiation is altered by rain water. They defined a scattering index which is mainly a function of the polarisation difference of the 85 GHz brightness temperatures. The rain intensity is directly related to this index. The coefficients in this relationship are empirically determined including a threshold for the scattering index which has to be exceeded if rain water is present.

Bauer (1992) used a radiative transfer model to simulate brightness temperatures of the SSM/I channels. Input of model are radiosonde profiles and a great variability of rain rates. Multiple regression analysis applied to the rain rates and the simulated T_B yields his algorithm. Since small rain are overestimated, Bauer defines an external threshold of 0.3 mm/h, estimated rain rates lower than these

Algorithm	Mode	SSM/I channels	Method	Threshold
Petty/Katsaros (1990)	scattering polarization	85 H+V(19 H+V, 22 V, 37 H+V)	theoretical, empirical	internal
Bauer (1992)	mixed	19 H+V, 22 V, 85 V	multiple regression of simulated data (schematic clouds)	external
Liu/Curry (1992)	mixed	19 H, 85 H	theoretical, empirical	external determined for each region
Prabhakara et al. (1992)	emission	37 H	statistical analysis of rain fall data, empirical	internal
Adler et al. (1993)	scattering	85 H (19 H+V, 22 V, 37 H)	regression of simulated data (cloud model)	external, decision tree for each pixel

Tab. 1 Algorithms included in the intercomparison study
Mode = basic physical process of the algorithm (mixed = emission and scattering); H, V = horizontal, vertical polarization.

values are set to zero.

Liu and Curry (1992) use also the results of a radiation transfer model. They deduced from these results that the difference of the emission dependent 19 GHz T_B and the 85 GHz T_B which depends mainly on the scattering effect can be related to the rain rates. The coefficients of their algorithm are empirically determined. SSM/I observations of a cloud free ocean are used to derive a rain/no rain threshold on a statistical basis.

Prabhakara et al. (1992) performed a statistical analysis of rain fall data and brightness temperature. The statistical results (e.g. cumulative frequency distributions) are similar. Their algorithm is developed on the basis of these comparison results. Empirically determined coefficients depend on total

precipitable water, which has to be retrieved by SSM/I observations in addition. A minimum 37 GHz T_B is determined depending on the total precipitable water, it serves as a rain fall threshold of the algorithm.
Adler et al. (1993) evaluated results of their numerical cloud model and could show that the 85 GHz T_B is linearly related to rain intensity. If rain fall increases 85 GHz T_B decreases because of the increase of the scattering effect, mainly of ice particles. A decision tree programme based on 19, 22 and 37 GHz T_B is applied to distinguish between rain and no-rain pixels.

Since we had no in-situ rainfall data available, the intercomparison is indirect. SSM/I data of November 1987 over the Atlantic Ocean were applied to the five algorithms. Rain intensities were calculated and intercompared. A validation was not possible, rather the intercomparison study shows differences between the different algorithms. The calculated rain intensities were spatially averaged for each 0.5° x 0.5° longitude, latitude area and each satellite overpass. These spatial means of every day of the month were then summed up to give the monthly value.
The Fig. 2 a - c show scatter diagrams of the monthly values of the four other algorithms against the scattering algorithm of Petty and Katsaros for three climatic regions (each dot represents a monthly value of a 0.5° x 0.5° longitude, latitude area). It is interesting that the estimated rainfall agrees best in mid latitudes. The largest differences occur in the tropics. The two scattering algorithms (Petty and Adler) show in general the smallest differences, even between these results large biases are found, even in the tropics, where the scattering on ice clouds dominates. The largest differences are found between the scattering and the emission (Prabhakara et al.) algorithms in the tropics. In general the Prabhakara et al. algorithm shows the greatest scatter with respect to the others.
With this study it is not possible to analyse the causes for the different behaviour of the five algorithms because the structure of the raining systems in each region is not known in detail. The partition into three climatic regions asssumes that the rain clouds are different: high convective clouds within the ITCZ (tropics), broken (may be warm) clouds in the subtropics and warm and cold front rain systems (mid-latitudes). This assumption is true in a climatic mean, how

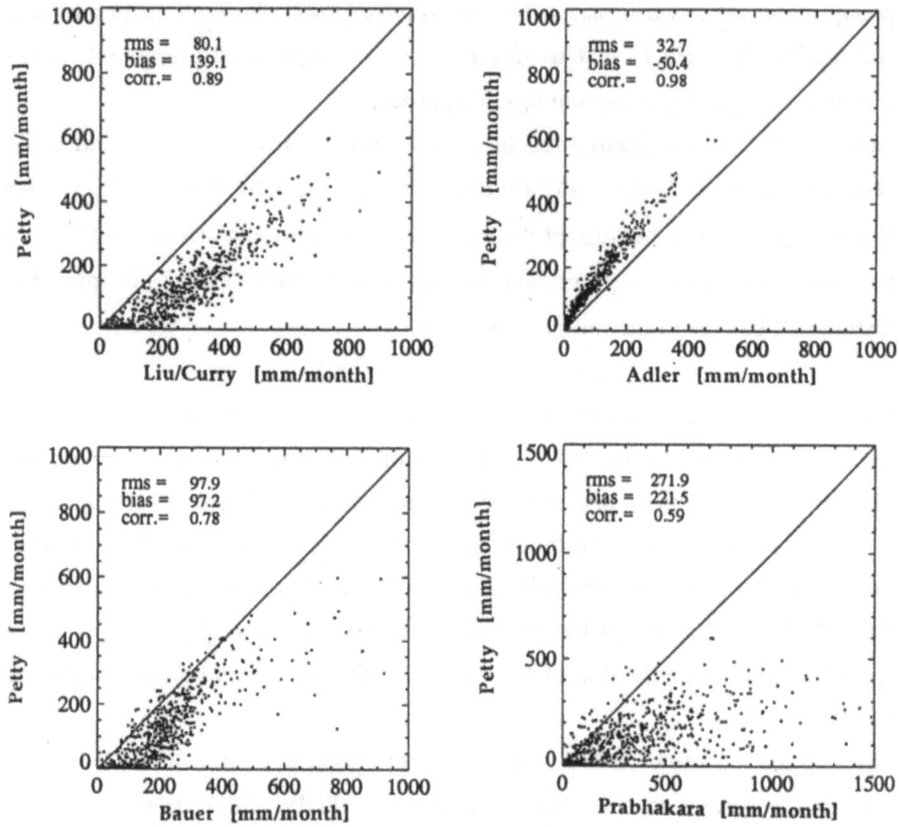

Fig. 2 Intercomparison of the five rainfall algorithms. The scatter plots show the November 1987 means for each 0.5° x 0.5° longitude, latitute area over the North Atlantic Ocean.

a) Tropics (01 - 08° N, 20 - 50° W)

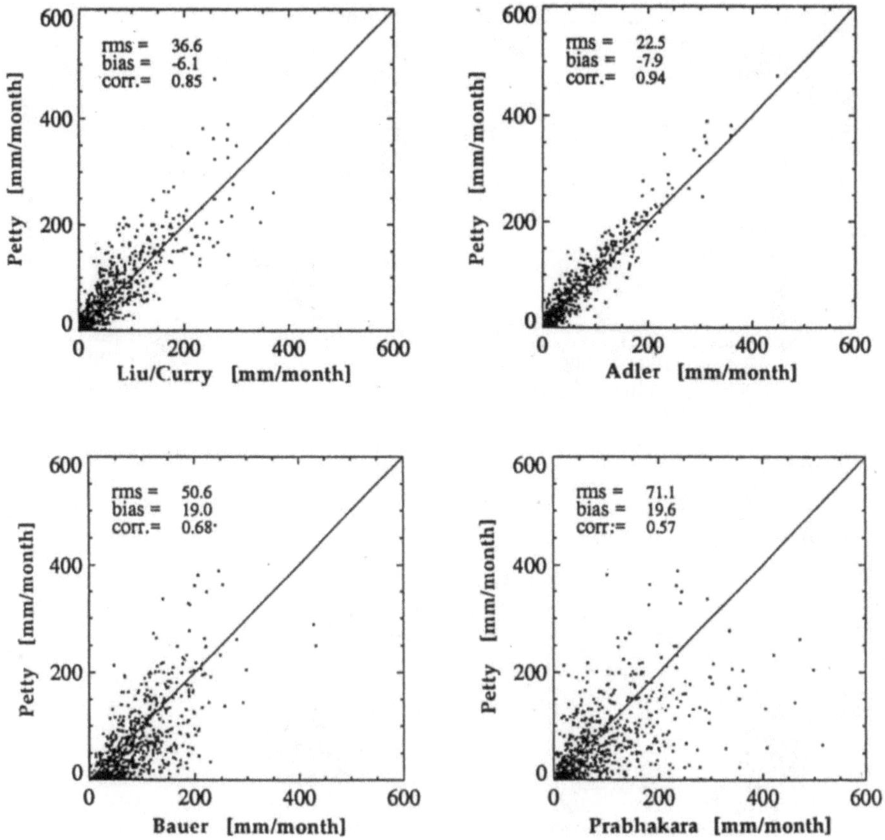

Fig. 2 continued

b) Subtropics (30 - 37° N, 30 - 60° W)

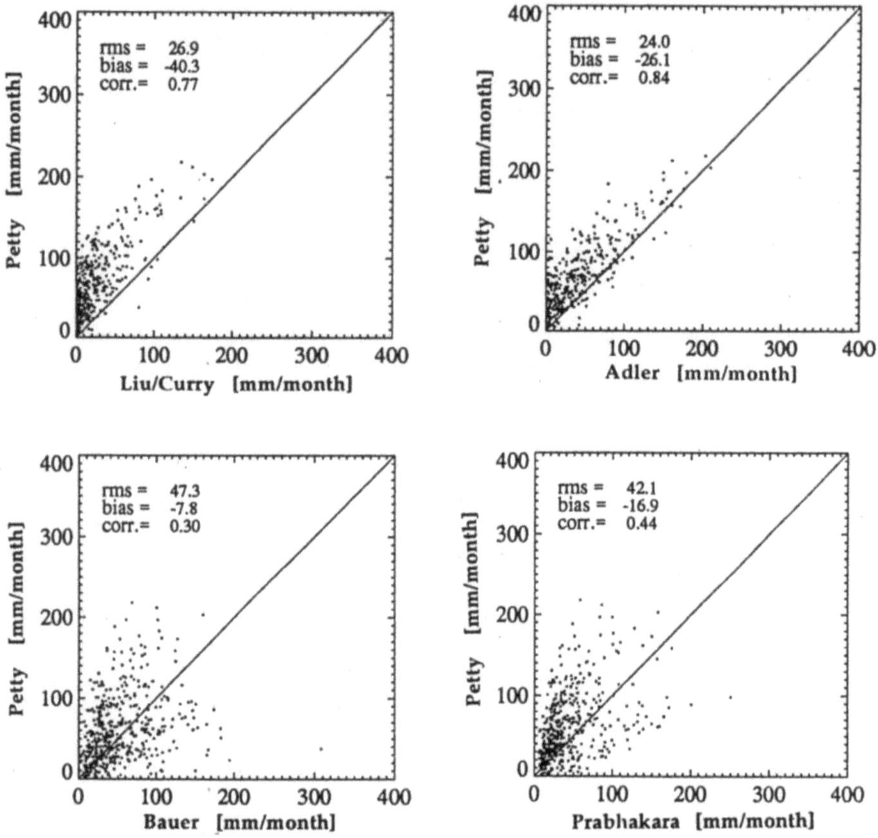

Fig. 2 continued
c) Mid-Latitudes (50 - 57° N, 00 - 30° W)

different the cloud systems are in this actual case of November 1987 is not known. In order to understand and to explain their behaviour we plan to apply these algorithms to the results of our numerical cloud model which gives the vertical distribution of rain water, cloud water and ice.

For further intercomparisons the mean rainfall maps of the Atlantic Ocean were constructed for November 1987 (Thomas, 1993). All algorithms reproduce of course the general rainfall distribution i.e., maximum within the ITCZ and heavy rain in the storm track regions of the midlatitudes and very little rain within the subtropics. (An algorithm, which cannot reproduce these general structures, is completely useless.) The differences of the absolute amounts are of interest. The purpose of these maps is, to demonstrate how large the differences of a monthly mean rainfall map can be, if one selects just one (published) algorithm. Within the ITCZ the Prabhakara et al. algorithm gives maximum values of about 1000 mm / month whereas Adler et al. determined less than 500mm / month at maximum. Similar differences are observed over the Gulf Stream region. In the Subtropics, in particular over the South Atlantic Ocean, the two scattering algorithms (Adler et al., Petty/Katsaros) give large areas without any rainfall, in contrast Bauer and Prabhakara et al. found nearly no rain-free areas. That is mainly the effect of the different thresholds. The results in the storm track regions agree fairly well at least in the Northern Hemisphere. Very large differences, however, are found in the Southern Hemisphere.

As mentioned above, we cannot assess which algorithm gives the best results, because we had no in-situ measurements. However, the 1st Algorithm Intercomparison of the GPCP (Lee et al., 1991) provided rainfall data (surface measurement and radar) over and around Japan for two time periods of 1989: June and July 15 - August 15 to compare with the estimated values. Four of the above algorithms were applied to this comparison: Adler et al., Petty/Katsaros, Liu/Curry, Prabhakara et al. Though we do not know whether any changes have been carried out to the published version of the algorithms, we assume that these are the same versions we applied.

Very large differences are obtained, for single values of each 1.25° x 1.25° latitude/longitude region deviations of more than 100% are not seldom. Even this intercomparison project does not allow to assess which algorithm works best, differences from June (frontal precipitation) to July/August (tropical convective rainfall) are large.

5. Conclusion

Our intercomparison shows, the results of the five algorithms do not support each other. The differences are very large. The 1st GPCP Algorithm Intercomparison Project did not yield more clarity. In order to calculate monthly mean rainfall or compare SSM/I retrieved rainfall with the amount measured at the ground, the sampling problem has to be considered. With one DMSP satellite a certain area is viewed only two times a day, in low latitudes even only every fourth day. As a solution to this problem one may consider either to have more satellites with microwave radiometer on board or to use additional data from other satellites with higher viewing frequencies but different channel observations (e.g. geostationay satellites). More research is needed in this direction. One conclusion from the 1st GPCP Project is that for a test phase a period of one month is too short to validate an algorithm sufficiently. During a month the rain process may be not very different so that an algorithm can be suitable to this special process, but totally fail in other situations.

The great variety of the rain processes e.g., with and without ice phases, with high amount of cloud water above the rain or not, leads to the conclusion that one algorithm emission or scattering type cannot estimate the rain intensities in all possible situations. The solution is not a mixed algorithm per se, but a decision tree algorithm, with two main decisions to carry out:

1. Is there rain within this pixel?
2. What type of rain and therefore what type of algorithm has to be applied?

The combination of numerical cloud models with radiative transfer models is a very helpful tool for the development of such algorithms. Additional information about the atmosphere and the vertical structure of the rain cloud is probably necessary.

Also the statistical inversion technique should be investigated in more detail. In order to understand and improve existing algorithms, further intensive intercomparison campaigns are needed.

REFERENCES

Adler R.F., Yeh H.-Y.M., Prasad N., Tao W.K., Simpson J. (1991): Microwave simulations of a tropical rainfall system with a three-dimensional cloud model. J. Appl. Meteor. 30: 924 - 953.

Adler R.F., Negri A.J., Keehn P.R., Hakkarinen I.M. (1993): Estimation of monthly rainfall over Japan and surrounding waters from a combination of low-orbit microwave and geosynchronous IR data. J. Appl. Meteor. 32: 335 - 356.

Bauer, P. (1992): Wasserdampf, Gesamtwasser und Niederschlagsraten aus Daten passiver Mikrowellenradiometer über dem Ozean. Forschungsbericht DLR-FB 92-37, 122 pp.

Hollinger, J.P., Lo, R., Poe, G., Savage, R., Peirce, J. (1987): Special Sensor Microwave/Imager User's Guide. Naval Research Laboratory, Washington D.C.

Kummerow C. and Giglio L. (1994): A satellite passive microwave technique for estimating rainfall and rainfall structure. Part I: Algorithm description. J. Appl. Meteor., 33, 3-18.

Lee T.H., Janowiak J.E., Arkin P.A. (1991): Atlas of Products from the Algorithm Intercomparison Project 1: Japan and surrounding oceanic regions. Univ. Corp. Atm. Sci., 131 pp.

Liu G. and Curry J.A. (1992): Retrieval of precipitation from satellite microwave measurement using both emission and scattering. J. Geophys. Res. 97: D 9959 - 9974.

Mugnai A. and Smith E.A. (1988:) Radiative transfer to space through a precipitating cloud at multiple microwave frequencies. Part I: Model Description. J. Appl. Meteor. 27: 1055 - 1073.

Petty G.W. (1994): Physical retrievals of over-ocean rain rate from multichannel microwave imagery. Part I: Theoretical characteristics of normalized polarization and scattering indices. (Subitted to Metor. Atm. Physics).

Petty G.W. and Katsaros K.B. (1990): New geophysical algorithms for the special sensor microwave imager. Proc 5th Conf. Sat. Meteor. Ocean. London, UK: 247- 251.

Prabhakara C., Dalu G., Liberti G.L., Nucciarone J.J., Suhasini R. (1992): Rainfall estimation over oceans from SMMR and SSM/I microwave data. J. Appl. Meteor. 31: 532 - 552.

Spencer R.W., Goodman H.M., Hood R.E. (1989): Precipitation retrieval over land and ocean with the SSM/I: Identification and characteristics of the scattering signal. J. Atm. Ocean. Technology 6: 254 - 273.

Thomas, C. (1993): Vergleich von Regenalgorithmen des Special Sensor Microwave/Imager. Master Thesis at the Institut für Meereskunde, University Kiel, 100 p.

Ulaby F.T., Moore R.K., Fung A.K. (1981: Microwave Remote Sensing. Vol I: Microwave remote sensing fundamentals and radiometry. Artech House Inc, Norwood 456 pp.

Wu R. and Weinman J. A. (1984): Microwave radiances from precipitating clouds containing aspherical ice, combined phase and liquid hydrometeors. J. Geophys. Res. 89: 7170 - 7178.

COMBINATION OF SATELLITE MICROWAVE AND INFRARED MEASUREMENTS FOR RAINFALL ESTIMATION

Isabelle Jobard and Michel Desbois
Laboratoire de Météorologie Dynamique du CNRS
Ecole Polytechnique
91128 Palaiseau Cédex, France

ABSTRACT

Satellite-based rainfall estimations are obtained from either microwave or infrared techniques but all these techniques have their own inherent limitations. Microwave radiances have a good physical connection with rainrate but the microwave observations from a single low-orbit satellite, (one or at most two images per day) are inadequate for the detection of rapidly changing precipitation distributions. Infrared images from geosynchronous-orbit satellites have high spatial and high temporal resolutions, but they do not have a direct physical connection with the instantaneous rainrate. However, infrared techniques provide reliable estimation of the rainfall amount accumulated during long time periods and averaged over large areas.

To overcome the limitations of these techniques, different methods for estimating precipitation at various time and space scales, which combine low-orbit microwave and geostationary infrared satellite data, are developed. The Rain and Cloud Classification (RACC) method, based on an automatic classification procedure, is described; the method is applied to two case-studies corresponding to different climatic regions: tropical land in West Africa and sub-tropical Japan and surrounding oceanic regions, for which the infrared data from geostationary satellites (Météosat, GMS) and the microwave data from the SSM/I radiometer on board a U.S. DMSP polar satellite are used. The comparisons of monthly and daily rainfall estimates with surface rainfall validation data are presented. The results show that the combined microwave-infrared technique performed well.

KEYWORDS

Rainfall, Satellite, microwave, infrared, precipitation algorithm

NATO ASI Series, Vol. I 26
Global Precipitations and Climate Change
Edited by M. Desbois and F. Désalmand

1. Introduction and background

The different techniques for satellite remote sensing of rainfall use either microwave (MW) or infrared (IR) radiometric obervations. Arkin and Ardanuy (1989) made a review of the numerous techniques for estimating climatic-scale precipitation from space and divided them into two classes : "indirect" estimates of rainfall based on observations of clouds in infrared (and visible) imagery and "direct" estimates of rainfall based on observations of the radiative effects of precipitation-sized hydrometeors in microwave imagery. The limitations of the IR and MW techniques are summarized below.

The main shortcoming of the infrared techniques is that there is not a direct physical connection between infrared brightness temperatures of a cloud top and the surface rain; nevertheless the infrared techniques, based on observations from geosynchronous satellites giving images with both a high spatial resolution and a good time sampling, provide reliable estimation of the rainfall amounts accumulated during long time periods and averaged over large areas as was shown for example, by Arkin and Meisner (1987) using the GOES precipitation index (GPI) or by Jobard and Desbois (1992) using a simple thresholding technique with Meteosat IR data.

The microwave measurements from passive radiometers are adequate for the estimation of the instantaneous rain-rate because the rain-rate is well related to the brightness temperatures of the different channels which measure either the scattering by hydrometeors of high frequency radiation or the absorption by the precipitation particles of low frequency radiation. Over the ocean, the rain-rate is also related to the difference between the vertically polarized and horizontally polarized radiations. However for the estimation of rainfall accumulated over long time periods, the microwave techniques face difficulties due to the very poor time sampling from a single low-orbit satellite (one or two observations per day). Another shortcoming is the problem of incomplete beam filling by the active rain cells due to the large size of the microwave footprints (37 km x 29 km for the 37 GHz channel, for example).

In spite of the inadequate time sampling, many microwave algorithms are now developed for application to monthly rainfall estimation; they are based on different methods (empirical techniques, classification, more or less sophisticated simulation,...) using the various operable SSM/I channels and calibrated from different climatic regions. An intercomparison of some MW algorithms applied over ocean is presented in the same volume (see Ruprecht and Thomas).

Two international programmes for algorithm intercomparisons are being conducted to improve the estimation of monthly precipitation. The WCRP Global Precipitation Climatology Project (GPCP) has organized three satellite Algorithm Intercomparison Projects (AIP) for different climatic areas and periods : AIP-1 around Japan between June and August 1989 (see Lee *et al.*, 1991 and Arkin and Xie, 1994), AIP-2 over UK and North-West Europe between February and April 1991 (see WCRP, 1994), and AIP-3 over the TOGA-COARE region (western equatorial Pacific Ocean) between November 1992 and February 1993. The Precipitation (algorithm) Intercomparison Project (PIP-1) run by the WetNet Precipitation Working Group compared a number of passive microwave techniques for global monthly rainfall monitoring by SSM/I for the period August to November 1987. The results of this project are discussed in Barrett (1994).

The first conclusions drawn from these intercomparison experiments indicate that no single algorithm out-performs the others. The study of the results of the comparisons with validation data sets will help to select the several algorithms or type of algorithms which perform best in different regions (land or ocean, tropics or mid-latitudes, ...) or over different types of surface, and will certainly lead to the refinement of the existing methods.

In the context of these increasing efforts to develop satellite-based rainfall algorithms, our approach for improving the climatic-scale rainfall estimates, is to combine MW and IR techniques to overcome the limitations inherent to these techniques. The next section deals with methods which combine geostationary IR and low-orbit MW satellite data and presents our Rain And Cloud Classification (RACC) method for estimating cumulated rainfall. The RACC method was applied for two case studies, a tropical land over West Africa and over Japan and surrounding oceanic regions (AIP-1 data set). The results are discussed in section 3.

2. Microwave-Infrared combining techniques

Rainfall, especially in tropical regions, is characterized by high spatial and temporal variability. For this reason it is of paramount importance to estimate rainfall from the satellite observations which have frequent sampling and good spatial resolution. This favours the IR techniques which use hourly or semi-hourly images with 5 km pixels. As indicated above, the information retrieved from the IR radiances is only related to the clouds, while the instantaneous rain-rate can be estimated from MW radiances. Methods combining IR and MW information can take advantage of both techniques. Two different approaches combining the IR and MW data are described below.

The method to estimate mean monthly rainfall proposed by Adler *et al.* (1993) is based on the idea of using the MW estimates to calibrate or adjust the monthly geosynchronous IR estimates. They use separately the low-orbit MW data set and the geostationary IR data set; with a MW technique they compute a MW monthly estimate and with an IR technique, they compute an IR monthly estimate from all the hourly images and a second IR monthly estimate from the hourly images contemporaneouswith the SSM/I sampling. By dividing the MW estimate by the second (subset) IR estimate for each grid box, they derive calibration factors which are then used to multiply the (full sampling) IR monthly estimate at each grid box to produce the combined MW-IR monthly rainfall estimates.

They applied this method using two different IR techniques combined with a scattering-based MW algorithm, to the AIP-1 data set (described in the following section) from the Japan region. They compared the monthly rainfall from the ground truth validation data set to the monthly estimates obtained from the MW-IR combined method and also from the MW technique and from the two different IR techniques alone. Their conclusion is that the monthly estimates resulting from the MW-IR combined method appear to be better than either the MW or the IR estimates individually, except when the MW based technique alone produces poor results. This is the case when orographic warm rain systems occurred during the July-August period, because the chosen MW algorithm fails to produce the good instantaneous rain-rate in that case. But we think that it is worth considering that the poor results for monthly estimates obtained from the MW technique may be due also to the very poor sampling of the MW observations which is approximately once a day for the studied July-August period.

Our approach to combine MW and geostationary IR satellite observations is based on an automatic classification procedure. The Rain And Cloud Classification (RACC) method is presented in detail in Jobard and Desbois (1994) and the flow-chart describing the method is shown in Figure 1. The low-orbit MW data set and the geostationary IR data set are used together for the learning phase of the classification process which uses a dynamic clustering technique. The learning phase, by combining several parameters, selects homogeneous classes characterizing different types of clouds. Each class is defined by the center of gravity of the cluster for each parameter. Three parameters are used : two parameters derived from the IR images, one 'spectral' parameter which is the IR brightness temperature, and one 'spatial' parameter which is the local standard deviation computed from the eight neighbors of the pixel in the IR image; this parameter, called "IR-Variance", characterizes the spatial structure of the cloud. This parameter is used here in order to discriminate better the highly convectively active parts of the clouds, supposed to present large temperature inhomogeneities near the convection towers, from less active areas where the cloud top temperature is more uniform. The third parameter, henceforth referred to as the MW parameter, is derived from the MW SSM/I images using a simple MW algorithm. For the studies presented in the next section, three different MW parameters were used. Over Africa (land) the brightness temperature of the horizontally polarized channel at 86-GHz was used. For the study concerning the Japan region, the scattering index proposed by Grody (1991) was used for land and ocean areas; and in addition a polarization difference at 37-GHz was used for the ocean only.

When combining MW and IR satellite data, it is necessary to use the same grid for the MW image and for the coincident IR image. The size of the box of the common grid is that of a pixel of the IR image. The method of interpolation used to obtain the MW image at a better spatial resolution than the resolution of the different channels used by the MW algorithm, is explained in detail in Boitel and Jobard (1994).

The application phase of the classification process assigns each pixel to the class to which it is the closest and produces a "classified" image. Note that this image is different from the one that would have been obtained by thresholding applied to any of the parameters separately. The value of the MW parameter of the center of gravity is used to attribute a Rain Factor to each

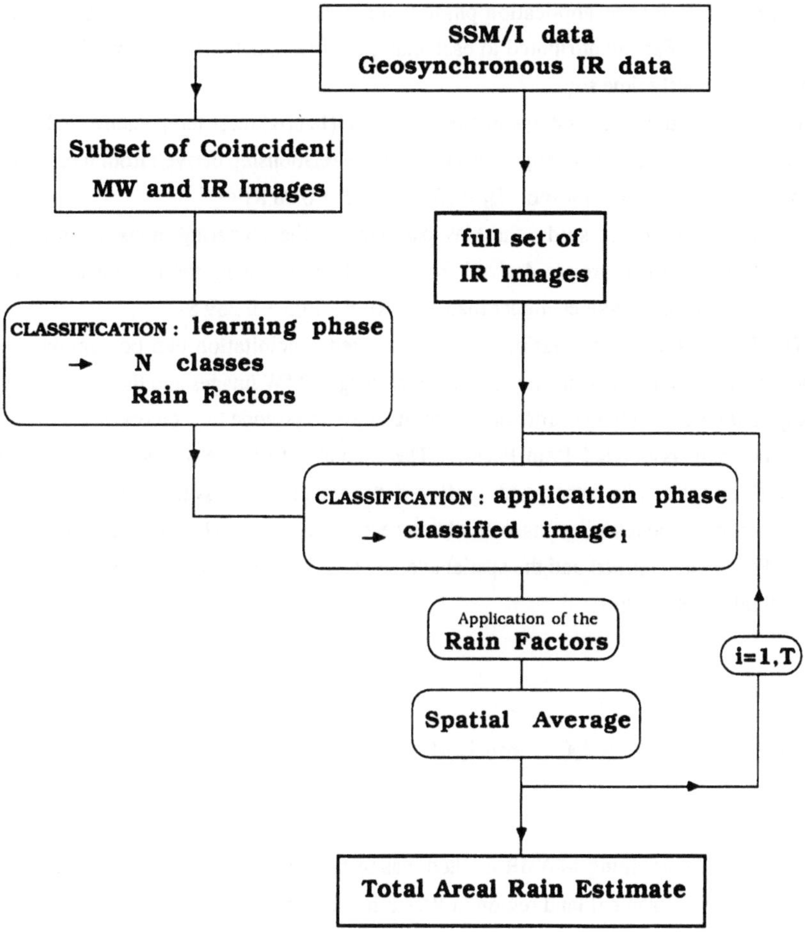

Figure 1 : Flow-chart of the RACC method.

class, so that each class represents a different type of precipitating or non-precipitating cloud area. Thus, by applying the Rain Factors to the classified image, an image of the rain estimates is obtained. This procedure can only be used on the one or two occasions each day when coincident MW and IR data are available. To take advantage of the half-hourly or hourly IR data which can track the rapid changes in the cloud and rainfall patterns, it is necessary to adapt the classification algorithm described above. Hence the classification application phase uses only the two available IR parameters (IR and IR-variance images). In this case, the classified image is computed using the same IR centers of gravity of the classes that were determined by the learning phase which used all three parameters. One therefore assumes that the combination of the two IR parameters is able in some way to simulate the combination of the three parameters, so that the classes that have been defined as rain classes from the MW parameter are considered as the rain classes for the application phase using only the two IR parameters. Further, by applying the Rain Factors attributed to each class, a rain estimate image is obtained for each hourly or semi-hourly time step.

For the land study, the MW parameter is the 86 GHz brightness temperature, and the Rain Factors are calculated using an empirical T_B - rain rate relationship derived from measurements over the the African validation site: $T_B = 253 - 7.0\,RF + 0.08\,RF^2$.

For the Japan region study, the MW parameter is the scattering index SI proposed by Grody and the Rain Factors are derived from the SI values, using the relationship given in Alam *et al.*(1993) for SI values higher than 10 : $RF = -1.65 + 0.289\,SI$.

The RACC method for estimating accumulated precipitation can be summarized as follows: in a first step, only a subset of coincident images (MW images and their corresponding IR images) are used in the classification learning phase to provide the centers of gravity of the classes and their associated Rain Factors. The second step is the application phase: each available IR image and its corresponding IR-variance image, are classified using the centers of gravity defined by the learning phase, to obtain semi-hourly or hourly classified images. Then the Rain Factors are applied and the spatial average and time integration are computed to give the total areal rain estimates.

3. Results of the RACC method

The RACC combining MW-IR method was tested for two studies concerning different climatic regimes : a tropical land region in Western Africa and a subtropical land and oceanic region around Japan. A summary of the results is presented in this section.

a. Tropical Africa

The surface precipitation measurements were collected over the EPSAT-NIGER validation site. The EPSAT (Estimation of Precipitation by Satellite) programme has been designed to improve the understanding of the precipitation systems of Sudano-Sahelian Africa and to develop operational rainfall estimation algorithms for this region (see Lebel *et al.*, 1992). We used the ground rainfall data given by the dense network of raingauge recorders implemented in the 1°x1° square site near Niamey (Niger) located at 13°N and 2°E, during a two-month period (from 6th July to 6th September 1989) corresponding to the period producing a major part of the total precipitation of the raining season (371 mm out of 540 mm). The rain fields were computed on a grid having a 15 km x 15 km mesh, using a kriging method for the interpolation of the 78 point measurements produced by the raingauge network. We used the full set of half-hourly IR images from Meteosat and the MW images from the 86 GHz (horizontal polarization) channel of SSM/I on board the DMSP-F8 satellite, for the application of the RACC method.

We have computed the correlation between the measured areal surface precipitations and both the rain estimates obtained from the RACC method and from a simple IR thresholding method (the IR rain estimates is proportional to the fraction of the area covered by pixels at a brightness temperature lower than a given threshold, using different thresholds from -30°C to -60°C).

The correlation coefficients obtained for different area scales (from 15x15 km^2 to 120x120 km^2) and for different time integrations (from hourly to daily accumulation) are reported in Jobard and Desbois (1994). They show that there is a significant improvement of the correlation for all the area and time scales, for the rain estimates obtained with the RACC method, when compared to the correlation obtained with the IR estimates. Figures 2(a) and 2(b) show the scatterplots and statistics for the daily rainfall in the 120x120 km^2 area, for the 26 days of the study period which had rain, for the RACC method in Fig. 2(a) and for the IR method using an IR temperature threshold of -40°C in Fig. 2(b). The combined MW-IR method has better or similar error statistics with the correlation coefficient ($r = 0.79$), the bias (Est.-Obs.= +5.9 mm) and the RMS error (12.9 mm) than the IR technique with the correlation coefficient ($r = 0.73$), the bias (Est.-Obs.= -1.5 mm) and the RMS error (13.0 mm). Moreover, it is seen on the scatterplot of Fig. 2(b) that the IR technique, using a single threshold, fails to retrieve correctly high rainfall values because, when the whole area is covered by cold clouds, it produces a "saturated" estimation.

b. Japan region

For the sub-tropical region study, we used the validation dataset provided for the AIP-1 campaign conducted over the area represented in Figure 3, from 127°E-145°E and 27°N-45°N,

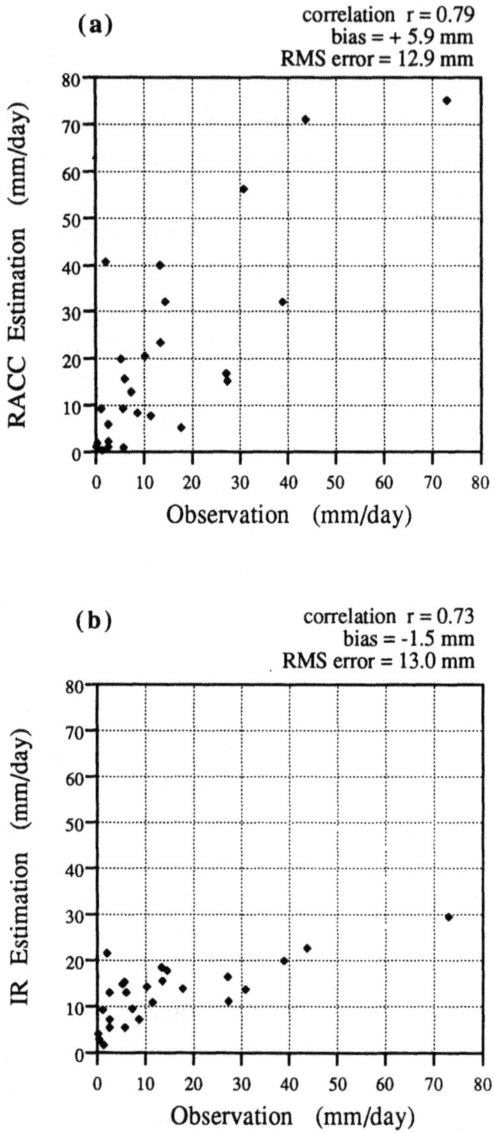

Figure 2 : Scatterplots and statistics for the daily rainfall over the Niger validation site (120x120 km^2 area), for the 26 days with rain, estimated from the RACC method in Fig. 2(a) and from the IR method using an IR temperature threshold of -40°C in Fig. 2(b).

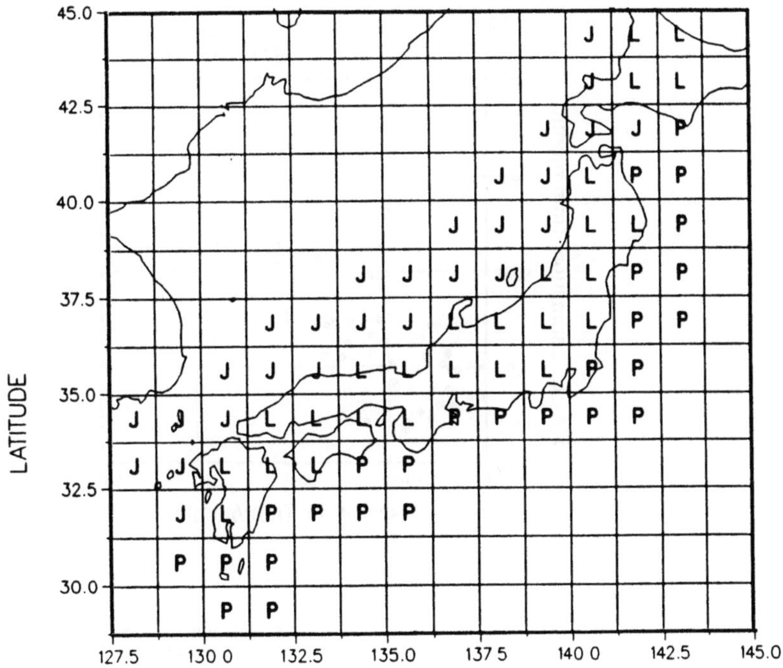

Figure 3 : Area covered during the AIP-1 campaign. The validation dataset covers the 1.25°x1.25° boxes marked with P (Pacific ocean) or J (Sea of Japan) or L (land). (from Lee *et al.* ,1991).

covering Japan and surrounding oceanic regions, during the periods 1-30 June and 15 July-15 August 1989. The dataset includes the hourly IR images from the GMS geostationary satellite, the SSM/I MW images from the polar-orbiting DMSP-F8 satellite and the validation dataset consisting of a composite of rainfall estimates from 15 precipitation radars and the Japanese AMeDAS network made of over 1300 automated rain gauges. We applied the satellite-based methods (RACC method and IR technique) to produce the daily and monthly rain estimates areally-averaged for the 1.25°x1.25° boxes represented in Fig. 3. The 79 boxes are divided in 52 "ocean" boxes and 27 "land" boxes.

The full results concerning this study are presented in details in Maia and Jobard (1994). Only the main results for the monthly estimates are presented in the following. The Figures 4 and Figures 5 show the scatterplots and statistics for the monthly rainfall estimates for the 79 land and ocean boxes, respectively for the period 1-30 June (Figures 4) and for the period 15 July-15 August (Figures 5). Estimates are from the RACC method in Figs. (a) and from the IR method using an IR temperature threshold of -38°C in Figs. (b). (This is the 235 K threshold used for the GPI of Arkin and Meisner, 1987).

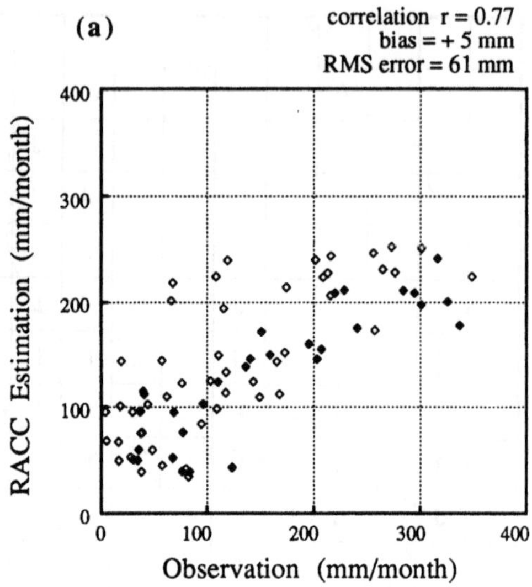

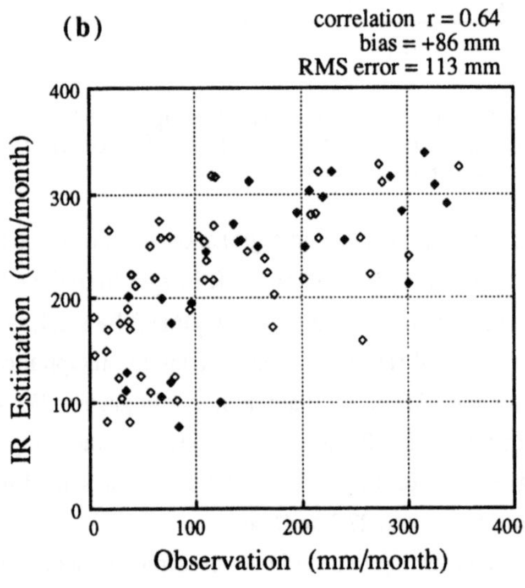

Figure 4: Scatterplots and statistics for the monthly rainfall estimates for the 79 land and ocean boxes, for the period 1-30 June, from the RACC method in Fig. 4(a) and from the IR method using an IR temperature threshold of 235 K in Fig. 4(b). (◆ : land, ◇ : ocean)

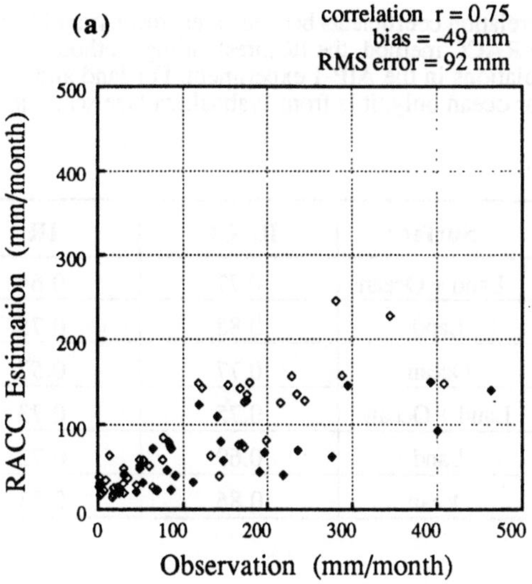

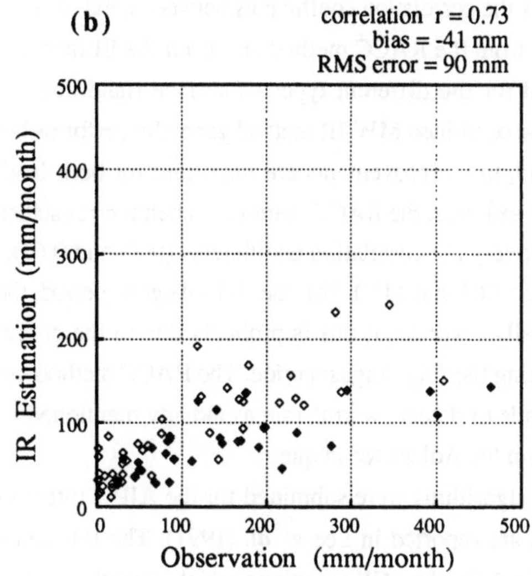

Figure 5 : Scatterplots and statistics for the monthly rainfall estimates for the 79 land and ocean boxes, for the period 15 July-15 August, from the RACC method in (a) and from the IR method using an IR temperature threshold of 235 K in (b). (◆ : land, ◇ : ocean)

Table 1 : Correlation coefficients between areal mean monthly rainfall validation data and estimates from the RACC method, the IR thresholding method and the MW technique which gave the best correlations in the AIP-1 experiment. For land and ocean the MW algorithm is from Grody and for ocean only, it is from Prabhakara (see scatterplots and statistics errors in Lee *et al.*, 1991).

Period	Surface	RACC	IR	MW
1-30 June	Land + Ocean	0.77	0.64	0.48
id	Land	0.88	0.75	-
id	Ocean	0.77	0.57	0.79
15 July -15 Aug.	Land + Ocean	0.75	0.73	0.22
id	Land	0.69	0.72	-
id	Ocean	0.86	0.83	0.54

Table 1 shows the correlation coefficients between areal mean monthly rainfall validation data and estimates from the RACC method and from the IR thresholding method, for the two month-periods and for the different types of surface (land and ocean, land, ocean). These results show that the combined MW-IR method generally performs better than the IR technique. For the June period, the improvements are important for both land and ocean; for land and ocean together, for example, the RACC method has better error statistics than the IR technique considering respectively the correlation coefficients (0.77 and 0.64), the biases (+ 5 and + 86) and the RMS error (61 and 113). For the July-August period, the improvements are less significant, especially over land; this is probably due to the orographic warm rain systems which occurred during the July-August period. The RACC method uses a scattering-based MW algorithm which fails to detect "warm" rain as already mentioned in section 2 for the similar results obtained from the Adler's technique.

Several MW algorithms were submitted for the AIP-1 intercomparison; the scatterplots and error statistics are reported in Lee *et al.* (1991). The last column of Table 1 shows the correlation coefficient for the MW technique which gave the best results for the two month-periods. For land and ocean, the MW algorithm is from Grody and for ocean only, it is from Prabhakara. The coefficients in Table 1 indicate clearly that the combined MW-IR method performs much better than the MW techniques for the estimation of monthly rainfall, in nearly all cases. The poor time sampling of the MW dataset might be a major reason.

4. Conclusion

Two different methods for estimating accumulated rainfall by combining geostationary satellite IR data and polar-orbiting satellite MW data have been presented. The common aim of these approaches is to take advantage of the strengths of the MW techniques and the IR techniques. In the studies presented in this paper, where the MW-IR combining methods were applied, it is shown that in most cases, they produced improved estimates compared to the MW estimates or the IR estimates individually. These results are encouraging and they indicate that the MW-IR approaches deserve to be developed because they can alleviate some difficulties of the other techniques, especially the problem of the poor sampling of the MW satellite data for the MW techniques.

***Acknowledgments*:** We thank ORSTOM which is in charge of the EPSAT Niger raingauge network, for providing us with the data used in this study.

REFERENCES

Adler, R.F., A.J. Negri, P.R. Keehn and I.M. Hakkarinen, 1993. Estimation of monthly rainfall over Japan and surrounding waters from a combination of low-orbit microwave and geosynchronous IR data. J. Appl. Meteor., Vol. 32: 335-356.

Allam, R., G. Holpin, P. Jackson and G.L. Liberti, 1993. Second Algorithm Intercomparison Project of the Global Precipitation Climatology Project AIP-2: Algorithms Review. Pre-Workshop Report, UK Met. Office, Reading : ALG 27 pp.

Arkin, P.A. and P.E. Ardanuy, 1989. Estimating climatic-scale precipitation from space: a review. J. Climate, Vol. 2: 1229-1238

Arkin, P.A. and B.N. Meisner, 1987. The relationship between large scale convective rainfall and cold cloud over the western hemisphere during 1982-1984. Mon.Weather Rev., Vol. 115: 51-74.

Arkin, P.A. and P.P. Xie, 1994. The Global Precipitation Climatology Project; First Algorithm Intercomparison Project. Bull. Amer. Meteor. Soc., Vol. 75, N° 3, 401-419.

Barrett, E.C., 1994. The First WetNet Precipitation Intercomparison Project (PIP-1): Interpretation of results. Special Edition, Remote Sensing Reviews: (in press).

Boitel, C. and I. Jobard, 1994. Acquisition de données-satellite et traitement d'images-satellite appliqués aux mesures du radiomètre microonde SSM/I. Note interne LMD N° 189, L.M.D./C.N.R.S., Palaiseau: 86 pp.

Grody, N.C., 1991. Classification of snow cover and precipitation using the Special Sensor Microwave Imager. J. Geophys. Res., Vol. 96, N° D4: 7423-7435.

Jobard, I. and M. Desbois, 1992. Remote sensing of rainfall over the tropical Africa using Meteosat IR imagery : Sensitivity to time and space averaging. International Journal of Remote Sensing, Vol. 13, N°14: 2683-2700.

Jobard, I. and M. Desbois, 1994. Satellite estimation of the tropical precipitation using the METEOSAT and SSM/I data. Atmospheric Research, Vol. 34/1-4: 285-298.

Lebel, T., H. Sauvageot, M. Hoepffner, M. Desbois, B. Guillot and P. Hubert, 1992. Rainfall estimation in the Sahel : the EPSAT-NIGER experiment. Hydrological Sciences - Journal des Sciences Hydrologiques, 37, 3: 201-215.

Lee, T.H., J.E. Janowiak and P.A. Arkin, 1991. Atlas of products from the algorithm intercomparison project 1 : Japan and surrounding oceanic regions June-August 1989. University Corporation for the Atmospheric Research, available from Climate Analysis Center, NOAA, Washington, 131 pp.

Maïa, A. and I. Jobard, 1994. Application d'une méthode d'estimation des précipitations combinant les donées-satellite Microondes et Infrarouges sur la région du Japon. Note interne L.M.D./C.N.R.S.

Ruprecht, E. and C. Thomas, 1994. Use of Microwave Radiometry to estimate Precipitation: Physical Basis and Intercomparison of Algorithms. Same volume.

W.C.R.P., 1994. Workshop report on the Global Precipitation Climatology Project (GPCP) Second Algorithm Intercomparison Project (AIP-2). (23-26 August 1993, U.K.), 13 pp.

SATELLITE DERIVED RAINFALL ESTIMATES OVER AFRICA : THEIR VALIDATION AND USE IN MONITORING CLIMATE CHANGE.

George Dugdale

Department of Meteorology

University of Reading

2 Earley Gate

Reading RG6 2AU U.K.

ABSTRACT

Experience in the estimation of rainfall using thermal infra-red data over tropical Africa has given insight into the problems of making and validating satellite rainfall estimates. The spatial and temporal variations in the rainfall mechanisms over Africa are demonstrated to be large. Rainfall algorithms that are mainly empirical and are tuned to perform in a particular climatic regime may be unsuitable for monitoring climate change. The high spatial variability of tropical rainfall combined with the sparse gauge network makes the validation of rainfall estimates difficult. Validation must take account of the non representative nature of gauge measurements and should be on the basis of identifying changes from climatologically expected rainfall. It is suggested that a robust multispectral rainfall algorithm will require physically based relationships between different cloud structures and observed radiances to be applied, probably in conjunction with GCM outputs which should define the weightings to be given to the different channels.

KEYWORDS

satellite, rainfall, Africa, spatial variability, storm structure, empiricism, validation.

1. Introduction

Satellite remote sensing offers global coverage and frequent observations so would appear to be an ideal way of monitoring global climate. The question that has to be answered is: to what extent can the

NATO ASI Series, Vol. I 26
Global Precipitations and Climate Change
Edited by M. Desbois and F. Désalmand

radiances measured by the satellite be quantitativly related to the variable of interest? In this case the variable is the mean rainfall on time scales of about a month and on space scales of the order of a square degree together with the statistical variability within these scales.

Many satellite based rainfall estimation methods have been developed and practiced to a greater or lesser extent over the last twenty years. (See, for example, Liberti 1993, Barrett 1990, Arkin and Arduanuy 1989 and Engman and Gurney 1991). Generally the operational techniques have been tuned to give the maximum useful information for a particular application and climatic region. These may be unsuitable for use in climate studies. During the last ten years the Department of Meteorology at the University of Reading has acquired experience of rainfall estimation over Africa with real time operational activities covering tropical Africa north of the equator since 1988. In 1991 these activities extended to southern Africa. These rainfall estimates have been based on a simple temperature thresholding of the thermal infra-red images from Meteosat and have be designed to identify periods when vegetation may be stressed. Both the temperature threshold and the calibration vary in space and with season (Milford & Dugdale, 1990). While more advanced techniques using multispectral microwave data will become an operational reality within the next few years some of the lessons learned with the simpler techniques may help us to assess the suitability of satellite based rainfall estimates for use in climate change studies and to help us to optimise the technology for this purpose.

Some aspects of the satellite estimates and of the rainfall characteristics that have been included in the Reading work are:

i) the variable relationship between the radiometric data and the rainfall,

ii) the influence of sampling interval on the rainfall estimates,

iii) the spatial variability of rainfall and its importance in validating satellite estimates.

While the findings have been derived from or applied to the use of thermal infra-red data to obtain the rainfall estimates, similar considerations will apply to most radiometrically based rainfall estimates including satellite borne radar. Particular attention is given to aspects relevant to the use of a satellite based rainfall estimation technique suitable for identifying changes in precipitation patterns associated with climate change. The requirements for this application are in some ways different from those needed to monitor rainfall for immediate applications in that care must be taken not to impose past climatology on the satellite estimates.

2. Storm cloud patterns over the Sahel

Figure 1 shows, in a simplified form, the monthly average pattern of storm clouds over Africa from 6 to 19 degrees north during July, August and September over the six year period 1987 to 1992. The figure was derived from hourly full resolution Meteosat images. Each image was compared to a threshold corresponding to -50°C; pixels colder than that threshold scored one, other pixels scored zero. The monthly accumulated pixel score is called the Cold Cloud Duration (CCD). The image is very similar to that produced by the GOES Precipitation Index (GPI) except that the GPI uses three hourly data and a temperature threshold of -38°C. (Arkin & Meisner, 1987). At first sight figure 1 could be mistaken for a rainfall map. The main features one would expect from rainfall climatology are displayed. A maximum extends across the continent corresponding roughly to the expected mean position of the Inter Tropical Convergence Zone. Local maxima also conform in a general manner to the rainfall; mainly being associated with the major topographic features of the highlands of Ethiopia, Jebel Mara, Cameroon, Jos Plateau and the Foutal Djala with an extension across the notoriously wet Guinea coast. The highlands are identified in figure 2. The rainfall map for the same period is shown in figure 3. Owing to the sparsity of raingauge data the rainfall contours should themselves be used with caution. While there is overall correspondence between the two sets of isopleths there are also significant differences. In particular, the west-east maximum of the CCD image is displaced northwards relative to that of the rainfall, also the maxima of CCD over high ground tend to occur to the west of the rainfall maxima and of associated hills. The most important deficiency of the CCD data is its failure to identify the very rainy south western coastal area. Also, the ratio of rainfall to CCD generally increases from north to south.

The differences noted above between the CCD and the surface rainfall measurements are due to the changing structure of the rainclouds as one moves away from the coast of west Africa and to the persistence of cirrus after storms have ceased to be active. The cirrus is carried west in

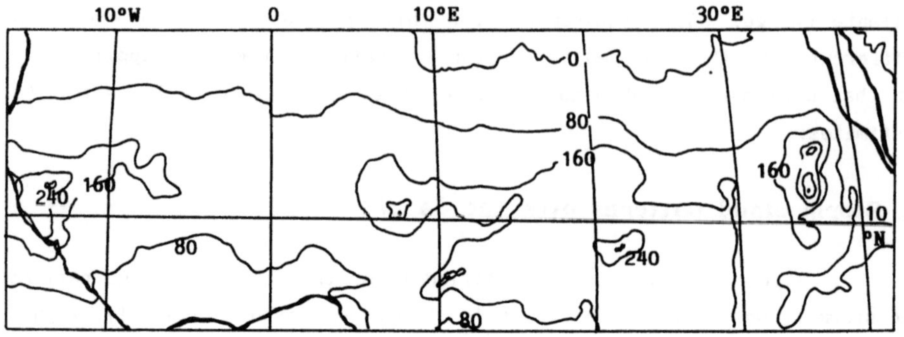

Fig.1: Average monthly duration (hours) of clouds with tops colder than -50°C during the months of July, August and September from 1987 to 1992. (After Chapa et al 1992)

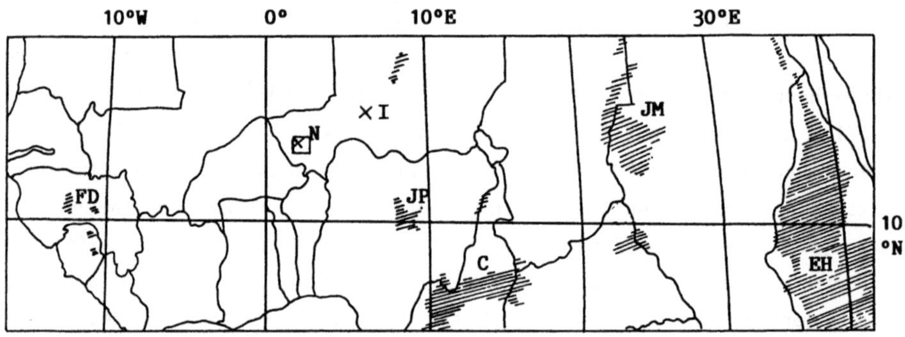

Fig.2: Africa from 6 to 19°N. Showing land above 900m (hatched) and regions referred to in the text.

KEY: EH - Ethiopian highlands JM - Jebel Mara C - Cameroon
JP - Jos Plateau FD - Foutal Djala I - Ibecetene
N - Niamey

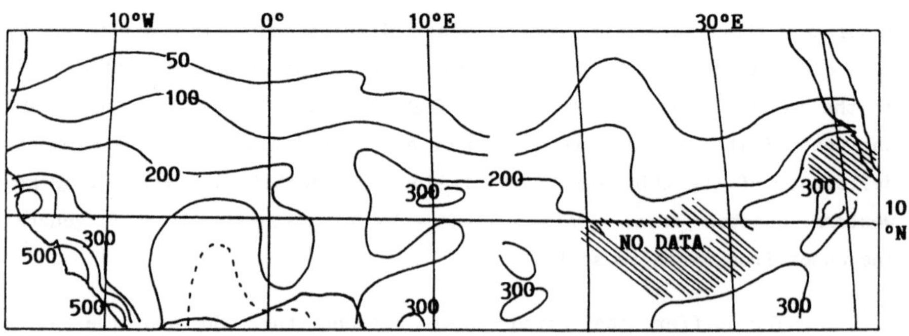

Fig.3: Average monthly rainfall (mm) for the months of July, August and September from 1987 to 1992.

the prevailing high level east winds. In the microwave bands the cirrus could perhaps be eliminated from the estimates but latitudinal changes in storm structure would not necessarily result in increased scattering being associated with increased rainfall as the ice and water particle size spectra and the sub cloud evaporation also change.

Figure 4 illustrates well the latitudinal varying structure of the storms over a north-south strip five degrees of longitude wide. This strip, from the meridian to five degrees east, is chosen because it is free of major topographic features. The cold cloud duration shows typical north south features with the CCD's at both the -40°C and -60°C thresholds showing minima in the north and south. The maximum in the -60°C threshold is found to the north of that at -40°C. The rainfall maximum is situated to the south of both the CCD maxima. It is seen that to the south of CCD maxima there is a zone in which the rain is increasing while the CCD is decreasing at each threshold. This corresponds to a zone where the anvil rainfall contributes an increasing fraction of the total rainfall from storms and where rain of non-convective origin contributes significantly to the total. (Eldridge, 1957). The pattern illustrated here occurs over much of the Sahel and moves north and south seasonally.

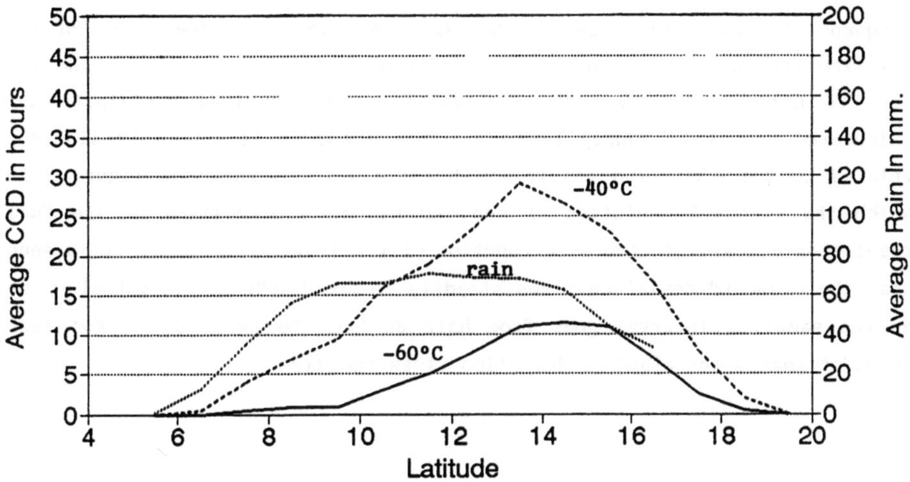

Fig.4: A north-south cross section from 5°N to 20°N through the ITCZ showing the duration of cold cloud at two temperature thresholds and the rainfall averaged over strip from 0 to 5°E, in the period August 1 to 10, 1988. (Private communication from Dr. V. Thorne)

In order to estimate rainfall from the CCD different groups have used different calibration techniques to represent the seasonally and latitudinally varying rainfall characteristics. These involve the inclusion of latitudinal effect either explicitly (the Lannion method) or implicitly (TAMSAT). In the TAMSAT method (Milford & Dugdale 1991) the influence of latitude is included by calibrating in narrow latitudinal belts on the basis of archived contemporaneous rainfall and CCD data. The inclusion of this climatological feature in the calibration of the estimation technique ensures that the estimates will, over a period, correlate well with the raingauge measured rainfall. But, it may prevent the methodologies from identifying unusual situations. More importantly for the purposes of climate modelling the inclusion of past climatic influences in the calibration technique may mask any actual future climate change.

3. Influence of sampling interval on rainfall estimates

Microwave data will only be available initially from polar orbiting satellites hence offering two snapshots per day per satellite of each point on the surface of the earth. Over areas where much of the rainfall comes from short lived convective storms the influence on the estimates of the frequency of observation needs to be taken into account. The effect of the sampling frequency on measured CCD has been studied by regressing the ten day CCD calculated from reduced temporal resolution images against that calculated from half hourly imagery. This standard deviation is linearly related to the CCD. The result of this analysis is shown in figure 5 in which the standard deviation of the CCD has been normalised by the full resolution CCD and plotted as a function of the image frequency. It is seen that the standard deviation rises steadily to about 33% of the actual value as the sampling interval goes from half an hour to two hours. For less frequent images the standard deviation rises rapidly.

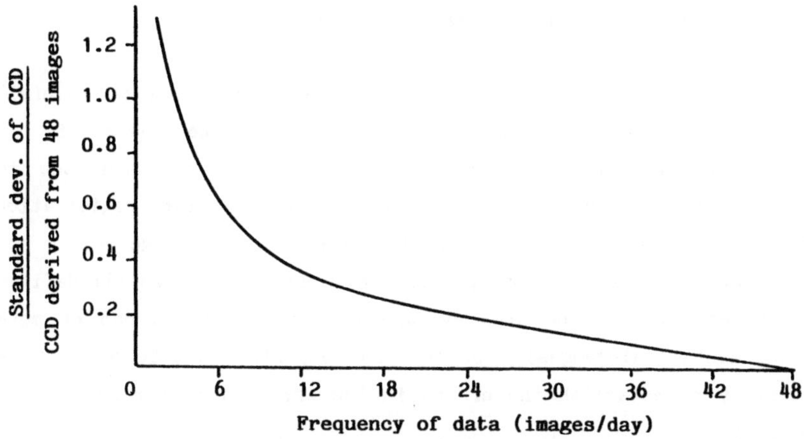

Fig.5: The influence of sampling frequency of satellite imagery on the calculated CCD, expressed as the normalised standard deviation of pixel CCD from the CCD calculated with half hourly data.

Despite the above discussion on the frequency of sampling if the occurrence and structure of the storms were independent of the time of day then their mean characteristics and the mean rainfall may be determined from a long series of single daily satellite images. However both the features of the rain producing system and their frequency vary diurnally and these variations themselves change with location and season. The diurnal variation of rainfall with location is illustrated in figure 6, which shows the patterns over the central Sahel and over coastal Sierra Leone in July. The diurnal effects are almost completely out of phase. Any change in regional climatology could influence these diurnal patterns.

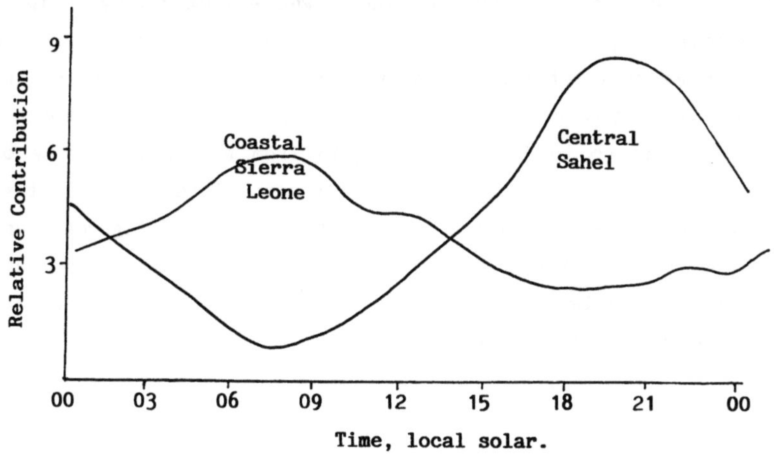

Fig.6: Diurnal pattern of rainfall over the central Sahel and over the coast of Sierra Leone in July.

4. The validation of satellite rainfall estimates

As with all satellite inferred data the validation of rainfall estimates presents a major problem in that there is no true measure of areal values on the ground. The problem is compounded when one is referring to period averages because the relative and absolute registration of the sequence of images used in the satellite estimate is generally not known. We therefore wish to compare a rainfall estimate for an ill defined area with an estimate of the area average rainfall based on one or more point measurements (raingauges). Before any comparison can be made it is first necessary to establish the errors in the area rainfall based on the raingauge data.

Area rainfall from point measurements

The systematic errors in raingauge data may be quite large but methods for correcting these in terms of gauge exposure and local climate are available (Sevruk 1989, for instance). However these errors may be small when compared to the errors in using gauges to estimate area rainfall, particularly in situations where most of the rainfall comes from convective rainfall which has high spatial variability.

The variability of rainfall over different time and space scales in the Sahel are illustrated in figures 7 and 8. Figure 7 shows the rainfall measured at points separated by 2km covering a square 10km x 10km. Figure 7a shows a typical daily rainfall while figure 7b gives a seasonal total for 1985. Even the seasonal total shows rainfall changing by a factor of two within the square. Figure 8, after Taupin et al (1992), shows the monthly variability on scales up to 100 x100km; again from an intense array of raingauges. The site here is some 250km south west of the previous test site in an area of heavier rainfall. Even a cursory study of figures 7 and 8 shows that it is not appropriate to compare single point gauge data directly with the satellite area averages. The sites from which these data were obtained are indicated in figure 2.

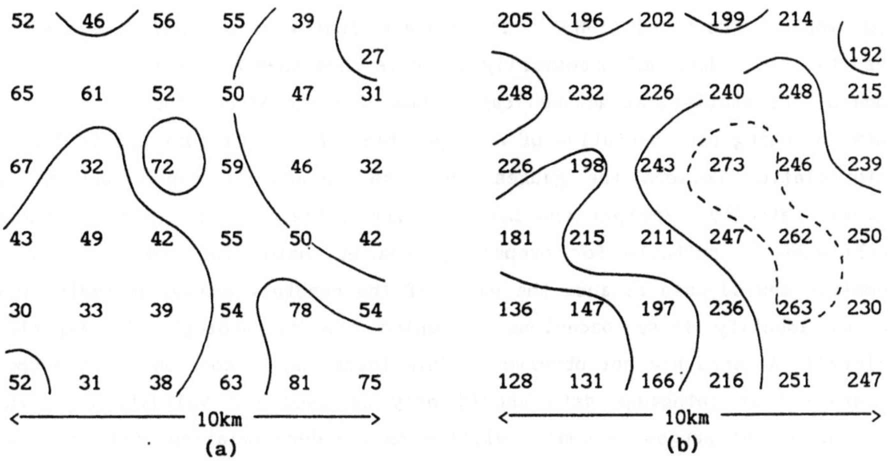

Fig.7: Rainfall (mm) measured at raingauges sited at 2km centres in a 10 x 10km square at Ibecetene (15.3°N, 5.3°E): a) A typical daily rainfall (20 July 1985); b) Total June, July, August and September 1985.

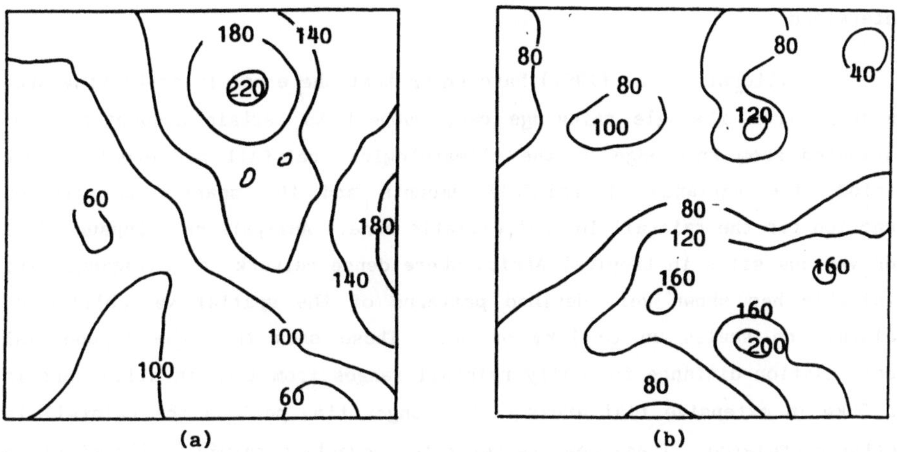

Fig.8: Rainfall measured by a dense network of raingauges in a 100x100km square centred at 13.5°N, 2.5°E. a) May 1991, and b) July 1991 (after Taupin et al, 1992).

For many purposes area average rainfall is calculated from gauge measurements using one of the many available statistical methods such as those described in the World Meteorological Organisation's Guide to

Hydrological Practices (1983). It is found that most methods give similar results. This does not necessarily indicate that they are accurate but that none of the statistical methods can account for the errors due to the point data not being representative of a large area. Also, for some applications, interpolation between the gauges may take account of topography or of climatologically expected spatial rainfall patterns. Neither of these techniques is suitable for preparing area estimates for comparison with remotely sensed data because the value of the remotely sensed estimates may be to identify those occasions on which the climatologically expected rainfall patterns are not observed. This leads one to conclude that mapped rainfall from raingauge data should only be used for validation if the spacing of the gauges is small relative to the decorrelation distance. If this condition is satisfied the errors in the gauge based estimates can be assessed and direct comparisons made with the satellite estimates. If the gauge spacing is large compared to the decorrelation distance, as is usually the case over Africa, then only the satellite data for the pixel corresponding to the position of the gauge can be compared to the gauge reading and the error in the area estimate from the gauge must be determined.

Flitcroft et al (1991) have shown that the error in estimating area rainfall from a single raingauge can, subject to certain assumptions, be estimated from knowledge of the climatological rainfall per event or per period, the variance of rainfall amounts and the spatial correlation function for the rainfall in that climatic zone. Analysis of raingauge data for various sites in tropical Africa where dense networks of raingauges are available has shown well defined patterns of the spatial variability of rainfall on scales up to 15km or so. These show that the exponential decorrelation distance for daily rainfall ranges from 15km in hilly terrain to 30km in extensive flat areas. For longer time periods the rainfall is better correlated. These results refer to convective rainfall, though often from extensive cloud clusters. Calculations have been made of the errors in estimating the area rainfall for a typical Meteosat pixel based on a raingauge reading at the centre of the nominal position of the pixel centre. Figure 9 illustrates the typical root mean standard error relative to the mean rainfall per event for a range of periods and terrain types.

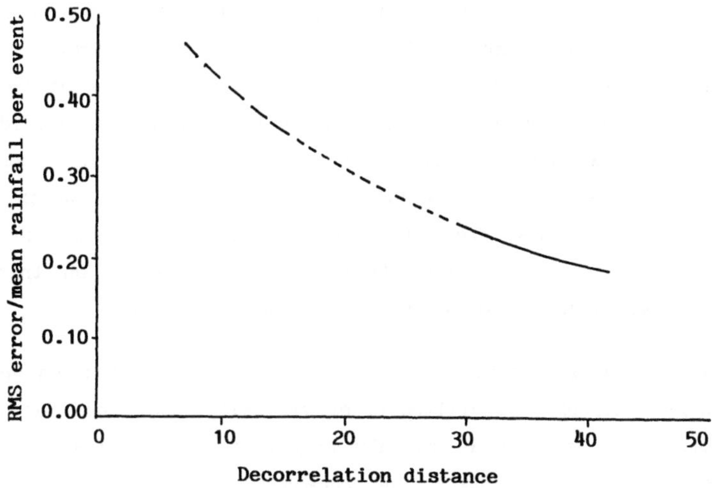

Fig.9: Typical root mean square errors of area rainfall estimated from a raingauge for the area of a Meteosat pixel. (based on Flitcroft et al 1991).

— — — daily rainfall over hilly terrain
- - - - - - - daily rainfall over level terrain
———— ten-day rainfall over level terrain.

The results of such calculations allow us to attribute a fraction of the variance between satellite and gauge data to the errors in estimating the pixel rainfall from raingauges. However, a large set of gauge data is needed to arrive at a statistically significant validation.

5. Concluding comments

Methods for estimating rainfall using thermal infra-red satellite data rely on empirical relationships between surface raingauge observations and satellite measured radiances. These relationships are normally only valid for the particular type of cloud system for which the relationship was established. Multispectral techniques employing microwave and visible bands include radiances more directly linked to cloud water properties. Hence these are, in principle, capable of more robust and portable algorithms. However, most current algorithms using such wavebands have little physical basis but are also rely on empirical relationships which depend on cloud structure being maintained. These, as with the thermal infra-red techniques, may be useful locally but not for climate change studies. Even those algorithms which have a stronger physical basis tend to use idealised cloud models and assume that the cloud viewed corresponds to that model. Perhaps what is needed are physically based algorithms in which general circulation models (GCM's) are used to define the environment within which clouds form and hence the cloud type to be expected. On this basis multispectral signatures could be interpreted appropriately.

Within the foreseeable future only the thermal infra-red detectors carried on geostationary satellites offer virtually continuous surveillance of the globe. So, in areas of rapidly varying rainfall and strong diurnal influences, it will continue to be necessary to use this data as an interpolation tool between passes of the polar orbiting satellites. Proposals to use the microwave data to give frequent calibration of the thermal infra-red may be realizable but will have to take account of the diurnal changes in storm cloud structure.

The precipitation indices are excellent tools for monitoring the changes in cirrus climatology over low latitudes where most of the high cloud originates from convection. But, as shown in section 2 above such changes cannot always be directly related to changes in rainfall, particularly over continental areas.

The validation of satellite based rainfall estimates over areas without radar and with poor raingauge networks requires particular care. The gauge data itself needs correction and interpretation before being used as an area estimate (Flitcroft et al, 1989). Also, validation of satellite estimates should be on the basis of the algorithm's ability to detect changes from climatology, not the ability to reflect the currently established rainfall climatology.

Current methods of rainfall estimation, while useful in some applications, are probably only marginally applicable to monitoring rainfall changes associated with climate change. This is particularly unfortunate over the oceans and over the data sparse tropics where there is little probability of achieving good coverage of reliable conventional rainfall data. A major effort should be made to develop true physically based algorithms which will exploit information from global scale, mesoscale and cloud models to interpret the satellite measured radiances. Can we look forward to radiances being assimilated directly into forecasting GCM's to improve their rainfall analyses?

Acknowledgements The various experimental results quoted in this report have come from the combined work of the TAMSAT group at the University of Reading; the principal researchers being I.D. Flitcroft, V.D. McDougall, J.R. Milford & V. Thorne. The research has been funded by the Overseas Development Administration of the United Kingdom, the United Nations Food & Agriculture Organisation and the European Community.

REFERENCES

Arkin, P.A., P.E. Arduanuy, 1989: Estimating climatic-scale precipitation from space: A review. J.Climate, 2, 1229-1238.

Arkin, P.A. and B.N Meisner, 1987: The relationship between large scale convective rainfall and cold cloud over western hemisphere during 1982-1984. Monthly Weather Review, 115, 51-74.

Barrett, E.C.,(ed), 1990: Satellite remote sensing for operational hydrology. Remote sensing reviews 4 (2) 223-474.

Chapa, S.R., J.R. Milford, G. Dugdale, 1993: Climatology of satellite derived cold cloud duration over sub-Saharan Africa. Proc. 9th Meteosat Scientific Users' Meeting Locarno Switz 14-18 Sept 1992. EUMETSAT. pp 245-250.

Eldridge, R.H., 1957: A synoptic study of West African disturbance lines. Q.J.R. Met Soc 83: 303-314

Engman, E.T. and R.S. Gurney, 1991: Remote Sensing in hydrology. Chapman & Hall, London.

Flitcroft, I.D., J.R. Milford and G. Dugdale, 1989: Relating point to area average rainfall in semi-arid West Africa and the implications for rainfall estimates derived from satellite data. J.of Applied Met 28: 252-266.

Flitcroft, I.D., G. Dugdale and J.R. Milford, 1991: Relating point rainfall and area average rainfall over the Sahel. Proc Int workshop on Precipitation Measurement. WMO/IAHS/ETH 4-7 Dec 1989 St Moritz. Hydrological Processes Vol.5, John Wiley & Sons Ltd.

Liberti, G.L., 1993: Precipitation estimation with AVHRR data: a review. Proc. AVHRR Users' Meeting Belgirate Italy. EUMETSAT. In press.

Milford, J.R. and G. Dugdale, 1990: Estimation of rainfall using geostationary satellite data. In Applications of Remote Sensing in Agriculture, Proc. 48th Easter School in Agricultural Science Univ. Nottingham April 1989. Butterworth, London.

Sevruk, B., 1989: Wind induced measurement error for high intensity rains. Proc WMO/IAHS/ETH Symposium on Precipitation Measurement St.Moritz Dec.1989. Swiss Federal Inst. of Technology.

Taupin, J.D., T. Lebel, H.F. Cazenave, F. Chiron, R. Gathelier, M. Greard, R. Gaulde, J. Kong and T. Valero, 1992: EPSAT-NIGER CAMPAGNE 1991, ORSTOM/DMN Montpelier.

World Meteorological Organisation (1983) Guide to Hydrometeorological Practices. WMO. No.168, Geneva.

ON THE USE OF THE NORMALIZED DIFFERENCE VEGETATION INDEX AS AN INDICATOR OF RAINFALL

Sharon E. Nicholson
Department of Meteorology
Florida State University
Tallahassee, FL 32306 (USA)

ABSTRACT

The remotely sensed Normalized Difference Vegetation Index, or NDVI, is being used operationally to evaluate rainfall and vegetation growth. It shows a good relationship to these variables in relatively dry locations, but the index saturates under humid conditions. Over much of Africa, the saturation takes place where annual rainfall is on the order of 1000 to 1200 mm. Above this threshold, NDVI cannot be used with any degree of confidence to assess either rainfall or vegetation.

KEYWORDS

vegetation index, rainfall, rainfall estimation, remote sensing, Africa, land surface processes, NDVI

1. Introduction

The Normalized Difference Vegetation Index has been used operationally to monitor vegetation and some water balance parameters. The assumption is made that it is a good indicator of "green-ness" and, in a more general sense, rainfall. However, this assumption is quite faulty. For example, in the arid and semi-arid regions where it is commonlyused, the distribution of rainfall is nearly as important as total rainfall. Also, a number of non-environmental factors affect the index. In this paper, the limitations of NDVI as an indicator of vegetation growth and rainfall are demonstrated, using a number of locations in Africa.

The index is calculated from Advanced Very High Resolution Radiometer (AVHRR) data from the NOAA polar orbiting satellites. It is defined as the difference in reflectances between channels 1 (visible) and 2 (near infrared), divided by the sum of reflectances in these

NATO ASI Series, Vol. I 26
Global Precipitations and Climate Change
Edited by M. Desbois and F. Désalmand

two channels. In theory, NDVI is most directly a measure of photosynthetic activity (Sellers, 1985; Tucker and Sellers, 1986). It has been shown to correlate with a number of variables related to bulk photosynthesis: green leaf biomass, leaf area index (LAI), total dry matter accumulation and annual net primary productivity (Curran, 1980; Tucker et al., 1981, 1983, 1985; Asrar et al., 1984). It also shows a relationship to rainfall patterns and interannual variability (Tucker et al., 1991; Nicholson et al., 1990). In most cases, these correlations have been demonstrated in regions of subhumid or semi-arid climate.

In practice, the measured values of NDVI are "corrupted" by a number of factors, including atmospheric effects (especially aerosols and humidity), viewing and illumination geometry, sensor calibration and drift, and the spectral signatures of soils underlying the vegetation cover. Also, in humid regions cloud containmination can be a problem (Derrien *et al.*, 1992). As a result, the variance in NDVI explained by changes in vegetation cover is considerably less than 100 %. Nevertheless, the normalization of the index helps to minimize these extraneous factors.

This study utilizes monthly rainfall totals and monthly composited NDVI (maximum value within the month) for the period 1983 to 1991. NDVI data were provided and processed by the GIMMS group at the Goddard Space Flight Center. NDVI is calculated from NOAA-AVHRR global area coverage (GAC) data, with a resolution of 4 km. The data are mapped to an equal-area projection with a grid cell of approximately 7.6 km (Tucker *et al.*, 1991) and the grid cell closet to each rainfall station is utilized. A cloud mask is applied and daily values are formed into monthly composite images by using for each pixel the maximum NDVI within the compositing period (Holben, 1986). This minimizes the effects of atmosphere, scan angle and cloud contamination. The definition of NDVI as a ratio minimizes the influence of varing solar zenith angles and surface topography on the derived data set.

In this paper, we examine the limits of the usefulness of NDVI through a number of analyses of the relationship between NDVI and rainfall. The results presented demonstrate the large scatter in the relationships, the variability in the relationships resulting from differences in vegeation and soil type, and the saturation of the index at relatively high values of NDVI and rainfall. The analyses do not directly evaluate the quality of the NDVI/vegetation growth relationship, but the results have important implications for this.

2. General NDVI-rainfall relationships

The relationship between NDVI and rainfall has been studied in a series of papers (Malo and Nicholson, 1990; Nicholson et al., 1990; Davenport and Nicholson, 1993; Farrar and Nicholson, 1993). For the Sahel region of West Africa (Fig. 1), a clear linear relationship exists between mean annually integrated NDVI and mean annual rainfall and the spatial patterns of both variables are remarkably similar (Fig. 3). When the values for individual years are considered, there is considerably more uncertainty in the relationship (Fig. 2).

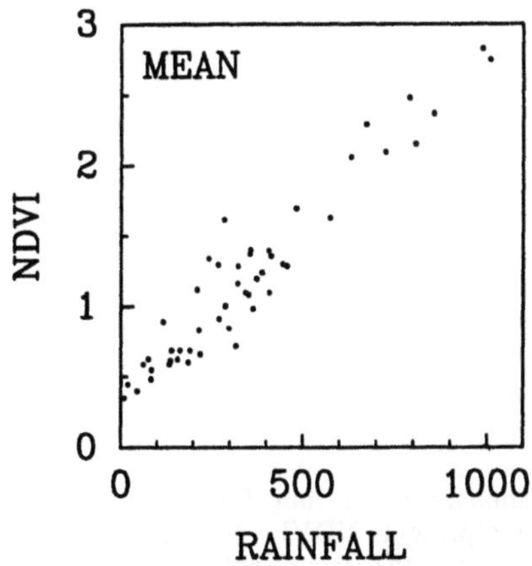

Fig. 1. Scatter diagram of mean annual integrated NDVI vs. mean annual rainfall (each points represents a four-year mean at one station) (from Malo and Nicholson, 1990).

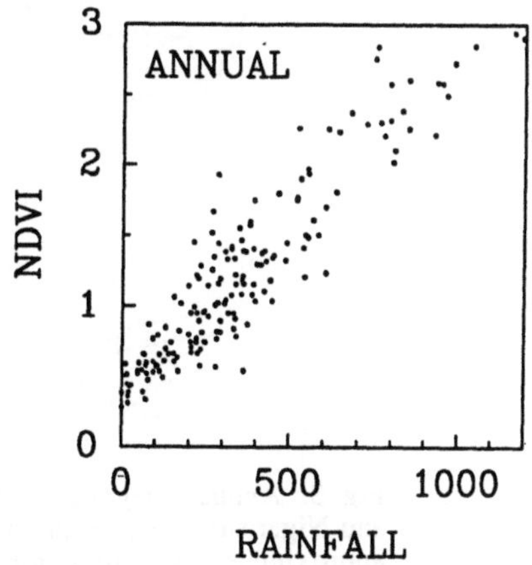

Fig. 2. Scatter diagram of annual integrated NDVI vs. annual rainfall (each point represents one station-year) (from Malo and Nicholson, 1990).

In East Africa, the mean spatial patterns of NDVI also reflect mean rainfall (Fig. 4). However, the ratio between these two variables is quite different than in the Sahel A mean annuallly integrated NDVI of 1 corresponds to roughly 200 mm of rainfall per year in both the Sahel and in East Africa. But a value of 2 corresponds to roughly 400 and 500 mm, respectively, in these two regions and a value of 3 corresponds to roughly 600 and 1100 in the two regions.

In East Africa, in contrast to the Sahel, the relationship is more log-linear than linear. As rainfall continues to increase, NDVI increases progressively more slowly and tapers off to a near-constant value (Fig. 5). This is hereforth referred to as "saturation" of the index. When the analysis in Fig. 1 is performed using monthly data for Sahelian stations, a similar saturation is suggested. In both East and West Africa, the threshold where it occurs appears to be on the order of 1000 to 1200 mm per year or about 200 mm per month (Nicholson et al., 1990).

In Southern Africa, the situation is similar but the threshold appears to be lower. Fig. 6 shows the comparison between NDVI and rainfall. NDVI tapers off at about 800 mm/an. Over some soil types in Botswana, the thresholds appear to be as low as 500 mm/an and 50 mm/mo (Farrar and Nicholson, 1993).

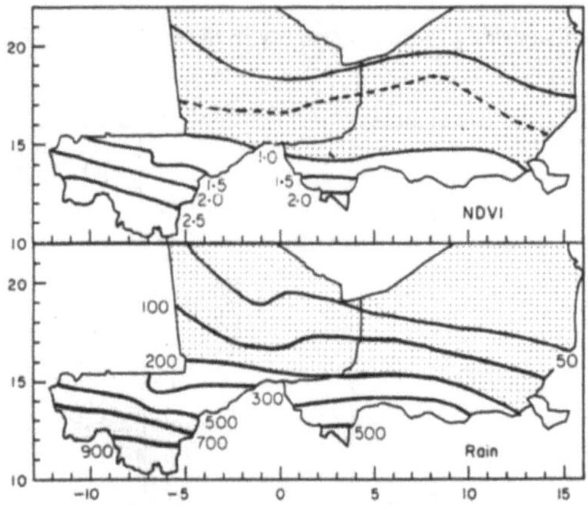

Fig. 3. Annual integrated NDVI and annual rainfall (mm) for Mali and Niger (four-year mean, dashed line represents NDVI = 0.6, the approximate lower limit for vegetation growth) (from Malo and Nicholson 1990).

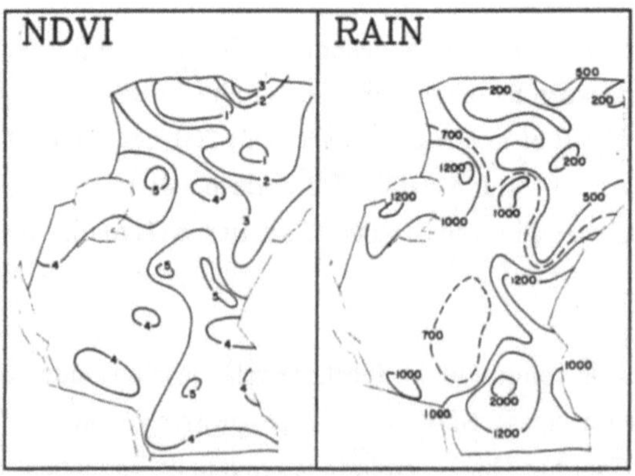

Fig. 4. As in Figure 3, but for East Africa (from Davenport and Nicholson, 1993).

3. NDVI and inter-annual variability of rainfall

Mean patterns of NDVI bear a close relationship to rainfall patterns, both spatially and temporally. However, the sensitivity of NDVI to interannual variations in rainfall remains to be established. There appears to be a good relationship in relatively dry locations, with rainfall well below the annual and monthly thresholds for saturation, but at wetter locations the phenology of NDVI is almost in variant from year-to-year despite large fluctuations in rainfall (Fig. 7).

This problem is further demonstrated using data for East Africa for three contrasting years (Fig. 8). The patterns of NDVI strongly reflect those of rainfall in central and northern regions, which are relatively dry. In the more humid forest regions to the south, no relationship is apparent. For example, NDVI actually increases during the year 1984/85, when rainfall is low compared to the two previous years.

The correlation between rainfall and NDVI also suggests a saturation of the index (Fig. 9). In Southern Africa monthly rainfall is well correlated with integrated NDVI, but much of the correlation is due to similar annual cycles. When the data are deseasoned by subtracting the monthly mean, correlations are high in relatively dry regions but decrease markedly as mean annual rainfall approaches 500 mm and become generally insignificant when rainfall exceeds about 700 mm. This provides further evidence of a saturation of NDVI at about 700 to 800 mm over Southern Africa.

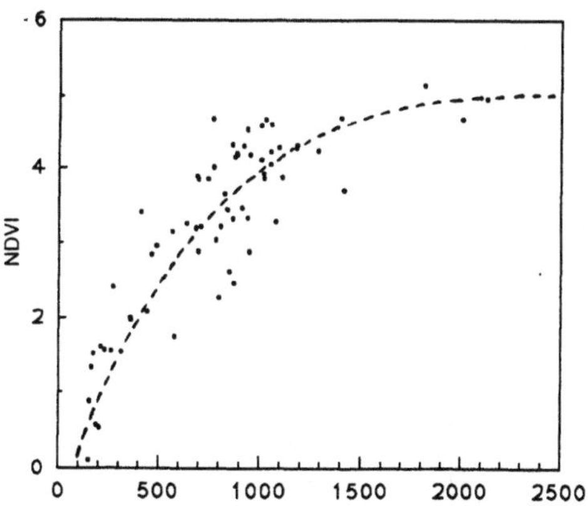

Fig. 5. Scatter diagram of annual integrated NDVI and mean annual rainfall (three-year average) for 65 stations in East Africa (dashed line is approximate best fit of data) (from Davenport and Nicholson, 1993).

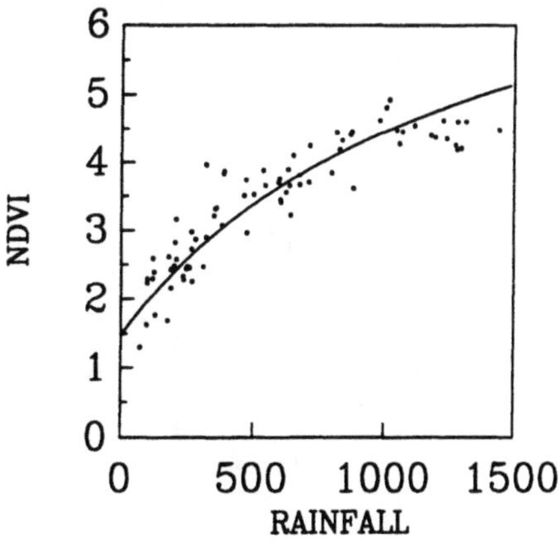

Fig. 6. As in figure 1, but for stations along a wet-to-dry transect in Southern Africa, extending from Botswana northeastward through Zambia and Zimbabwe (from P. Santos, unpublished).

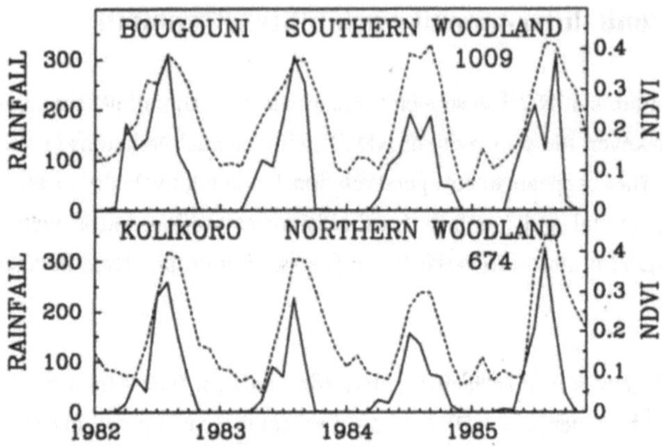

Fig. 7. Time series of monthly composites of NDVI and rainfall for two stations in West Africa (thick line, NDVI; dashed line, three-month moving average of rainfall in the concurrent plus two previous months; mean annual rainfall is indicated in the upper left).

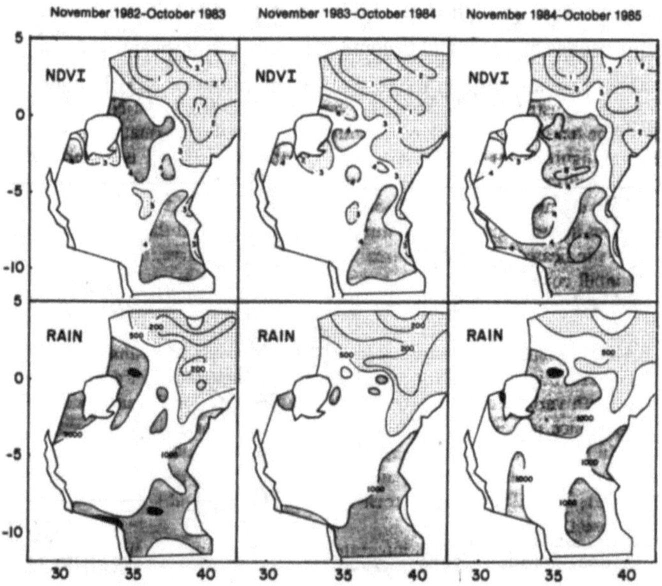

Fig. 8. Annual integrated NDVI and annual rainfall for the years 1983, 1984 and 1985 (stipling, rainfall below 500 mm/an and NDVI below 3; shading, rainfall above 1000 mm/an and NDVI greater than 4; dark shading, rainfall above 2000 mm/an (from Davenport and Nicholson, 1993).

4. NDVI-rainfall relationships by vegetation class

Fig. 10 illustrates the NDVI/rainfall relationship for several vegetation types in West Africa. To quantify this relationship, the ratio of individual values of the variables is utilized. The relationships have been assessed for three sectors of Africa: areas of East, West and Southern Africa which include large areas of savanna and woodland. The results are summarized in Tab. 1.

These results suggest two major conclusions: that the NDVI/rainfall relationship is vegetation specific and that it differs greatly from one region to another. In East Africa, the ratios vary from 3.5 for both the semi-desert shrub and the wet miombo woodland regions to 5.6 for thickets. In the Sahel, for southern and northern woodlands (which are analogous to wet and dry miombo), the ratios are 2.8 and 3.5, respectively, compared to 3.5 and 4.3 for the corresponding vegetation types in East Africa. In Botswana, in the mopane woodland, the ratio is 7.0; it is nearly as high in the transition zone and drier thornveld. Thus, the values of NDVI have much different meanings in the three locations and for different vegetation types. Differences can be due to a variety of factors, including percent cover, underlying soils, efficiency of evaporation in the various climate regimes, water-use-efficiency of various species, and soil fertility.

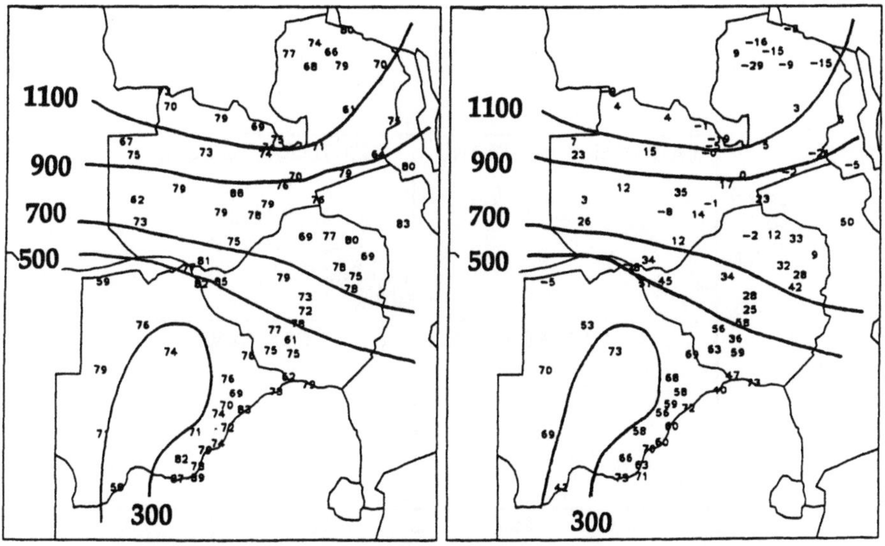

Fig. 9. Correlation between rainfall and NDVI over Southern Africa during the period 1983 to 1991. Rainfall during a given month is correlated with the average NDVI for that month plus the two preceding months. On the left, raw rainfall and NDVI data are correlated; on the right, both NDVI and rainfall data have been deseasoned prior to correlation. Annual rainfall isohyets in mm are indicated.

Tab. 1 Relationship between NDVI and rainfall for various vegetation formations in the West African SAHEL, East Africa and Botswana.

Vegetation	Annual NDVI	Annual Rainfall	NDVI/RAIN
	EAST AFRICA		
Wet Miombo	3.7	1102	3.5
Dry Miombo	3.8	948	4.3
Thicket	3.9	700	5.6
Shrub-Thicket Mosaic	3.3	546	5.1
Semi-Desert Shrub	1.1	297	3.5
	SAHEL		
Southern Woodland	2.6	951	2.8
Northern Woodland	1.5	460	3.5
Grassland/ Woodland	0.9	249	3.9
	BOTSWANA		
Mopane Woodland	2.7	381	7.0
Transition Zone	2.3	350	6.5
Kalahari Thornveld	2.2	333	6.6

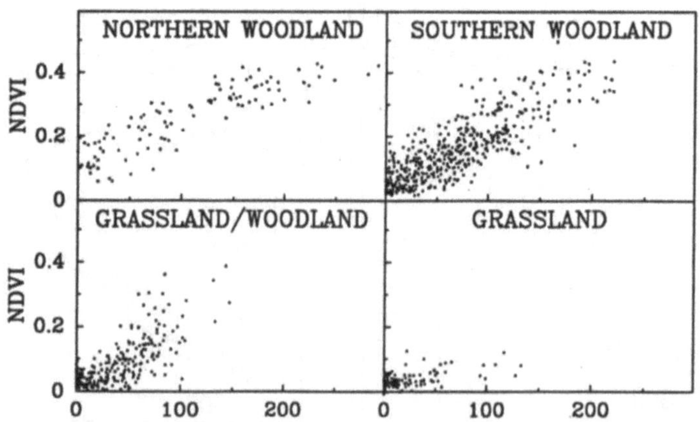

Fig. 10. Scatter diagrams of NDVI and rainfall (mm/mo) for four vegetation formations in West Africa (from Malo and Nicholson, 1990).

5. NDVI-rainfall relationships by soil type

Soil affects NDVI in a number of ways. The spectral signature of the underlying soil has a significant effect on the value of NDVI (Huete, 1989), but the soil also determines the rate of vegetation growth as a result of differences in fertility and water-holding and infiltration capacity. Fig. 11 shows monthly time series of NDVI and rainfall for two stations (Palapye and Tshane) on different types of sandy soils (arenosols) and for one station (Pandamatenga) in a region with vertisols (clay-rich soils) and one (Kanye) in a region with primarily cambisols (fine-to-medium-grained). Mean annual rainfall is comparable at Palapye and Tshane and at Kanye and Pandamatenga, so that differences in NDVI reflect soil effects (Note: in this region soil types are closely linked to vegetation types as well).

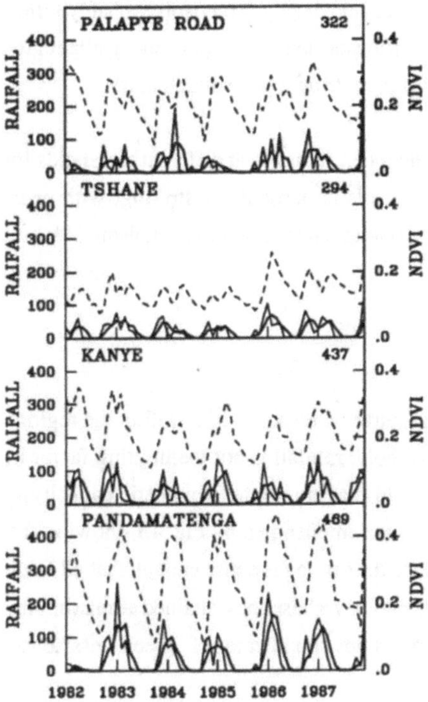

Fig. 11. As in 8, but for four stations in Botswana (Southern Africa)

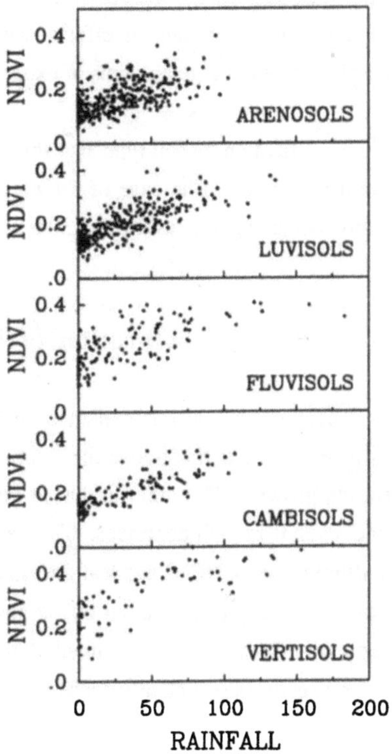

Fig. 12. As in 10, but for stations representing five soil types in Botswana (from Farrar and Nicholson, 1993).

Tab. 2 Ratio of NDVI to rainfall for six soil types in Botswana

	NDVI/RAIN
Arenosols	.66
Luvisols	.67
Fluvisols	.82
Cambisols	.70
Vertisols	.71
Solonchaks	.61

Fig. 12 shows scatter diagrams of NDVI and rainfall for five soil types in Botswana. Systematic differences are apparent. The ratio of NDVI to rainfall varies from .66 for arenosols to .82 for fluvisols (Tab. 2). With the exception of the fluvisols, the ratio appears to increase with decreasing sand content. The high values calculated for fluvisols probably reflect station location, rather than soil effects. Most of the fluvisols are in the Okavango Delta region of the country, with large amounts of exogenous water contributing to soil moisture.

Stratification by soil type also shows that the index saturates at different thresholds for different soil types. The case of the clay-rich vertisols is particularly striking, with near-saturation oocurring with as little as 30 mm/mo or 500/mm/an (Farrar and Nicholson, 1993).

6. Reasons for saturation of NDVI

There are two posible explanations for the saturation response of NDVI in the regions investigated. One is that above the saturation threshold, rainfall is not the limiting factor in vegetation growth. The other is that at this threshold maximum photosynthetic capacity is reached, even if growth continues. This latter effect is demonstrated by Fig. 13, showing the relationship between NDVI and leaf-area index. Once the canopy is dense enough that all PAR (photosynthetically active radiation) is absorbed, photosynthesis has attained a maximum, although growth can still be sustained. Thus, NDVI is not an indicator of "green-ness" once the canopy becomes sufficiently dense.

From this exercise, it is impossible to determine which explanation is most likely. Probably both factors play a role. However, in West Africa the 1000-1200 mm threshold is close to that above which NDVI reflects evapotranspiration, rather than soil moisture, indicating a change from a water-limited to an atmosphere-limited system.

7. Summary and Conclusions

The mean temporal and spatial patterns of NDVI closely resemble those of rainfall. In relatively dry locations, NDVI tends to increase linearly with rainfall. In more humid regions, the index saturation and approaches a constant value as rainfall continues to increase. Because of this saturation, the use of NDVI as an indicator of either rainfall or greenness is limited to relatively dry regions. The threshold for saturation is not constant. In East and West Africa, in is on the order of 200 mm/mo or 1000-1200 mm/an. In Southern Africa, it is generally on the order of 800 mm, but as low as 30-50 mm/mo and 500/an on clay-rich soils. If a constant threshold of 1000 mm were assumed, this would imply that NDVI is a poor indicator over about one third of the African continent (Fig. 14).

Both the saturation threshold and the quantitative relationship between NDVI and rainfall is dependent on vegetation class and soil type, as well as being regionally specific. This limits the degree to which NDVI can be used to monitor vegetation or rainfall, particularly interannual variability.

***Acknowledgements*:** This research was partially supported by NASA grants NAG-5-764, NAG-5-1587, NGT-70062, NGT-70273, and by a grant from AID. C. J. Tucker, of NASA, and Steve Prince, of the University of Maryland, provided NDVI data.

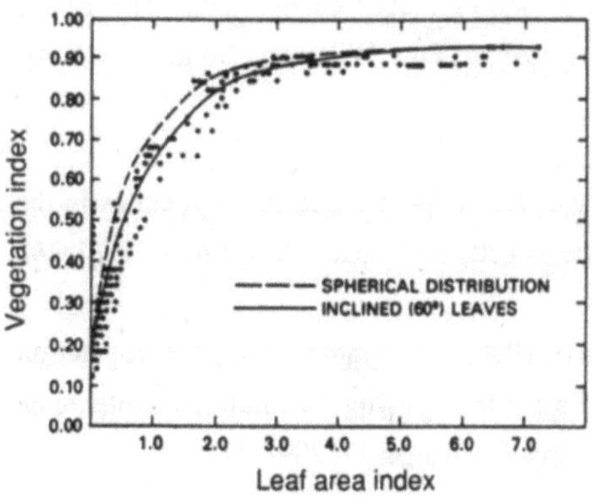

Fig. 13. Observed and predicted values of NDVI as a function of leaf area index for a wheat canopy (from Sellers, 1985; Asrar et al., 1984).

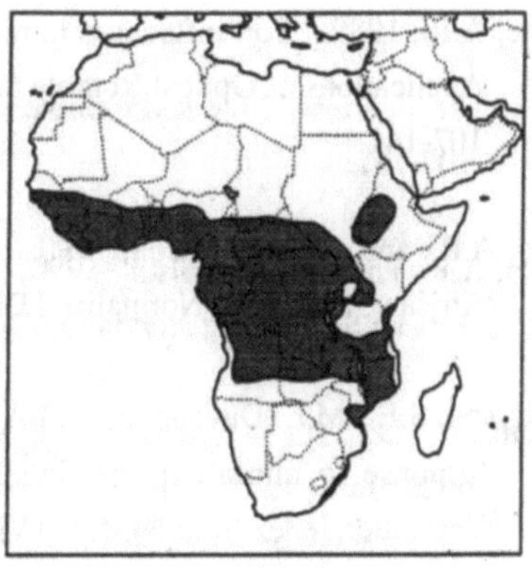

Fig. 14. Map of Africa, indicating areas receiving more than 1000 mm of rainfall annually, in the mean.

REFERENCES

Asrar, G., M. Fuchs, E.T. Kanemasu and J.L. Hatfield, 1984: Estimating absorbed photosynthetic radiation and leaf area index from spectral reflectance in wheat. Agron J., 76, 300-306.

Curran, P.J., 1980: Multispectral remote sensing of vegetation amount. Prog. Physical Geog., 4, 315-341.

Davenport, M.L., and S.E. Nicholson, 1993: On the relation between rainfall and the Normalized Difference Vegetation Index for diverse vegetation types in East Africa. Int. J. Remote Sens., 14, 2369-2389.

Derrien, M., B. Farki , H. Le Gleau and A. Sairouni, 1992: Vegetation cover mapping over France using NOAA-11/AVHRR. Int. J. Remote Sens., 13, 1787-1795.

Farrar, T.J., S.E. Nicholson and A.R. Lare, 1993: The influence of soil type on the relationships between NDVI, rainfall and soil moisture in semi-arid Botswana. Part I. Relationship to rainfall, Remote Sensing of Environment.

Holben, B.N., 1986: Characteristics of maximum-value composite images from temporal AVHRR data. Int. J.Remote Sens., 7, 1395-1416.

Huete, A.R., 1989: Soil influences in remotely sensed vegetation canopy data. Theory and Applications of Optical Remote Sensing. G. Asrar, Ed.: Wiley and Sons, New York, 107-141.

Malo, A.R., and S.E. Nicholson, 1990: A study of rainfall and vegetation dynamics in the African Sahel using Normalized Difference Vegetation Index. J. Arid Envir., 19, 1-24.

Nicholson, S.E., M.L. Davenport, and A.R. Malo, 1990: A comparison of the vegetation response to rainfall in the Sahel and East Africa, using Normalized Difference Vegetation Index from NOAA AVHRR. Climatic Change, 17, 209-241.

Sellers, P.J., 1985: Canopy reflectance, photosynthesis, and transpiration. Int. J. Remote Sens., 6, 1335-1372.

Tucker, C.J., B.N. Holben, J.H. Elgin, and J.E. McMurtrey, 1981: Remote sensing of total dry-matter accumulation in winter wheat. Remote Sens. Environ., 11, 171-189.

Tucker, C.J., C. Vanpraet, E. Boerwinkel, and A. Gaston, 1983: Satellite remote sensing of total dry matter production in the Senegalese Sahel. Remote Sens. Environ., 13, 461-474.

Tucker, C.J., C.L. Vanpraet, M.J. Sharman, and G. Van Ittersum, 1985: Satellite remote sensing of total herbaceous biomass production in the Senegalese Sahel: 1980-1984. Remote Sens. Environ., 17, 233-249.

Tucker, C.J., H.E. Dregne, and W.W. Newcomb, 1991: Expansion and contraction of the Sahara Desert from 1980 to 1990. Science, 253, 299-301.

Tucker, C.J., and P.J. Sellers, 1986: Satellite remote sensing of primary production. Int. J. Remote Sens., 7, 1395-1416.

VALIDATION AND INITIALIZATION OF NUMERICAL MODELS

IMPROVED PRECIPITATION FORECAST SKILL FROM THE USE OF PHYSICAL INITIALIZATION

T.N. Krishnamurti, G.D. Rohaly and H.S. Bedi
Department of Meteorology
Florida State University
Tallahassee, FL 32306-3034

ABSTRACT

This study explores the impact of physical initialization on the numerical weather prediction of tropical rainfall. An important finding of this study is related to the point correlations of the model-based rainfall at the initial time and at the end of a one day forecast which are significantly improved with the use of physical initialization within the tropics. Physical initialization refers to the use of a number of 'reverse' algorithms during an assimilation phase of the model forecast. The goal is to improve the definition of the initial state by the assimilation of proposed or currently available surface and satellite-based observations during a pre-integration phase of a forecast using a global spectral model.

The 'observed' rainrates are obtained from algorithms that translate satellite-based measurements of outgoing longwave radiation (OLR) and radiances for an array of microwave frequencies. In addition, the available rain gauge records over the land area are incorporated to define the 'observed' rainfall over the gaussian transform grid squares of a global spectral model at a high resolution (T106). Thus the rainfall measurements are averages over roughly 100 × 100 square kilometers and 7.5 minutes (the time step of the spectral model). It is for this averaged representation that we are able to demonstrate a very marked skill in nowcasting and one day forecasting of tropical rainfall. The monthly mean rainfall climatology, thus obtained, nearly replicates the rainfall analyses provided to the physical initialization.

NATO ASI Series, Vol. I 26
Global Precipitations and Climate Change
Edited by M. Desbois and F. Désalmand

This study also addresses a comparison of two different rainrate algorithms for the microwave measurements from a polar orbiting satellite. We note somewhat excessive rain for one algorithm as compared to the other; however, the physical initialization appears to accept both rainrates well. This calls for a more reliable measure of observed rain.

1. Introduction

The sparsity of conventional observations over the tropics contributes to the low skill in the numerical weather prediction of the low latitudes. Given the satellite-based estimates of rainrates, it is possible to invoke a physical initialization procedure within an assimilation phase of a model forecast to enhance the information content of the initial state. The term "physical initialization" was first introduced by Krishnamurti et al. (1984) to use satellite-based rainrates thus augmenting the data content of the tropical latitudes. Currently, the satellite-based rainrates and the outgoing long wave radiation are used to initialize the surface fluxes, the model rainrates and the model-based outgoing long wave radiation (Krishnamurti et al., 1991, 1993).

It should be possible to improve the nowcasting and short range forecasting skill of rainfall considerably by implementing the aforementioned physical initialization. The purpose of this paper is to demonstrate the nowcasting and forecasting skill from a large number of experiments. Our definition of skill will be based on a direct calculation of point correlations of the model and analysis rainrates. The "observed rain" is defined as a mix of: algorithms based on microwave, outgoing long wave radiation and raingauge data. The tropical latitudes between approximately 30°S and 30°N are considered. We have prepared analyses of the daily 'observed' rainfall distributions for the month of October, 1991. This rainfall was analyzed directly on to the transform grid squares of the global model at the horizontal resolution T106.

The Special Sensor Microwave Imager (SSM/I) includes vertically and horizontally polarized radiation sensors at 19.3, 37.0 and 85.5 Gigahertz (GHz). The fourth frequency at 22.2 GHz only includes the capability to receive vertically polarized radiation.

It is perhaps premature to qualify the superiority of one rainrate algorithm as compared to another at the present time. Several algorithm intercomparison exercises are currently under way. In this context, we shall examine the impact on physical initialization of two such algorithms, Olson et al. (1990) and Kummerow and Giglio (1994).

The Florida State University (FSU) global spectral model has been used in our earlier studies on physical initialization, Krishnamurti et al. (1991, 1993). The same model at the resolution T106 has been used to perform the experiments carried out here.

2. The Sequence of Computations

Lacking a unified column model that integrates all of the physical processes (i.e. the surface flux parameterization, the cumulus parameterization and the radiative transfer scheme), the current model has separate computational algorithms for these components. The sequence of computations within the physical initialization modifies the moisture variable separately to satisfy each requirement, i.e. fluxes, rainrate and the earth radiation budget.

Figure 1 presents a schematic outline of these computations. The sequence of reverse algorithms are: reverse similarity followed by OLR matching and then the reverse cumulus parameterization. The time interpolated rainrates are an input to the reverse similarity and the reverse cumulus parameterization algorithms. The satellite-based OLR enters the matching algorithm. A one time step, forward integration follows the execution of the reverse cumulus parameterization algorithm. Since the reverse cumulus and the normal cumulus are carefully designed to be almost completely reversible, the 'observed' rainrate is nearly returned by the model at the end of the time step. Thus, an accumulation over 24 hours of assimilated rainfall is nearly provided by the model. It is this feature that provides us with a mean nowcasting skill of approximately 85% as

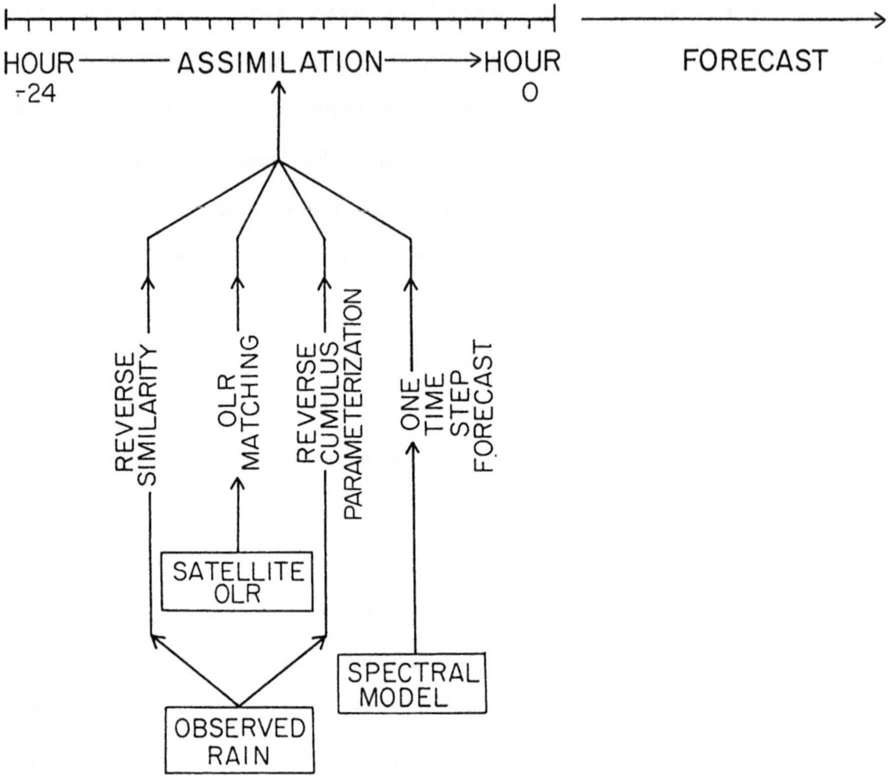

Figure 1. An outline of the physical initialization within the assimilation phase of the global model. The package of reverse similarity, OLR matching, reverse cumulus parameterization slides from left (hour - 24) to right (hour 0) during this initialization period.

measured by point correlation over the model gaussian grid box of roughly 100 × 100 square kilometers.

3. Forecast skill from Physical Initialization

We shall show the results of the physical initialization and one day forecasts based on this improved initial state for a large sample of experiments. In Figure 2 we illustrate the rainfall predictive skill in the global tropics for days 0 and 1 when the physical initialization was invoked. Also shown is the rainfall forecast skill for day 1 of the control experiments where the physical initialization was not used. Results of 31 forecasts are shown; these results show a marked skill at day 0 and day 1 for the physically initialized experiments. These skills are substantially above those of the control run. The observed rainfall is the 24 hour rainfall total ending at day 0 or 1 of the forecasts. This is based on the mix of OLR, SSM/I and raingauge data following Gairola and Krishnamurti (1992).

Perhaps the most spectacular result is the nowcasting skill from the physical initialization. The mean skill, as measured by point correlation, for the entire month of October, 1991 is around 0.85. The mean one day forecast skill for the physically initialized experiment is around 0.55. The mean skill of the one day forecast of the control run (without physical initialization) is around 0.40. The one day skill with the physical initialization far exceeds the skill of operational forecasts which are of the order of 0.20 - 0.40. This demonstrates a major improvement in the one day rainfall forecast over the global tropics from physical initialization. The one day forecast skill of persistence was around 0.30 which is well below that stated above.

4. Monthly Mean Rainfall

Since the physical initialization forces the analysis of the daily rain towards the observed amount to an accuracy of nearly 85% (i.e. point correlation of 0.85), it is not surprising that one can attain a rather high skill for the monthly total rainfall by simply adding a month of daily observations. There is a very good agreement between the

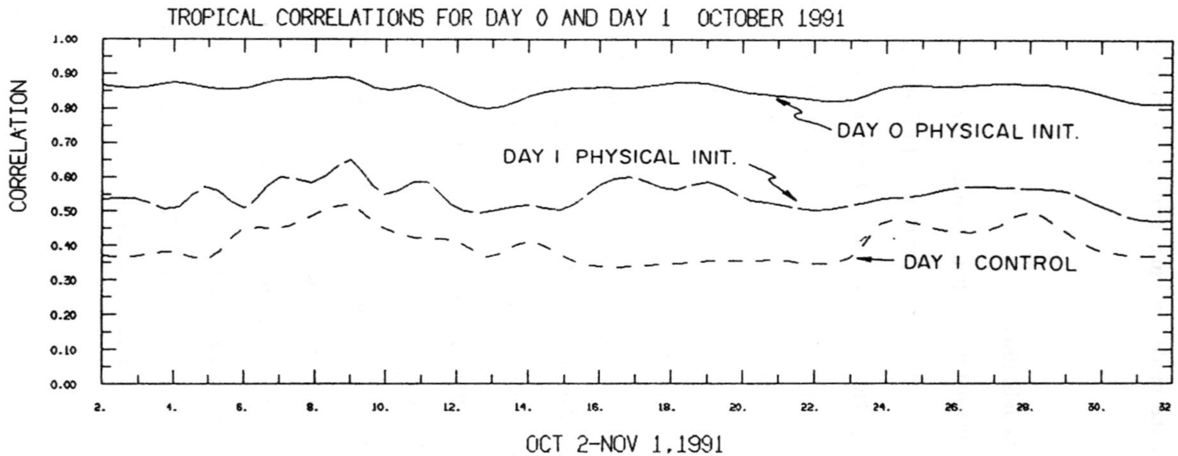

Figure 2. Correlation of 'observed', based on FSU rainfall analyses and the predicted rainfall. Top line is for the day 0 skill with physical initialization, the middle (long dashed) line shows the day 1 skill with physical initialization and the bottom (short dashed) line shows the day 1 skill based on the control run without physical initialization. Abscissa shows days of the month.

'observed' and predicted monthly total rainfall for October, 1991. Even smaller scale features of the rainfall maxima over most regions have been reproduced by the proposed initialization.

The proposed physical initialization was only invoked over the tropics, i.e. approximately 30S to 30N. We note a lack of agreement between the 'observed' and the physically initialized monthly rainfall over the South Pacific Convergence Zone (SPCZ) east of Australia. That region of discrepancy is due to the 29.719S boundary for the physical initialization. It implies that convective rain was occurring along the SPCZ near 30S and the model was not able to recover it because of interaction with midlatitude systems which were not physically initialized due to the latitudinal cut off. This perhaps calls for the need for the extension of the boundaries of physical initialization beyond the tropics.

We calculated the difference between the predicted and the observed monthly mean values (expressed in the units of mm/day). The difference between the observed and the predicted fields are less than ±2 mm/day over most regions. The maximum difference over some isolated points is of the order of ±6 mm/day. The largest discrepancies are noted over Brazil, Africa and over the maritime continent (near Indonesia). The SPCZ near 30S has already been addressed.

The coverage of the SSM/I instrument may be one reason why the discrepancies arise. The areas where the maximum differences occur are the areas which have the highest total rainfall for October, 1991. This can be seen in the Global Precipitation Climatology Project (GPCP) precipitation estimates of monthly accumulated rainfall. These estimates are computed using techniques developed by Arkin (1979), Arkin and Meisner (1987) and Janowiak and Arkin (1991) which uses a temperature threshold to produce a rainfall analysis. It appears that the SSM/I experiences small data voids between the satellite swaths. These data voids are virtually nonexistent when two or more satellites are providing observations; however, for this time period the Defense Meteorology Space Program (DMSP) F10 satellite SSM/I was not providing a steady stream of data, and the 85 GHz channels were faulty on the DMSP F8 satellite.

5. Forecast skill beyond 1 day

The skills for days 0 and 1 for the physical initialization are of the order of 0.9 and 0.55 respectively. The day 1 skill with physical initialization is much higher than the control; note there was no rainfall for the control at day 0. This is an important result of the present study. However, the skill of forecasts for days 2, 3 and 4 do not exhibit any appreciable advantage from the inclusion of physical initialization. Physical initialization does not substantially modify the rotational part of the wind. The data void tropical oceans do have a poor description of the basic flow field on the synoptic and the mesoscale thus we are not able to enhance the forecast skill beyond the control experiment. The 2, 3 and 4 day skills are comparable to the day 0 and day 1 skill of operational forecast centers.

Another factor that leads to a loss of skill by day 2 is the phase speed errors of the tropical disturbances. Although the storms are very well retained later in the forecast, the slight phase errors in the storm motion result in an effective doubling of the penalty for the forecast skill scores. The 'observed' and the predicted rainfall maxima are sufficiently separated and large errors occur over the respective locations. Reduction of such errors will require a better control on the phase speed error of the disturbances. To some extent that can be achieved from a further increase of resolution of the global model. However, the sparsity of overall tropical observations may place a limit on the reduction of such errors.

6. A Comparison of the Kummerow and Olson Rainrate Algorithms

Sample rainrate descriptions for September 6, 1987 for the Kummerow and the Olson rainrate algorithms are illustrated in Figures 3a and 3b, respectively. This example pertains to a triple typhoon event over the western Pacific Ocean described by Krishnamurti et al. (1993). The scale identifies the rainfall in the units of mm/day. This

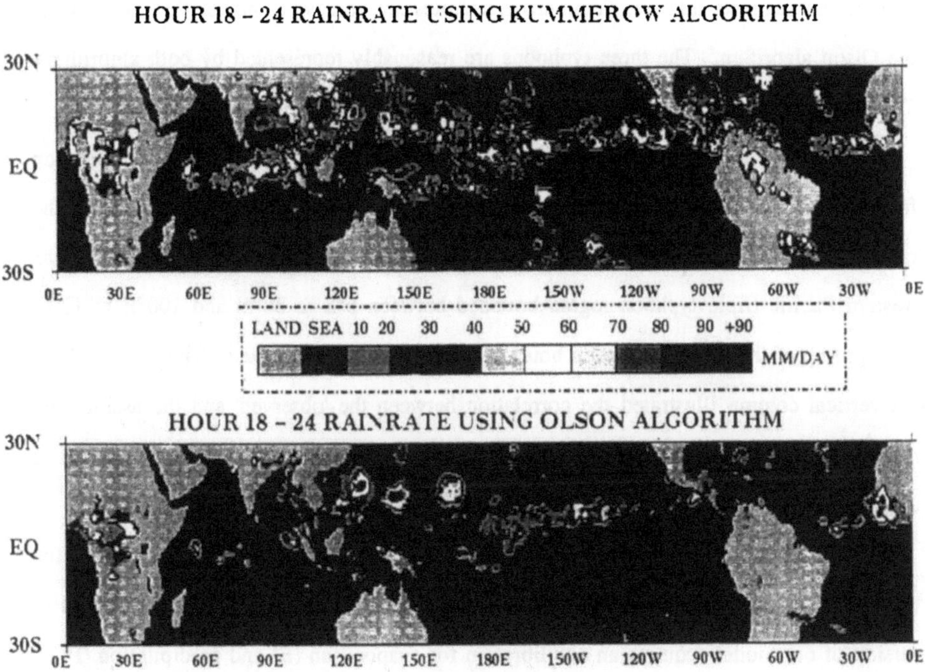

FIGURE 3A & 3B

Figure 3. Physically initialized mean rainrate between Sept. 6, 1987, 600 UTC and Sept. 6, 1987, 12 UTC for two rainrate algorithms. a) Kummerow and Giglio, b) Olson.

comparison suggests that the Kummerow rainrates are somewhat higher than those from the Olson algorithm. The three typhoons are reasonably represented by both algorithms. The African monsoon rainfall appears to be stronger for the Olson algorithm.

We have illustrated the spin-up of the rainfall during the physical initialization from the use of the two algorithms in Table 1. This table illustrates the spin-up of the rainfall over the entire tropical belt between approximately 30°S and 30°N and for the western Pacific triple typhoon region bounded between 5°S to 30°N and 100°E to 180°. The spin-up of the rainfall between hours -18 and hour 0 is illustrated. The entries along the vertical column illustrated the correlation between the 'observed' and the assimilated rainfall for the respective rainrate algorithms. It is apparent that the final correlation obtained is higher for the Olson as compared to the Kummerow algorithm.

The reason for this difference lies in part to the equilibration level of rainfall for the T106 resolution of our global spectral model. During a ten day forecast the T106 version of our model acquires an equilibration for evaporation (E) and precipitation (P) at a level of roughly 3.25 mm/day. If the assimilation for a particular algorithm yields values of P far different from this equilibration level then the model has difficulties adjusting during and after the physical initialization. The Kummerow algorithm produces an unusually large mean value (≈ 8.5 mm/day) for the tropicals, whereas the Olson algorithm has a value (≈ 4.2 mm/day) somewhat closer to the equilibration value for the T106 version of our model. The Pacific typhoon belt experiences an unusually high value of wide spread rain from the Kummerow algorithm, seen in Table 1. It appears from this analysis that care should be exercised in the development of rainrate algorithms to assure that global rainfall rates are reasonable from the global hydrological perspective.

It should also be noted that the physical initialization does seem to nearly accept whatever rainrates we have been able to provide to it. Thus the initialization is a strong function of the rainrate algorithms which the model is being forced to adjust toward via these reverse algorithms. This brings up the need for a more reliable global distribution of rainrates.

Table 1: An illustration of rainfall spin-up during physical initialization expressed in point correlation for the global tropics (top) and Pacific typhoon belt (bottom).

Global Tropical Belt (30°S to 30°N)

Olson Rain in Analysis	≈ 4.20 mm / day
Kummerow Rain in Analysis	≈ 8.50 mm / day

Hour	Olson	Kummerow
-18	0.71	0.67
-12	0.82	0.77
-6	0.85	0.77
0	0.85	0.75

Pacific Typhoon Belt (5°S to 30°N, 100°E to 180°)

Olson Rain in Analysis	≈ 9.00 mm / day
Kummerow Rain in Analysis	≈ 15.00 mm / day

Hour	Olson	Kummerow
-18	0.80	0.72
-12	0.92	0.81
-6	0.91	0.80
0	0.91	0.80

6.1 Comparison of Prediction using the Olson and the Kummerow Algorithms

The triple typhoon event was subjected to physical initialization using the Olson and the Kummerow rainrate algorithms. Separate prediction experiments were carried out with these respective algorithms at the horizontal resolution T106. The starting date for these experiments was September 5, 1987 at 12UTC. The experiment based on the Olson algorithm is described in some detail in Krishnamurti et al. (1993). The three typhoons over the western Pacific ocean (from west to east) were Gerald, Freda and Holly. Holly had become a supertyphoon by September 8th and essentially moved northwards. Gerald moved northwestwards whereas the central storm Freda moved slowly westward during the first three days.

Overall the track forecasts were handled better with the initialization based on the Olson algorithm. The excessive rain from the use of the Kummerow algorithm was noted over the entire central Pacific ocean. The central and the easternmost typhoons were nearly quasistationary. For both of the rainrate specifications the storms were maintained during the forecasts (which were carried out through 5 days). In contrast for a control experiment where no physical initialization was invoked the storms weakened considerably and Freda and Holly dissipated by day 2 of the forecast, Krishnamurti et al. (1993). Overall both forecasts with the rainrate initialization appear quite impressive when compared to the control forecast.

6.2 Predicted Rainrates from the use of the Olson and Kummerow Algorithm

We shall present the different 24 hour rainfall accumulations from the experiments and compare those to an analysis of rainfall which is valid on day 4 of the forecast. The four panels of Figure 4 illustrate these different forecasts in mm/day.

Figure 4a shows the results of the control experiment where no physical initialization was invoked. What is conspicuous is the absence of heavy rainfall maxima for the central and the easternmost typhoons Freda and Holly. These storms had essentially dissipated by day 2 of the forecast in the control run. The forecast of rain

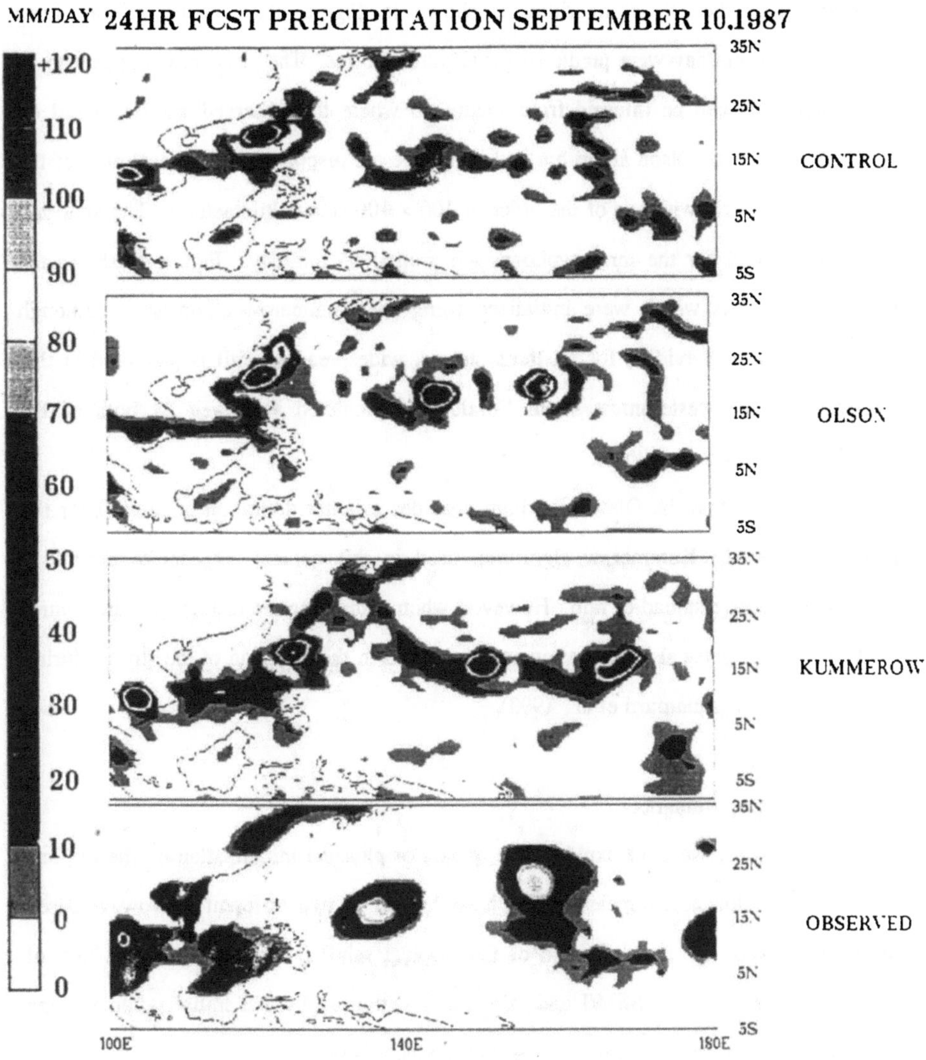

Figure 4. Forecast 24 hour rainfall totals for day 1 to day 2 for a) the control experiment, b)forecast physically initialized using the Olson algorithm, c) forecast physically initialized using the Kummerow algorithm and d) FSU analysis 'observed' rainfall based on OLR, raingauge and SSM/I via Olson algorithm. Units are mm/day. Color bar shown on left.

initialized with the Olson algorithm is shown in Figure 4b. Intense rainfall rates exceeding 110 mm/day were predicted for all three storms. The phase errors appear to be quite small, this can be inferred from Figure 4d where the 'observed' rainfall based on OLR, SSM/I via the Olson algorithm and raingauge are displayed (at the resolution of the T106 transform grid which is of the order of 100 × 100 square kilometers). The analyzed maximum rainfall for the three typhoons are around 85 mm/day. Figure 4c shows the results of forecasts which were initialized using the Kummerow algorithm. Although rainfall maxima are evident for all three storms, widespread rainfall is predicted in this experiment. The westernmost storm Gerald was predicted very well by both of the rainfall initialization algorithms.

It appears that the Olson algorithm provides a better definition of rainfall for the three typhoons, the Kummerow algorithm, used in this context, appears to provide a somewhat excessive spread of rain. However, when comparing the results with the control it is evident that physical initialization appears to aid in the retention of the storms during these forecasts (Krishnamurti et al., 1993).

7. Concluding Remarks

In this study we have explored the impact of physical initialization on the tropical rainfall assimilation and forecasts. We have looked at two different microwave-based rainrate algorithms for the definition of the tropical rainfall. Introduction of a mix of measurements from OLR, SSM/I and raingauges within a physical initialization scheme has improved the nowcasting and the one day forecasting skill.

Using a month long sample of assimilation-prediction experiments a point correlation of roughly 0.85 for the mean nowcasting skill of rainfall was demonstrated in this study. The observed and the predicted rainfall are averages over the transform grid squares of a global model which was run at a horizontal resolution of T106 (i.e. transform grid separation of roughly 100 × 100 square kilometers in the tropics). The mean correlation for day 1 of the forecast is around 0.55 and exceeds the current operational capability. However as one proceeds beyond day 1 the skill falls off rapidly towards the operational levels. This loss of skill beyond day 1 is attributed to a lack of adequate observations to define the rotational part of the wind on synoptic and subsynoptic scales. The physical initialization does enhance the divergent part of wind consistent with the rainfall description but does not have a large impact on the rotational part of the wind due to the Newtonian relaxation.

Another possible factor for the loss of skill may arise from small phase errors in the forecast of tropical disturbances. Although heavy rain is predicted by this procedure the phase errors in the position of typhoons appear to contribute to substantial errors of the rainfall.

Physical initialization using two different rainrate algorithms illustrates different amounts of tropical rain, one being somewhat higher than the other. Physical initialization appears to accept these different rainfall totals in the description of the initial state. Thus we are led to conclude that physical initialization can only be as accurate as the rainfall we provide to it.

The resolution of the present study, T106, does not permit us to examine meso-convective disturbances. We are currently extending the physical initialization to very high resolution global models such as T213. Even at this resolution the meso-convective disturbances whose scales are of the order of a few hundred kilometers are barely resolved. The resolution of the SSM/I microwave channels varies from roughly 30 to 60 kilometers. Physical initialization at a high resolution is able to nearly return the 'observed' rainrate over the transform grid squares. The spin-up during physical initialization modifies the vertical structure of humidity and in the evolution of stronger divergence (and possibly of smaller scale motions) and heating fields locally. Lacking mesoscale data we must rely on the physical initialization to generate the structure of the meso-convective systems. This shall be one of the major efforts for numerical weather prediction in the future.

Acknowledgements

Financial support for this work was received from NSF grant ATM 8812053, NASA grant NAG 5-1595, ONR grant N 00014-93-1-0243, and NASA grant NAG 8-914. Computations for this work were carried out on the National Center for Atmospheric Research (NCAR) supercomputer CRAY/YMP 8/64; NCAR is sponsored by the National Science Foundation.

REFERENCES

Arkin, P.A. and B.N. Meisner, 1987: The relationship between large-scale convective rainfall and cold cloud over the Western Hemisphere during 1982-84. *Mon. Wea. Rev.*, **115**, 51-74.

Arkin, P.A., 1979: The relationship between the fractional coverage of high cloud and rainfall accumulations during GATE over the B-scale array. *Mon. Wea. Rev.*, **107**, 1382-1387.

Gairola, R.K., and T.N. Krishnamurti, 1992: Rain rates based on SSM/I, OLR and raingauge data sets. *J. Meteorol. Atmos. Phys.*, **50**, 165-174.

Janowiak, J.E. and P.A. Arkin, 1991: Rainfall variations in the tropics during 1986-89, as estimated from observations of cloud top temperatures. *J. Geophys. Res.*, **96** Supl., 3359-3373.

Krishnamurti, T.N., H.S. Bedi and K. Ingles, 1993: Physical initialization using the SSM/I rain rates. *Tellus*, **45A**, 247-269.

Krishnamurti, T.N., J. Xue, H.S. Bedi, K. Ingles and D. Oosterhof, 1991: Physical initialization for numerical weather prediction over the tropics. *Tellus*, **43AB**, 53-81.

Krishnamurti, T.N., K. Ingles, S. Cocke, R. Pasch and T. Kitade, 1984: Details of low latitude medium range numerical weather prediction using a global spectral model II. Effect of orography and physical initialization. *J. Meteor. Soc. Japan*, **62**, 613-649.

Kummerow, C. and L. Gilio, 1994: A passive microwave technique for estimating rainfall and vertical structure information from space. Part 1: Algorithm description. *J. App. Meteor.*, **33**, 3-18.

Olson, W.S., F.S. Fontaine, W.L. Smith and R.H. Achtor, 1990: Recommended algorithms for the retrieval of rainfall rates in the tropics using SSM/I (DMSP-F8) Manuscript. University of Wisconsin, Madison, 10 pp.

FUTURE CLOUD REPRESENTATION IN THE ECMWF MODEL

Graeme Kelly, Michael Tiedtke, Martin Miller and Jean-Jacques Morcrette.

ECMWF
Reading
UK

ABSTRACT

A prognostic scheme for stratiform and convective clouds is developed for the model. The time evolution of clouds is defined through the large-scale budget equations for cloud water and cloud air. The scheme has been tested in the ECMWF global model and compared with the current operational scheme. The representation of the cloud process has a strong effect on the hydrological cycle and the maintenance of the tropospheric water vapour cycle.

KEYWORD

Clouds, Numerical Models, Convection, Radiation, Hydrological cycle.

INTRODUCTION

Parametrization of cloud physical processes and radiative transfer in clouds has been a major research area at ECMWF during the past years. Modifications previously made to cloud parametrization in the operational model (e.g. in cumulus convection in 1985 and 1989 and in radiative transfer in 1985, 1989 and 1991) have improved the model performance in various aspects including forecast quality. However, cloud processes are still represented poorly in the ECMWF model as indeed they are in other models. Most of the difficulties are of a fundamental nature, although some are more specific such as the specification of cloud optical properties, which are still rather uncertain. The most obvious inadequacy is the inconsistent and incomplete treatment of cloud processes which is most evident in the decoupling of the radiation cloud scheme from the rest of the model's physics. There have been various attempts in recent years to develop unified cloud schemes for large scale models (e.g. Sundqvist, 1988; Smith, 1990; Le Treut and Li, 1988) but these schemes are still rather incomplete and inconsistent in various aspects (e.g. the treatment of convective clouds) and contain undesirable diagnostic features (e.g. for specifying cloud cover and cloud water content). The prognostic cloud scheme developed recently at ECMWF overcomes some of these shortcomings by adopting a model approach where all cloud-related processes are treated in a fully consistent way. The scheme is described in detail in Tiedtke 1993.

Initially, validation of the scheme has concentrated mainly on general circulation studies in order to see how well the major cloud fields are reproduced with the new scheme. Recently we

NATO ASI Series, Vol. I 26
Global Precipitations and Climate Change
Edited by M. Desbois and F. Désalmand

started to validate the scheme in synoptic case studies. Results summarized in Section 3 show that the new scheme produces more realistic cloud fields as well as a more realistic hydrological cycle. However, the interaction of the model-produced clouds with the radiative transfer parametrization needs further work as the net radiative cooling is excessive. Parametrization of radiative transfer in clouds is discussed in Section 4. There is also the initial value problem; presently cloud variables are initialized to zero. As a consequence spin-up problems are encountered. Both aspects, the excessive radiative cooling and the spin-up, are likely to affect forecast quality with the new cloud scheme and might be responsible for the negative impact found at resolution T213, although T106 forecasts are improved with the new cloud scheme.

2. DESIGN OF ECMWF UNIFIED CLOUD SCHEME

Here we summarize the basic features of the new cloud scheme and refer to Tiedtke (1993) for details. The scheme considers explicitly the sources and sinks of cloud water and cloud air model processes relevant for formation of cloud liquid water are: moist adiabatic lifting in large scale ascent, in cumulus updrafts, in convective updrafts in well-mixed boundary layers and diabatic cooling, clouds dissipate through: evaporation in moist adiabatic descent (large scale and/or cumulus-induced), diabatic heating, turbulent mixing with unsaturated environmental air and precipitation processes. The following indicates where each of these processes are represented:

Cloud formation:

i) Formation of stratiform clouds by large scale ascent and/or diabatic cooling (Tiedtke 1993,Section 2.4)
ii) Formation of convective clouds by detrainment of cloud air from convective updrafts (Tiedtke 1993, Section 2.2)
iii) Formation of cloud at the top of convective boundary layers (Tiedtke 1993, Section 2.3)

Cloud dissipation:

i) Evaporation by moist adiabatic descent (large-scale and/or cumulus-induced subsidence) and diabatic heating (Tiedtke 1993, Section 2.5)
ii) Turbulent mixing of unsaturated environmental air (Tiedtke 1993, Section 2.5)
iii) Entrainment processes at top of boundary layer clouds (Tiedtke 1993, Section 2.3)
iv) Precipitation processes (Sundqvist, 1988, Tiedtke 1993, Section 2.6, Heymsfield and Donner, 1990)

It should be noted that the scheme has two prognostic equations and, in contrast to other prognostic schemes, predicts cloud water liquid (CLW) as well as cloud cover. The equation for fractional cloud cover is derived from the large-scale budget equation for cloud air and contains source and sink terms due to cloud formation and cloud dissipation. Three principles, consistency, completeness and accuracy, guided the design of the new scheme.

Consistency: The new cloud scheme provides a large step towards a consistent treatment of cloud processes in large scale models, as cloud formation and dissipation are linked directly to model processes. The physical basis is that formation and evaporation of cloud water is determined by the time change of the saturation water vapour content. This physical concept is applied equally to stratiform and convective clouds. With regard to cumulus convection we note that part of the cloud water condensed in updrafts immediately falls out as rain while the remainder is transported upwards and detrained into the environmental air, thereby contributing to cloud formation. Although the physical concept for the new cloud scheme is sound and internally consistent, a fully consistent representation of all cloud related processes has not yet been obtained with the ECMWF model. This is because supersaturation is produced by non-physical processes such as spectral truncation. Condensation produced by non physical processes can be large as it amounts on average to 15% of the total condensation in stratiform clouds at resolution T213L31 and about twice as much at T63L31. The main culprit is probably the spectral truncation since diabatic processes are calculated on the full computational grid but are fitted with a coarser spectral resolution. A more consistent treatment might be obtained if the physics were calculated on a coarser grid on which tendencies could be reproduced after transformation to spectral space and back.

Completeness: Although the original motivation for developing the new cloud scheme has been to obtain a model-consistent treatment of cloud processes, the more apparent benefit is that of a more complete representation of cloud processes than with the ECMWF operational cloud scheme. The increased complexity is most evident in the model's hydrological cycle, e.g. formation and storage of cloud liquid water with the possibility of re-evaporation, introduction of partial cloudiness, formation of anvil and cirrus clouds by convection and the representation of boundary layer clouds. The coupling of cloud formation to the model's cumulus convection scheme represents an important extension of the present cumulus parametrization with benefits not only for the simulation of anvil and cirrus clouds due to cumulus convection but also for the overall hydrological cycle (Tiedtke 1993, Section 4.3 a, Fig. 8).

Accuracy: Accuracy of the prognostic scheme depends on the representation of the processes for formation and dissipation of clouds. Formation of clouds is determined by model processes such as large-scale ascent, cumulus convection, radiative cooling and boundary layer turbulence. Therefore, errors in simulated cloudiness may first of all indicate errors and deficiencies in those processes. Additional errors originate from the uncertainty of the tunable parameters for the precipitation processes, the turbulent diffusion coefficient for mixing of cloud air and unsaturated environmental air and the assumption for the critical humidity above which stratiform clouds are assumed to form, which at present is between 80% and 100% depending on height above ground.

3. VALIDATION OF NEW CLOUD SCHEME

3.1 Cloud cover

Initially we have verified cloudiness in extended integrations by comparing time averaged model-produced cloud cover against satellite observations from ISCCP and climate data from surface observations. Recently, we have also started to evaluate cloudiness synoptically in short range forecasts at resolution T213.

3.1.1 *Extended integrations*

The extended integrations show (see Tiedtke (1993) for details):

i) Total cloud cover appears realistic as the major cloud fields are reproduced: i.e. along the ITCZ, extratropical storm-tracks and over the cold water off the subtropical west coasts of the continents.

ii) High level clouds appear excessive in comparison to ISCCP, but model clouds are optically thin to a large extent as indicated by values of effective cloudiness (Tiedtke 1993, Fig. 2). Recent satellite observations employing multi-spectral techniques indicate widespread occurrence of thin cirrus which support the model's large cloud cover. The excess in high level cloudiness in tropical regions is smaller with the new scheme than the operational scheme.

iii) Extension of the model's boundary layer scheme to include the formation of clouds at the top of convective boundary layers has been successful in reproducing maritime stratocumulus fields to the west of the continents. Also, the transition to the cumulus regime downstream of the stratocumulus clouds is realistically captured by the model (Tiedtke 1993, Fig. 3).

iv) Low and middle level cloudiness in tropical regions is increased with the new cloud scheme but is still much lower than observed. Insufficient low level clouds in tropical regions indicate deficiencies in the convection scheme, in particular in the representation of the transports by shallow cumuli in the presence of deeper clouds. It is important to realize that one of the basic parametrization assumptions in the ECMWF convection scheme is that of a unimodal cloud distribution whereas observations typically show bimodal distributions with two maxima in entrainment/detrainment rates at the level of zero-buoyancy and just above the boundary layer (e.g. Yanai et al., 1973; Nitta 1975).

Fig. 1 compares the evolution histograms of the long-wave window brightness temperature for a 7-day period in June 1987, produced by T106 L31 simulations with the operational (SFF) and the new (SAF) cloud scheme over 10° by 10° limited areas. Over Central Africa, the operational cloud scheme with its convective cloud directly linked to the convective activity of the model shows a regular pattern of development followed by dissipation of the convective towers. By contrast, the new cloud scheme does not display such a regular pattern as it maintains cloud water in stratiform anvils even after convection has switched off. Although histograms have not yet been produced for observed radiances we are certain that the new scheme is more realistic

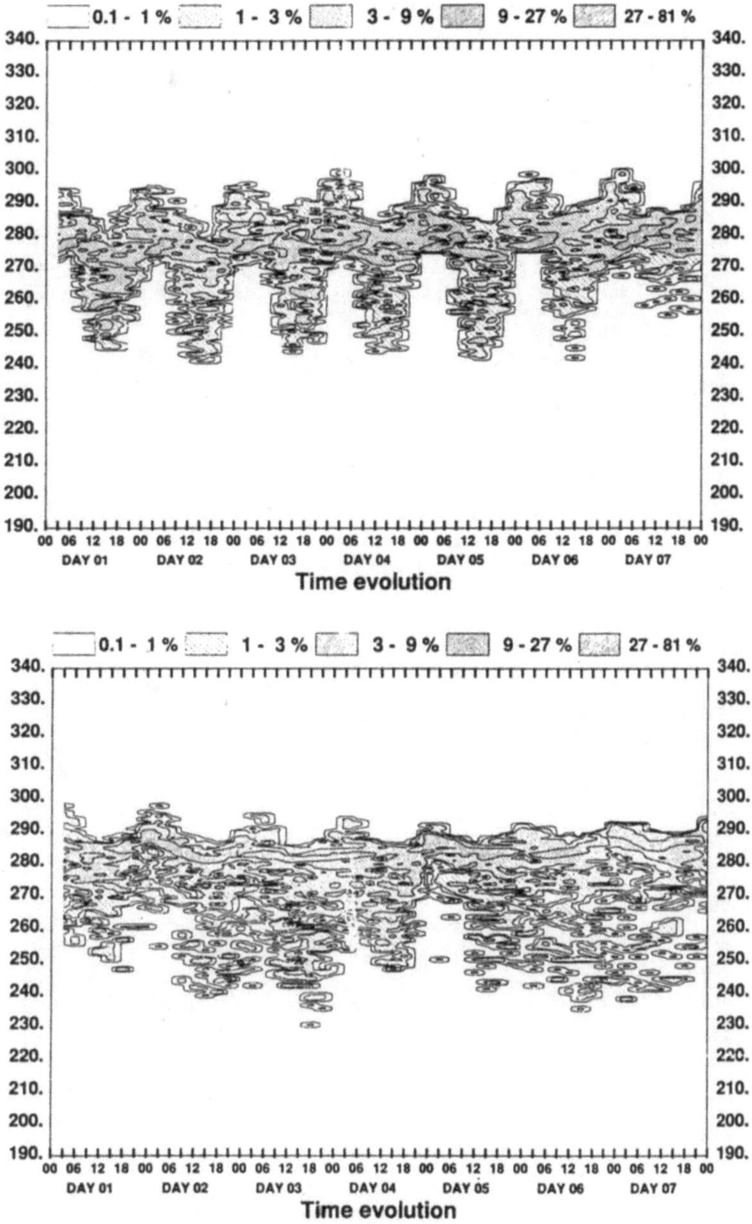

Fig. 1a) Evolution histogram of the longwave window brightness temperature over Central Africa limited area (7.9N - 2.2S, 21.4E - 31.5E) as simulated in T63 L31 forecasts with the operational scheme (top) and new prognostic scheme (bottom).

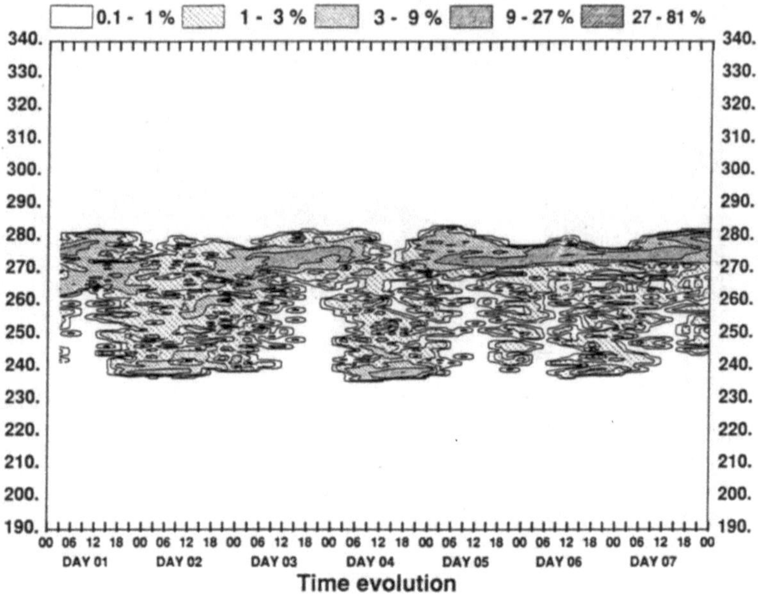

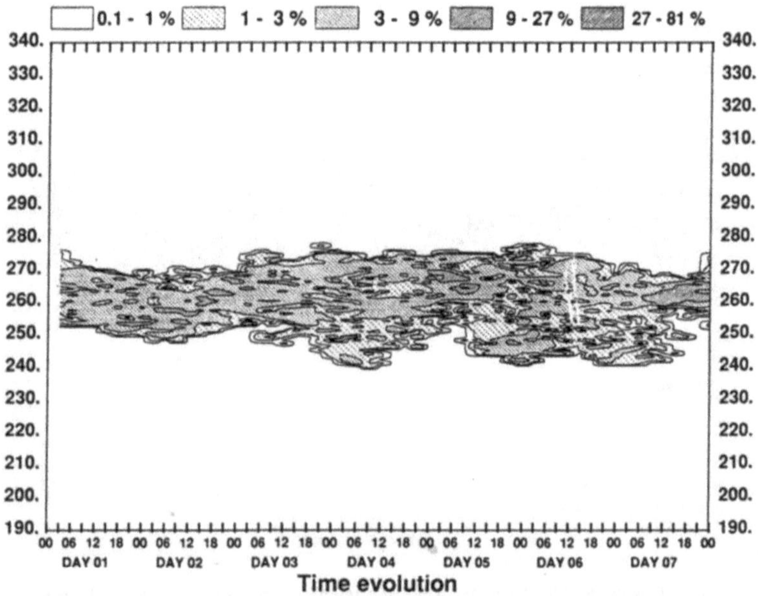

Fig. 1b) As Fig. 1a but for the Gough Island area (39.4S - 49.5S, 15.7W - 5.6W).

as cloudiness over tropical continents is typically largest in the late afternoon and at night and smallest in the morning. The change in the lifetime of the clouds has a strong impact on the radiation available at the surface and so on the diurnal cycle of the surface temperature. In the Gough Island area, the new cloud scheme produces more clouds (the signature of the surface is not seen any more on the histogram) and puts the maximum of the cloud water higher up; however, the effective cloud top appears slightly lower than with the operational cloud scheme.

3.1.2 *Synoptic case studies*

The new cloud scheme and the operational scheme have been validated in short range forecasts (36-72h) at resolution T213 in five winter cases over Europe and the Atlantic (i.e. 15-18 Jan. 1993: 14 Feb. 1993). Predicted cloud fields have been compared with cloud structures in METEOSAT IR imagery and surface synoptic observations. The skill of the operational scheme (as a reference) in representing clouds in the various flow regimes during the periods is summarized in Table 1.

Cloud regime	**Horizontal distribution of cloudiness**	**Vertical distribution of cloudiness**
Cold front	Realistic, but overestimated cloud fraction in the upper layers	Cloud amount increases height
Behind the cold front	Underestimated, both over land and sea	Mostly low clouds
Warm front	Underestimated	Mostly low clouds
Occlusion and hammer head region	Realistic in occlusion, but underestimated in the upper hammer head region	
Anticyclone	Underestimation of low clouds. Spurious low clouds over land due to unrealistic inversions. Overestimation of clouds associated with inactive cold fronts	

Table 1: Summary of the performance of the cloudiness forecast with the operational ECMWF scheme for two winter cases

The horizontal extent of total cloudiness associated with the large scale lifting of moist air along fronts, the occlusion and the "hammer head" region (Thorncroft et al. 1993) appear quite realistic. However, the cross (cold) frontal extent of high cloud cover HCC (450-100hPa) is generally too large with the operational scheme . Inactive fronts in anticyclonic regions are also associated with excessive HCC due to poor representation of the cloud decay process. On the other hand, the HCC in the "hammerhead" region is underestimated. The horizontal extent of high clouds in the above mentioned regions is more realistic with the new scheme apart from the "hammer head" region, which is even more deficient. This deficiency may be reduced when advection of cloud variables is included in the scheme. The differences between the two schemes are also reflected in the vertical distribution of the cloud cover. The operational scheme seems to have a quite systematic increase of cloud cover with height, especially in regions with large scale ascent. The cloudiness produced by the new scheme is more diverse and there is no systematic change of cloud cover with height except in the occlusion and "hammer head" region. The low cloud cover LCC (1000 - 800hPa) behind cold fronts (convective over sea and stratiform over land) and along warm fronts is systematically underestimated with the operational cloud scheme . The new scheme produces much more realistic clouds in these conditions and this applies equally to land and sea. However, both schemes often produce spurious clouds in connection with unrealistic thermal states (i.e. too strong low level inversions) in anticyclones over land.

In summary we find that the new cloud scheme has reduced significantly the deficiencies of the operational scheme for almost all the cloud regimes contained in Table 1. The only exception is the horizontal extent of clouds in the "hammer head" region which is too small.

In order to have an objective evaluation we have also calculated bias (Fig. 2a), root mean square errors (Fig. 2b) and correlation coefficients (Fig. 2c) for verifications of the two schemes against SYNOP cloud observations over Europe. The evaluation confirms that the new scheme produces smaller errors in terms of each of these verifications than the operational scheme.

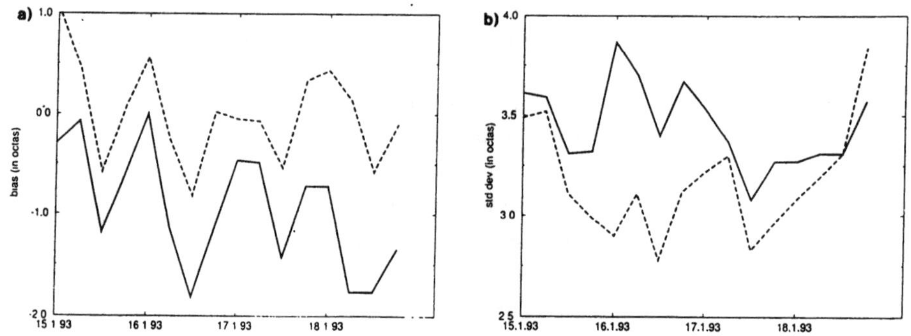

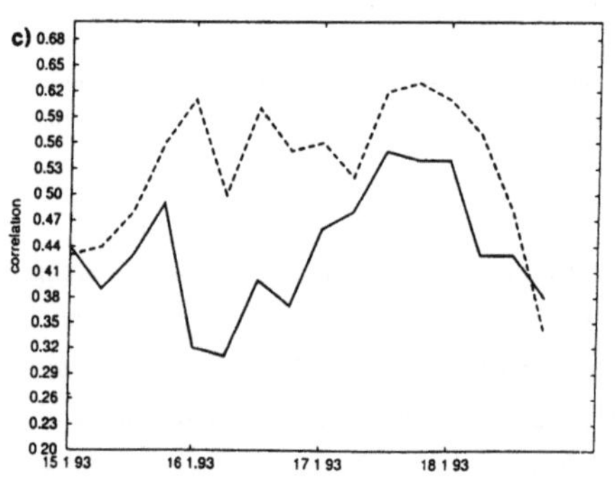

Fig. 2 a) Averaged differences (bias) between predicted and surface observed fractional cloud cover (in octas). Solid and dashed lines represent the operational and new cloud scheme, respectively. The horizontal axis represents the verification time (each day at 00UTC is tick-marked). The four observation times each day (0, 6, 12 and 18UTC) verify the 36h, 42h, 48h and 54h forecasts respectively.
b) Same as a), except for averaged standard deviation.
c) Same as a), except for averaged correlation.

3.2 Cloud water content

Verification of cloud liquid water content (CLW) is difficult because of its large variability in space and time and the lack of adequate observational data. Microwave measurements from satellite (SSM/I) provide estimates of vertically integrated amounts averaged over areas of 10 by 10. Comparison of model-produced values with observed monthly mean values (July 1987) indicate that the new cloud scheme reproduces the maxima associated with clouds in extratropical disturbances and along the ITCZ (Tiedtke 1993, Fig. 4). However, model values along the ITCZ are underestimated which indicates again the lack of shallow cumuli in the ECMWF convection scheme. In fact, model values become more realistic when shallow cumuli are represented by means of a bimodal distribution of clouds as outlined above.

Cloud water content is specified in the operational cloud scheme as a given fraction of the saturation mixing ratio, which makes it highly dependent on temperature. This parametrization produces unrealistic vertical profiles of CLW (i.e. with maximum values within the boundary layer) and fails to reproduce the variability of CLW as a function of the strength of vertical lifting. The new cloud scheme is superior in both aspects (Tiedtke 1993, Section 4.3 b and c). The reproduction of the variability of CLW is particularly important for cirrus clouds as it determines whether they are transmissive or opaque.

3.3 Hydrological cycle

The introduction of cloud liquid water content as an additional prognostic variable constitutes a major change to the model's physics. The hydrological cycle becomes more complex and more realistic as additional cloud processes are represented (Tiedtke 1993, Section 4.3). Inspection of the model's global hydrological cycle (Tiedtke 1993, Fig. 8) reveals the dominant role of cumulus convection for the time evolution of cloud fields, which clearly emphasizes the need for coupling the cloud scheme to the model's cumulus parametrization. As a direct result of the coupling, precipitation characteristics in tropical regions have changed towards more realistic conditions (Tiedtke 1993, Fig. 9): a significant proportion of the rain falls from stratiform anvils as observed (e.g. in GATE). The new cloud parametrization also has a strong effect on the maintenance of the tropospheric water vapour content. From zonally averaged mean values of extended integrations it appears that the water vapour content is reduced significantly, especially in two regions: in the tropics, because cloud water detrained from cumulus updrafts can precipitate out more efficiently and in the extratropics because precipitation can occur before grid points become saturated. The change in moisture content is likely to have important implications for the assimilation of moisture and for short and medium range forecasts.

3.4 Thermal state - cold temperature bias

All experiments performed to date with the new cloud scheme show a large bias towards too low temperatures at all latitudes.

The excessive cooling in tropical regions was partly due to the lack of an important process in the convection scheme, i.e. the latent heat release as cloud liquid water freezes when advected upwards within cumulus updrafts. When freezing of cloud water is introduced the cold bias in tropical regions reduces considerably (Fig. 3).

The remaining cold bias is strongly linked to the larger radiative cooling with the new cloud scheme.

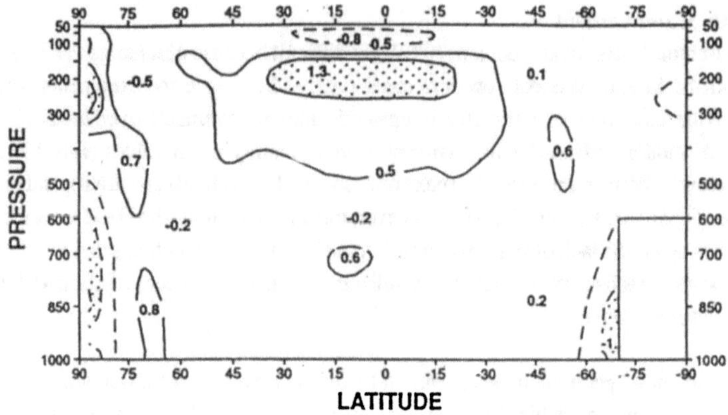

Fig. 3 30 day mean differences in T due to latent heat release by freezing of cloud water in cumulus updrafts (from integrations a) without latent heat release and b) with latent heat release included, cy44 T63L31).

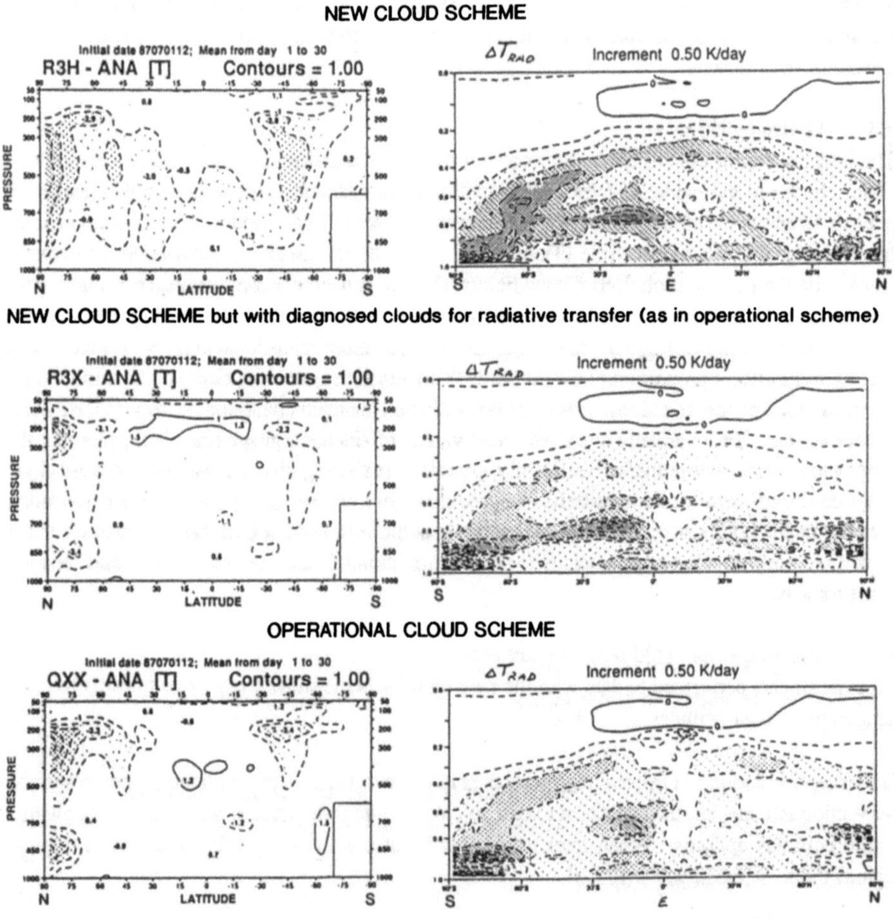

Fig. 4 30 day mean T errors (left) and radiative heating (right) for forecasts with cy44, T63L31.

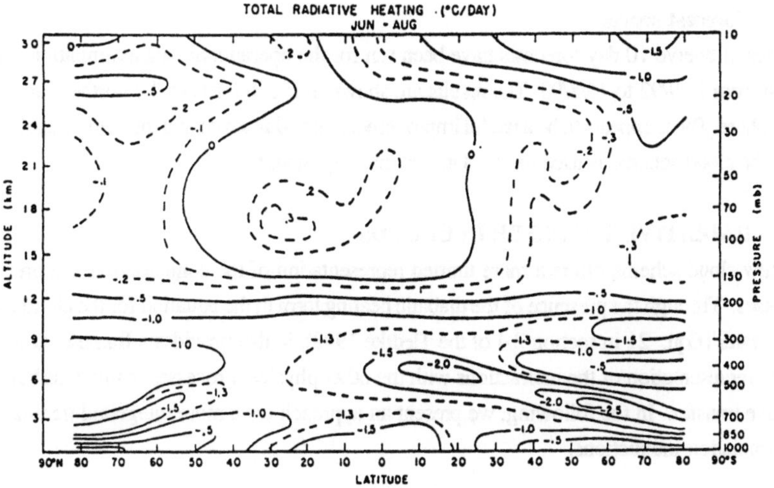

Fig. 5 Zonally averaged heating rates calculated by Dopplick for summer season June-August (input is observed cloud cover).

This is evident from Fig. 4 which shows temperature errors and radiative cooling rates. Note the larger tropospheric T-bias at almost all latitudes and the larger radiative cooling rates. The overall radiative cooling with the new cloud scheme is certainly excessive since the globally averaged value exceeds 120W/m^2 which is too large by 10 to 15 W/m^2. However the vertical profiles appear more realistic with the new scheme reflecting the more realistic cloud fields. For comparison we show Dopplick's estimates (Fig. 5) for London's observed values of cloud cover (Dopplick, 1972, 1979; London, 1957)

The link of the cold bias to the radiative transfer parametrization is further established by an additional experiment where the cloud scheme was not interfaced with the radiation scheme but cloud variables were defined in the same way as in the operational model. The cold bias disappeared as the radiative cooling rates became similar to those of the control experiment (Fig. 4 middle).

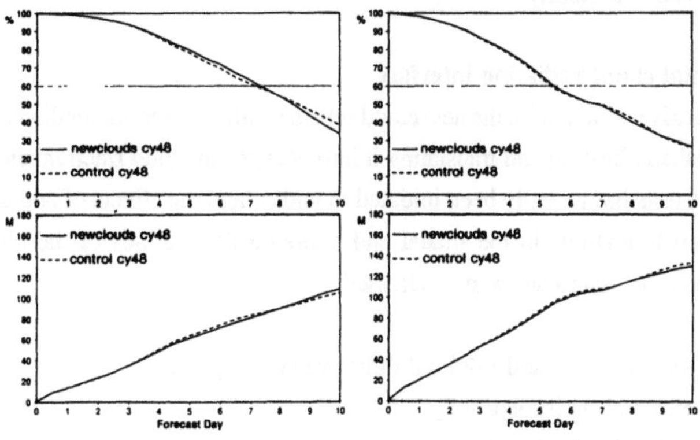

Fig. 6 Forecast scores comparing the current ECMWF model and the new cloud scheme. Forecasts scores less than the 60% are not considered useful.

3.5 Forecast scores.

A series of twelve 10 day forecasts have been run for the operational data assimilation - one each month from 15/9/92 to 15/8/93. The results are shown in Fig. 6 for both the northern and southern hemisphere. There appears to be a useful improvement up to day seven and this will possibly improve when the cloud scheme is used by the data assimilating model.

4. RADIATIVE TRANSFER IN CLOUDS

The new cloud scheme offers a more unified representation of the condensation and precipitation processes. Through the presence of the diabatic heating term in the equation for the cloud cover and cloud water (Eqn. 21 in section 2.4 of the Tiedtke 1993), it also provides a framework for a more unified representation of the interactions with the other physical processes, namely turbulence and radiative transfer. In the following, we present an approach for a more integrated treatment of the cloud-radiation interactions.

4.1. Present operational cloud-radiation interface

Since the introduction of the diurnal cycle of radiation in December 1983, the ECMWF operational forecast model has included a full radiative transfer calculation once every 3 hours (and once every hour during the first 6 hours of the forecast since May 1989). For any time step in between, radiative tendencies and fluxes are interpolated, accounting only for the change in temperature in the long-wave and the change in solar zenith angle in the short-wave part of the spectrum. Furthermore, due to the cost of the radiative computations, even the full radiative computations are performed on a sampled grid (1 point out of 4 on a latitude row). The latter feature has the effect of horizontally smoothing, in the zonal direction, the cloud distribution and hence its radiative effects. The former keeps the cloud distribution fixed between two full radiation computations (for three hours). Up to now, the outputs of the radiation scheme have been grid-averaged net radiative cooling and fluxes, and these quantities are inputs for the thermodynamic equation and the surface energy balance equation, respectively. These are adequate for a diagnostic cloud scheme whose main predictor is the grid-averaged relative humidity.

4.2. Experimental cloud-radiation interface

Such features are clearly restrictive for the new cloud scheme with its potential for directly responding to any change in diabatic heating and translating it into changes in cloud fraction and cloud water loading. A series of tests has recently been initiated to understand the effects of and to improve on the above mentioned limitations in the spatial and temporal descriptions of the cloud-radiation interactions. A number of issues are being addressed:

- grid-averaged vs. clear-sky and in-cloud radiative cooling rates;
- effects of the temporal interpolation;
- effects of the spatial interpolation;
- relative contribution of the short-wave and long-wave radiation to the in-cloud net radiative cooling.

These four points are being studied but results are presently not available. For this ongoing study, the operational radiation scheme has been extended to provide, at the time of the full radiation computations, equivalent clear-sky upward and downward long-wave emissivities from which clear-sky long-wave fluxes can be efficiently reconstructed at every time step and for every grid point, accounting for any change in temperature. When dealing with the radiation transfer within clouds, a number of simplifications and approximations can be made that lead to simplified versions of both the long-wave and short-wave radiation transfer codes. These stripped-down versions are then being used to compute the radiation within the clouds at every time step and for every grid point.

5. CONCLUSIONS

The new cloud scheme differs from conventional schemes in important aspects. By using a model consistent approach we have linked the formation of clouds to processes represented in large-scale models. The relevant processes are: moist adiabatic lifting due to grid scale and sub-grid scale circulations (i.e. large scale ascent, cumulus updrafts and boundary layer turbulence) and diabatic cooling, by radiation for example. We note that mesoscale circulations are not represented in large-scale models and therefore do not contribute to cloud formation. The new scheme compares well with the ECMWF operational cloud scheme both in extended and short range forecasts. This result has been expected in view of the model's ability to reproduce realistically the various synoptic regimes where clouds are abundant (e.g. extra tropical fronts, tropical convection, convection behind cold fronts etc.). However, validation of the scheme has revealed insufficient cloudiness in some regions, thus indicating model deficiencies. We have already identified two areas and introduced modifications to the model's physics:

a) In order to represent marine stratocumulus clouds we have extended the boundary layer scheme by including condensation within convective updrafts above the lifting condensation level;

b) We have modified the cumulus convection scheme towards a bimodal cloud distribution by including the transports by very shallow cumuli (this modification will be implemented in cy48 of the operational model). Another area, discussed below, is the radiative transfer in clouds and the interface with the cloud scheme. We expect that validation of the cloud scheme will expose more model problems in future which will eventually lead to a more realistic model.

Since formation of clouds is determined by model processes, our direct parametrization of the formation process could be kept to a minimum, i.e. we have only introduced a humidity threshold value where clouds begin to form. Dissipation of clouds by precipitation processes and turbulent mixing with unsaturated environmental air is less certain since we must rely on very crude parametrizations. Both aspects will be subject of extensive future research. The parametrization of precipitation processes needs to be validated with observational data as well as results from explicit cloud models. The present precipitation scheme (Sundqvist, 1988) distinguishes diagnostically between warm clouds, mixed water/ice clouds (Bergeron-Findeisen process) and ice clouds but in future a parametrization with comprehensive cloud physics similar to that used in mesoscale models (with water and ice as separate prognostic variables) may be required.

The cloud scheme is still incomplete with regard to the treatment of turbulent transports related to cloud processes. So far we have accounted only for entrainment processes at the top of boundary layer clouds. The parametrization must be extended to the general case, but "double counting " is to be avoided for convective situations.

Preliminary results using a more unified representation of the cloud-radiation interactions indicate that the new cloud scheme is sensitive to the details of this representation. An extensive validation of the various aspects of both the new cloud scheme and the radiation parametrization is more important than before. Time-averaged behaviour and synoptic assessment of the performance of the scheme in terms of cloud cover, cloud water and associated radiation fluxes are important aspects of the validation. Effort will also be directed at testing the scheme with respect to the temporal and spatial variability of the cloud fields it produces. The sensitivity of the prognosed cloud parameters to the microphysical parametrizations embedded in the scheme will also be explored. It is presently believed that the radiation package is responsible for the cold temperature bias seen with the new cloud scheme. The three-dimensional distributions of cloud cover and cloud water are believed to be more realistic than those produced by the operational scheme. The new scheme shows large changes in cloud cover and vertical profiles of liquid water with more medium-level clouds associated with larger water contents. These changes are directly felt by the radiation scheme which associates larger LW cooling (because of their higher LW emissivity) to these more radiatively active medium-level clouds. Starting with the operational introduction of ice/mixed cloud optical properties in February 1993, work is continuing on a reassessment of the relationships between the cloud water content and the cloud optical properties used in the radiation scheme.

References

Dopplick, T.G., 1972: Radiative heating of the global atmosphere. J. Atmos. Sci., **29**, 1278-1294.

Dopplick, T.G., 1979: Radiative heating of the global atmosphere. Corrigendum. J. Atmos. Sci., **36**,1812-1817.

Heymsfield A. J. and L. J. Donner, 1990: J. Atmos. Sci. , **47**, 1865-1877.

Le Treut, H., and Z. X.. Li, 1988: Using METEOSAT data to validate a prognostic cloud water scheme. Atmos. Res., **21**, 273-292.

London, J., 1957: A study of the atmospheric heat balance. Final Rep. AFC-TR-57-287, OTS PB 129551. Dept. Meteor.& Oceanography., New York University, 99 pp.

Nitta, T., 1975: Observational determination of cloud mass-flux distributions. J. Atmos. Sci., **32**, 73-92.

Smith, R. N. B., 1990: A scheme for predicting layer clouds and their water content in a general circulation model. Q. J. Roy. Meteor. Soc., **116**, 435-460.

Sundqvist, 1988: Parametrization of condensation and associated clouds in models for weather prediction and general circulation simulation. In: Physically-based Modelling and Simulation of Climate and Climatic Change, ed. M. E. Schlesinger, Kluwer, 433-461.

Tiedtke, M., 1993: Representation of Clouds in Large-Scale Models. Mon. Wea. Rev., **121**,3040-3061.

Thorncroft, C. D., B. J. Hoskins and M. E. McIntyre, 1993: Two paradigms of baroclinic-wave life-cycle behaviour. Quart. J. Roy. Meteor. Soc., **119**, 17-55.

Yanai, M., S. Esbensen and J.-H. Chu, 1973: Determination of bulk properties of tropical cloud clusters from large-scale heat and moisture budgets. J. Atmos. Sci., **30**, 611-627.

ANALYSIS AND PREDICTION OF RAINFALL IN THE TROPICS

M.J. Manton

Bureau of Meteorology Research Centre

GPO Box 1289K

Melbourne VIC 3001

Australia

ABSTRACT

The representation of rainfall in global atmospheric models remains a significant source of uncertainty in the prediction of both weather and climate, particularly in the tropics where diabatic heating is a dominant process. Moreover, the measurement of rainfall from both conventional surface techniques and satellite remote sensing is difficult, especially over the oceans. It is therefore expected that optimal estimates of rainfall for analysis and prediction will come from data assimilation systems, in which the complementary features of models and measurements are combined. To support this activity, it is necessary for rainfall observations from both surface raingauges and satellites to be made readily available in real time for incorporation in operational assimilation systems.

KEYWORDS

Climate modelling, numerical weather prediction, rainfall analysis, data assimilation.

1. Introduction

Climatological analyses of rainfall are fundamental to efforts to improve our understanding of the global climate system and to support economic and social activities, such as the agriculture and tourism industries. These analyses are also used to develop and validate global climate models and also to initialize numerical weather prediction models. It is therefore important for us to develop techniques to improve the analysis of rainfall on regional and global scales.

All general circulation models (GCMs) have systematic errors, especially in the representation of precipitation. Moreover the prediction of precipitation in GCMs tends to be sensitive to the detailed form of the various physical parameterizations in the models.

NATO ASI Series, Vol. I 26
Global Precipitations and Climate Change
Edited by M. Desbois and F. Désalmand

However, despite these errors and sensitivities, GCMs can produce quite good simulations of features, such as monsoons, that are dependent upon the diabatic forcing from precipitation.

While GCMs have some problem areas, observations of precipitation also have some significant limitations. These limitations occur especially over the oceans where there are few direct or even reliable measurements. Although satellites provide extensive geographical coverage, the estimates of rainfall from them are indirect and they have only a limited base on physical understanding.

As with other parameters that are difficult to observe and simulate, such as the surface fluxes over the ocean, the best estimates of precipitation are likely to come from assimilation methods, where observations and models complement each other to produce optimal estimates. In this way, the uncertainties in each component are minimized.

2. Uncertainties in models and observations

The GCM of the Bureau of Meteorology Research Centre (BMRC) is described by Hart et al. (1990). It is currently used for both numerical weather prediction and climate simulations, including climate change experiments. The GCM has participated in a number of international intercomparison projects, such as the Atmospheric Model Intercomparison Project (AMIP) described by Gates (1992). The performance of the model is therefore well documented, and it is found to be quite satisfactory for the simulation of many aspects of the climate system.

Like all other GCMs, the BMRC model has some systematic errors in its simulated climate. For example, it is found that the simulation of rainfall over the Philippines region is over-estimated when compared with the climatology of Jaeger (1976). This apparently local error leads to larger-scale impacts, so that the Aleutian low is somewhat too deep in winter.

If one considers the zonally averaged precipitation then such local anomalies are smoothed and the precipitation distribution is found to be reasonably well represented by the model. On the other hand, the BMRC model appears to underestimate the summer rainfall in the storm tracks of the southern hemisphere, when compared to Jaeger (1976). This feature is common to most models (Gates et al., 1990), suggesting that the fault may lie more with the "observed" climatology than with the models. We note that the lack of direct observations of rainfall in the Southern Ocean makes it very difficult to derive accurate climatological estimates in this region. Thus it must be appreciated that uncertainties arise not only with climate models but also with climate observations.

In addition to systematic errors, climate models also have sensitivities to changes in the representation of various sub-grid-scale processes. These sensitivities are not necessarily an indication of a fault, rather their existence needs to be taken into account when considering the predictions of models. For example, Miller et al. (1991) find that model simulations are sensitive to the manner in which the surface fluxes over the ocean are represented at low wind speeds. This feature is also found by Frederiksen et al. (1992) in simulations of rainfall during the Indian monsoon with the BMRC model. Such a sensitivity is not unexpected for a monsoon simulation in which the moisture is derived mainly from the ocean. However, it does emphasize the need for great care in the specification of all physical processes, even those that are thought to be well understood.

A process that is not well represented in models is cumulus convection, although it plays a dominant role in the determination of the overall radiation balance and of the thermal structure of the troposphere. Changing the cumulus parameterization in the BMRC model from the classical Kuo (1974) to the more recent mass-flux scheme of Tiedtke (1989) has some impact on the global climate, but the main impact is at regional scales. For example, the precipitation anomaly around the Philippines is reduced when the mass-flux scheme is used. These regional differences are even more apparent when different parameterizations are used in $2xCO_2$ experiments.

3. Model simulations of boundary-value problems

Although GCMs are far from perfect, they are capable of simulating many significant aspects of the climate system that depend upon precipitation. The Atmospheric Model Intercomparison Project (AMIP), a project of the World Climate Research Programme (WCRP), has been a valuable means of testing the ability of current models to simulate the present climate under strictly controlled boundary conditions (Gates, 1992). The BMRC model was one of the thirty GCMs participating in AMIP, in which the sea-surface temperature (SST) and sea-ice distributions were specified for the decade of the 1980s.

The El Nino-Southern Oscillation (ENSO) is a phenomenon that is closely linked to the SST of the Pacific Ocean and which influences the precipitation distribution over large regions of the tropics and sub-tropics. The Southern Oscillation Index (SOI) is the normalized pressure difference between Tahiti and Darwin (Troup, 1967), and it is a measure of the strength of the atmospheric circulation associated with ENSO. Analysis of the results of the AMIP integration of the BMRC model shows that the SOI is well simulated; in particular, the ENSO events of 1982-83 and 1986-87 are clearly represented by large negative values of the SOI.

It is found that the Australian monsoon is well simulated by the BMRC model throughout the AMIP integration. The accuracy of the simulation is shown by comparison of the monsoon for 1981-82, which was a normal monsoon period, with that for 1982-83, which was the ENSO year. For 1981-82, the model yields widespread rain and westerly winds across northern Australia for the December to February period. In contrast, the simulation for the summer of 1982-83 gives heavy rains north of Australia and the westerly winds are very weak. Thus the large interannual variation of the monsoon is well represented by the model during the extended AMIP integration. (It is interesting to note that shorter-term integrations generally do not have such success at simulating the monsoon and its associated rainfall. This emphasizes the need for careful dynamic and thermodynamic balances in model simulations of large-scale precipitation events.)

The success of the AMIP simulation of the Australian monsoon, which is dependent upon the SST distribution, suggests that the seasonal rainfall distribution across much of Australia may be amenable to prediction from a GCM when the global SST is specified appropriately. Initial work with the BMRC model shows that regional rainfall can be simulated reasonably well when the observed SST distribution is used, and this gives some confidence that GCM-based seasonal predictions could be developed over the next few years.

4. Model simulations of initial-value problems

The GCM results in Section 3 correspond to climate simulations where the boundary conditions (i.e. the SSTs) are dominant constraints. It is also useful to consider weather prediction problems where the initial conditions are important to the prediction of rainfall.

The prediction of rainfall from a GCM days or weeks ahead is extremely sensitive to the initial conditions, particularly to the initial distribution of moisture. The lack of conventional observing systems in the tropics and in the southern hemisphere makes it very difficult to provide an accurate estimate of the initial distribution of moisture for weather prediction. This also implies that accurate climatologies of moisture (and other meteorological variables) cannot be readily obtained in these regions. Soundings derived from the Tiros Vertical Operational Sounder (TOVS) on the NOAA polar-orbiting satellites provide additional information that is important for numerical weather prediction in the southern hemisphere, particularly for the thermal structure of the atmosphere.

However, the TOVS data are not sufficient in the tropics, particularly since their coverage and quality are limited in cloudy areas. Work has been carried out in Japan and Australia on the use of infrared imagery from the Geostationary Meteorological Satellite (GMS) to infer the moisture structure over the tropical oceans at a resolution of about 50 km. The scheme is based on the identification of relationships between moisture profiles and cloud types (Mills & Davidson, 1987). Nineteen categories of distinct cloud type are selected, based on the parameters of cloud top height, cloud cover, and variance of cloud height. Tests with the BMRC numerical model show that these "bogus" moisture data are needed in order to predict the onset of the Australian monsoon a few days ahead (Puri & Davidson, 1992).

The imagery from GMS can also be used in other ways to improve the initialization of numerical prediction in the tropics so that spin-up can be reduced. The threshold method of Arkin (1979) can be used to estimate the rainfall rate and hence the initial distribution of diabatic heating. Davidson & Puri (1992) show that weather prediction in the tropics is improved by dynamical nudging, a form of data assimilation in which an integration is commenced at least 12 hours prior to the base analysis time. During the initial period, the model is nudged towards the target analysis and the diabatic heating from the cloud parameterization scheme is replaced by the heating inferred from the GMS imagery. This technique leads not only to improved 24-hour and 48-hour predictions of the large-scale flow but also to the reduction of spin-up so that the precipitation field is dynamically consistent throughout the integration from the base time.

5. Conclusions

It is clear that all GCMs have systematic errors and sensitivities to the representation of processes that are not well understood. Many of these errors and sensitivities are associated with precipitation processes. Despite these problems, numerical models can produce useful simulations of precipitation-dependent phenomena, such as the ENSO and the monsoon, provided that the initial and boundary conditions are specified accurately.

Conventional meteorological observations are neither uniformly distributed in space and time nor uniform in their accuracy. Satellite-based observations can give good coverage, but they are indirect and so they need considerable calibration to convert them to meteorological data. The uncertainties in both models and observations imply that the best estimates of fields like rainfall will be obtained by combining the observations with models through data assimilation.

It follows that operational centres that carry out routine numerical weather prediction should be encouraged to validate the precipitation fields estimated from their global models. This activity is currently being supported by the WCRP Working Group on Numerical Experimentation (WGNE). Such validation requires the development of routine analyses of rainfall observations. In turn, this requires access to real-time global rainfall data. Therefore significant progress on the analysis and prediction of rainfall will depend upon the encouragement of national meteorological services to include daily rainfall data on the Global Telecommunications System.

Since the optimal estimate of fields like rainfall is expected to be developed from data assimilation and these systems tend to be operated in real time, it is important that all relevant data should be made available in real time, if possible. This requirement applies especially to satellite data which generally provide much more comprehensive spatial coverage than conventional observations.

REFERENCES

Arkin, P.A. 1979: The relationship between fractional coverage of high cloud and rainfall accumulations during GATE over the B-scale array. Mon. Weath. Rev., 107, 1382-1387.

Davidson, N.E. & Puri, K. 1992: Tropical prediction using dynamical nudging, satellite-defined convective heat sources, and a cyclone bogus. Mon. Weath. Rev., 120, 2502-2522.

Frederiksen, C.S., Puri, K. & Balgovind, R.C. 1992: Summer monsoon simulation with the Bureau of Meteorology Research Centre global spectral model: MONEG experimentation. Proc. MONEG Workshop, 21-24 October 1991, WMO/TD No. 470, 247-250.

Gates, W.L., Rowntree, P.R. & Zeng, Q.C. 1990: Validation of climate models. In "Climate Change: The IPCC Scientific Assessment", Eds Houghton, J.T., Jenkins, G.J. & Ephraums, J.J., 93-130.

Gates, W.L. 1992: AMIP: the Atmospheric Model Intercomparison Project. Bull. Amer. Meteor. Soc., 73, 1962-1970.

Hart, T.L., Bourke, W., McAvaney, B., Forgan, B.W. & McGregor, J.L. 1990: Atmospheric general circulation simulations with the BMRC global spectral model: impact of revised physical paramaterizations. J. Climate, 3: 436-459.

Jaeger, L. 1976: Monatskarten des Niederschlags fur die ganze Erde. Bericht Deutscher Wetterdienst, 18, Nr. 139, 38pp.

Kuo, H.L. 1974: Further studies of the parameterization of the influence of cumulus convection on large-scale flow. J. Atmos. Sci., 31, 1232-1240.

Miller, M.J., Beljaars, A.C.M. & Palmer, T.N. 1991: The sensitivity of the ECMWF model to parameterization of evaporation from tropical oceans. J. Climate, 5, 418-434.

Mills, G.A. & Davidson, N.E. 1987: Tropospheric moisture profiles from digital IR satellite imagery: system description and analysis/forecast impact. Aust. Meteor.Mag., 35, 109-118.

Puri, K. 1992: Modelling studies of the Australian monsoon. Proc. MONEG Workshop, 21-24 October 1991, WMO/TD NO. 470, 133-136.

Puri, K. & Davidson, N.E. 1992: The use of infrared satellite cloud imagery data as proxy data for moisture and diabatic heating in data assimilation. Mon. Weath. Rev., 120, 2329-2341.

Tiedtke, M. 1989: A comprehensive mass flux scheme for cumulus parameterization in large-scale models. Mon. Weath. Rev., 117, 1779-1800.

Troup, A.J. 1967: Opposition of anomalies of upper tropospheric winds at Singapore and Canton Island. Aust. Meteor. Mag., 15, 32-37.

UNDERSTANDING PERTURBATIONS OF THE INTERTROPICAL CONVERGENCE ZONE OVER TROPICAL CONTINENTS

Kerry H. Cook
Atmospheric Sciences Program
1110 Bradfield Hall
Cornell University
Ithaca, NY 14853, USA

ABSTRACT

Repeated integrations of GCMs with simple boundary conditions and the diagnostic application of a linear primitive equation model are used to understand the physical mechanisms that lead to the observed perturbation of the intertropical convergence zone (ITCZ) in summer. Two examples are considered.

One application addresses the role of surface drying in determining the longitudinal precipitation gradient across South America and Africa in austral summer. Different mechanisms are responsible for the response in different regions, leading to a longitudinal dependence in the response to surface drying.

A second application concerns the role of surface roughness in establishing the West African precipitation climatology. Frictional convergence and vorticity advection are found to force anomalous low-level convergence in this region in boreal summer.

KEYWORDS

tropical precipitation/ African drought/ South American climate/ land surface-atmosphere interactions/ tropical climatology/ African precipitation/ intertropical convergence zone

1. Introduction

In this paper we address some fundamental questions of how certain features of the climatological precipitation field are established. Here, the focus is on the tropics and, in particular, the perturbation of the intertropical convergence zone (ITCZ) near the African and South American continents in summer.

The stationary features of the observed precipitation climatology suggest that (semi)permanent features of the surface influence rainfall distributions. An approach for using climate models to investigate the mechanisms that lead to this influence is outlined in

NATO ASI Series, Vol. I 26
Global Precipitations and Climate Change
Edited by M. Desbois and F. Désalmand

the following section. Then, two features of the large-scale precipitation climatology are investigated, namely, the precipitation gradient across Africa and South America in austral summer, and the precipitation enhancement north of the Gulf of Guinea in boreal summer.

2. Method

General circulation models (GCMs) are, in a sense, a compilation of our knowledge of the climate system. Perhaps the most familiar applications of these models involve attempts to replicate the present day climate with as much accuracy as possible, and to simulate other climate states due to changes in atmospheric greenhouse gas concentrations or ice age conditions. This study revolves around the use of a GCM, but in a different way. The goal is not to reproduce a realistic climate, but to use the model to reveal the physical mechanisms responsible for a particular feature of the observed climate.

The GCM used here is a version of the model developed and maintained by the Climate Dynamics Group at NOAA's Geophysical Fluid Dynamics Laboratory (GFDL). The GCM is used with low resolution for the examples below; the spectral representation of latitudinal and longitudinal dependence is limited to 15 waves, amounting to a horizontal resolution of 4.5° latitude and 7.5° longitude on the transform grid. The model is used with 9 vertical levels in one example, and 14 in another. The reader is referred to Washington and Parkinson (1986) for a complete description of this class of climate model. Here we will note only that the model is used with simple parameterizations for surface hydrology (the "bucket" parameterization) and convection ("convective adjustment"). These simple treatments are necessary for the aim of the work to analyze physical mechanisms cleanly, but it will also be important to consider the possible dependence of the results on these (and other) physical parameterizations since unrealistic parameterizations can obscure realistic behavior.

To use the GCM to investigate physical mechanisms associated with the perturbation of the large-scale tropical precipitation climatology, we take the following steps:

1. Identify a feature in the observations.
2. Identify the feature in the GCM with realistic boundary conditions. If the fully-complex GCM replicates (or approximates) the observed feature, we assume it is the same feature as in the observations in the sense that the causes are the same. (This might not always be a good assumption.)
3. Hypothesize about how the feature may be related to the model's boundary conditions. (Ask: What attribute(s) of the surface may be causing the perturbation?)
4. Design and run GCM experiments with simplified boundary conditions to isolate the relationship between the surface feature and the precipitation perturbation. It is

important to choose problems that are appropriate for the model, and to keep in mind the model's strengths and weaknesses when designing experiments.

5. Analyze the GCM results to understand mechanisms. (Ask: How does this attribute of the surface modify the precipitation field?)

When compared with observational analyses, analysis of output has the advantage of completeness, with many fields that are difficult to observe (like mixing ratio or diabatic heating rates) available. If the GCM has realistic boundary conditions, however, the fields are so complex that it is generally not possible to extract clear cause-and-effect relationships. When simplified boundary conditions are used in the model, connections are cleaner (but at the expense of realism, of course). However, even with the simplest boundary conditions, the GCM is still based on the fully non-linear primitive equations and has complete physics (e.g., radiation, hydrology), so the results can be complex.

In the next section, the GCM analysis is aided by the diagnostic application of a steady state, linear primitive equation model. This model (Ting and Held, 1990) is based on the linearized form of the GCM's governing equations. The linear model is dry, and has no "physics" (e.g., no radiation calculation, clouds, surface heat budget). Given zonal-mean wind and temperature fields and a three-dimensional diabatic heating distribution, the linear model solves for the deviation from the zonal mean (perturbations) of the wind and temperature fields. This model is an excellent tool for understanding the response of the atmosphere to a given forcing under certain conditions, but is not useful for investigating the mutual adjustment between diabatic heating fields and atmospheric motion. To use the linear model, it is first necessary to demonstrate that it is applicable by showing that the precipitation perturbation in the GCM is closely related to the (linear) response to the total forcing field (diabatic heating, transient eddies, topography). This is not always the case, but when it is the linear model is a powerful tool that can reveal exactly how a precipitation perturbation is forced (see next section).

With an understanding of physical mechanisms from simple GCM experiments, a less restricted representation of the climate system must be rebuilt in the model to appreciate the role of the isolated physical mechanisms when competing processes are included.

Two examples are given. In each, GCM experiments with simple representations of land surfaces are compared with all-ocean control integrations. The control integrations provide zonally-uniform climatologies, with seasonality defined by the specification of solar insolation and zonally-uniform sea surface temperatures. The control climatologies (as well as the experiments) are formed by averaging over the last 1800 model days of a 2000-day integration beginning from an isothermal atmosphere at rest. (The length of the integrations was chosen to insure statistical robustness of the precipitation anomalies; the 2 900-day means are virtually identical to the 1800-day climatologies in all cases.) In all cases, clouds are fixed, hemispherically-symmetric and zonally uniform; ozone concentrations are symmetrized; and sea ice is not allowed to form. In this way, the complexity of the model climate system is reduced to highlight only certain relationships, and allow in-depth diagnosis.

Despite the simplicity of the boundary conditions, the control integration provides a reasonable zonal-mean climate (Cook and Gnanadesikan, 1991). Figure 1 shows the precipitation from the all-ocean control integration for boreal summer. The fundamental features of the zonal-mean precipitation field are reproduced, with a tropical maximum in the summer hemisphere, subtropical minima, and a secondary maximum in the mid-latitudes of the winter hemisphere. The precipitation maximum associated with the intertropical convergence zone (ITCZ) occurs at about 7° latitude in the summer hemisphere, coincident with the SST maximum. The maximum values of 6mm/day are a little weak compared with observations, but this may be partially attributable to the simple boundary conditions and the low horizontal resolution. The zonal uniformity of this very noisy field indicates that the 1800-day mean is sufficient for establishing a quasi-steady climate state.

In the experiments described below, continental surfaces are introduced as the only source of longitudinal structure in the model. The analysis of the experiments reveals how the precipitation perturbations are forced.

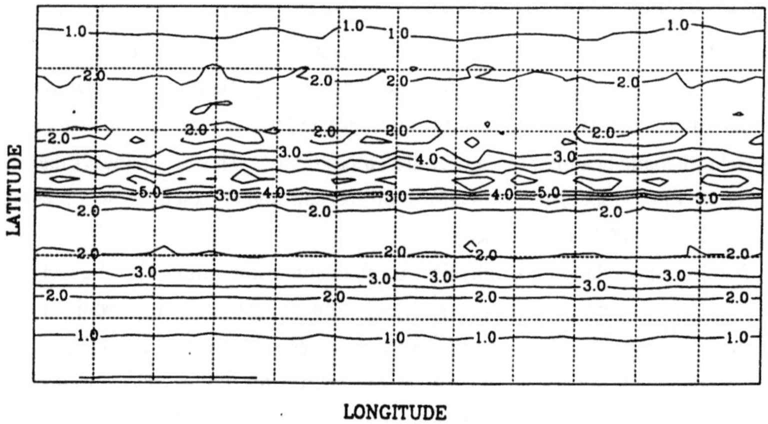

Figure 1. Precipitation (mm/day) from the all-ocean control GCM integration.

3. African and South American precipitation in austral summer

Figure 2 shows the observed precipitation climatology for austral summer. Here, Legates and Willmott's (1990) precipitation climatology on a 0.5° square grid is averaged over December, January, and February, and interpolated to the transform grid of the low-resolution GCM. The Mollweide projection emphasizes the tropics, where the precipitation field has a similar structure crossing South America and Africa; precipitation is enhanced over the eastern portions of both continents, with decreasing precipitation rates to the west. These longitudinal gradients could be related to a number of surface features, including climatological sea surface temperatures and land surface characteristics. For this study we single out one factor, and try to understand the role of surface drying in establishing this feature.

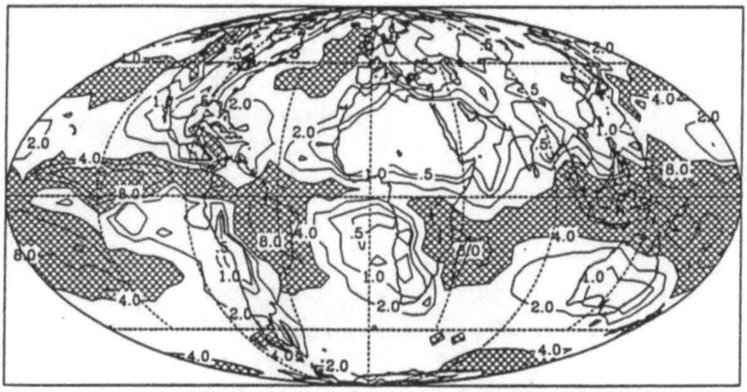

Figure 2. Observed DJF precipitation from Legates and Willmott (1991).

In a series of 5 experiments, longitudinal structure is introduced into the GCM by specifying a simple land surface. The continental surface is 60° wide in longitude, and extends to about 33° north and south latitude. It has the same surface roughness and albedo as the ocean surface. In the first experiment, surface wetness is prescribed at saturation, so the only difference between the continental surface and the ocean surface in the model is that the land surface can change temperature in response to the surface heat budget. In four additional experiments, the surface wetness of the continental surface is prescribed at progressively drier values: "wet", "intermediate", "dry", and "very dry".
Figure 3 shows the response of the precipitation field (differences from the zonal mean from the all-ocean control in Fig. 1) for three experiments; Fig. 3a is for the saturated land surface, and Fig. 3b and c are for the "intermediate" (roughly savanna conditions) and very dry (desert) values of surface wetness, respectively. Over the saturated surface (Fig. 3a),

there is no net loss of rainfall onto the continent, but the precipitation is shifted to the southeast. On the east coast, well into the summer hemisphere, rainfall is enhanced. With intermediate surface wetness (Fig. 3b), the precipitation enhancement in the continental interior largely disappears. When the surface is very dry (Fig. 3c), the negative precipitation anomaly spreads throughout the continental interior of both hemispheres but, again, there is increased precipitation on the east coast of the summer hemisphere. In summary, as the land surface becomes drier, precipitation rates in the continental interior decrease, but precipitation is enhanced on the east coast of the summer hemisphere in all cases. This response leads to an increase in the longitudinal precipitation gradient with increased surface drying.

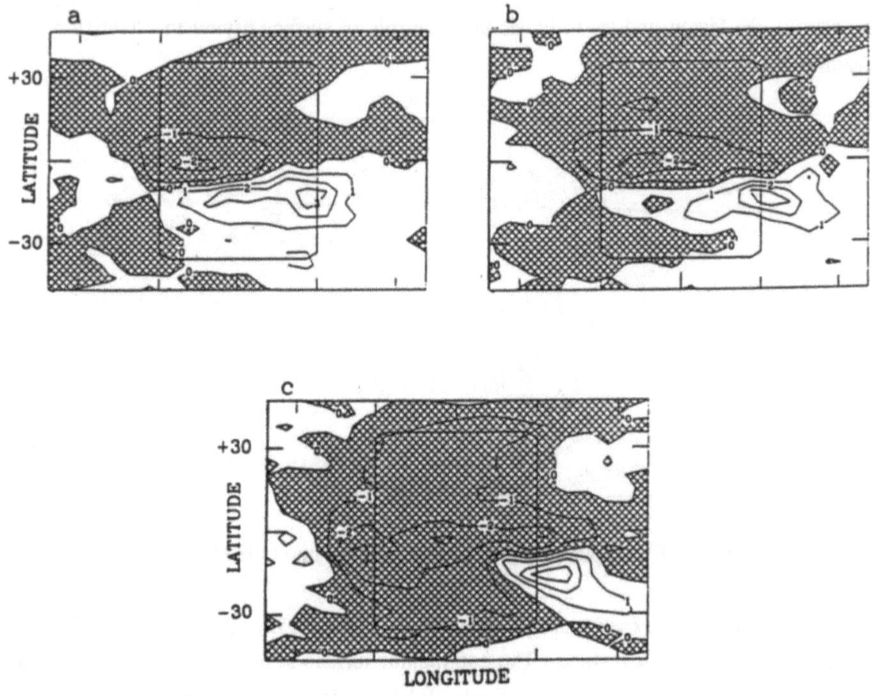

Figure 3. Precipitation perturbation (mm/day) over the (a) saturated (b) intermediate wetness and (c) dry continental surface.

The precipitation perturbation in the GCM is closely related to the perturbation of the low-level (below 680 mb) moisture convergence field and not to the perturbation of the evaporation field. Further, the structure of the anomalous moisture convergence is very similar to that of the wind convergence, indicating that an understanding of the response of the lower troposphere to surface conditions is central to understanding the precipitation perturbation. This is as far as the GCM output can take us in the analysis, for there is no way of separating the effects of individual forcing functions; for this we need the linear model.

For the wet and intermediate soil moisture cases, the linear model validates as an appropriate diagnostic tool. Application of this model shows that different physical mechanisms are operating in different regions. In the continental interior and west, there are 2 competing processes, both of which operate through the warming of the surface air:

I. warmer surface -> increased dry convection -> increased low-level convergence -> increased precipitation

II. warmer surface -> reduced relative humidity -> reduced low-level condensational heating -> reduced low-level convergence -> reduced precipitation

Over the saturated continent (Fig. 3a), mechanism I dominates. As the surface becomes drier, mechanism II becomes increasingly important and leads to precipitation decreases over drier surfaces. For the very driest case, many of the elements seem to be the same, but there is the added complication of structure introduced into the moisture convergence field associated with strong surface moisture gradients. A diagnostic model with explicit water vapor transport is needed to treat this case.

In the east, where strong precipitation persists despite surface drying, the precipitation anomaly occurs due to a different mechanism. Through application of the linear model, we can demonstrate that the low-level (moisture) convergence is driven by condensational heating in the middle and upper troposphere (approximately between 500 and 200 mb). Away from the surface, diabatic heating is balanced by adiabatic cooling (rising motion) throughout the tropics, and this forces convergence below by continuity. With an adequate moisture supply from "upstream", this precipitation maximum is independent of the surface wetness field.

Given this level of understanding of how the precipitation perturbation is forced, the next step is to build in more realism and investigate the effects of the shapes and locations of the continents. Figure 4 shows the precipitation perturbation that is generated over more realistic (but still flat) representations of Africa and South America. In these experiments, the surface wetness is fixed at the intermediate value, so comparison with Fig.3b is appropriate. As in the observations, the precipitation gradient is more zonally-oriented over south America than over Africa. Unlike the observations, the longitudinal precipitation gradient across Africa is larger than across South America. This discrepancy may point to an important role of the Andes in enhancing the dryness of the west. Another model feature not seen in the observations is the northwest-southeast tilt of the positive precipitation anomaly over South America. This feature is a topic for future study.

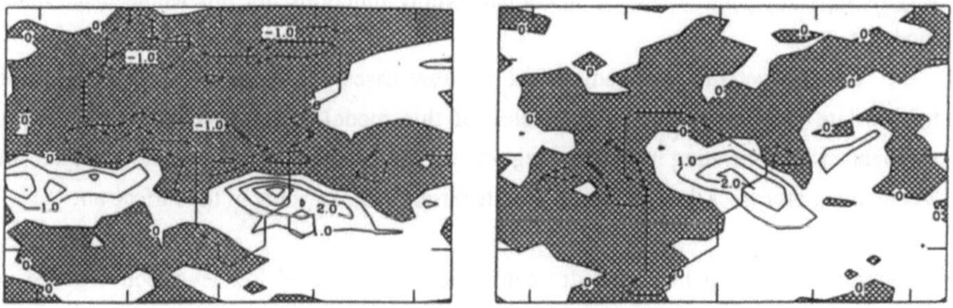

Figure 4. Precipitation anomaly (mm/day) over Africa and South America in the GCM.

4. Precipitation north of the Gulf of Guinea

In contrast to the austral summer case, positive longitudinal precipitation gradients do not occur across Africa and South America in Northern Hemisphere summer, as shown in Figure 5. Over Africa, precipitation increases from east to west at the latitude of the ITCZ. The role of the Gulf of Guinea as a source of moisture for northwestern Africa has been discussed in the literature, for example, in the study by Cadet and Nnoli (1987). Here, we focus on one aspect of the problem and ask: What is the role of land surface roughness in establishing this precipitation maximum?

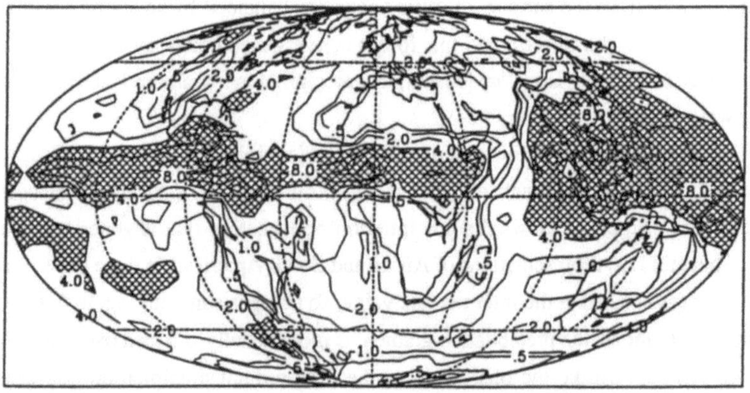

Figure 5. Observed JJA precipitation from Legates and Willmott (1991).

The direct effects of increasing surface roughness in the model are an enhancement of the momentum, sensible heating, and moisture fluxes from the surface to the atmosphere. The bulk aerodynamic parameterization is used. For example, in the horizontal equations of motion,

$$\frac{d\bar{v}}{dt} + f\ \bar{k} \times \bar{v} = -\nabla\phi + \bar{F} \tag{1}$$

where $\bar{v}$ is the horizontal wind vector, f the Coriolis parameter, $\bar{k}$ the unit vector in the vertical direction, and ϕ is the geopotential. The "frictional" term, F, is given by

$$\bar{F} = \bar{F}_x + \bar{F}_y \tag{2}$$

where subscripts x and y indicate the longitudinal and latitudinal directions and

$$\left|\bar{F}_x\right| = \frac{1}{\rho}\frac{\partial}{\partial z}(\rho C_D V u) \tag{3}$$

$$\left|\bar{F}_y\right| = \frac{1}{\rho}\frac{\partial}{\partial z}(\rho C_D V v) \tag{4}$$

Here, r is the density of air and V the magnitude of the surface wind with components u and v. C_D is called the drag coefficient, and it is commonly set equal to 0.001 for ocean surfaces and 0.003 for land. Fluxes of moisture and sensible heat are also linearly proportional to C_D in the model.

To examine the effects of the larger C_D over land on West African precipitation, the results from two GCM experiments are examined. In one, a realistically-shaped African continent is used, but the continent is flat, with fixed intermediate soil wetness and the same surface roughness and albedo as the ocean surface. In a second experiment, the continent is given increased surface roughness, representing the effects of a soil or vegetated surface. Anomalies are calculated with reference to an all-ocean control with solar insolation and sea surface temperatures representing July conditions.

Figures 6a and b show the precipitation perturbations over the "smooth" and "rough" continents, respectively. In both cases, the pronounced precipitation maximum on the east coast is joined by a second maximum north of the Gulf of Guinea. It is weak over the smooth African continent, but reaches magnitudes over 2 mm/day over the rough continent. The precipitation maximum on the east coast is also increased over the rough land surface.

To isolate the effects of enhanced momentum transport on the precipitation field, the GCM was integrated with C_D increased only for momentum fluxes over the continent; C_D

was held at 0.001 everywhere for the moisture and sensible heat flux parameterizations. The result is a precipitation anomaly that is very similar to that shown in Fig. 6b (not shown).

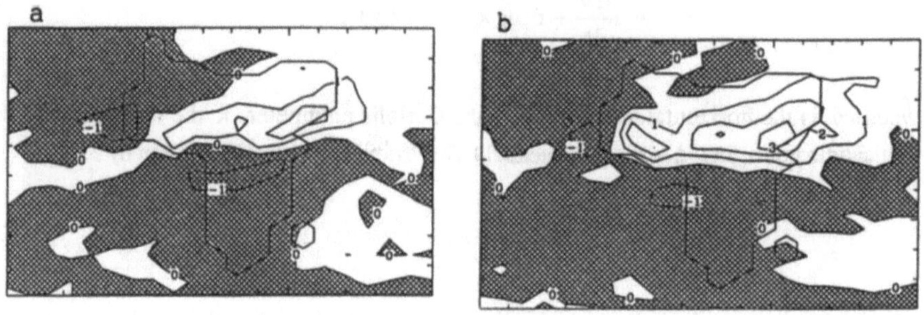

Figure 6. Precipitation anomaly (mm/day) over the "smooth" and "rough" African continent.

As in the case discussed in the previous section, the precipitation perturbation is found to be closely related to the perturbation of the low-level winds. To understand the connection between increased surface roughness and low-level convergence, consider the vorticity equation, which can be derived from Eq. (1):

$$\frac{\partial \zeta}{\partial t} = -\bar{v} \bullet \nabla(\zeta + f) - (\zeta + f)\nabla \bullet \bar{v} + \bar{k} \bullet \left[\frac{\partial \bar{v}}{\partial p} \times \nabla\omega\right] + \left(\frac{\partial F_x}{\partial x} + \frac{\partial F_y}{\partial y}\right) \tag{5}$$

Averaging over the climatology, we must have $\partial z/\partial t = 0$, where the overbar indicates the time-mean. Each term in the vorticity equation can be calculated from the GCM output. The tilting term is found to be small, so Eq. (5) can be approximated as

$$-(\zeta + f)\nabla \bullet \bar{v} \cong \bar{v} \bullet \nabla(\zeta + f) - \left(\frac{\partial F_x}{\partial x} + \frac{\partial F_y}{\partial y}\right) \tag{6}$$

and we can view the horizontal convergence as being associated with the advection of vorticity (first term RHS) and frictional convergence (second term RHS).

Eq. (6) shows how gradients in the surface roughness can lead to convergence. Frictional convergence has been discussed in the literature (Bryson and Kuhn, 1961; Hastenrath, 1991) as possibly being an important mechanism for convergence in coastal and we can view the horizontal convergence as being associated with the advection of vorticity (first term RHS) and frictional convergence (second term RHS).

Eq. (6) shows how gradients in the surface roughness can lead to convergence. Frictional convergence has been discussed in the literature (Bryson and Kuhn, 1961; Hastenrath, 1991) as possibly being an important mechanism for convergence in coastal regions. If the climatology did not change when the surface roughness increased, then we would have $F_{x,y}$ varying only with C_D. Then, north of the Gulf of Guinea, $\partial F_x/\partial y > 0$, which would force positive convergence. On the east coast, $\partial F_y/\partial x < 0$, also forcing convergence.

Examination of the GCM output allows a quantitative assessment of the importance of this mechanism. While frictional convergence does operate in the GCM, and supports convergence, it is not the only mechanism that leads to the precipitation anomalies. The indirect effects of increasing CD on land are important for the frictional convergence term, i.e., changes in the low-level winds. In addition, the GCM results indicate that differences in the vorticity advection, especially the meridional advection of planetary vorticity, play a central role in the precipitation enhancement. Cook (1993) explores these results in more detail.

Despite the similarity between the precipitation fields in the rough-land-surface and the momentum-alone-enhanced cases, enhancing surface fluxes of sensible heat and moisture alone (without momentum fluxes enhanced) also leads to precipitation increases north of the Gulf of Guinea. Thus we cannot conclude that the momentum fluxes alone controll precipitation in this region.

4. Conclusions

Some of the physical mechanisms responsible for the large-scale observed precipitation climatology over tropical continents are investigated using climate models, including a GCM and a linear, primitive equation model.

The influence of surface wetness on the longitudinal precipitation gradient across South America and Africa in austral summer is explored. While precipitation rates in the continental interiors and on the west coasts are sensitive to surface wetness, precipitation in the east is not. The differences in sensitivity arise because different processes are responsible for the influence of surface wetness in different regions.

The strength of the modeled precipitation maximum north of the Gulf of Guinea in boreal summer is shown to be related, at least in part, to the land/sea surface roughness contrast. Three processes can be distinguished. One is the direct effect of a surface roughness gradient in causing frictional convergence. Additional low-level moisture convergence occurs due to the response of the model climate to the roughness gradient. This response leads to additional frictional convergence and convergence associated with the advection of vorticity into the region.

These results rely on GCM experiments with simple boundary conditions, designed to isolate certain aspects of land surface/atmosphere interactions. The relevance of the mechanisms identified here for more realistic situations remains to be shown. Further, the dependence of the results on certain physical parameterizations should be explored.

This study is an example of studying model results at the process level. Because of the problems associated with defining the present day precipitation climatology, and because the processes that maintain climate may not be the same as those important for climate change, a perfect replication of the observed precipitation field by a climate model does not guarantee a reliable prediction of future precipitation distributions. For these reasons, it is suggested that models and observations be studied and compared at the process level.

REFERENCES

Bryson, RA, and P.M. Kuhn, 1961 : Stress-differential induced divergence with application to littoral precipitation. Erdkunde, 15, 187-294.

Cadet, D, and N. Nnoli, 1987 : Water vapor transport over Africa and the Atlantic Ocean during summer 1979. Quart. J. Roy. Meteor. Soc., 113, 581-602.

Cook, K.H. and A. Gnanadesikan, 1991 : Effects of saturated and dry land surfaces on the tropical circulation and precipitation in a general circulation model. J. Climate, 4, 873-889.

Cook, K.H., 1993 : Forcing of the African monsoon : Role of land surface conditions. In preparation.

Hastenrath, S., 1991 : Climate Dynamics of the Tropics. Kluwer Academic Publishers, Dordrecht, The Netherlands.

Legates, D.R., and C.J. Willmott, 1990 : Mean seasonal and spatial variability in gauge-corrected, global precipitation. Int. J. Climatol., 10 , 111-127.

Ting, M.F., and I. M. Held, 1990 : The stationary wave response to a tropical SST anomaly in a idealized GCM, J. Atmos. Sci., 21, 2546-2566.

Washington, W.M., and C.L. Parkinson, 1986 : An Introduction to Three-Dimensional Climate Modeling. University Science Books, Mill Valley, CA.

THE HYDROLOGICAL CYCLE IN THE ECHAM3 SIMULATIONS OF THE ATMOSPHERIC CIRCULATION

K. Arpe, L. Bengtsson, L. Dümenil and E. Roeckner
Max-Planck Institut für Meteorologie
Bundesstraße 55
20146 Hamburg
Germany

ABSTRACT
The ability of the ECHAM3 model to simulated the hydrological cycle of the atmosphere is investigated. The impact of horizontal resolution is studied. It is shown that the validation is severely hampered by the large uncertainty of observational data. It is tried to minimize this problem by applying a wide range of validation methods and data, e.g. using run-off data from large catchment areas. Mostly the model simulations lie within the range of realizations of climatological estimates though some model deficiencies become obvious.

KEYWORDS
Hydrological cycle, precipitation, rain, snow, run off, model, climate simulations, model validation, ECHAM, model resolution, general circulation model.

1. Introduction

The ECHAM3 model originates from the ECMWF model and still resembles in the formulation of the dynamical part the version of their cycle 36 while the parametrization of sub-grid scale processes have mainly been developed in Hamburg. For a more detailed description of the model and improvements in respect to earlier versions of the ECHAM models see Roeckner et al (1992). In this study we analyse the impact of model resolution on the hydrological cycle.

The hydrological cycle is a key process in the atmosphere and it is essential that this is well simulated in the model. It is intended to use the model for simulating future climates, e.g. to show the impact of increased CO2, and for that it is needed to demonstrate that the model is able to simulate the climatological quantities which are important for the people, i.e. precipitation, snow amount, surface temperature and the river run-off of the present climate.

NATO ASI Series, Vol. I 26
Global Precipitations and Climate Change
Edited by M. Desbois and F. Désalmand

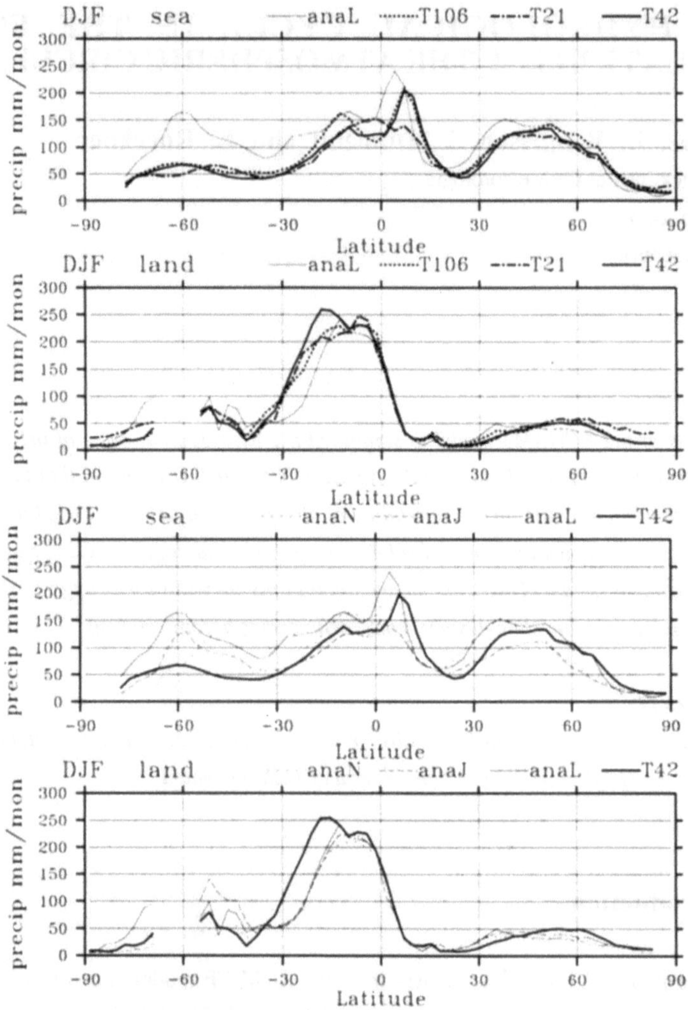

Fig. 1: Zonal mean precipitation separately for sea and land points for the season DJF. In the upper panels simulations with the ECHAM3 T21, T42 and T106 models are compared with climatological estimates by Legates and Willmott (1990). In the lower panels the T42 model results are compared with 3 climatological estimates: anaJ=Jaeger (1976), anaL=Legates and Willmott (1990), anaN=NASA, which is a blend of estimates from MSU observations over oceans by Spencer (1993) and from conventional observations over land by Schemm et al (1992).

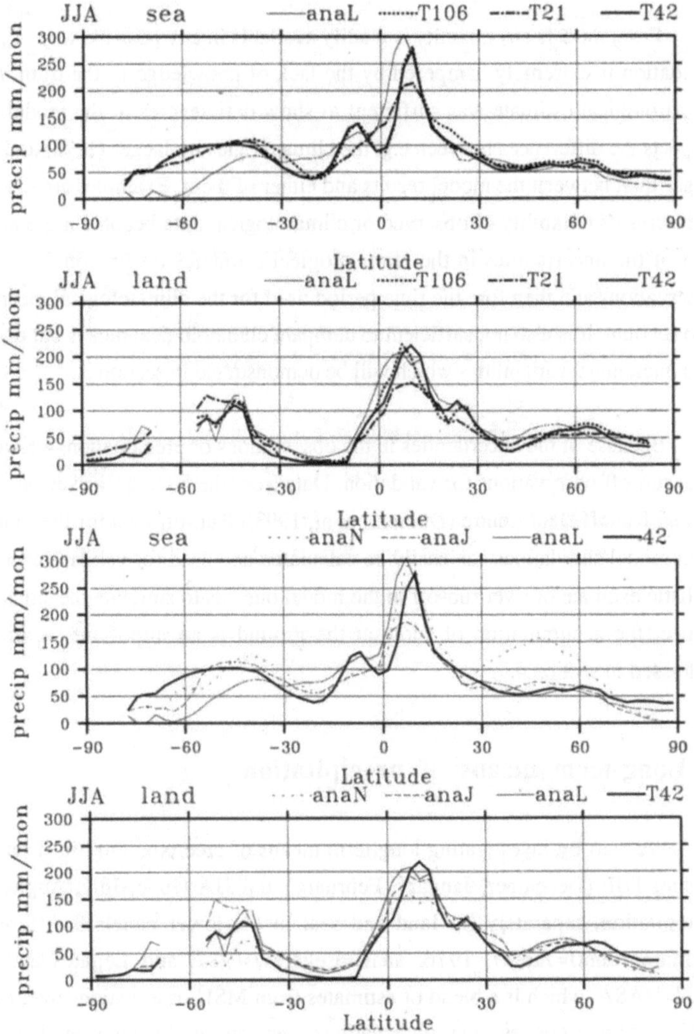

Fig. 2: As Fig. 1 but JJA.

From models any quantity is readily available in any possible way of presentation but the validation is extremely hampered by the lack of knowledge of the truth. In earlier days any climatological estimate was sufficient to show deficiencies of the models but now in many respects the differences between e.g. the climatologies by Jaeger (1976) and Legates (1987) are larger than between the model results and either of them. Examples are discussed in section 2. Measures of reliability of observed or climatological data become more and more important. Part of the uncertainties in the climatological estimates result from long term trends in the observations and therefore the time period used for the climatological means have to be taken into account. It is also not sufficient to compare climatological means but one has to investigate also interannual variabilities which will be demonstrated in section 3.

Because of the uncertainties in the observations of precipitation, we are trying to use also river run-off observations for validation. Data from the 50 largest rivers are available from the Global Runoff Data Centre (Dümenil et al, 1993). Run-off data for the same catchment areas from model simulations can easily be calculated and used for validation (see section 5). For a realistic estimate of river run-off in the model one has to simulate the storage of water. In this respect the accumulation of snow at the ground is an important factor, a topic which is addressed in section 4.

2. Long-term means of precipitation

We start by investigating long term means of precipitation. Fig. 1 and 2 display for the seasons DJF (December, January, February) and JJA (June, July, August) zonal means of precipitation, separately for land and sea. In the lower panels 3 different climatological estimates (anaJ=Jaeger, 1976, anaL=Legates, 1987, and Legates and Willmott, 1990, anaN=NASA, which is a blend of estimates from MSU observations over oceans by Spencer, 1993, and from conventional observations over land by Schemm et al, 1992) are compared with 30 year means from the ECHAM3 T42 model, in the upper panel the Legates climatology is compared with 3 different resolution runs with the ECHAM3 model. The T106 results consist of 6 years of simulation for DJF and of 5 years for JJA while T21 and T42 results consist out of 30 years, all using climatological SST, which is a 10 year mean of the AMIP data set (Reynolds, 1988). Over the oceans, especially during JJA there is a marked relative minimum at the equator in the T42 and T106 experiments which is not present in the T21 model nor in the climatological estimates. However, it is difficult to judge if this minimum is unrealistic because of the large uncertainty in the observations. We can be more sure that the northern ITCZ over the oceans is simulated too far north in both seasons because all climatological estimates agree in this respect. In the belt 20° to 30° S over land during DJF the models simulate consistently more precipitation than the climatological estimates, a problem over Australia and South Africa which will be discussed below. Over the southern extra-tropical oceans during DJF the climatological estimates suggest clearly larger precipitation amounts than the model simulations. Though the climatological estimates agree with each other, many doubts about the accuracy in this respect have been raised.

On an global average the models show a steady increase of precipitation with resolution but still stay within the range of the climatological estimates. The annual global mean values are as follows: Legates: 95, Jaeger: 79, NASA: 102, T21: 81, T42: 85, T106: 89 mm/month. The Legates data have been corrected for observational errors. If one would apply the same correction to Jaeger's data the spread between them would decrease as the correction always leads to an increase, strongest at high elevations and in snow conditions.

Fig. 3 shows mean maps of precipitation for DJF for 3 simulations with different resolutions. For validation 3 climatologies (Legates, Jaeger and NASA) are given. Because of the large uncertainty in the observations, the simulations are mostly within the range of climatological estimates and there are only few points where the model deviates from the climatologies significantly. Over subtropical oceanic upwelling areas the models show much larger areas with low precipitation amounts than the estimates by Legates or Jaeger but the estimates from MSU observations are quite near to the model results. In the zonal means over oceans a relative minimum at the equator in the model simulations had been shown and here it becomes obvious that this results mainly from the Pacific where the models look unrealistic. There is excessive precipitation over Australia and South Africa, a problem which was introduced into the simulations when changing the parameterization of convection from a Kuo (1974) to a mass-flux scheme (Tiedtke, 1989) though this may not neccessarily be the cause of the problem. For the latter problem improvements with increasing resolution can clearly be seen.

During winter all climatological estimates agree that there should be a precipitation maximum over the eastern Mediterranean though with quite different amplitudes. The higher resolution models agree reasonably well concerning the pattern but are on the low side even when compared with the lowest climatological estimate by NASA. The T21 model gives an unrealistic distribution of precipitation. The relatively low simulated precipitation values over the Mediterranean is connected with an eastward shift and strengthening of the Azores anticyclone, as discussed by Roeckner et al (1992).

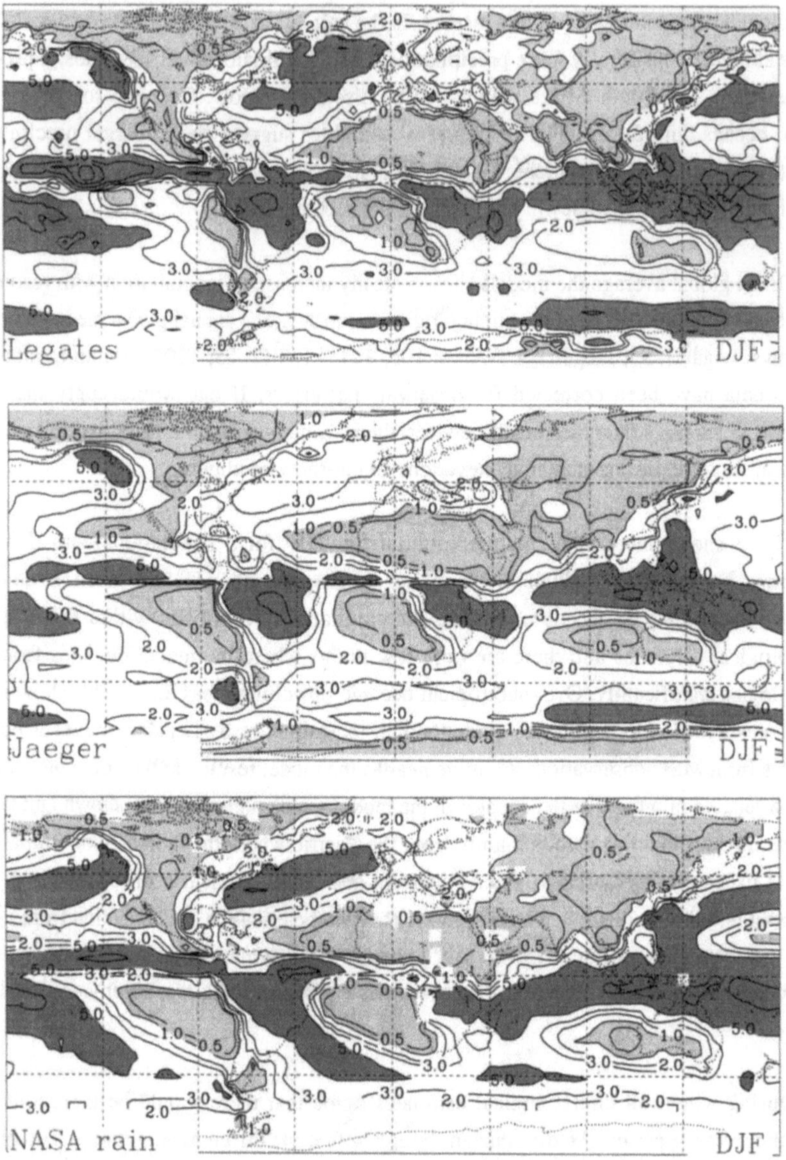

Fig. 3: Maps of precipitation during DJF in 3 climatological estimates and the ECHAM3 model simulations with T21, T42 and T106 resolution. Contours at 0.5, 1, 2, 3, 5, 10, 20 mm/day. Light shading for < 1 mm/day, heavy shading for > 5 mm/day.

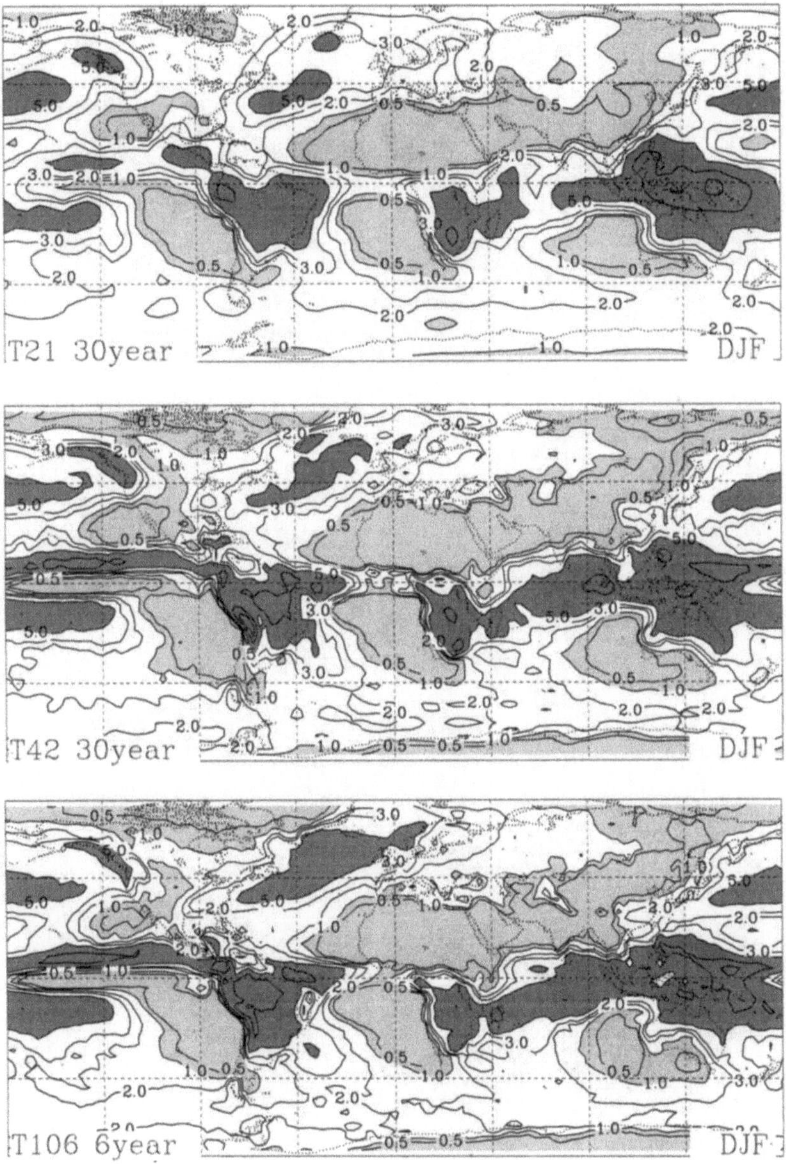
T21 30year
DJF
T42 30year
DJF
T106 6year
DJF

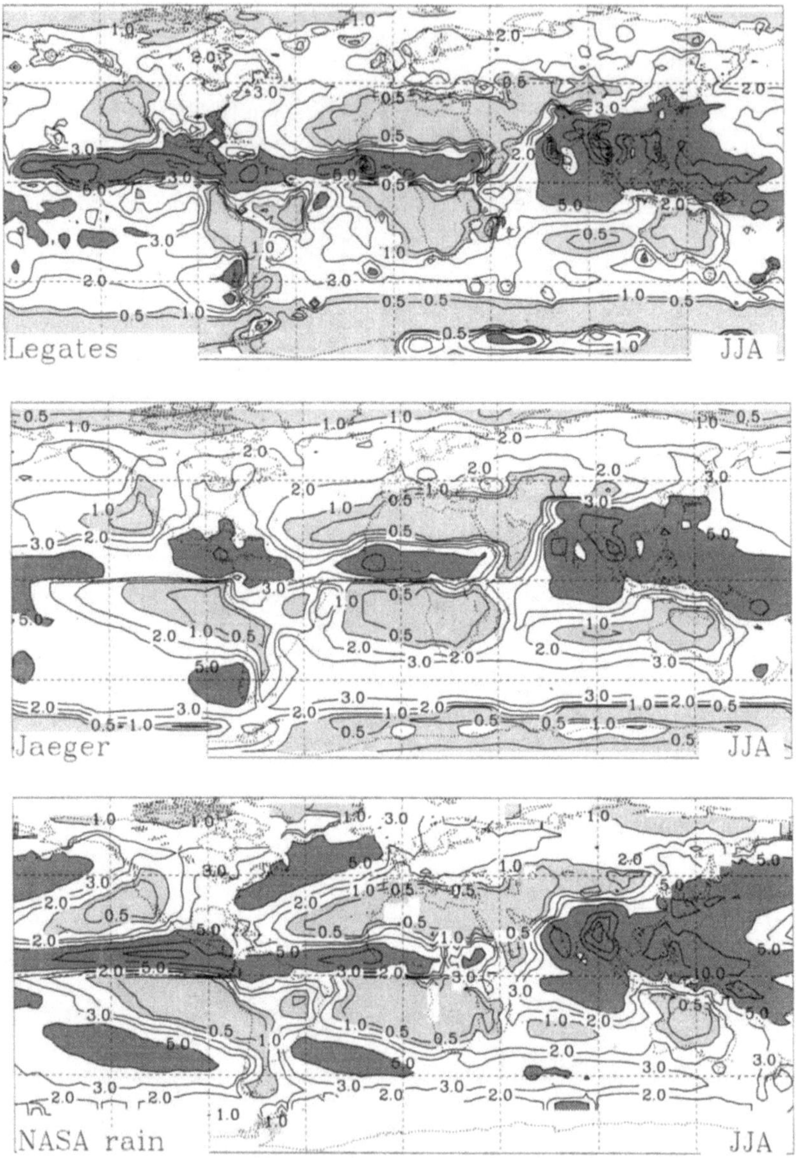

Fig. 4: As Fig. 3 but JJA.

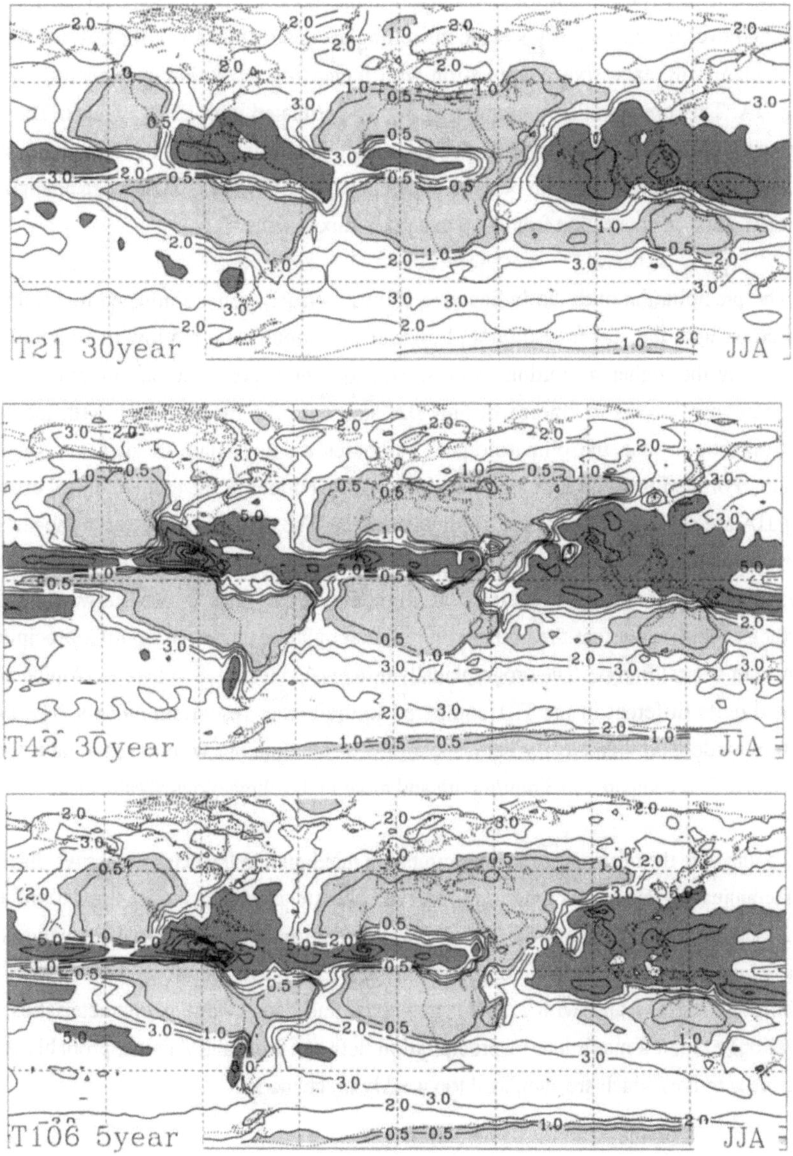
T21 30year
JJA
T42 30year
JJA
T106 5year
JJA

Fig. 4 shows global maps for JJA. Further to the too low precipitation over the equatorial Pacific and over the subtropical oceanic upwelling areas already known for DJF we find that there is increasing excessive precipitation with increasing resolution over central America during JJA. This precipitation maximum is highly correlated with ENSO events. The better performance of the T21 model is partly due to the fact that there is no land in central America at this resolution. Also this problem became apparent in the ECHAM models when changing the parameterization of convection from Kuo to a mass-flux scheme.

The precipitation over the Indian area shows a clearly better simulated distribution of precipitation with the higher resolution, however the amounts are too low in all simulations especially by the higher resolution models. The equatorial precipitation minimum, already discussed above, seems to emerge during JJA also over the Indian Ocean with increasing resolution though also the climatological estimates display some of such an equatorial minimum. For the European area the precipitation distributions of the climatological estimates during JJA are remarkably different from each other. The differences between Jaeger's and Legates' climatologies result partly from applying a correction of observational data which always mean an increase, especially at high elevations and under snow conditions. Nevertheless they agree that the higher resolution simulations provide too little precipitation over most of central Europe. The orographically influenced precipitation over the Alpine area is simulated quite different in the T21 compared to the higher resolution models. The higher resolution models provide clearly better positions of maxima but with too low values, i.e. 2 mm/month or less compared to 3 mm/month and more in the climatological estimates.

To illustrate the missing summer precipitation maximum in the extra-tropics we display in Fig.5 the annual cycles of precipitation (observed climate as well as model simulation) at the two stations Berlin and Chicago. Station observation data are more accurate than analyses data but suffer naturally from lack of representativeness. We have therefore checked also with neighbouring stations and with area mean values and found similar characteristics. The observations exhibit a maximum of precipitation between June and August, probably due to convective activities which are simulated too weakly in the model.

3. The interannual variability of precipitation

We investigated further the impact of Sea Surface Temperature (SST) anomalies on the variability of precipitation by running several experiments with different initial states but with the same observed SST. Monthly observed means for the period 1979 to 1992, prepared by Reynolds (1988), have been used. For the inner tropics the impact of the SST turned out to be

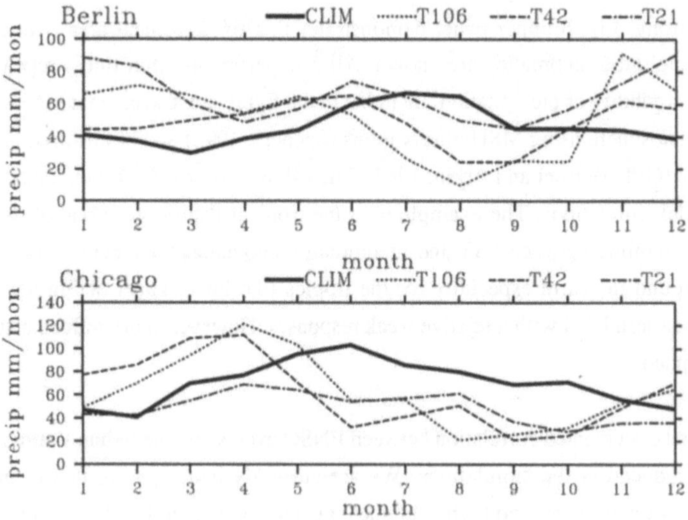

Fig. 5: Annual cycle of precipitation for the stations Berlin and Chicago. The climatological values are station observations.

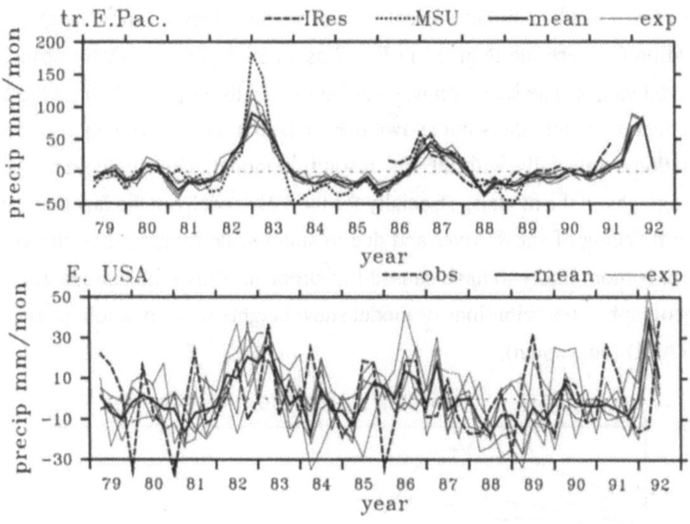

Fig. 6: Seasonal mean precipitation averaged over the eastern tropical Pacific (10°N-10°S, 90°-160°W) and over the eastern USA (30°-45°N, 70°-102°W) in 5 simulations with observed monthly mean SST for the period 1979 to 1992. The heavy line represents the average of the 5 experiments. Dotted and dashed lines represent estimates of the true precipitation.

extremely strong. Fig. 6 upper panel demonstrates this for a tropical area over the eastern Pacific. Precipitation anomalies are shown. All 5 experiments (thin lines) repeat the same interannual variability of precipitation. Included in the figure are also estimates of precipitation of the real atmosphere using MSU observations (Spencer, 1993) and of infra-red observations from satellite (GPI, Meisner and Arkin, 1987). They show a very similar interannual variability like the model simulations. The example over the tropical Pacific gave one of the strongest signal but a coupling between SST and precipitation anomalies have been found over many places all around the earth especially for the tropics (the lower panel of Fig.6 shows as an example the eastern USA with a relative weak response). However, more refined analyses have still to be applied.

The well documented correlation between ENSO events and the Indian summer monsoon was not reproduced in the simulations. We are currently investigating this in more detail because it is not well understood whether such a correlation should really exist in the model simulations.

4. Precipitation in the form of snow and its melting

For only few Canadian climatological stations we had observed snow heights available of which the station Coppermine is given in Fig. 7 as an example to compare observation with model simulated values. The selection was guided by the discussion below on the MacKenzie river. For Coppermine and others not shown one finds that the snow is melting too late in the model simulations, especially with the low resolution model. This could partly be due to an unrealistic orography in the models, especially for the T21 model, but the fact that the change of the albedo with ageing of snow cover and due to small-scale roughness of the ground is not parameterized, is more likely to have caused this problem. This problem can also be seen by comparing geographical distributions of model snow heights with climatological estimates by NASA and RAND (not shown).

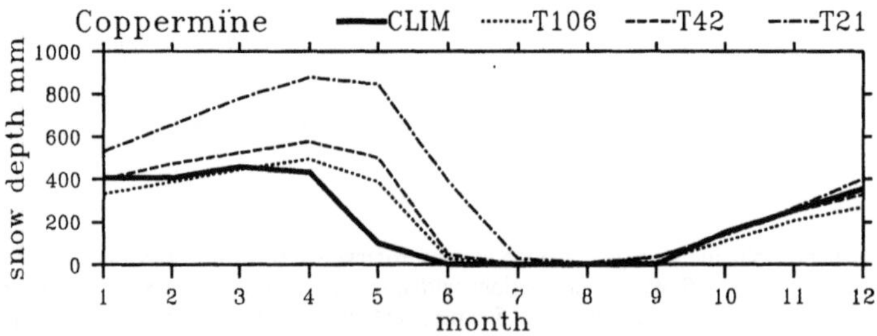

Fig. 7: Annual cycle of snow heights at the station Coppermine (Potter, 1965).

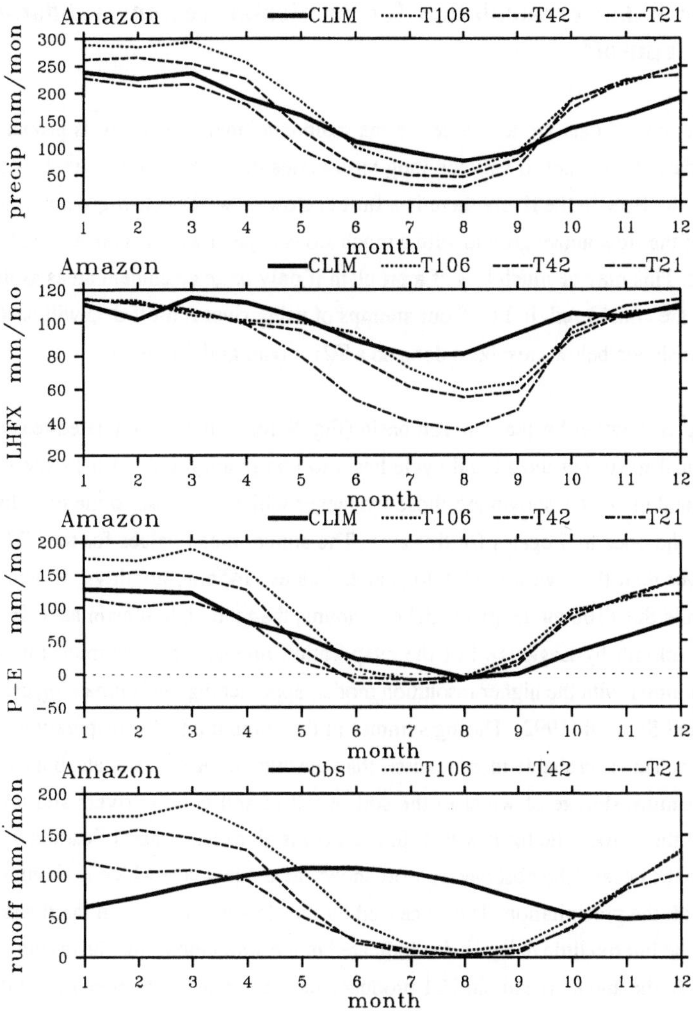

Fig. 8: Annaul cycle of precipitation, evaporation, their difference and water run-off for the Amazon catchment area.

5. Water balance calculation for particular regions validation using river run-off

River run-off data are available for the main catchment areas and is probably the best source of data if we want to validate long term series of large scale means. It is said that in addition to the flow in the rivers there is a further flow of water below ground in the order of one third of the flow above ground with a much slower speed which is not covered by the run-off data set. This may be much less of a problem if only large catchment areas as the Amazon river basin are considered. In Fig. 8 our attemps of using river run-off data will be shown. The calculations shown below have been done on a T21 model grid throughout.

The precipitation for the Amazon basin (Fig. 8, top panel) is simulated reasonably well for the annual mean but the annual cycle has a too large amplitude in the experiments. The climatological estimates shown are those by Jaeger which are close to the ones by Nasa but lower than the ones by Legates for this area. The annual mean values for the T21 model are clearly lower than the lowest climatological estimates (145mm/month vs. 153mm/month by NASA) while the T106 model gives higher amounts than the highest estimate (190mm/month vs. 171mm/month by Legates). For the evaporation one sees a clear trend towards higher values in summer with the higher resolution model, approaching the "climatological estimates" by Mintz and Serafini (1992). During summer in the simulations the evaporation exceeds the precipitation. This is probably more realistic than the climatological estimate, but the model has only a maximum storage of water in the soil of 0.2 m and none in rivers and lakes and we believe that this is too little for this area. In the models all excess water is leading to immediate run-off into the ocean. The observed run-off shows an annual cycle which is about 4 month out of phase with the precipitation. This is caused by the slowness of flow in the rivers and in the ground. Some but too little delay is also indicated in the simulations due the drying and wetting of the soil. In the annual mean the T21 produces too little run-off in agreement with too little precipitation already mentioned above. Because of the lack of supply of water by the rivers in the simulations, the ground can run dry in summer and this leads to excessively high temperatures for the Amazon basin.

Over the MacKenzie basin in north-western Canada (not shown) the T21 model produces by far too much precipitation. The precipitation in the higher resolution model is still too large though much better and with a reasonable annual cycle. During summer there is considerable evaporation in this area which is in the higher resolution models larger than the precipitation, however not as large as in the "climatological estimates" by Mintz and Serafini (1992). The latter looks unrealistic as the evaporation nearly cancels the precipitation for an annual mean.

The run-off in this area is dominated by the melting of snow. Therefore the maximum run-off in the simulations can be found in spring and early summer. However, the observations have a maximum run-off in June - July, i.e. 1 month later than the simulations, suggesting that the snow-melting is simulated too early. However as shown above in Fig. 7 for Coppermine in the surrounding of the MacKenzie river, the melting in the simulation is generally delayed by one month. The higher resolution models perform better because of the better simulation of precipitation. The slowness of flow in the river and slow melting of ice in the ground and lakes must be an important factor which has to be taken into account in the models.

6. Conclusions

When comparing maps of long term mean precipitation in the model simulations with estimates of the true precipitation one finds generally that model results lie mostly well within the range of realisations. Simulations of the atmosphere with general circulation models have now reached a high level of quality so that a detailed validation is severely hampered by the lack of observational data, their uncertainty and/or their unrepresentativeness. At least we do not know these characteristics in all cases.

Though the range of realisations of the "truth" is quite large, there is clearly too much precipitation simulated over Australia and South Africa during DJF, connected with too hot surfaces. An improvement with increased resolution has been noticed. There is also too much precipitation over central America during JJA. Both these problems were introduced or enhanced with the introduction of the massflux scheme for parameterizing convective precipitation and are common to other models as well. Over oceanic upwelling areas, e.g. over the Pacific off Peru the model simulates much dryer climate than the estimates by Legates and Willmott (1990). Different estimates of climatological means differ considerably in this respect leaving the model simulation within their range of realisation. Any monthly mean of model simulation can hardly be judged to be unrealistic, the long term mean errors mentioned above are a result of many extreme though possible situations of the same kind. The model produces a strong relative minimum of precipitation at the equator with main contributions from the central Pacific.

We found river discharge to be valuable independent observations for model validation and intend to make more use of these data. For a better use of these data but also for improving the model performance, the storage of water and its transport has to be parameterized in the model. On the whole the T106 and T42 version of the ECHAM3 model give results which are much more similar to each other than to the T21 version.

Acknowledgments: The authors are indebted to U. Schlese (DKRZ) for carrying out the T106 experiments and to the Swiss Climate Computing Center which has been providing computational resources for this experiment under a joint co-operative program between the Max-Planck-Institut für Meteorologie in Hamburg and the Eidgenössische Technische Hochschule in Zürich.

REFERENCES

Dümenil, L., K. Isele, H.-J. Liebscher, U. Schröder, M. Schumacher, K. Wilke (1993): Discharge data from 50 selected rivers for GCM validation. Max-Planck-Institut für Meteorologie Report No. 100, 62pp.

Jaeger, L. (1976): Monatskarten des Niederschlags für die ganze Erde. Ber. d. Dt. Wetterdienstes 139, Offenbach.

Kuo, H.L. (1974): Further studies of the parameterization of the influence of cumulus convection on large-scale flow. J. Atm. Sci., 31, 1232-1240.

Legates, D.R. (1987): A climatology of global precipitation. Publ. Climatol. 40 (1), Newark, Delaware, 85pp.

Legates, D.R.and C.D. Willmott (1990): Mean seasonal and spatial variability in gauge-corrected global precipitation. Internat. J. Climatol. 9, 111-127.

Meisner, B.N. and Arkin, P. A. (1987): Spatial and annual variations in the diurnal cycle of large-scale tropical convective cloudiness and precipitation. Mon. Wea. Rev. 115, 2009-2032.

Mintz, Y. and Y.V. Serafini (1992): A global monthly climatology of soil moisture and water balance. Climate Dynamics, 8, 13-27.

Potter, J.G. (1965): Snow cover. Climatological studies No.3, Canada, Department of Transport, Meteorological Branch, Toronto 1965, 69pp.

Reynolds, R.W. (1988): A real-time global sea surface temperature analysis. J. Climate, 1, 75-86.

Roeckner, E., K. Arpe, L. Bengtsson, S. Brinkop, L. Dümenil, M. Esch, E. Kirk, F. Lunkeit, M. Ponater, B. Rockel, R. Sausen, U. Schlese, S. Schubert and M. Windelband (1992): Simulation of the present-day climate with the ECHAM model: Impact of model physics and resolution. Max-Planck-Institut für Meteorologie Report No. 93, 171 pp.

Schemm, J., S. Schubert, J. Terry and S. Bloom (1992): Estimates of monthly mean soil moisture for 1979-1989. NASA Technical Memo. 104571, 262pp.

Spencer, R.W. (1993): Global oceanic precipitation from the MSU during 1979-1991 and comparisons to other climatologies. J. Climate, 6, 1301-1326.

Tiedtke, M. (1989): A comprehensive massflux scheme for cumulus parametrization in large-scale models. Mon. Wea. Rev., 117, 1777-1798.

THE PARAMETERIZATION OF THE PRECIPITATION PROCESS IN THE LMD GENERAL CIRCULATION MODEL

H. Le Treut, Z.X. Li, and O. Boucher

Laboratoire de Météorologie Dynamique du CNRS,
Ecole Normale Supérieure, 24 rue Lhomond, 75005 Paris, France

ABSTRACT:
The LMD atmospheric General Circulation Model includes a cloud water budget equation in which the condensation of water vapour and the precipitation of cloud water are treated as separate processes. Such a distinction has become a standard feature of many models in the recent years. We show here that it has little impact on the simulated monthly mean precipitation. It does however affect strongly the water vapour mixing ratio, the cloud water content and the cloud distribution. If we submit the model to an external perturbation, such as an increase in greenhouse effect, these parameters cause powerful feedback processes, which may affect the mean precipitation: the parameterization of the precipitation process then appears as a key process determining the amplitude of the climate response.

KEYWORDS:
General Circulation Model; Cloud parameterization; Climate feedback

1. Introduction

For many years most atmospheric General Circulation Models (GCMs) have assumed that water vapour, once condensed, is immediately released as precipitation. During the last decade however many GCMs have begun to include in their formulation a budget equation for cloud condensed water. Water vapour condensation and precipitation are then distinct processes, condensation being the source of cloud water, while precipitation, which transforms cloud water into rain water, is its main sink.

The aim of this short paper is to examine how this distinction may affect the simulated precipitation and cloudiness fields, as far as the simulation of the mean climate is concerned, or in the case of a climate perturbation. For this purpose we use different simulations in which various representations of the precipitation processes are introduced. A more complete description of these simulations is available elsewhere (Li and Le Treut, 1992; Le Treut *et al.*, 1994; Boucher *et al.*, 1994) and only a few essential features are reviewed here.

2. Model description: the parameterization of precipitation

The general formulation of the LMD GCM has been originally described by Sadourny and Laval (1984). The dynamics is formulated in finite differences. The resolution of the model used in this paper is a low one with 48 points in the latitudinal direction, 36 points

NATO ASI Series, Vol. I 26
Global Precipitations and Climate Change
Edited by M. Desbois and F. Désalmand

in the meridional direction and 11 vertical levels. The model includes a complete set of physical parameterizations. The condensation of water vapour is parameterized differently for convective and stratiform clouds. For the convection, a moist adiabatic adjustment scheme is used to eliminate the atmospheric conditional instability and a Kuo-type scheme is then sequentially applied over the boundary layer to take into account the large-scale convergence. The vertically integrated amount of condensed water is then distributed at different levels in proportional to the local decrease of water vapour due to convection process. Stratiform condensation is parameterized using a simple statistical scheme, which allows it to start before saturation, by taking into account the inhomogeneity of the water vapour mixing ratio within a grid-box.

After a budget equation for cloud water was introduced in the model, the precipitation process has been parameterized through three different approaches of increasing complexity. In the present text they are referred to as parameterizations A, B and C.

(i) The approach A simply consists in prescribing a threshold for the precipitation of cloud water: the water in excess of this threshold is precipitated to the ground. It is, however, necessary to distinguish between warm and cold clouds and a lower threshold was used for clouds whose top layer was below -10°C. This is a very simple way to take into account the Bergeron process, by which ice clouds precipitate more efficiently. A validation of this first scheme using the ISCCP data over the Meteosat area gave good results (Le Treut and Li, 1988).

(ii) Deficiencies of the parameterization soon began to appear, however. The comparison with SMMR or SMMI estimates showed that the integrated cloud water content was too low. Comparison to ERBE data showed that the longwave cloud forcing was too large, especially at high latitudes. The discontinuities introduced by the thresholds were also constituting a serious problem for further comparisons with observed data (Yu, 1993) and climate sensitivity studies (Li and Le Treut, 1992). A second version of the model was therefore developed. We call it here version B and it is referenced in Le Treut *et al.* (1994) as version LMD4^{MO}. It includes a refined parameterization of the precipitation scheme. In particular the scheme of Sundqvist (1981) was introduced for warm clouds. It allows a smoother transition from non-precipitating clouds to thicker precipitating clouds. Ice clouds are treated through an ice sedimentation mechanism (Star and Cox, 1985; Heymsfield and Donner, 1990). The rate of ice cloud precipitation then depends on the terminal velocity of the falling crystals and the rate of condensed water removal is proportional to the cloud condensed water itself. Finally, a linear transition between ice and water clouds is also defined for temperature between 0°C and -15°C.

(iii) In a third step (referred here to as scheme C and described in more detail in Boucher *et al.*, 1994) a new parameterization of water clouds was introduced, to take into account the role of cloud condensation nuclei, and the possible change in their concentration. In this approach, which is adapted from a one-dimensional modelling study by Liou and Ou (1989), the concentration of cloud droplets N is prescribed. The prediction of the amount of condensed water makes it then possible to determine a mean radius of the cloud droplet size distribution. This is done by using the following equation:

$$\frac{4}{3}\pi r_3^3 \rho_{water} N = l\rho_{air} \qquad (1)$$

where r_3 is the mean volume cloud droplet radius, ρ_{water} and ρ_{air} represent the mass

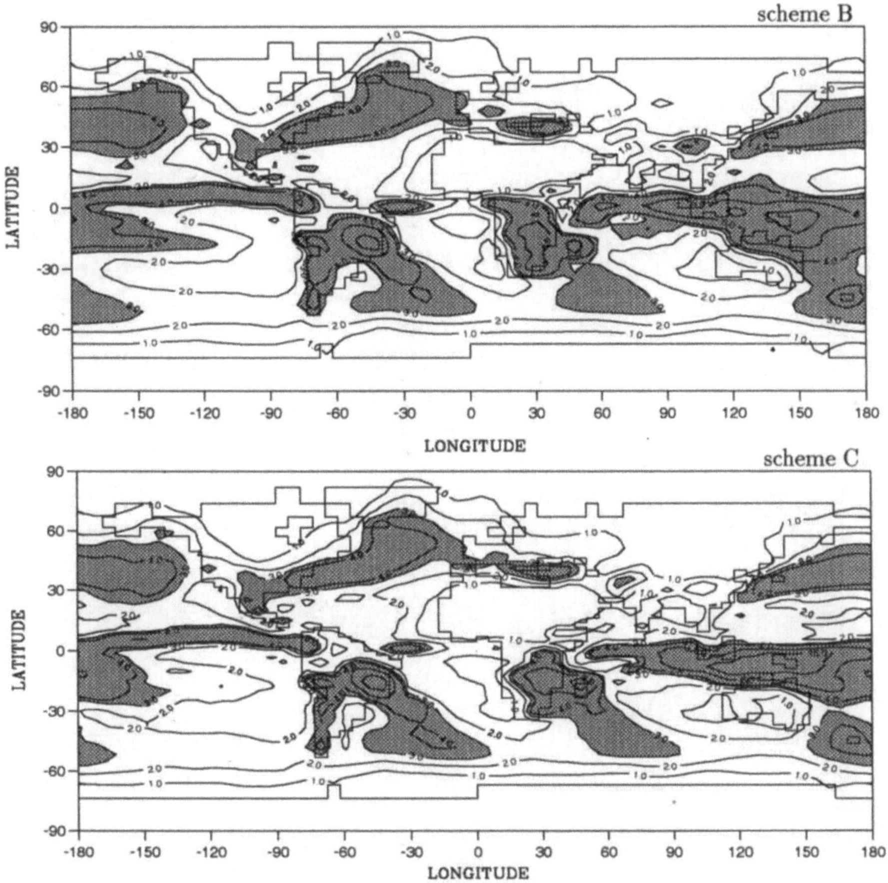

Figure 1: *Monthly mean precipitation (mm day^{-1}) for January: experiment B (top), experiment C (bottom).*

per unit volume of water and air, respectively, l is the cloud water mixing ratio (within the cloud). We further assume that the different equivalent radii of the distribution, the average value $r = r_1$, or the higher orders equivalent radii r_2, r_3, r_4, vary proportionally. We can then use the mean radius to parameterize the precipitation process itself and to estimate the cloud optical thickness. In particular, the autoconversion rate of cloud water into precipitation water has the form:

$$A = Clr^4 NH(r - r_0), \tag{2}$$

where C is a constant and H the Heavyside function to specify a threshold cloud droplet radius r_0 at which precipitation can start.

The parameterization also includes the role of cloud water collection by rain water, and the reevaporation of falling rain. This is done by introducing a budget equation for rain water.

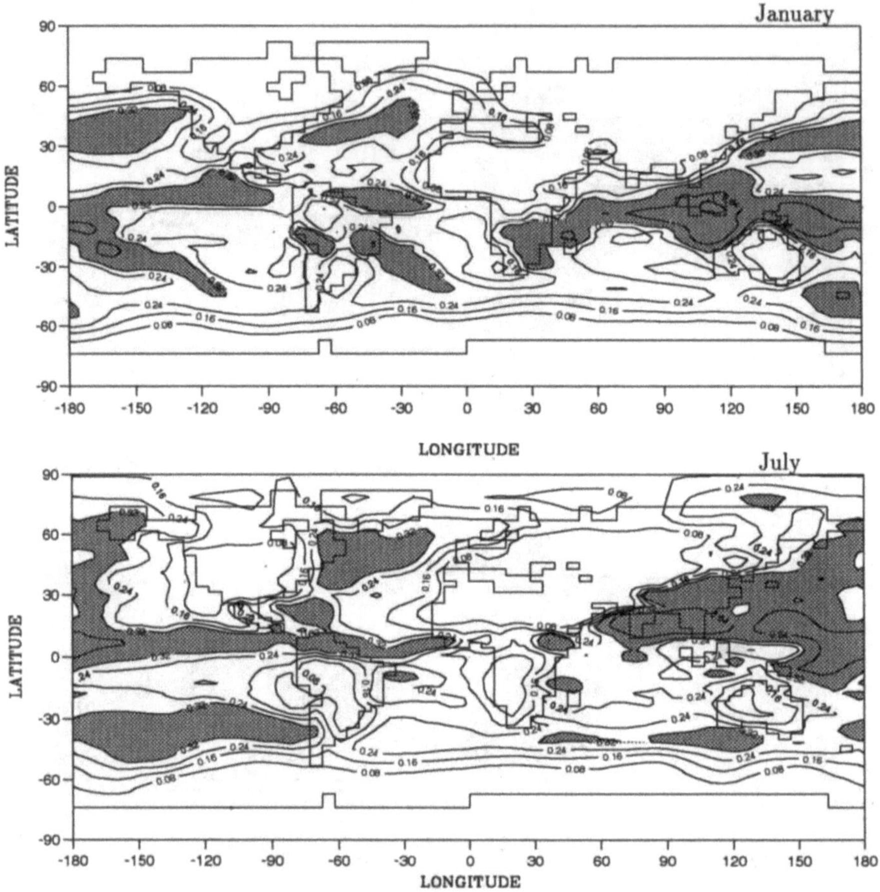

Figure 2: *Vertically integrated cloud liquid water content (g m^{-2}) simulated in the experiment C for January (top) and July (bottom).*

This parameterization C was introduced in order to be able to use observed correlations between N and the aerosol concentration, and study such effects as the role of man-made sulphate aerosols (Boucher and Rodhe, 1994). In the present paper we examine more fundamentally how the introduction of this new mechanism modifies the model physics.

3. Results

3.1. Simulation of the mean climate

The role of parameterizations B and C can be compared in the experiments reported by Boucher *et al.* (1994). We have displayed in Figures 1(a) and 1(b) respectively the mean precipitation fields obtained for a month of January using those two parameterizations. The very drastic modification of the representation of the precipitation process is not reflected in the results, which are very similar in the two cases. The differences are

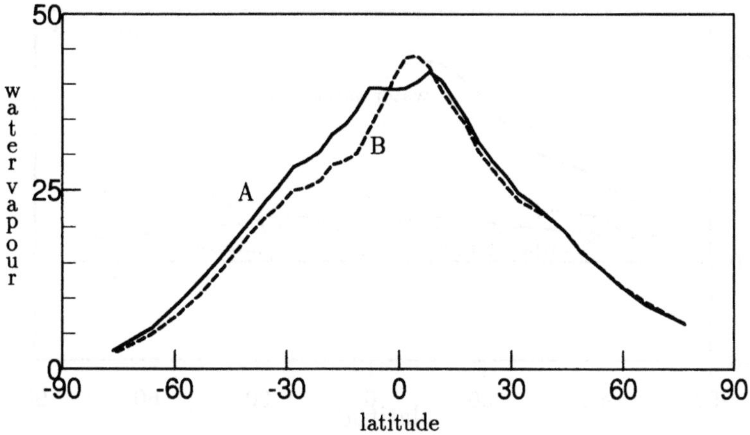

Figure 3: *Vertically integrated water vapour content (kg m^{-2}) in function of latitude for respectively the experiment A and the experiment B.*

not significant, as was shown using a set of 5 simulations differing through their initial state only. This indicates that for monthly mean conditions, the supply of water vapour through large-scale convergence is much more important for the determination of the mean precipitation that the details of the precipitation onset, because the cloud water reservoir is generally much smaller than this convergence term. This is confirmed by Figure 2 which shows the vertically integrated cloud water amount, as simulated using parameterization C. The orders of magnitude - and the location of the main precipitation areas - are in good agreement with the SSMI data of Wentz (1989), and they illustrate the order of magnitude of the cloud water reservoir.

Of course there is an indirect impact of the cloud microphysical parameterizations. One of these effects is illustrated in Figure 3 by the comparison of two experiments using parameterizations A and B, presented in Le Treut *et al.* (1994). We show the zonally averaged difference between the mean water vapour distribution of those two experiments: when using parameterization B the amount of water vapour in the ITCZ region is significantly higher, with a corresponding decrease in the subtropics. This most certainly results from some indirect radiative effect, through which convection is enhanced. Note that in those experiments the atmospheric model is coupled with a slab ocean model, and this may enhance the radiative feedbacks with the surface.

3.2. Climate sensitivity

The parameterizations A and B were used for two simulations of the climate response to a doubling of the atmospheric CO_2 content. A slight additional difference between the two simulations concerned the parameterization of the cloud optical properties. The globally averaged surface warming was reasonably similar in the two cases: 3.9°C for parameterization A and 3.6°C for parameterization B. But the regional response was largely modified. Indeed a study of the main feedback mechanisms explaining the model response shows that this similar global warming occurs through different feedback processes. The

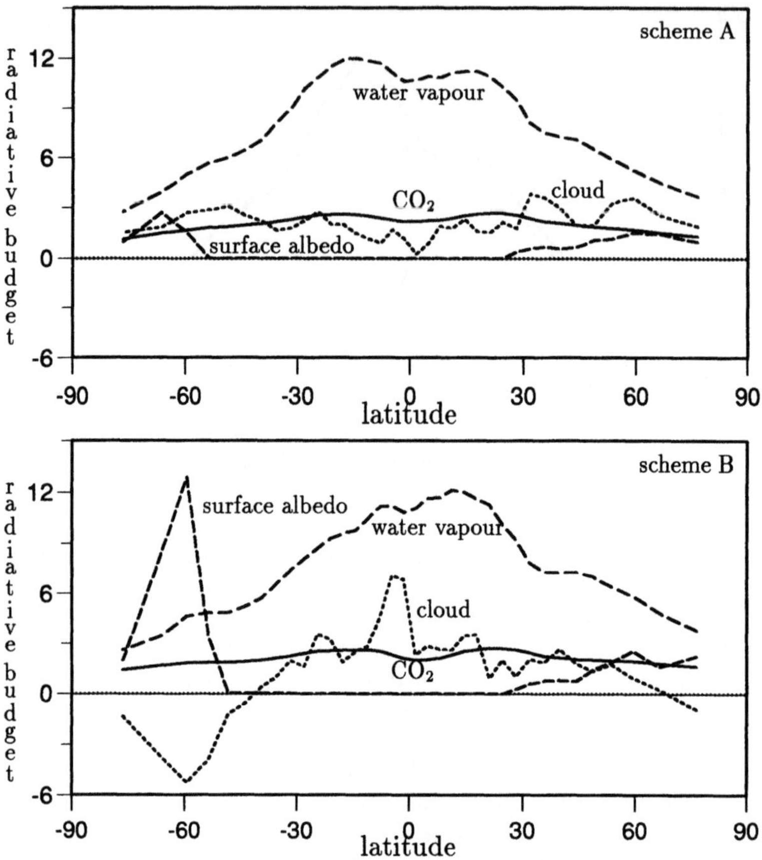

Figure 4: *Zonal means ($W\ m^{-2}$) of the direct forcing and the feedbacks at the top of the atmosphere for the experiment A (top) and the experiment B (bottom).*

notion of feedback is defined by considering the modification of the radiation budget at the top of the atmosphere between the $2\times CO_2$ and the $1\times CO_2$ integrations. Off-line radiative computations allows us to determine what is the part of this change which may be attributed to the change of a given parameter: water vapour, cloudiness or surface albedo. As these parameters do not vary independently and the response of the model is not linear, such a computation can only be indicative. But experience shows that it is useful and informative. Figures 4(a) and 4(b) display these "feedback terms" as a function of latitude for experiments A and B respectively.

The water vapour feedback is very strong in the two experiments. It is however slightly modified from one experiment to the other. The noticeable modification is a decrease in the subtropical regions of the Southern Hemisphere for the experiment B. This is consistent with the fact that in this experiment the water vapour content is diminished in the subtropical regions and the water vapour increase in the $2\times CO_2$ simulation is also diminished.

The cloud feedback is largely modified from A to B. In the experiment A, it is almost uniformly positive over the globe. In the experiment B, it is positive in the tropical regions, but it becomes negative at high latitudes. As explained in Le Treut *et al.* (1994), this feature is strongly related to the cloud parameterization and the changes of cloud properties at different layers of the atmosphere. The total cloud feedback effect is diminished in the experiment B, which is compensated by a larger value of the surface albedo feedback.

4. Conclusions

The conversion of cloud water into rain water is a microphysical process, which begins to be parameterized crudely in large-scale models. It may not affect directly the mean large-scale precipitation amount over a month, but the indirect effects may be important. A good validation of the parameterization of this process is essential for the continuous development of GCMs. We may note that satellite estimates of the rainfall distribution, cloud optical properties or cloud droplet radii are obtained indirectly and also depend on the microphysics of the precipitation process. Evaluating a physically consistent model by comparison to different types of satellite data may help both the model design and the development of satellite retrieval systems.

REFERENCES

Boucher, O., H. Le Treut and M. Baker, 1994: Precipitation and Radiation Modelling in a GCM: Introduction of Cloud Microphysical Processes, *J. Geophys. Res.*, submitted.

Boucher, O., and H. Rodhe, 1994: The Sulfate-CCN-Cloud Albedo Effect: A Sensitivity Study, Report CM-83, Dept. of Meteor., Univ. of Stockholm.

Heymsfield, A.J., and L.J. Donner, 1990: A scheme for parameterizing ice-cloud water content in general circulation models. *J. Atmos. Sci.*, **47**, 1865-1877.

Le Treut, H. and Z.X. Li, 1988: Using Meteosat data to validate a cloud prognostic scheme, *Atmos. Res.*, **21**, 273-292.

Le Treut, H., Z.X. Li and M. Forichon, 1994: Sensitivity of the LMD General Circulation Model to greenhouse forcing associated with two different cloud water parameterizations, *J. of Climate*, submitted.

Li, Z.X., and H. Le Treut, 1992: Cloud-radiation feedbacks in a general circulation model and their dependence on cloud modelling assumptions. *Climate Dyn.*, **7**, 133-139.

Liou, K.N. and S.C. Ou, 1989: The role of cloud microphysical processes in climate: an assessment from a one-dimensional perspective, *J. Geophys. Res.*, **94**, 8599-8607.

Sadourny, R., and K. Laval, 1984: January and July performance of the LMD general circulation model. *New Perspectives in Climate Modelling*, A. Berger, C. Nicolis, Eds., Elsevier, 173-198.

Star, D.O'C., and S.K. Cox, 1985: Cirrus clouds. Part I: a cirrus cloud model. *J. Atmos. Sci.*, **42**, 2663-2681.

Sundqvist, H., 1981: Prediction of stratiform clouds: results of a 5-day forecast with a global model. *Tellus*, **33**, 242-253.

Yu, W., 1993: Etude comparative des propriétés de la couverture nuageuse observée par satellites et simulée par un modèle du climat, Thèse de Doctorat, LMD, France.

Wentz, F.J., 1989: User's manual, SSM/I Geophysical tapes, Remote sensing systems, Tech Rep. No. 060989, 16pp.

VALIDATION OF LARGE-SCALE PRECIPITATION FIELDS IN GENERAL CIRCULATION MODELS

Mike Hulme

Climatic Research Unit
School of Environmental Sciences
University of East Anglia
Norwich NR4 7TJ, UK

ABSTRACT

How well do the current generation of General Circulation Models (GCMs) simulate present-day climate? Answering this question is an important stage in the improvement of our ability to model global climate change. The answer to the question will also determine the confidence that national and international bodies might place both in the projections of climate change generated by these models and in the regional climate change impact studies which use model output. This paper quantitatively evaluates the performance of a number of model precipitation climatologies, using four 'independent' observed climatologies. Three of the observed climatologies consist of mean monthly fields (JAEGER, LEGATES and IIASA), while the fourth is a new land-only time series of monthly precipitation from 1900 to 1992 (CRU0092). Comparisons are undertaken of global means and of the spatial pattern of the field of mean monthly precipitation. A distinction is made between global and "terrestrial" precipitation fields. CRU0092 includes interannual variability and therefore allows other validation exercises to be undertaken. The importance of the observed reference period for model validation is assessed, and observed and modelled global precipitation sensitivities are compared. These validation exercises reveal markedly different model performances and show that uncertainty about the 'true' observed precipitation hinder the task of model validation. Validation analyses such as these, together with more detailed validation of the interannual variability and daily precipitation characteristics of model output, must remain an essential dimension both of 'greenhouse climate' predictions and of climate change impact analyses which rely upon GCM experiments.

KEYWORDS

Global precipitation / Climate models / Global warming / Climate change / Model validation / Precipitation observations

NATO ASI Series, Vol. I 26
Global Precipitations and Climate Change
Edited by M. Desbois and F. Désalmand

1 Introduction

Precipitation is probably the most difficult atmospheric processes to model and therefore to predict. The occurrence and intensity of precipitation is controlled both by cloud microphysics and particle growth and also by the regional patterns of pressure, humidity and temperature (Chahine, 1992). In General Circulation Models (GCMs) which are used to perform climate change experiments, these problems are compounded by the coarse spatial scale over which these models are currently integrated (between 2.5° and 8° longitude/latitude). Most of the atmospheric processes which control precipitation occur at space scales smaller than these. It is primarily for these reasons that the prediction by GCMs of future precipitation changes which may result from elevated greenhouse gas concentrations are highly uncertain on most spatial and temporal scales. Yet it is at small spatial scales - a few 10kms to a few 100kms - that the impact of precipitation changes on a wide range of human and natural activities will be most obviously felt. Attempts to assess these impacts using results from GCM experiments (e.g. on crop growth, Rosenzweig and Parry, 1994; on vegetation, Vloedbeld and Leemans, 1993) are therefore seriously undermined by the unreliability of the precipitation predictions used. Even more precarious become the attempts by environmental economists to convert such impact assessments into monetary values for the purposes of benefit/cost analyses (e.g. Hope *et al.*, 1993). It is therefore essential that with each potential improvement in climate models that an evaluation is made of how well the model is simulating present-day precipitation at a variety of space and time-scales.

To perform this task satisfactorily requires good quality data sets of observed precipitation at this same range of space and time-scales. As with the modelling of precipitation, however, the measurement of precipitation is beset with problems. Precipitation falling on the ground is one of the hardest climate variables to measure accurately (see Sevruk, this volume). Problems with gauge design, microclimate around the measuring site and the heterogeneity of precipitation in space make both small-scale and large-scale empirical estimates of precipitation subject to great uncertainty (see Rudolf *et al.*, this volume).

A number of extensive GCM precipitation validation exercises have been undertaken (see Hulme, 1991; Legates, 1991; Skelly, 1993; and many others) which have addressed different aspects of the overall precipitation validation issue. In this chapter I will extend the intercomparison of model and observed precipitation climatologies at global scales performed in Hulme (1991). An improved observed precipitation climatology for global land areas has been constructed for this purpose and consists of gridded estimates of monthly precipitation from 1900 to 1992 at two different resolutions. This new data set is referred to as CRU0092 and a brief description of its construction is provided. Also used in this type of analysis for the first time is the precipitation climatology of Leemans and Cramer (1991). In addition to the six GCM experiments used in Hulme (1991), an additional high resolution equilibrium experiment is included (the Canadian Climate Centre) and also results from two transient experiments (performed at the Max Planck Institute and the Hadley Centre). Three aspects of model precipitation simulations are evaluated:

- global-mean estimates of current precipitation and of the precipitation sensitivity to global warming;
- the large scale spatial patterns of precipitation;
- the importance of the reference period of the observed climatology to the validation exercise.

The observed and model data sets used in the analysis are first described, including an extended description of the new CRU0092 data set. The results of the validation analyses are then presented and the relevance of the results discussed.

2 The Data Sets

Observed Data

Four observed precipitation climatologies are used in this analysis. Three of these have been described elsewhere (Jaeger, 1976; Legates and Willmott, 1990; Leemans and Cramer, 1991) and are here referred to as JAEGER, LEGATES (and LEG-COR, a "corrected" version of LEGATES) and IIASA. Their key characteristics are summarised in Table 1. The fourth climatology is a new one constructed from the station data set described in Eischeid *et al.* (1991) and subsequently updated in the Climatic Research Unit (see Hulme *et al.*, 1993). This latter climatology is referred to as CRU0092 and consists of monthly gridded time series of precipitation for global land areas between 1900 and 1992. CRU0092 has not previously been described in the literature therefore a brief summary is given here.

The construction of CRU0092 broadly follows that used by Hulme (1992) to compile a gridded precipitation climatology for the period 1951 to 1980 at 5° resolution. Station time series are selected from the master data set of over 8300 stations on the basis of the completeness of their records in a specified reference period. In constructing CRU0092, two partly overlapping reference periods were used, 1931-70 (3919 stations selected) and 1951-90 (3926 stations). 2854 of these stations were in common, thus a total of 4991 station time series from the master data set were used. For each of these two selected station data sets, the station time series were gridded at 2.5° latitude by 3.75° longitude using Thiessen weighting. Missing monthly values were estimated using an anomaly interpolation method described in Hulme (1992), but with two improvements. First, when a station possessed a missing data value the search for nearby valid data was allowed to extend outside the relevant grid cell and was constrained only by a circular search radius. This was chosen to be 278km, i.e., about 2.5° latitude/longitude at the Equator. The ten nearest stations, within the search radius, which possessed data for the month in question were used to calculate a precipitation anomaly. This anomaly was used to estimate a monthly value for the missing station and was weighted, not only according to the inverse of distance as in Hulme (1992), but also with an angular weighting to account for direction. This latter procedure follows that used in SPHEREMAP (Legates and Willmott, 1990; also the GPCC, see Rudolf *et al.*, this volume) and ensures that directional bias in the resulting area-averaged precipitation anomaly is minimised.

	"Corrected"?	Land & Ocean?	Time-dependent?	Time series?
JAEGER	N	Y	N (?)	N
LEGATES	N	Y	N	N
LEG-COR	Y	Y	N	N
IIASA	N	N	Y (?)	N
CRU0092	N	N	Y	Y

Table 1: Summary characteristics of the observed precipitation climatologies used. "Corrected" means adjustments for gauge undercatch have been applied; "time-dependent" means only station means from fixed reference periods have been used in constructing the climatology. JAEGER and IIASA only loosely use fixed reference periods.

The above procedure resulted in two global gridded precipitation data sets between 1900 and 1992, one based on stations possessing good records between 1931-70 and the other on stations with good records between 1951-90. Because of the varying geographic distributions of these stations, time series for some grid cells existed only in one or the other of these two versions and even where grid cell time series existed in both, the length of the gridded record could vary. To maximise the geographic coverage and temporal length of the final version of CRU0092, these two gridded data sets were merged. Where grid cell time series existed in both versions, the 1951-90 data were preferred to the 1931-70 data. For years where only 1931-70 data existed, these grid cell data were re-scaled to represent 1951-90 values using the respective mean and variance ratios of the two versions based on the common 1951-70 period (for which period, by definition, the data were virtually complete in both versions). This merging resulted in a more complete CRU0092 data set, both in time and space, than would have been achieved using a single reference period for station selection. This is illustrated graphically in Figure 1 where one can see the better coverage of the 1931-70 version in the earlier decades of the 20th century and the 1951-90 version in later decades. A total of 1310 grid cells in the merged CRU0092 possess monthly precipitation time series at least 30 years in length, compared to only 1037 cells in the 1931-70 version and 1191 in the 1951-90 version. A total of 115 grid cells possess complete time series from January 1900 to December 1992, distributed mostly in western Europe and eastern North America. The spatial distribution of the grid cells with nearly complete data in three selected decades is shown in Figure 2. Best coverage is achieved in the decade 1951-60 when 1280 grid cells possessed data for more than 90% of months.

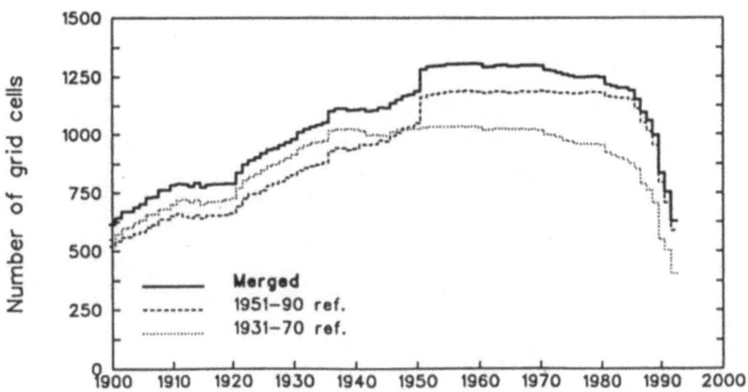

Figure 1: Annual total of 2.5° by 3.75° grid cells (maximum = 72 x 96 = 6912) with complete monthly precipitation estimates from three versions of CRU0092: using 1931-70 as reference period; using 1951-90 as reference period; and the final merged version.

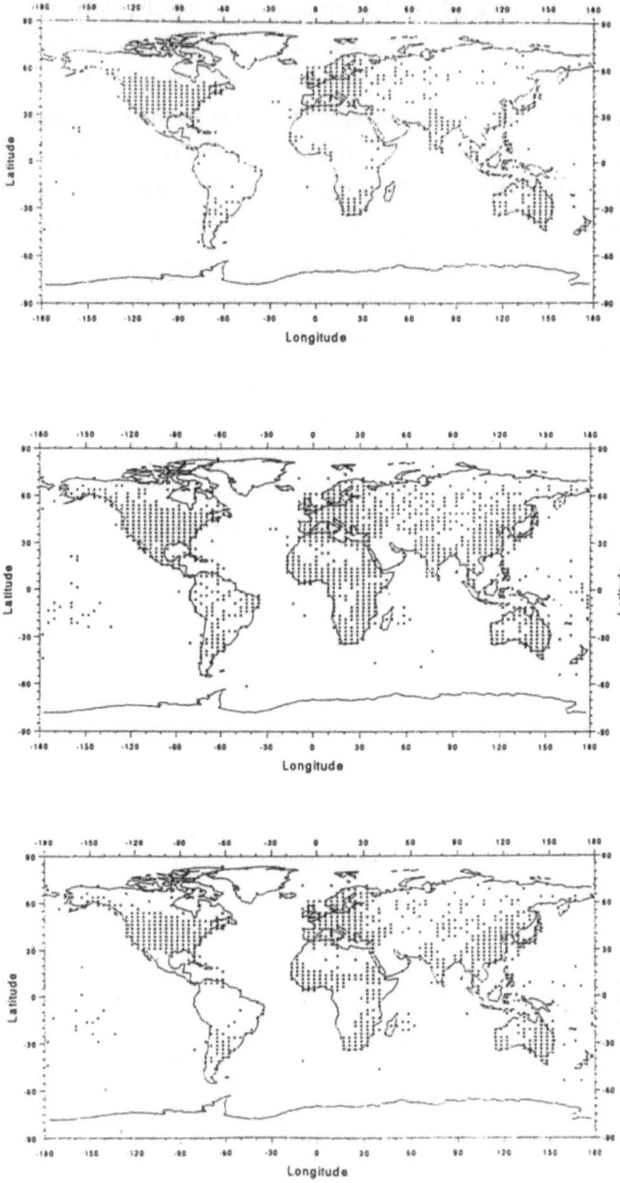

Figure 2: Distribution of 2.5° by 3.75° grid cells in CRU0092 with >90% of months possessing precipitation estimates for three decades: (top) 1901-10 (n = 618), (middle) 1951-60 (n = 1280) and (bottom) 1981-90 (n = 913).

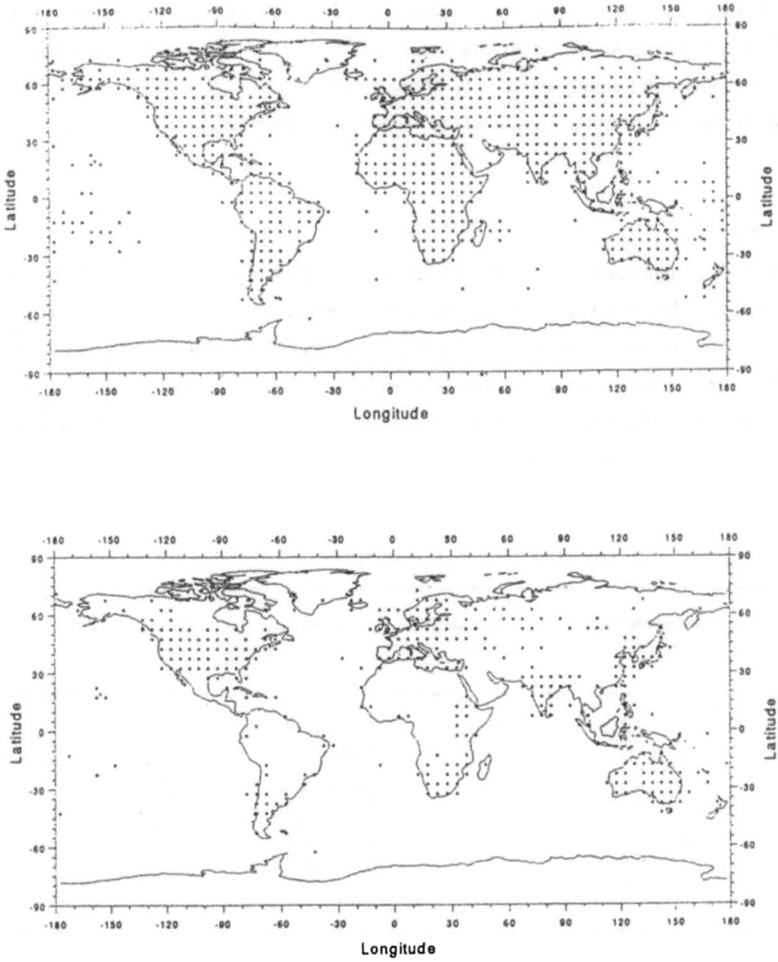

Figure 3: Distribution of 5° by 5° grid cells (maximum = 2592) in CRU0092 which possess (top) a calculated monthly precipitation time series (740 cells) and (bottom) 10-year mean monthly values for each of the 84 10-year climatologies used in the fixed-grid correlation experiment (334 cells).

The above description applies to the construction of the 2.5° by 3.75° version of CRU0092. This resolution was chosen to be consistent with that used in the UK Met. Office GCM, for whom the data set was originally constructed. Exactly the same methodology was followed in constructing a 5° by 5° resolution version of CRU0092 and it is this version which is used in the rest of the analysis described here. A total of 740 cells in this 5° version possess at least 30 years of monthly data (Figure 3), compared to only 595 in the CRU5180 data set described in Hulme (1992). The only terrestrial areas not covered by this new data set are Antarctica, parts of high latitude Canada and Russia, interior South America and Africa, the Gulf region and some of the Indonesian islands. These 740 cells, representing about 33% of the Earth's surface (cf. 28% in the Hulme (1992) version), define the "terrestrial" portion of the Earth which is analysed in later sections.

Model Data

Mean monthly precipitation fields from the control integrations of nine GCM experiments are used in the analysis. Seven of these experiments are equilibrium experiments and two (UKTR and ECHAM1-LSG; the latter performed at the Max Planck Institute for Meteorology) are transient experiments performed using a coupled ocean-atmosphere GCM. For the UKTR experiment monthly time series of precipitation for the 75-year duration of the control integration are also used. Table 2 summarises these model experiments and the acronyms which are subsequently used to refer to them.

Model	Spatial resolution Long.	Lat.	Atmospheric layers	Ocean Model	ΔT_{eq} (°C)	ΔP_{eq} (%)
CCC	3.75°	3.75°	10	Mixed layer	3.5	4
GISS	10°	7.8°	9	Mixed layer	4.2	11
GFDL	7.5°	4.5°	9	Mixed layer	4.0	9
LLNL	5°	4°	2	2-layer ocean	3.8	11
ECHAM1-LSG	5.6°	5.6°	19	11-layer ocean	1.6	3
OSU	5°	4°	2	Mixed layer	2.8	8
UKLO	7.5°	5°	11	Mixed layer	5.2	15
UKHI	3.75°	2.5°	11	Mixed layer	3.5	11
UKTR	3.75°	2.5°	11	17-layer ocean	2.3	4

Table 2: Summary characteristics of the nine GCM experiments used in the analysis. All model experiments are equilibrium experiments except for ECHAM1-LSG and UKTR. ΔT_{eq} and ΔP_{eq} are the *equilibrium* global-mean temperature and precipitation sensitivities for a CO_2-doubling, except for ECHAM1-LSG and UKTR for which the *realised* changes by model years 61-70 (ECHAM1-LSG, CO_2 increased by 115%) and 66-75 (UKTR, CO_2-doubled) are shown.

References:		
	CCC	Canadian Climate Centre: Boer *et al.* (1989)
	GISS	Goddard Institute for Space Studies: Hansen *et al.* (1984)
	GFDL	Geophysical Fluid Dynamics Lab.: Wetherald and Manabe (1986)
	LLNL	Lawrence Livermore National Lab.: Gates (pers. comm.)
	ECHAM1-LSG	Max Planck Institute (Geostrophic ocean): Cubasch *et al.* (1992)
	OSU	Oregon State University: Schlesinger and Zhao (1989)
	UKLO	UK Met. Office low resn.: Wilson and Mitchell (1987)
	UKHI	UK Met. Office high resn.: Mitchell *et al.* (1990)
	UKTR	UK Met. Office transient: Murphy (1994)

To ensure direct comparability between the model and observed fields all the model data sets were interpolated onto the 5° by 5° grid of CRU0092. This resolution is a reasonable compromise between the various resolutions of the GCMs used. For the model fields and for JAEGER this was accomplished using a Gaussian space-filtering technique (Santer, 1988). For LEGATES, LEG-COR and IIASA, originally constructed at a 0.5° resolution, the appropriate 100 grid cells were averaged using an area weighting.

3 Estimates of Global-Mean Precipitation

Global-mean annual precipitation has been estimated by various authors to lie between 750mm and 1150mm and terrestrial precipitation between 550 and 850mm (Willmott and Legates, 1991). Figure 4 indicates the global and "terrestrial" mean annual precipitation derived from the observed and GCM climatologies used in this study. GCM-estimated global precipitation ranges between 909mm (OSU) and 1135mm (GISS), with a median value of 1037mm (UKTR). The two observed climatologies straddle this range, JAEGER being slightly drier than LEGATES, but the gauge corrections of LEG-COR lead to a substantially higher estimate of 1139mm, higher than any GCM value. This latter result suggests either that the GCMs are somehow "tuned" to reproduce conventional uncorrected precipitation estimates, or that all the models are too dry for some reason, or that the corrections of LEG-COR are exaggerated.

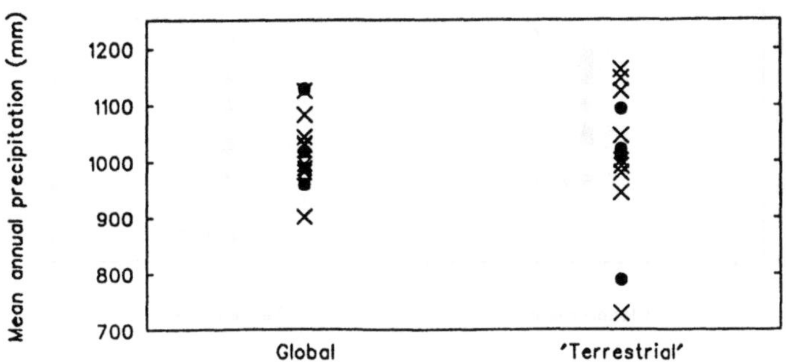

Figure 4: Estimates of global and "terrestrial" (i.e., only 740 grid cells) mean annual precipitation from model (X) and observed (•) climatologies.

The GCM estimates of "terrestrial" annual mean precipitation are more divergent, ranging between 737mm (OSU) and 1172mm (GISS). Again, the observed estimates fall within this range with LEGATES, JAEGER and CRU0092 all agreeing with each other to within 7mm (range 1015 to 1022mm). The "corrected" LEGATES is again substantially wetter, although now not as extreme as the wettest GCM. (The IIASA estimate of only 797mm is biased because the regridding procedure from 0.5° to 5° excluded a number of oceanic islands which had large precipitation totals. The full subset of 740 "terrestrial" cells could not therefore be used for IIASA).

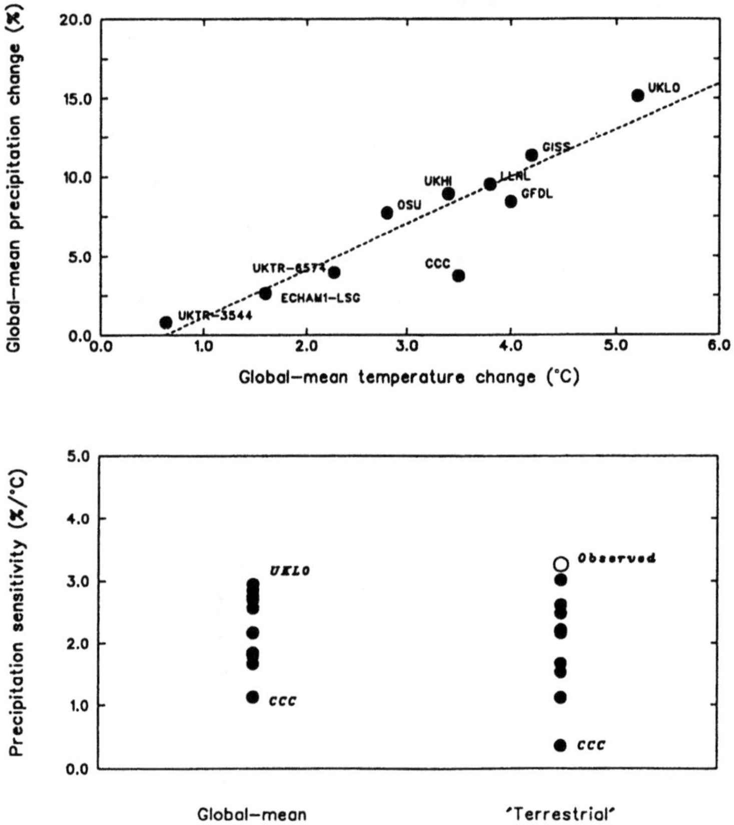

Figure 5: (top) Global-mean precipitation change as a function of global warming as simulated by nine GCM experiments; (bottom) The distribution of global and "terrestrial" precipitation sensitivities to global warming as simulated by nine GCM experiments. The observed estimate derives from the CRU0092 data set described here and the global temperature time series described in Folland *et al.* (1992).

GCM-estimated global-mean precipitation sensitivity to global warming ranges between about 1.5%/°C and 3%/°C (Mitchell *et al.*, 1990). This range derives from equilibrium experiments, but remains valid for the more recent transient GCM experiments also, as indicated by the values in Figure 5 from UKTR and ECHAM1-LSG. The CCC equilibrium experiment produced the most anomalous result generating a sensitivity of only just over 1%. By analysing the CRU0092 data set in conjunction with the global-mean surface air temperature record used by IPCC (Folland *et al.*, 1992), it is possible to make an empirical estimate of the "terrestrial" precipitation sensitivity to warming over the last 93 years. Figure 6 shows the time-evolution of this sensitivity between 1900 and 1992 plotted as 30-year filtered anomaly values. While there has been considerable decadal-scale variability in this sensitivity, over the duration of the observed record the "terrestrial" surface has got wetter by about 1.6%, while the world has warmed by about 0.5°C. The resulting sensitivity of 3.2%/°C places this observed estimate just outside the top end of the modelled range of "terrestrial" sensitivities (Figure 5). This observed sensitivity may be somewhat overestimated, however, because of the potential wetting bias in terrestrial-mean precipitation which has been introduced since the 1950s and 1960s by high latitude gauge changes. These have almost certainly introduced to these latitudes a wetting bias in recent decades (Groisman *et al.*, 1992) which will contaminate the "terrestrial"-mean precipitation record. With this adjustment in mind, the observed precipitation sensitivity to global warming probably falls within the modelled range.

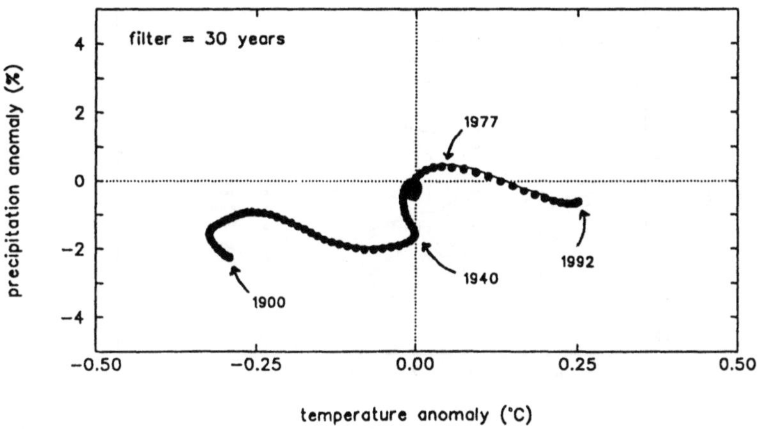

Figure 6: Time evolution of the "terrestrial"-mean annual precipitation anomaly versus the global-mean annual temperature anomaly between 1900 and 1992 (values filtered using a 30-point low-pass filter).

4 Spatial Patterns of Precipitation

A further comparison which can be made between model and observed precipitation climatologies is to examine their spatial pattern correlation coefficients. Examining the annual cycle of the mean monthly pattern correlation coefficients provides a useful relative indicator of model performance and, when these are compared with correlations among observed climatologies, this provides some indication of how substantial model errors might be.

Using Fixed Period Observed Climatologies

Figure 7 shows these annual cycles with respect to the global field of LEGATES and to the "terrestrial" field of CRU0092. For the purposes of these pattern correlations based on fixed periods, mean fields for 1951-80 were extracted from CRU0092. Figure 7 indicates a number of features:

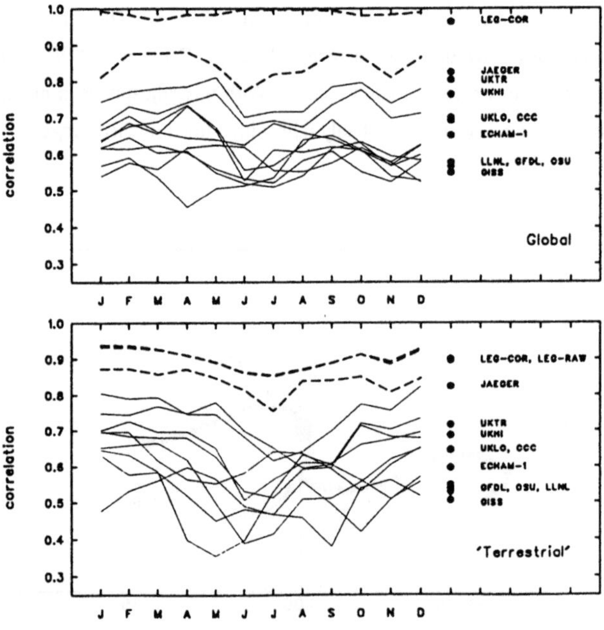

Figure 7: Monthly pattern correlations between the mean precipitation fields of (top) LEGATES and nine model control integrations for the global surface and (bottom) CRU0092 and nine model control integrations for the "terrestrial" surface. The dashed lines refer to the equivalent observed-observed correlations. The dots on the right-hand side of the plots indicate the average of the 12 monthly coefficients for each respective model (or observed) climatology.

- the observed climatologies correlate only at between 0.8 and 0.95; the poorest such correlation is 0.78 (JAEGER and CRU0092) in July for "terrestrial" areas.

- there is a wide diversity of model-observed correlations which, for the poorest model, fall to below 0.4 for "terrestrial" areas. This is a remarkably poor result.

- throughout the annual cycle there is a generally superior performance from the two high resolution UK Met. Office experiments, UKHI and UKTR. The best of these monthly correlations exceeds 0.8 in winter months and exceeds the degree of agreement between JAEGER and CRU0092 achieved in summer.

Using a Time-Varying Observed Climatology

The above correlation analysis used fixed period observed climatologies against which GCM precipitation fields were compared. In the case of CRU0092 this was a 1951-80 period, for JAEGER it was a quasi-1931-60 period and for LEGATES no fixed period was used, the climatology being "representative" of the entire 20th century. The time series nature of CRU0092 enables a further question to be explored: how dependent on the time period of the observed climatology is the model-observed pattern correlation coefficient? For example, is the model simulating a mean precipitation field which is more similar to the observed precipitation of 1951-80 than to 1921-50? This might be expected since many of the boundary conditions used to initialise GCMs are related to 1951-80 data.

To explore this question successive 10-year observed climatologies are calculated from CRU0092 starting with 1900-09 and ending with 1983-92, a total of 84 climatologies. Two methods were used to select the terrestrial grid cells which contributed to these climatologies. First, a fixed grid was used whereby only cells which could generate 10-year mean values for each of these 84 time periods were used (90% of months required data values for a 10-year mean to be calculated). A total of 334 out of the 740 CRU0092 grid cells met this criteria (see Figure 3). This fixed grid therefore defined the model cells which were used in the pattern correlation analysis. The second method allowed the distribution of used grid cells to vary from one 10-year period to the next. Thus as the data coverage improved over time the number of valid cells increased from 351 (1900-09) to a maximum of 735 (1957-66). In this varying-grid experiment the selected model cells also varied to follow the distribution of observed cells.

The model precipitation climatologies from the control integration of two GCM experiments and two seasons were chosen for this analysis: the UKHI and UKTR experiments for DJF and JJA seasons (only the UKHI results are shown here since UKTR showed similar sensitivities). The LEGATES climatology was also selected to identify the sensitivity of the observed-observed correlations to varying time periods. RMS errors with respect to CRU0092 were also calculated to provide an indication of absolute model errors. Figures 8 and 9 plot the various time series of these pattern correlation coefficients and RMS errors for JJA and DJF respectively from which a number of conclusions can be drawn:

- the agreement between LEGATES and CRU0092 does vary slightly, whether using a fixed or varying grid, depending on the CRU0092 decade. This is particularly so in the fixed grid experiment in JJA when RMS errors vary between 30% and 45%. The best agreement between CRU0092 and LEGATES occurs in the decade 1921-30 (JJA) and 1973-82 (DJF).

- in the fixed grid analysis, there is little sensitivity of the pattern correlations to the time period of the observed climatology, i.e., the relative pattern of "terrestrial" precipitation is changing little from decade to decade. A slight improvement in the agreement of the model DJF field with the observations is noticeable when the observed climatology is based on periods after about 1945, but the improvement is small (the RMS error falls from nearly 70% to just over 60%).

- in the variable grid analysis, there are quite noticeable improvements in the model performance as successively later decades are used. This is so particularly in summer when pattern correlations increase from about 0.55 to 0.65 and RMS errors fall from 90% to less than 80%. This is a result of the particular distribution of grid cells which are selected in any given 10-year period. Over time it would appear that areas of the world in which the models can simulate precipitation more faithfully are included in the observed climatology (e.g. Equatorial Africa and S.America), therefore resulting in better agreement with the observed climatology.

5 Conclusions

An improved gridded precipitation data set for global terrestrial areas (CRU0092) has been constructed yielding monthly time series from 1900 to 1992 at either 2.5° by 3.75° or 5° by 5° resolution. This gridded data set has been constructed from nearly 5000 individual station time series. I have used this new data set, together with existing mean global precipitation climatologies,

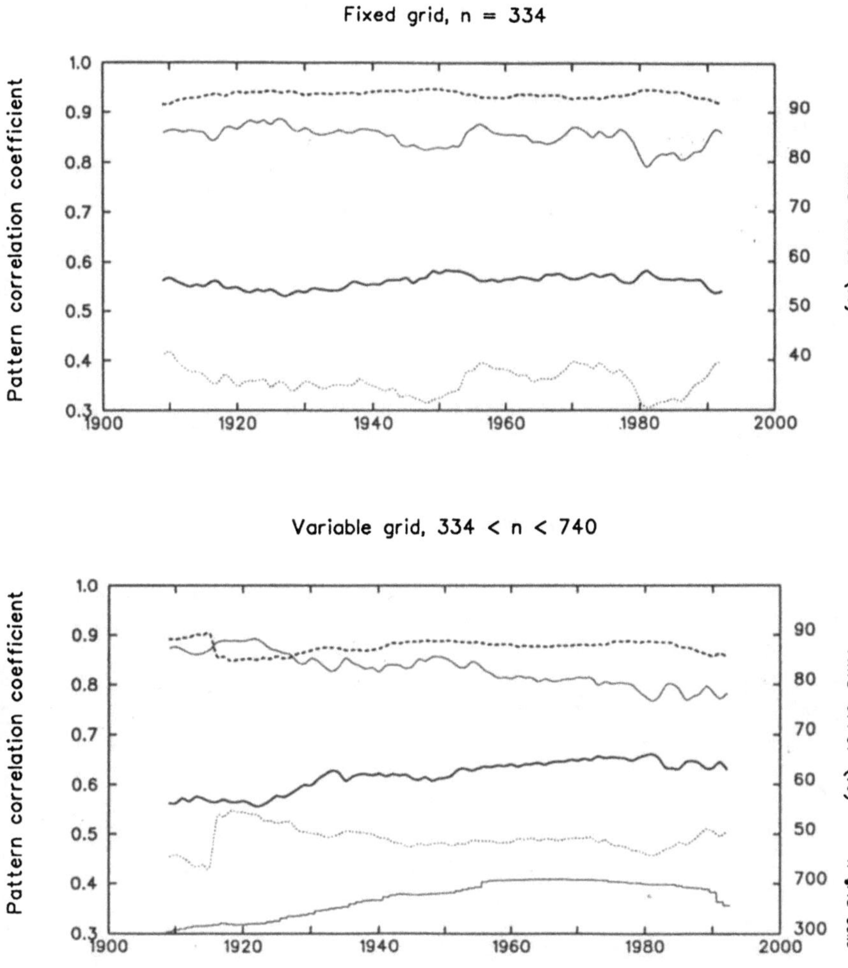

Figure 8: Model performance indicators of the similarity between mean JJA precipitation fields for LEGATES (dashed lines) and UKHI (continuous lines) versus successive 10-year climatologies from 1900-09 to 1983-92 derived from CRU0092. Thick lines indicate pattern correlation coefficients (left axis) and thin lines RMS errors (right axis). Top panel is for a fixed grid and bottom panel for a time-varying grid. The bottom panel also shows for the time-varying grid the frequency of grid cells used in each 10-year intercomparison.

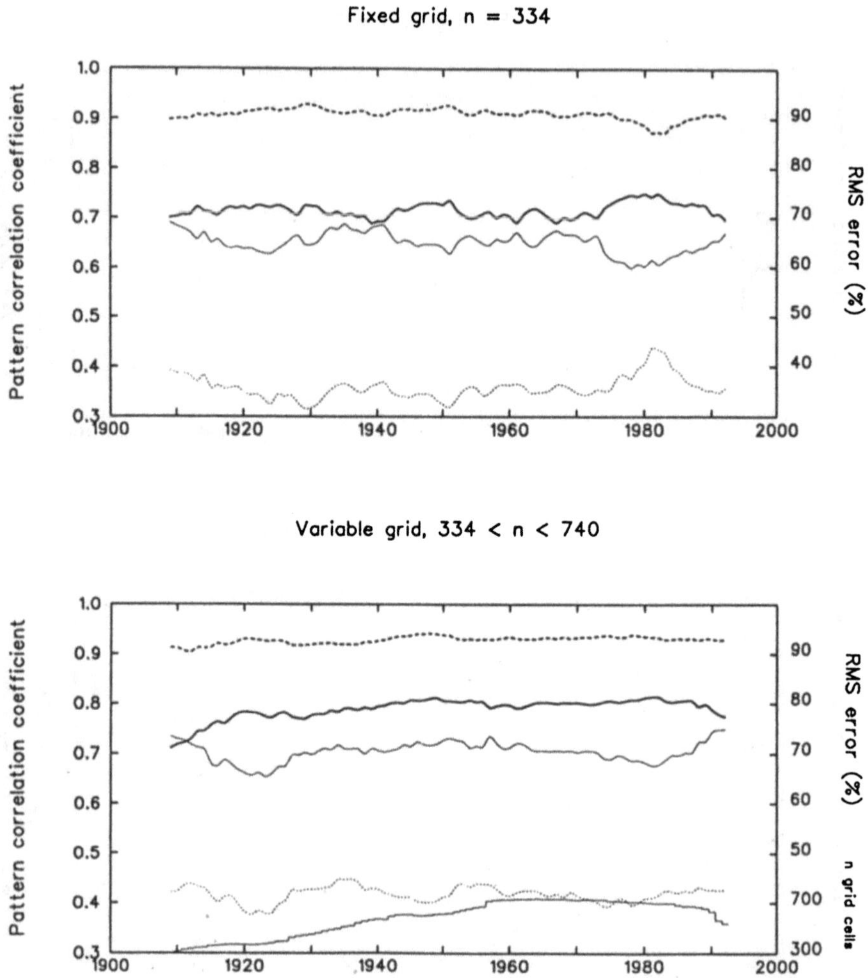

Figure 9: Model performance indicators of the similarity between mean DJF precipitation fields for LEGATES (dashed lines) and UKHI (continuous lines) versus successive 10-year climatologies from 1900-09 to 1983-92 derived from CRU0092. Thick lines indicate pattern correlation coefficients (left axis) and thin lines RMS errors (right axis). Top panel is for a fixed grid and bottom panel for a time-varying grid. The bottom panel also shows for the time-varying grid the frequency of grid cells used in each 10-year intercomparison.

to evaluate the large-scale precipitation fields from the control integrations of nine different GCM experiments, including two coupled ocean-atmosphere model simulations. Global-mean precipitation estimates and the terrestrial precipitation sensitivity to global warming have been compared for the observed and modelled climatologies. The observed "terrestrial" precipitation sensitivity of the last 93 years - about 3.2%/°C - is slightly higher than model estimates, but this may be due to a wetting bias introduced into the observed data through uncorrected gauge changes in the 1950s and 1960s at high latitudes in the Northern Hemisphere. Work is underway in the Climatic Research Unit to incorporate corrections for these gauge changes in Canada, Russia and Scandinavia into CRU0092.

The spatial patterns of large-scale precipitation fields have also been intercompared. Uncertainties in knowing the "true" observed precipitation field hinder the objective evaluation of model precipitation simulations. In this regard, the future precipitation climatology for 1986-95 being produced by the Global Precipitation Climatology Centre (GPCC; see Rudolf *et al.*, this volume) will be a valuable new resource for GCM validation. The difference in pattern correlation performance between an observed climatology and the best GCMs (UKHI and UKTR) in some months is not significantly greater than that between two independent observed climatologies (JAEGER and CRU0092). The sensitivity of this intercomparison exercise to the time period of the observed reference climatology has also been assessed. For a fixed grid the sensitivity is generally small, although measurable, indicating that the changes in terrestrial precipitation patterns from one 10-year period to another are not large enough to alter substantially the results of a model validation exercise. The greater sensitivity of the model performance indicators when using a time-varying grid results from the fact that model precipitation is closer to reality in some geographical regions than others, rather than to the time period of the observations per se.

Acknowledgements

The work described here has been supported by the UK Department of the Environment under Contract No. PECD/7/10/198. The modelling centres who supplied GCM output are thanked, as are Rik Leemans for the IIASA climatology and David Legates for the LEGATES and LEG-COR climatologies. The results from the UKTR experiment have been supplied through the Climate Impacts LINK Project at the Climatic Research Unit (UK Department of the Environment, Contract No. PECD/7/12/96).

References

Chahine,M.T., 1992: The hydrological cycle and its influence on climate, Nature, 359, 373-379.

Cubasch,U., K.Hasselmann, H.Höck, E.Maier-Reimer, U.Mikolajewicz, B.D.Santer and R.Sausen, 1992: Time-dependent greenhouse warming computations with a coupled ocean-atmosphere model, Climate Dynamics, 8, 55-69.

Eischeid,J.K., H.F.Diaz, R.S.Bradley, and P.D.Jones, 1991: A comprehensive precipitation data set for global land areas U.S.Department of Energy Report No. DOE/ER-69017T-H1, Washington DC, 81pp.

Folland.C.K., T.R.Karl, N.Nicholls, B.S.Nyenzi, D.E.Parker and K.Ya.Vinnikov, 1992: Observed climate change and variability, pp.136-170 in, Climate change 1992: the supplementary report to the IPCC scientific assessment (eds.) Houghton,J.T., Callander,B.A. and Varney,S.K., Cambridge Univ. Press, Cambridge, 200pp.

Gates,W.L., J.F.B.Mitchell, G.J.Boer, U.Cubasch and V.P.Meleshko, 1992: Climate modelling, climate prediction and model validation, pp.97-134 in, Climate change 1992: the supplementary report to the IPCC scientific assessment (eds.) Houghton,J.T., Callander,B.A. and Varney,S.K., Cambridge Univ. Press, Cambridge, 200pp.

Hansen,J., A.Lacis, D.Rind, L.Russell, P.Stone, I.Fung, R.Ruedy. and J.Lerner, 1984: Climate sensitivity analysis of feedback mechanisms, pp.130-163 in, Climate porcesses and climate sensitivity (eds.) Hansen,J. and Takahashi,T. Geophysical Monograph No.29, American Geophysical Union, Washington DC

Hope,C., J.Anderson and P.Wenman, 1993: Policy analysis of the greenhouse effect, Energy Policy, 21, 327-338.

Hulme,M., 1991: An intercomparison of model and observed global precipitation climatologies, Geophys. Res. Lett., 18, 1715-1718.

Hulme,M., 1992: A 1951-80 global land precipitation climatology for the evaluation of General Circulation Models, Climate Dynamics, 7, 57-72.

Hulme,M., K.R.Briffa, P.D.Jones and T.M.L.Wigley, 1992: GCM validation and climate change detection, Interim Report for the UK Department of the Environment, Contract No. PECD/7/12/78, Climatic Research Unit, Norwich, 48pp.

Jaeger,L., 1976: Monatskarten des Niederschlags fur die ganze Erde, Berichte des Deutschen Wetterdienstes, 18, No.139, Offenbach, FR Germany, 38pp.

Leemans,R. and W.Cramer, 1991: The IIASA database for mean monthly values of temperature, precipitation and cloudiness on a global terrestrial grid, IIASA Report RR-91-18, Laxenburg, 63pp.

Legates,D.R. and C.J.Willmott, 1990: Mean seasonal and spatial variability in gauge-corrected, global precipitation, Int. J. Climatol., 10, 111-128.

Legates,D.R. and C.J.Willmott, 1992: A comparison of GCM-simulated and observed mean January and July precipitation, Glob. Planetary Change, 5, 345-364.

McFarlene,N.A., G.J.Boer, J-P.Blanchet and M.Lazare, 1992: The Canadian Climate Centre second-generation General Circulation Model and its equilibrium climate, J. Climate, 5, 1013-1044.

Mitchell,J.F.B., S.Manabe, V.Meleshko and T.Tokioka, 1990: Equilibrium climate change - and its implications for the future, pp. 137-164 in, Climate change: the IPCC scientific assessment (eds.) Houghton,J.T., Jenkins,G.J. and Ephraums,J.J. Cambridge University Press, Cambridge, 364pp.

Murphy,J.M., 1994: Transient response of the Hadley Centre coupled ocean-atmosphere model to increasing carbon dioxide Part I. Control climate and flux correction, J. Climate (in press)

Rosenzweig,C. and M.L.Parry, 1994: Potential impact of climate change on world food supply Nature, 367, 133-138.

Rudolf,B., H.Hauschild, W.Rueth and U.Schneider, 1994: This volume

Santer,B.D., 1988: Regional validation of General Circulation Models, Climatic Research Unit Publication No.9, Climatic Research Unit, Norwich, 375pp.

Schlesinger,M.E. and Z-C.Zhao, 1989: Seasonal climate changes induced by doubled CO_2 as simulated by the OSU atmospheric GCM mixed layer ocean model, J. of Climate, 2, 459-495.

Sevruk,B., 1994: This volume

Skelly,W.C., 1993: A geostatistical approach to the interpolation and evaluation of General Circulation Model output for purposes of climate impacts research, Unpublished PhD Thesis, Macquarie University, New South Wales, 204pp.

Vloedbeld,M. and R.Leemans, 1993: Quantifying feedback processes in the response of the terrestrial carbon cycle to global change: the modelling approach of IMAGE-2, Water, Air and Soil Pollution, 70, 615-618.

Wetherald,R.T. and S.Manabe, 1986: An investigation of cloud cover change in response to thermal forcing, Climatic Change, 8, 5-23.

Willmott,C.J. and D.R.Legates, 1991: Rising estimates of terrestrial and global precipitation, Climate Research, 1, 179-186.

Wilson,C.A. and J.F.B.Mitchell, 1987: A doubled CO_2 climate sensitivity experiment with a global climate model including a simple ocean, J. Geophys. Research, 92, 13,315-13,343.

STATUS OF PRECIPITATION CLIMATOLOGIES AND CURRENT PROGRAMMES

THE GLOBAL PRECIPITATION CLIMATOLOGY PROJECT

Phillip A. Arkin
National Meteorological Center, National Weather Service
National Oceanic and Atmospheric Administration
Washington, DC 20233 (USA)

ABSTRACT
Modeling and observational research requirements have led the World Climate Research Programme to initiate the Global Precipitation Climatology Project (GPCP) with the goal of the production of a 10-year set of monthly analyses of area-averaged precipitation on a 2.5 ^{o} global grid using satellite estimates based on IR and microwave data, together with station observations. Its several principal centers each presently uses existing observations and techniques, including an IR-based tropical rainfall estimate and oceanic estimates derived from SSM/I observations as well as gauge observations, to produce rainfall analyses. A Calibration/Validation component ensures that the satellite estimates used in the project are of high quality by providing validation data sets and by encouraging the development of new observing technology. Examples of the satellite-based estimates are presented.

KEYWORDS
GPCP, global precipitation, rainfall analysis, satellite precipitation estimates, algorithm intercomparisons.

1. Introduction

Global analyses of areally averaged precipitation are required for a great number of climate modeling and observational research applications. Knowledge of the spatial and temporal distribution of large-scale precipitation is required for verification of climate models. Similarly, diagnostic studies of the climate system require such analyses to understand the relationships among surface boundary forcings and the large-scale circulation. However, the precipitation is extremely difficult to observe on these scales, and its global distribution is not well known. In this paper we will describe the Global Precipitation Climatology Project (GPCP), an international effort to produce a time series of global analyses of precipitation.

NATO ASI Series, Vol. I 26
Global Precipitations and Climate Change
Edited by M. Desbois and F. Désalmand

2. The Global Precipitation Climatology Project

The GPCP is an element of the World Climate Research Programme. Its goal is the production of a 10-year set of monthly analyses of area-averaged precipitation on a 2.5^{o} latitude x 2.5^{o} longitude grid. It uses rain gauge observations, together with satellite estimates based on IR and microwave data and ancillary information. The satellite estimates are derived from observations of the GOES (U. S.), GMS (Japan), and METEOSAT (European community) geostationary satellites and the operational NOAA polar orbiting satellites, along with the SSM/I data from the U. S. DMSP satellite.

The GPCP comprises two satellite components (fig. 1) - an IR component composed of the Geostationary Satellite Data Processing Centers (GSDPCs) and the Geostationary Satellite Precipitation Data Center (GSPDC), and a microwave component composed of the Polar Satellite Data Processing Center (PSDPC) and the Polar Satellite Precipitation Data Center (PSPDC) - combined with a Calibration/Validation Component and the Global Precipitation Climatology Centre (GPCC). In the following sections, we will briefly describe the IR and microwave precipitation estimation activities and show some examples. We will also discuss the calibration/validation activities. The GPCC will be addressed in the paper by Rudolf (this volume).

3. Satellite Estimates

Two sorts of satellite observations can be used to estimate precipitation. They are:

1. observations of the tops (and to some extent the thickness) of clouds, generally inferred from infrared (IR) temperature and/or visible brightness (often referred to jointly as vis/IR); and
2. observations of liquid and solid hydrometeors, inferred from their effects on microwave radiation relative to that emitted by the surface of the land or ocean.

These techniques have been labeled indirect and direct, respectively, by Arkin and Meisner (1987), and have been discussed there as well as in Arkin and Ardanuy (1989). In the GPCP, the GSPDC uses an IR-based indirect algorithm, while the PSPDC uses a microwave-based direct technique.

3.1 IR component

The IR estimates for the GPCP are produced using the GOES Precipitation Index (GPI, Arkin and Meisner, 1987), one of the simplest of the indirect satellite techniques. It is based

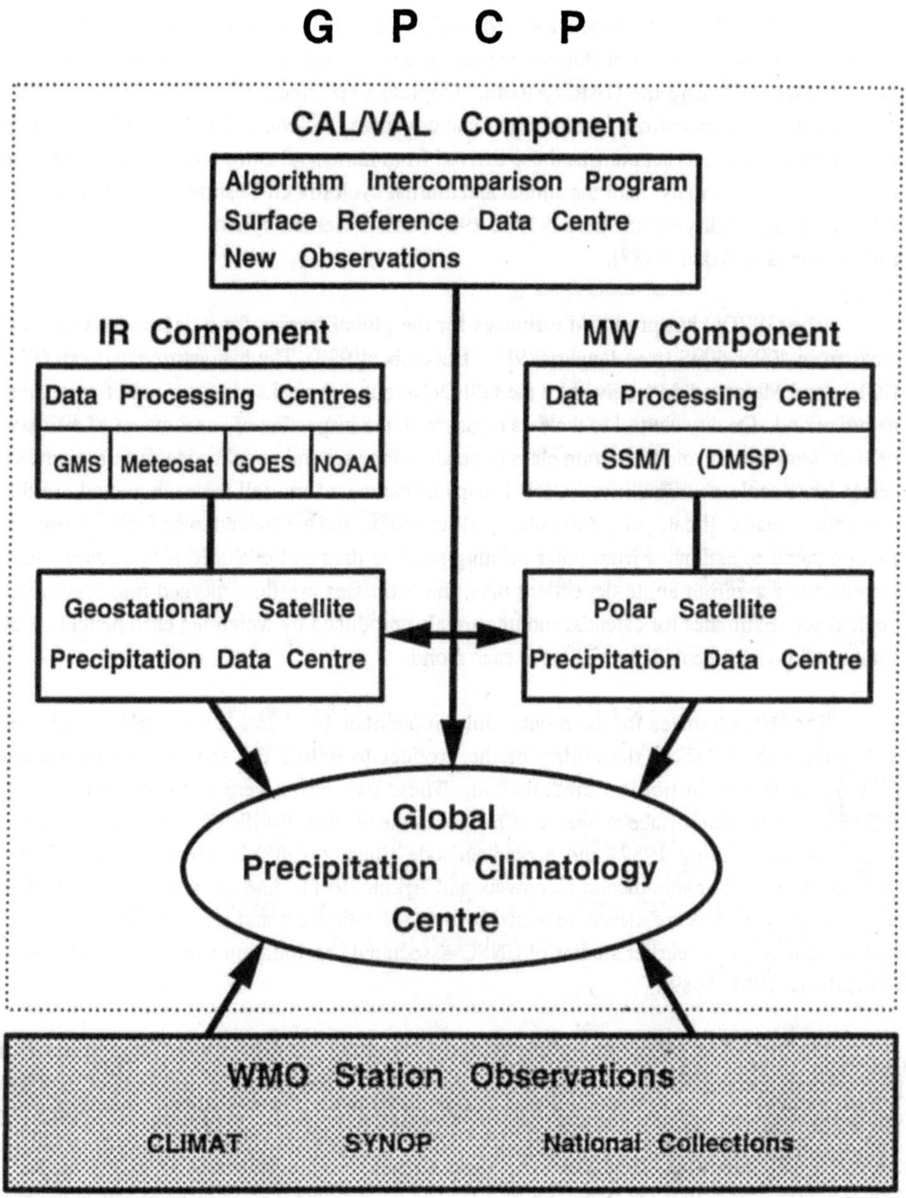

Fig. 1 : Components and data flow of the Global Precipitation Climatology Project.

upon the work of Arkin (1979) and Richards and Arkin (1981), who compared rainfall analyses derived from a combination of ship-borne radar and rain gauge observations with geostationary satellite IR data during the GARP Atlantic Tropical Experiment (GATE). They found that rainfall estimates made from just the fractional coverage of cloud colder that 235K over large areas appeared to be as accurate as those derived from more sophisticated algorithms (Richards and Arkin, 1981). Analyses of the annual and diurnal cycles in GPI-estimated rainfall and their interannual variability for the period 1982-1984 were presented by Arkin and Meisner (1987) and Meisner and Arkin (1987).

The GSPDC has produced estimates for the global tropics for 5-day periods and 2.5^o areas from 40°N-40°S from January 1986 - December 1993. The histogram data from GMS, GOES, and Meteosat are received by the GSPDC about 4-8 weeks after the end of each three-month period. Quality control of the data begins with the inspection of zonal means of estimated rainfall, temperature and histogram class populations for several zonal bands and time series to detect large-scale abnormalities. Global maps of estimated rainfall for each pentad are then inspected visually. If data for a particular satellite and pentad are judged to be faulty, those data are compared to estimates from polar orbiting satellite data and excluded if necessary. After correction of a zenith angle-dependent bias, the estimates are then merged into a complete tropical set. Estimates for calendar months are also produced by weighting each pentad by the number of days the pentad contributes to each month.

The GPI estimates for the months July-September 1987 and 1988, together with their difference (fig. 2), show the ability of the product to define the spatial distribution and interannual change in tropical precipitation. These two years were characterized by large differences in the sea surface temperature pattern in the tropical Pacific Ocean associated with a warm episode (during 1987) and a cold episode (during 1988) of the El Niño/Southern Oscillation (ENSO) phenomenon (Janowiak and Arkin, 1991). The extraordinary differences between the two years are clearly revealed by the GPI estimates, and are quite consistent with expectations based on earlier studies of ENSO-associated precipitation variations (Ropelewski and Halpert, 1987, 1989).

The IR-based estimates for the period 1986-1992 have been used in a number of presentations and publications. Janowiak and Arkin (1991) described the evolution of the large rainfall variations in the tropical Pacific associated with the evolution of ENSO (El Niño/Southern Oscillation) conditions from a warm episode during 1986-87 to the cold regime during 1988-89. Janowiak (1992) compared the estimates with published climatologies (Jaeger, 1976; Legates and Willmott, 1990) and used them to evaluate the skill of tropical oceanic precipitation forecasts by the National Meteorological Center and the European Center for Medium Range Weather Forecasts.

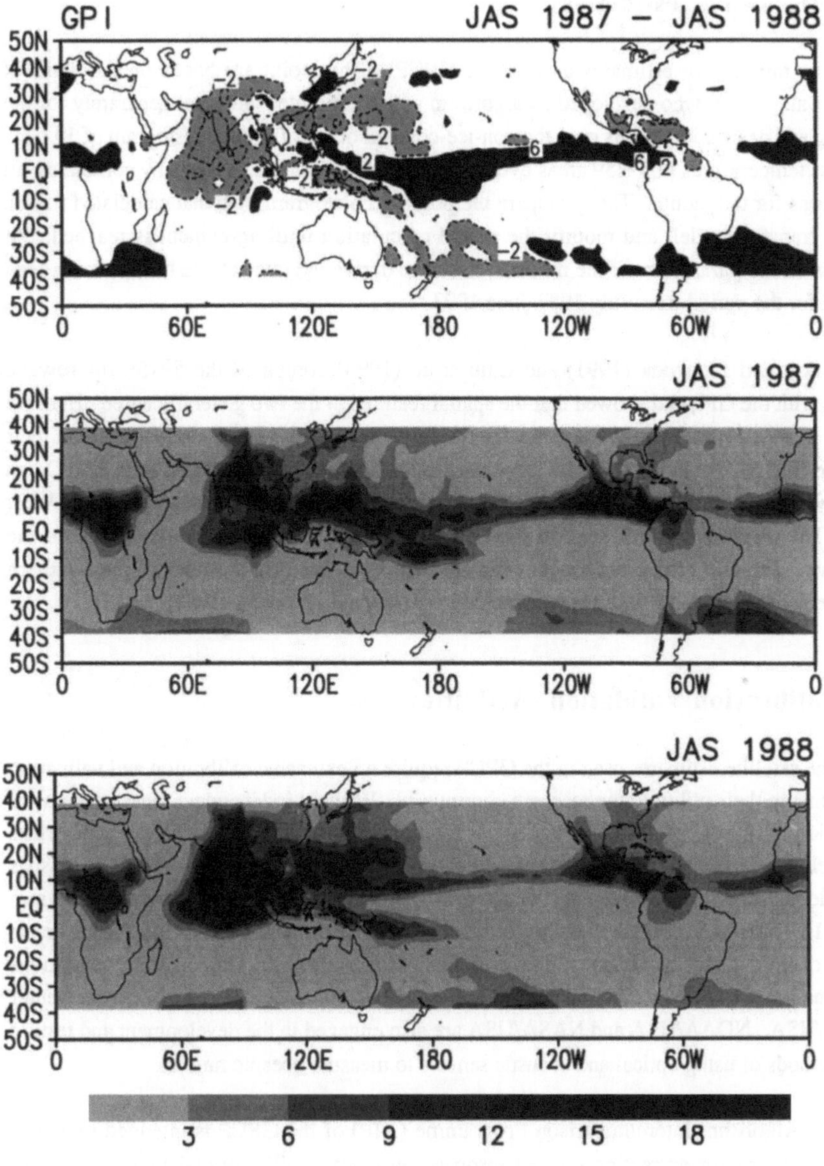

Fig. 2 : Precipitation estimated by the GPI for July-September 1987 (middle panel), July-September 1988 (bottom panel), and the difference (1987 minus 1988 - top panel) of the two. Contours on the difference plot are at +/- 2 mm/day and at intervals of 4 mm/day with negative contours dashed. Areas of difference greater/lesser than +/-2mm/day are indicated by heavy/light shading. The gray scale bar at the bottom of the figure indicates the values for the middle and lower panels.

3.2 Microwave estimates

The microwave estimates used in the GPCP to this point are based on the work of Wilheit et al. (1991), who developed an algorithm which estimates rain rates separately for the ascending and descending orbits over the non-ice-covered oceans from the histogram of 19 GHz brightness temperatures in $5^o x 5^o$ areas over a month, and for $2.5^o x 2.5^o$ areas by combining all observations for the month. They compare the observed histogram with that calculated from a radiative transfer model, and modify the model calculation until agreement is reached. At convergence, the parameters of the model are used to derive the rain rate. Estimates have been produced for the period from July 1987-June 1993.

Arkin and Janowiak (1991) and Chiu et al. (1993) compared the $5^o x 5^o$ microwave estimates with the GPI, and showed that the spatial features in the two generally agree. Here we show the $2.5^o x 2.5^o$ microwave estimates for the July-September 1987-1988 seasons and their difference (fig. 3). As in the GPI, the large interannual differences associated with ENSO can be clearly seen, although some disagreement is evident in the Indian Ocean and over West Africa. The tropical features seen in the GPI (fig. 2) are somewhat smoother than in the microwave. The GPI also gives higher estimated values in the mid-latitudes of the Southern Hemisphere. Both of these findings were noted by Arkin and Janowiak (1991).

4. Calibration/Validation Activities

The satellite estimates used in the GPCP require an extensive calibration and validation effort to ensure their utility to the research community. Without independent measurements of precipitation from different climatic regimes, it is not possible to understand the expected errors of the satellite estimates. The Calibration/Validation (Cal/Val) component of the GPCP consists of three elements: a Surface Reference Data Centre (SRDC), a technique development program and an Algorithm Intercomparison Programme. The SRDC uses observations from dense networks of rain gauges in various climatic regimes around the world to construct validation-quality time series of observed precipitation. It is operated by the National Climatic Data Center of NOAA/USA. NOAA/USA and NASA/USA are also engaged in the development and testing of new methods of using optical and acoustic sensors to measure oceanic rainfall.

The Algorithm Intercomparison Programme (AIP) of the GPCP is intended to assess current algorithms for extracting rainfall information from satellite digital imagery, and to develop new methods. It consists of a series of Algorithm Intercomparison Projects (AIP/1, AIP/2, etc.) conducted over differing climatic regions where good-quality, high-density in situ data (raingauge, digital radar) and coincident satellite visible/infrared and microwave digital

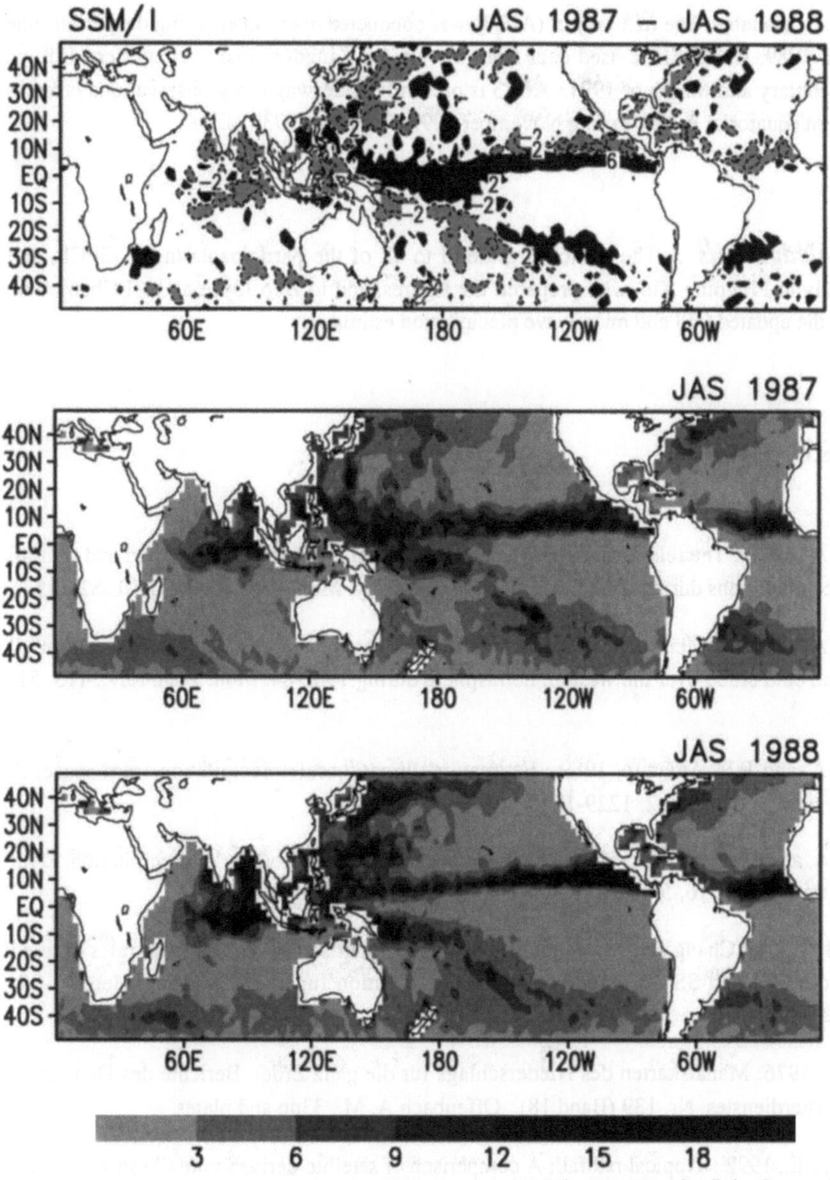

Fig. 3 : Precipitation estimated by the Wilheit et al. (1987) algorithm for July-September 1987 (middle panel), July-September 1988 (bottom panel), and the difference (1987 minus 1988 - top panel) of the two. Contours on the difference plot are at +/- 2 mm/day and at intervals of 4 mm/day with negative contours dashed. Areas of difference greater/lesser than +/- 2mm/day are indicated by heavy/light shading. The gray scale bar at the bottom of the figure indicates the values for the middle and lower panels.

imagery are available. The first project (AIP/1) was conducted over Japan during the period June - August, 1989, while AIP/2 used data from the United Kingdom and northwestern Europe during February and March of 1991. AIP/3 is presently underway using observations taken in the western equatorial Pacific during November 1992-February 1993.

Acknowledgments **:** The author is grateful to all of the participants in the GPCP, and particularly to Pingping Xie, who prepared the figures, and to Bob Joyce and Al Chang who provided the updated GPI and microwave precipitation estimates.

REFERENCES

Arkin, P.A., 1979 : The relationship between the fractional coverage of high cloud and rainfall accumulations during GATE over the B-scale array. Mon. Wea. Rev., 107, 1382-1387.

Arkin, P.A. and B.N. Meisner, 1987 : The relationship between large-scale convective rainfall and cold cloud over the western hemisphere during 1982-84. Mon. Wea. Rev., 115, 51-74.

Arkin, P.A. and P.E. Ardanuy, 1989 : Estimating climatic-scale precipitation from space: A review. J. Climate, 2, 1229-1238.

Arkin, P.A. and J.E. Janowiak, 1991 : Analyses of the global distribution of precipitation. Dyn. Atm. Oceans, 16, 5-16.

Chiu, L.S., A.T.C. Chang and J.E. Janowiak, 1993 : Comparison of monthly rain rates derived from GPI and SSM/I using probability distribution functions. J. Appl. Meteor., 32, 323-334.

Jaeger, L., 1976: Manatskarten des Niederschlags fur die ganz Erde. Berichte des Deutschen Wetterdienstes, Nr. 139 (Band 18). Offenbach A. M., 33pp and plates.

Janowiak, J.E., 1992 : Tropical rainfall: A comparison of satellite-derived rainfall estimates with model precipitation forecasts, climatologies and observations. Mon. Wea. Rev., 120, 448-462.

Janowiak, J.E. and P.A. Arkin, 1991 : Rainfall variations in the tropics during 1986-1989. J. Geophys. Res., 96(sup), 3359-3373

Legates, D.R. and C.J. Willmott, 1990 : Mean seasonal and spatial variability in gage-corrected, global precipitation. Int. J. Climate, 10, 111-127.

Meisner, B.N. and P.A. Arkin, 1987 : Spatial and annual variations in the diurnal cycle of large-scale tropical convective cloudiness and precipitation. Mon. Wea. Rev., 115, 2009-2032.

Richards, F. and P. Arkin, 1981 : On the relationship between satellite-observed cloud cover and precipitation. Mon. Wea. Rev., 109, 1081-1093.

Ropelewski, C.F. and M.S. Halpert, 1987 : Global and regional scale precipitation patterns associated with the El Niño/Southern Oscillation. Mon. Wea. Rev., 115, 1606-1626.

Ropelewski, C.F. and M.S. Halpert, 1989 : Precipitation patterns associated with the high index phase of the Southern Oscillation. J. Climate, 2, 268-284.

Wilheit, T.T., A.T.C. Chang and L.S. Chiu, 1991 : Retrieval of monthly rainfall indices from microwave radiometric measurements using probability distribution functions. J. Atmos. Ocean. Tech., 3, 118-137

GLOBAL AND TERRESTRIAL PRECIPITATION CLIMATOLOGIES: AN OVERVIEW AND SOME INTERCOMPARISONS

H. Hauschild, M. Reiss and B. Rudolf
Global Precipitation Climatology Centre (GPCC)
Deutscher Wetterdienst
Postfach 100465
D-63004 Offenbach am Main (FRG)

ABSTRACT
Over the years, a number of climatological precipitation data sets have been compiled from conventional rain-gauge measurements, mainly from land based stations. Besides, remotely sensed precipitation data and accumulated short-range forecasts from numerical weather prediction models are also used for studies of the global climate and its variability. The different data sets comprise station-related time-series, climatological normals or long-term means, isohyetal maps as well as global or continental gridded data. A first purpose of this paper is to give a brief but probably incomplete overview of existing sources of precipitation information. Second, some of the mostly used gridded climatologies including the new, still preliminary results of the GPCC are compared. A distinction is made between continental and oceanic areas. It is shown that the precipitation estimates are in reasonable agreement both in temporal and spatial variation, although some large regional differences were found having to be interpreted.

1 Introduction

Climate change is one of the most important environmental issues facing us. This includes both short-term climate variability and long-term trends. Precipitation is especially important as its spatial distribution identifies the regions of maximum latent heat release which is the major driving force of the atmospheric circulation. If observed precipitation data is used for the assessment of the earth's energy and water balances or for validation of climate models, the data need to be temporarily and spatially integrated, e.g. monthly mean on a grid area.

Large-scale area averages of precipitation have been estimated traditionally from conventional measurements representing points only in the spatial field. Beginning with the GATE results, precipitation has been estimated from remotely sensed space-based observations, which are mainly used for oceanic areas. Radar observations will not be treated in this paper since there are only few regions covered.

NATO ASI Series, Vol. I 26
Global Precipitations and Climate Change
Edited by M. Desbois and F. Désalmand

Long-term large-scale precipitation averages derived from historical data have been generally accepted by the climatological and hydrological research communities as an available approach to describe the statistical characteristics of the precipitation field. However, it was recognized that data of highest possible accuracy and covering the entire globe are required for the verification of climate models (WMO/ICSU, 1990). The GPCC has been established as the central operational element of the Global Precipitation Climatology Project (GPCP) in the framework of the World Climate Research Programme (WCRP) in order to continuously provide monthly gridded analyses of the global precipitation on an operational basis, at least for the period 1986 to 1995 (Arkin, this issue).

2 Collections of Conventional Precipitation Measurements at Stations

Conventional measurements continue to be regarded as the most reliable observation method, even if they are also contaminated by systematic and sampling errors (Rudolf et al., 1993, Sevruk, 1993). Thus, rain-gauge data are still the main basis for analysing the monthly precipitation totals over land.

The optimum station density to calculate reliable area-averaged monthly precipitation depths on a grid of 2.5° longitude by latitude lies between 10 to 40 stations per grid (2 to 8 stations per 10,000 km^2) depending on orographical structure and climatic conditions in the grid, resulting in a number of about 40,000 stations world-wide.

As a matter of fact, only for about 4,500 stations monthly precipitation totals can be obtained from SYNOP and CLIMAT messages, which are globally distributed via the World Weather Watch Global Telecommunication System (GTS) of the WMO. An extension of the data base is only possible by merging the different regional collections or by bilateral data exchange. However, several countries deliver the data combined with restrictions that the data should not be passed to other users. Therefore, some existing important data collections are not generally available for climate research, and also not for the GPCC, although this centre has an official international function.

2.1 Collections of Historical Data

Global and regional collections of station-related precipitation data are listed in Table 1. Besides the number of stations (maximum number in the data set, number varies from month to month for time series), the time period covered and the spatial coverage is included and some comments are given in the last column (e.g. if time series available or long-term means or normals only). Some of the global historical data collections are based on the same sources, but there are differences with regard to the grade of quality-control procedures applied.

The GPCC has access to most of the listed data sets. Information on further numerous existing climatological data sets is compiled by Jenne (1991) and Burtis (1992), both possibly updated.

The only data immediately available are the CLIMAT and SYNOP reports distributed globally and received via GTS at the Regional Telecommunication Hub (RTH) Offenbach. The resulting data collection (monthly precipitation totals for only about 4,500 stations) could be improved by merging the partly different data sets received at all of the WMO regional centres. The total number is estimated to reach 7,000 stations or more. The US National Weather Service (GTS-RTH Washington) provides data of about 5,000 stations which are partly complementing, partly overlapping the data received at Offenbach, and the merged data set contains about 6,000 stations. Other regional data centres do not pass the data (if collected), since an approval of the individual countries to pass their data to the GPCC is required. Connections via WMO-channels would be necessary.

Over many parts of the continents, the resulting station density is still insufficient with regard to the mentioned optimum data density and the requirement of world-wide 40,000 stations. By bilateral contacts supported by WMO-activities, the GPCC has acquired larger amounts of additional data. Up to now the national Meteorological Services (partly research institutes and private companies, too) of more than 60 countries have supplied data from totally about 30,000 stations. The global collections of Legates (1987) and Eischeid et al. (1991), the GHCN (Vose et al. 1992) data as well as the African data set of the FAO are also included. The GPCC data base is still increasing and the GPCC continues its efforts to fill some gaps still existing and to uphold the data flow also in future. However, with regard to the realization of the GCOS, an international convention on the exchange of data is inalienable.

3 Global and Terrestrial Isohyetal Maps and Gridded Data Sets

Terrestrial or continental analyses have been carried out in earlier times by drawing isohyets manually, which have been reproduced as maps. This method incorporates subjective interpretations based on the author's experience and knowledge of the regional precipitation patterns and orographic-climatological relationships. An overview of maps and atlases published is given in Table 2.

source	year of publication	period	number of stations	continent	comments
CCTA (Commission de Cooperation Technique en Afrique)	1961	30 years	720	Africa	digitized at the GPCC
WMO/UNESCO	1970	1931 - 1960	7,000	Europe	digitized at the GPCC
WMO/UNESCO	1975	1931 - 1960	1,700	South-America	digitized at the GPCC
WMO/UNESCO	1981	1931 - 1960	6,600	Asia	digitized at the GPCC
WMO/UNESCO	1979	1931 - 1960	3,000	North-America	digitized at the GPCC
Geiger	1965	1931 - 1960		global	mean annual
Shea (NCAR)	1986	1950 - 1979	3,600	global	mean monthly

Table 2: Summary of global and continual climatological precipitation maps (atlases, no pretension on completeness)

source	area	station number	period	comments
SYNOP	global	3,500	current	received via GTS, monthly totals calculated from short-term data
CLIMAT	global	1,500	current monthly	received via GTS, monthly data
Climate Analysis Centre (NOAA-NMC)	global	4,900	current monthly	provided to GPCC on tape, main source GTS
Monthly Climatic Data for the World (NOAA-NCDC)	global	1,500	current	monthly journal some months delay
World Weather Records (WWR)	global	1,800	1961 - 1980	decade books for the continents
World Monthly Surface Station Climatology (NCAR)	global	3,600	1700 - 1983	data sources: World Weather Rec., Monthly Clim. Data for the World
World Climate Disc (CRU and Chadwyck-Healey)	global	7,339	various best coverage 1951 - 1980	time series of monthly data, available on CD-Rom, data collection of CRU Norwich
Global Historical Climatic Network (NOAA-NCDC)	global	7,000	var. - 1990	continuously updated time series, monthly data
Legates (University of Oklahoma)	global	25,000	various 1920 - 1980	long-term monthly means
FGGE	global	24,000	Dec. 1978 - Nov. 1979	daily data
GPCC	global	30,000	1986, onwards	monthly data from GTS and bilateral conventions
WMO Drought Monitoring Centres (Nairobi and Harare)	regional Africa	200	1991, onwards	monthly journals
Food and Agriculture Organization (FAO)	regional Africa	300	1986 - 1991	more stations available, however, with incomplete observations
Nicholson (University of Florida, Tallahassee)	regional Africa	548	1900 - 1988	collected during travels
Colorado State University African Rainfall Database	regional Africa	720	various 1920 - 1988	data sources WMSSC, MCDW, Nicholson, AID, Lamb
Morrissey (University of Oklahoma)	Pacific	88	1971 - 1990	only existing rain-gauge data being representative for oceanic areas

Tab. 1: Summary of commonly known important precipitation data collections (rain-gauge measurements, no pretension on completeness)

2.2 GPCC Data Collection

The GPCC collects monthly precipitation totals for the period starting with January 1986, according to its international tasks (WMO/ICSU, 1990) and its corresponding personal capacity. However, apart from the continuous data series, climatological long-term means (normals) are also acquired and stored since these data provide important background information both for the quality-control of the continuous data as well as for the calculation of anomalies.

Virtually all climatological research today requires digital information suitable for processing by computers. To facilitate this use of the isohyetal analyses, several of the maps have been digitized and interpolated to a grid. For example, Jaeger's (1976) global precipitation climatology, which is still widely used, results from subjective visual interpolation of isohyetal maps from the available climatic atlases supplemented by some additional station-related data. Jaeger obtained his oceanic estimates from monthly percentage precipitation frequency maps given in the United States Marine Climatic Atlas (ref. Jaeger). These precipitation frequencies were applied to the map of mean annual precipitation from Geiger (1965) to generate monthly averages. The frequencies and Geiger's map have been interpolated onto the grid visually.

source	publication year	period	stations number over land	grid size	method	comments
Global						
Schutz, Gates (RAND)	1972	various	1,500	4x5°	visual interp.	based on Möller (1951)
Jaeger (DWD)	1976	various 1931 - 1960	from atlases	5°	visual interpolation	continents: atlases, ocean: Geiger and US-Marine atlases
Legates (University of Oklahoma)	1987	various 1920 - 1980	24,000	0,5°	objective analysis	ocean: Tucker method, south of 30° S Jaeger
Shea (NCAR)	1986	1950 - 1979	3,600	2,5°	obj. analysis	ocean: COADS data
Cogley, Briggs	?	1950 - 1979	?	1°	?	ref. Shea 1993 (NCAR)
Schemm, Terry, Schubert, Bloom (NASA)	1992	1979 - 1989	1,300	4x5°	objective analysis	merged rain-gauge and MSU satellite data
GPCC	1993	1987, 1988	7,000	2,5°	objective analysis	merged rain-gauge, IR and SSMI satellite data and model results
Terrestrial (continents only)						
Leemans, Cramer (IIASA)	1990	1931 - 1960	1,500	0,5°	obj. analysis	calendar months, year
Eischeid, Diaz, Bradley, Jones (DOE)	1991	1800 - 1980	5,328	4x5°	objective analysis	time series, monthly data
Hulme (CRU)	1991	1951 - 1980	4,700	5°	objective analysis	time series, monthly data: not all grids
Digitized atlases of (WMO/UNESCO)	in preparation	various 1931 - 1960	isohyets	0,5°	objective analysis	continents as Tab. 2: calendar months, year

Table 3: Summary of global and terrestrial gridded precipitation maps (digital data available, no pretension on completeness)

Most of the gridded data developed in the last ten years are derived by objective methods using directly the station-related observational data rather than by subjective interpolation of isohyetal maps. The examples given in Table 3 are based on different data collections and different interpolation approaches. Also the goals of the authors differed in setting up time series of analyses based on homogeneous but smaller input data sets (5°-grid, ca. 4,000 stations, Hulme, 1991) or providing a gridded global long-term climatology of highest possible spatial resolution (0.5°-grid, ca. 25,000 stations, Legates, 1987). Legates' results are criticized by some authors since data of very different periods are mixed so that spatial inhomogeneities can be caused by climatic trends.

The isohyetal maps of the WMO/UNESCO atlases have been digitized by the GPCC. The transfer onto a grid using an objective interpolation method is in preparation, so that comparison studies with other gridded precipitation data purely objectively derived will be possible.

Legates' data, both the station-related and the gridded, are available as measured as well as corrected with regard to the systematic gauge-measuring error. The assessment of the size of this error (even if it can be done only roughly) is very important if the observed data are used for validation of satellite based estimates or climate model results.

The authors of global data sets have used different sources of information on precipitation over the oceans. Legates for example included the estimations from ship weather observations published by Dorman and Bourke (1979, 1981), whereas Schemm et al. (1992) have based their oceanic data on MSU satellite observations and the GPCP is using infrared and microwave satellite data as well as numerical weather prediction results (Rudolf, 1993, GPCC, 1993).

4 Intercomparison of Gridded Global and Terrestrial Precipitation Climatologies

4.1 Remarks on the Data Sets Compared

Jaeger provided the classical and mostly used gridded precipitation set in 1976. The product was assembled from subjective estimates of mean monthly rainfall from contemporary rainfall atlases, supplemented by other data sets (period: 1931 to 1960) . Over the ocean, digitized values from subjectively analyzed isohyets from the U.S. Marine Climatic Atlas were used.

The intention of Legates (1987) was to realize a sufficient spatial station density by inclusion of all available rain-gauge data. Therefore, means over different periods, partly going back to the 19th century and partly of the current time, are mixed in the data base. The largest quantity of the data collection comprise in the period before 1960. Over the ocean, precipitation estimates are incorporated, which are based on synoptic weather observations from ships.

Legates gridded data are provided as two separate versions, "nc" derived from the station related data as measured, and "c" from the data corrected with regard to the systematic error from evaporation losses and rain drop and snow drift due to wind.

The estimates of Leemans and Cramer (1990) are based on the time series collected by Bradley et al. (1985). For this analysis, the data of the period 1931-1960 are preferred.

The data set from Hulme (1991) was developed especially for model validation and is based on quality-controlled data of homogeneous time series for the period 1951-1980. This climatology therefore covers only land areas (on a 5° grid), but grid cells without any station are not covered.

In order to obtain complete global products, the GPCC merges the data from differ ent observation techniques. In the version 001 (GPCC, 1993), a blending scheme is used, whereby the grid values from the different data sources are selected by predefined priorities depending on the geographical location: analyses from conventional gauge-measurements are exclusively used for all land grids, infrared-microwave-mixed estimates over tropical oceans, microwave-only estimates over the subtropical and mid-latitude ocean, and model results of the European Centre for Medium Range Weather Forecasts (ECMWF) over the high-latitude oceans, where no observed data are available. Up to now, preliminary results for the 24 months of 1987 and 1988 have been prepared. The continental results of these data set are based on the measurements from only about 6,700 stations. The additionally acquired data from about 30,000 stations, mentioned in paragraph 2.2 are not yet used, but a re-analysis based on all data is planned.

Additionally, the pure ECMWF model results are included in this intercomparison. The daily 12h to 36h precipitation forecasts are accumulated to monthly totals.

4.2 Intercomparison Results

An overall intercomparison is given by the annual global means, whereby the monthly gridded precipitation totals are accumulated for the year and weighted by the grid-area size which decreases with the latitude (Table 4).

Jaeger	985 mm
Legates (nc)	1,034 mm
Legates (c)	1,140 mm
GPCC (1987)	1,015 mm
GPCC (1988)	1,025 mm
ECMWF (1987)	1,205 mm
ECMWF (1988)	1,230 mm

Tab. 4: Annual global mean precipitation derived from monthly gridded data from different sources.

The lowest annual global mean is calculated for Jaeger with 985 mm, while Jaeger himself gives 1,000 mm in his original paper. The difference results from rounding before the accumulation on the year, which can cause an error up to about 2% and should be considered for the other annual data sets, too.

The highest over all precipitation results from the forecast model and is about 8% larger than the highest value of the observed data, which are corrected with regard to systematic measuring error. Arpe (1991) studied the characteristics of the ECMWF forecast model results and found a tendency of the contemporary model to overestimate convective rainfall.

If the systematic error is not considered, the annual global means from the climatologies based on observational data are on the same level with relative differences of less than 3% to the total mean (1,015 mm).

The overall mean of Legates' corrected data is about 10% larger than the result from his data as measured. Because the consideration of the systematic measuring error is inalienable, when measured data and model results or remote sensing data are being compared, some more detailed comparisons are made for the two Legates data sets. The continental zonal means are shown in Figure 1, separately for the northern summer and winter seasons. The systematic gauge-measuring error is estimated to be highest, when and where a large portion of snow occurs in the total precipitation. The seasonal zonal mean reaches its largest relative values with about 70% (absolute 15 mm/month) in the subpolar northern hemisphere and more than 100% (absolute 30 mm/month) over the Antarctica. The error, also influenced by characteristics of the measuring instrument used, could be quite higher for specific regions and months. In the tropics, the absolute error can also reach more than 10 mm/month, however, but the relative error is generally less than 5%.

The GPCC is planning to provide the gridded monthly precipitation with a supplement for the error size on each individual grid cell.

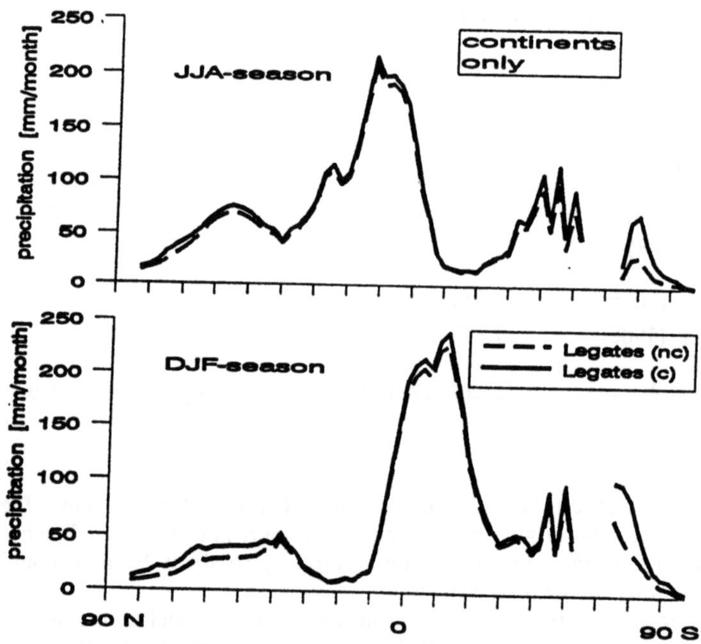

Fig. 1: Latitudinal profile of zonal-mean seasonal precipitation in mm/month exclusively for the continental areas, derived from Legates gridded data (nc: data as measured, c: corrected with regard to the systematic measuring error), above northern summer (June, July, August) and below northern winter (December, January, February).

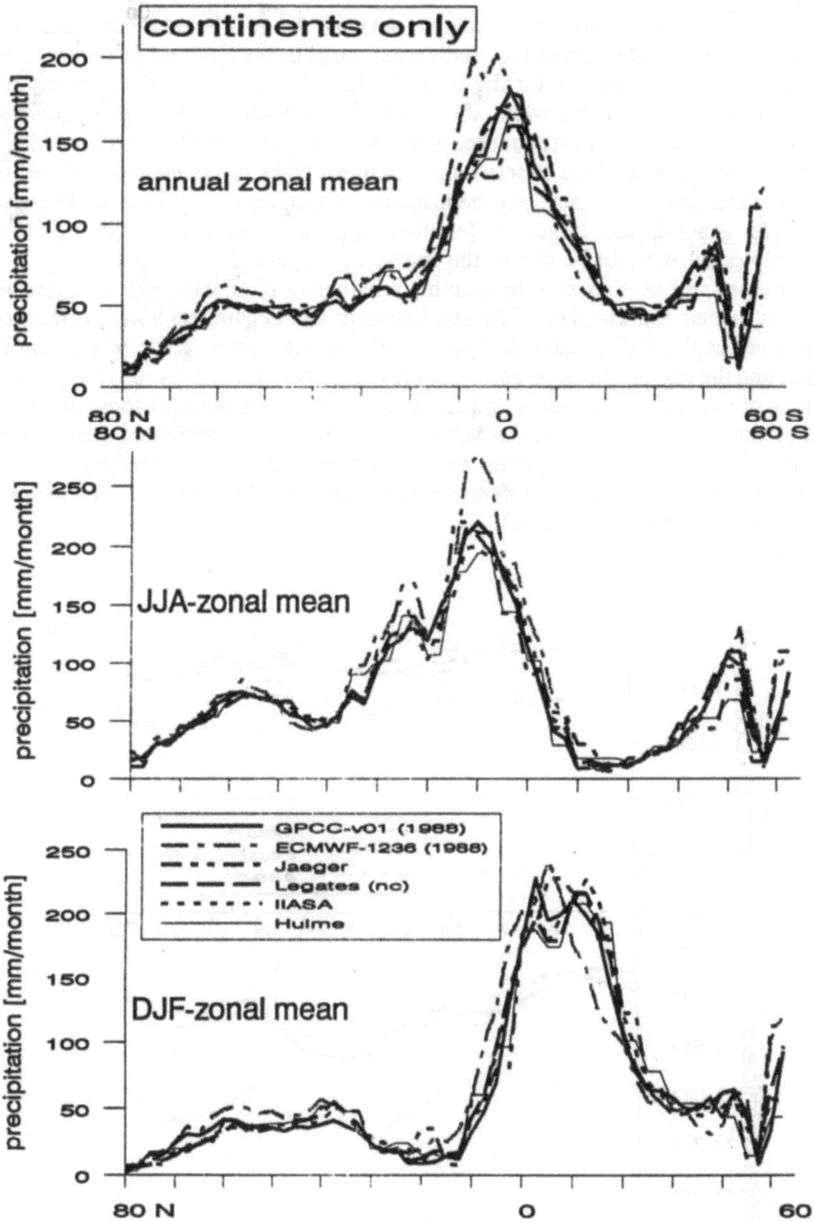

Fig. 2: Latitudinal profile of zonal-mean precipitation in mm/month exclusively for the continental areas, for the climatological data sets of Hulme, Jaeger, Leemans/Cramer (IIASA) and Legates, and the 1988 analyses of the GPCC and the ECMWF forecast results (upper panel annual mean, mid panel northern summer and lower panel northern winter).

The similarity of the climatological data sets and the mentioned model defect can also be seen by the comparison of the continental zonal means, plotted in Figure 2. For the northern summer as well as for the year, the ECMWF model yield the broadest ITCZ with highest precipitation, while the Hulme data indicate the lowest precipitation amounts in the tropics. Furthermore, there is a small shift in the location of the ITCZ between the model and the observation based estimates in the northern winter season. In arid zones and in the northern hemisphere subtropics all results are in good agreement. In general, the largest differences are found in the southern hemisphere extratropics, what might be due to the sparse coverage with measured data.

For those grid cells covered by monthly area-means of all data sets, a regression analysis was carried out. The data of Jaeger, Leemans and Legates each were compared separately versus the Hulme data. In Figure 3, the annual cycles of the correlation (upper panel) and the regression slope coefficient (lower panel) are displayed. The spatial distributions of the Hulme, Leemans and Legates climatologies seem to show nearly similar mean features. However, a annual variation is found with a minimum slope coefficient for July. Since the Jaeger, Leemans and Legates data show the same behavior, it is assumed, that it results from the different base periods used by Hulme (1951-1980) and the other authors (mainly older data).

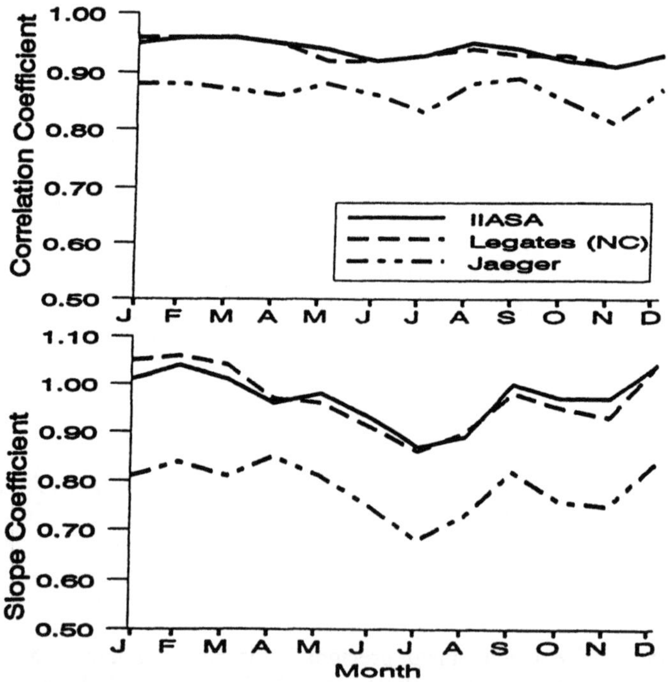

Fig. 3: Correlation and slope coefficient of a linear regression analysis for the gridded data of Jaeger, Leemans/Cramer (IIASA) and Legates (not corrected), each versus the data of Hulme.

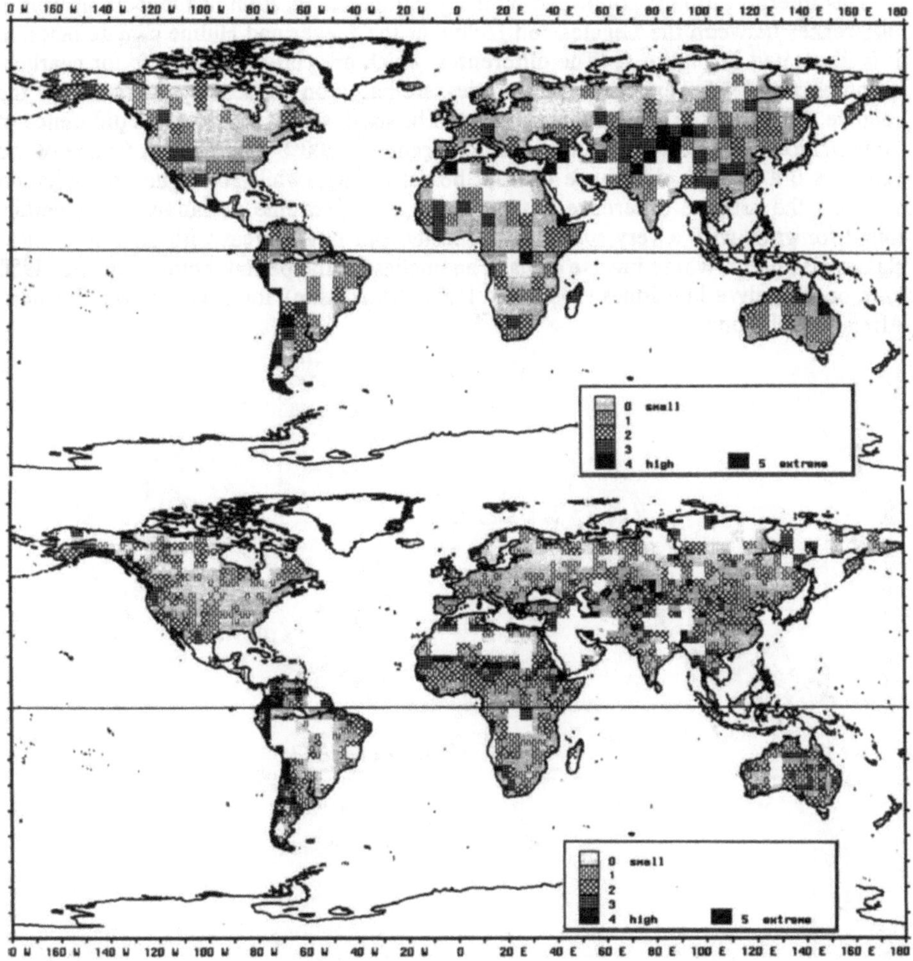

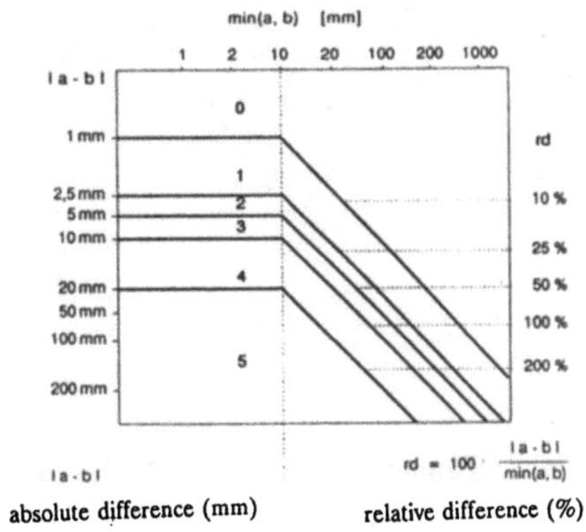

Fig. 4:

Classified differences as defined by the diagram below for the gridded annual global precipitation data of Jaeger and Hulme (upper panel) as well as of Legates and Hulme (lower panel).

This interpretation is confirmed by the geographical location of the most significant differences between the Legates and Hulme or the Jaeger and Hulme climatologies, as it is illustrated in Figure 4. (The differences, which are classified from "0" for nearly in accordance to "5" for extreme discrepancies, are based on the absolute as well as on the relative differences. This combination has to be used, since for arid areas difference of 10 mm/month is relevant but a relative difference of 100% might be not (if one of the values is 0 a relative difference can not be calculated), while for higher precipitation amounts the relative difference is most important.) Extreme deviations are generally found for grids with a very sparse data density. But larger areas with high difference-classes also occur, where inverse climatic anomalies for the periods before and after 1950 have been analyzed by Jones (1989) and Hulme (this issue), for example over Northern Africa, Sahel Zone.

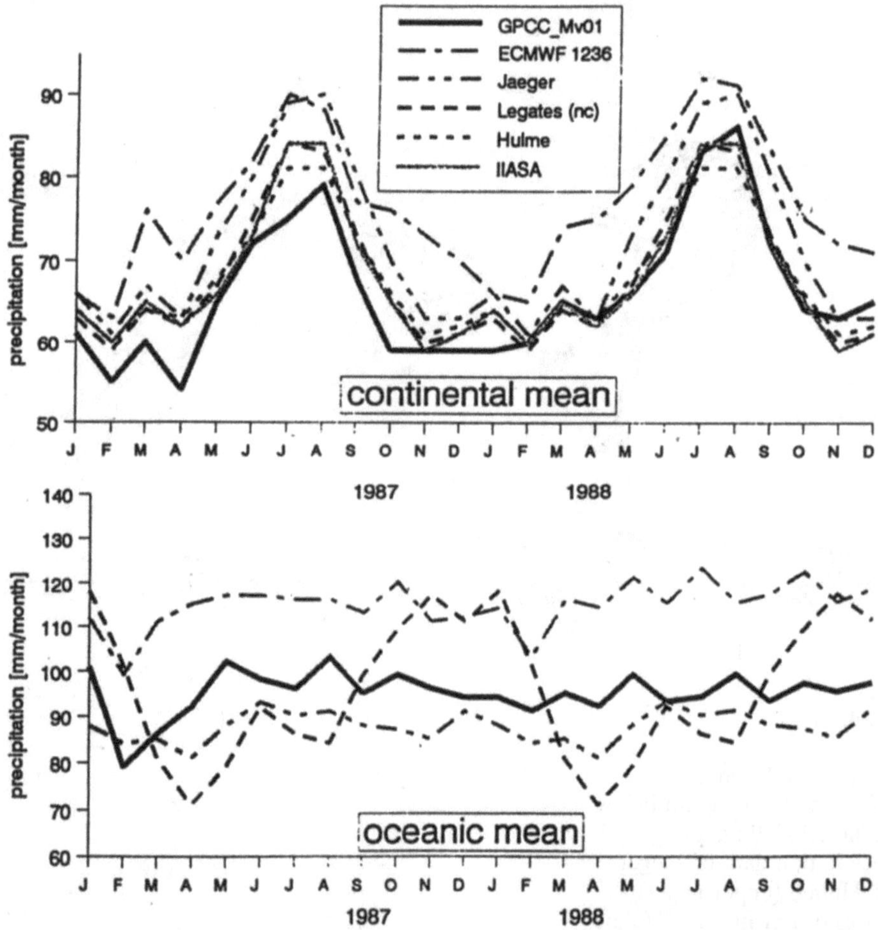

Fig. 5: Temporal variation of the large-areal mean precipitation (mm/month) for the climatologies of Hulme, Jaeger, Leemans/Cramer (IIASA), and Legates, as well as for the GPCC analyses and ECMWF model results (years 1987 and 1988).

Finally, the large-area means for the continental and oceanic areas are compared separately, whereby over land only grids containing results from all data sets are used in order to include also the Hulme data. Figure 5 shows the time series for January 1987 to December 1988 of the GPCC and ECMWF results, while the climatological estimates are periodically plotted. Over land, a strong intra-annual variation occurs in all data sets. Over ocean, only the Legates data indicate a pronounced variation over the year, which seems to be artificial with regard to the data used (mainly very rough estimates from ship weather observations) of the large rainfall amounts. Generally, the model results are clearly higher than the GPCC analyses by between 10 and 40%.

5 Conclusion

The collections of historical and current observed precipitation data provide an important information on the underlying mechanisms of climate dynamics and these data sets becomes more and more of interest in different research activities concerning climate variability and change.

Area-averaged data should be based on objective interpolation methods rather than on subjective interpretations. The incorporation of a sufficient concentration of station observations is necessary to produce reliable quantitative results over land areas, as is evidenced by the distribution of the differences between the presented gridded data sets. Furthermore, differences can be attributed to the different time periods which are represented by the different data sets. Therefore, data from homogeneous time series must be used for the analysis of climate variability or trends. For comparison of conventionally measured data and results of other observation techniques or models, and especially before merging of the different data, the systematic errors have to be corrected. Otherwise inaccuracies and inhomogeneities are systematically produced.

Oceanic and polar estimates still are the most uncertain part of any precipitation climatology. At present, precipitation over ocean can be estimated best by remote sensing techniques, even if the calibration based on reliable ground data is necessary. Further research on algorithms and intercomparison studies are necessary.

Acknowledgement: The authors would like to thank many individuals and institutions helping with valuable suggestions and providing data. The list would be to long to be presented here. The establishment and development of the GPCC is supported by the Federal Ministry of Research and Technology (BMFT), Germany, under Registration No. 07 KF 005.

REFERENCES:

Arkin, P. A. (1993): The Global Precipitation Climatology Project (this issue).

Arpe, K. (1991): The hydrological cycle in the ECMWF short range forecats. Dynamics of Atmospheres and Oceans, **16**, 33-60.

Bradley, R.S., P.M. Kelly, P.D. Jones, H.F. Diaz, and C. Goodess (1985): A Climatic Data Bank for the Northern Hemisphere Land Areas, 1851-1980. DOE Technical

Report No. TR017, US Department of Energy, Carbon Dioxide Research Division, Washington D.C..

Burtis, M.D. (1992): Catalog of data bases and reports. Oak Ridge National laboratory, ORNL/CDIAC-34/R2 No.3878, Tennessee.

Commission for Technical Co-cooperation in Africa South of the Sahara (1961): Climatological Atlas of Africa (CCTA/CSA Lagos, Nairobi). Printed by the Government Printer Pretoria, Rep. South Africa.

Dorman, C.E. and R.H. Bourke (1979): Precipitation over the Pacific Ocean, 30°S to 60°N.- Month. Weath. Rev. **107** (7), 896-910.

Dorman, C.E. and R.H. Bourke (1981): Precipitation over the Atlantic Ocean, 30°S to 70°N.- Month. Weath. Rev. **109** (3), 941-948.

Eischeid, J.K., H.F. Diaz, R.S. Bradley and P.D. Jones (1991): A Comprehensive Precipitation Data Set for Global Land Areas. U.S.Department of Energy Report No. DOE/ ER 69017T-H1,Washington D.C., 81 pp.

Geiger, R. (1965): Jährlicher Niederschlag. Wandkarte 1:30 Mill., Verlag Justus Perthes, Darmstadt.

GPCC (1993): Global Area-Mean Monthly Precipitation Totals on a 2.5° Grid for the Year 1988 (Preliminary Results, Derived from Rain-Gauge Measurements, Satellite Observations and Numerical Weather Prediction Results). Rep.-No. DWD/WZN-1993/07-1, Offenbach a.M., 16 pp and cover (to order from GPCC).

Hulme, M. (1991): An Intercomparison of Model and Observed Global Precipitation Climatologies. Geophys. Res. Lett., 18, 1715-1718.

Jaeger, L. (1976): Monatskarten des Niederschlags für die ganze Erde. Bericht des Deutschen Wetterdienstes, Nr. 139 (Band 18), 33 pp and plates.

Jenne, R.L. (1991): Data Available on CD-ROMs. National Center for Atmospheric Research, Scientific Computer Division; Boulder, Colorado.

Jones, P.D. (1989): Regional, Hemispheric and Global Scale Precipitation Measurements. Proc. WMO/IAHS/ETH Workshop on Precipitation Measurement, St. Moritz, 3-7 December 1989, 513-518.

Leemans, R. and W.P. Cramer (1990): The IIASA Database for Mean Monthly Values of Temperature, Precipitation and Cloudiness on a Global Terrestrial Grid. Working Paper WP-90-41, Biosphere Dynamics Project, International Institute for Applied Systems Analysis, Laxenburg, Austria, 62 pp.

Legates, D.R. (1987): A Climatology of Global Precipitation. Publ. in Climatology **40** (1), Newark, Delaware, 85 pp.

Morrissey, M.L. and J.S. Greene (1991): The Pacific Atoll Raingage Data Set. Planetary Geosciences Division Contribution/JIMAR No. 648, Honolulu, Hawaii.

Nicholson, S. (1988): Atlas of African Rainfall and Its Interannual Variability. Florida State University, Department of Meteorology, Tallahassee.

Rudolf, B. (1993): Management and Analysis of Precipitation Data on a Routine Basis. Proceedings Intern. WMO/IAHS/ETH Symp. on Precipitation and Evaporation, Slovak Hydromet Inst., Bratislava Sept. 1993, Vol. 1, 69-76.

Rudolf, B., W. Rueth and U. Schneider (1993): The Spatial Sampling Error of Areal Mean Monthly Precipitation Totals Analyzed from Gauge Measurements. Fourth International Conference on Precipitation; Hydrological and Meteorological Aspects of Rainfall Measurement and Predictability, Iowa City, USA, 80-82. (see also Rudolf et al., this issue.)

Schemm, J., S. Schubert and J. Terry (1992): Estimates of monthly mean soil moisture for 1979-1989. NASA Technical Memorandum 104571, GSFC, Greenbelt, Maryland.

Schutz, C. and W.L. Gates (1972-74): Global Climatic Data for the Surface: R-1029-ARPA (1972), R-915/1-ARPA (1972), R-1317-ARPA (1973), R-1425-ARPA (1974), RAND Corporation, Santa Monica.

Sevruk, B. (1982): Method of Correction for Systematic Error in Point Precipitation Measurement for Operational Use. WMO-No. 589, 91 pp. (see also Sevruk and Nespor, this issue.)

Shea, D.J. (1986): Climatology atlas: 1950-1979, Surface Air Temperature, Precipitation, Sea-Level Pressure, and Sea-Surface Temperature (45°S-60°N). NCAR Technical Note, NCAR/TN-269+STR, Boulder, Colorado, 35 pp and plates.

Shea, D.J. and R. Jenne (1993): NCAR's Historical and Climatological Data Base. In: Analysis Methods of Precipitation on a Global Scale. Report of a GEWEX Workshop organized by GPCC, WCRP-81, WMO/TD-No. 558, A/18-A/21.

Vose, R.S., R.L. Schmoyer, P.M. Steurer, T.C. Peterson, R. Heim, T.R. Karl and J.K. Eischeid (1992): The Global Historical Climatology Network: Long-Term Monthly Temperature, Precipitation, Sea Level Pressure, and Station Pressure Data. Oak Ridge National Laboratory, ORNL/CDIAC-53, 99 pp.

WMO/ICSU (1990): The Global Precipitation Climatology Project - Implementation and Data Management Plan. WMO/TD-No. 367, Geneva, 45 pp and 6 appendices.

WMO/UNESCO (1970): Climatic Atlas of Europe; Maps of Mean Temperature and Precipitation. Vol. I.

WMO/UNESCO (1975): Climatic Atlas of South America; Maps of Mean Temperature and Precipitation. Vol. I.

WMO/UNESCO (1979): Climatic Atlas of North and Central America; Maps of Mean Temperature and Precipitation. Vol. I.

WMO/UNESCO (1981): Climatic Atlas of Asia; Maps of Mean Temperature and Precipitation. Vol. I.

World Climate Disc, Global Climatic Change Data. Chadwyck-Healey Ltd., Cambridge/UK (1992).

PRECIPITATION DISTRIBUTIONS FROM COMBINED AIRBORNE MICROWAVE RADIOMETRIC/RADAR MEASUREMENTS AND GROUND BASED VLF SFERICS OBSERVATIONS IN SUPPORT OF NASA'S TROPICAL RAINFALL MEASURING MISSION

James A. Weinman
Microwave Sensors Branch
NASA Goddard Space Flight Center
Greenbelt, MD 20771

Jack L. Schols
Long S. Chiu
General Sciences Corp.
Laurel MD. 20707

ABSTRACT
A description of the Tropical Rainfall Measuring Mission (TRMM) program is presented. The main thrust of the TRMM program is a satellite that will carry the first rain measuring radar in space along with microwave radiometers, visible and infrared imagers and radiometers as well as a lightning imaging instrument. Measurements from nadir viewing airborne radars and microwave radiometers were analyzed to develop algorithms to retrieve precipitation characteristics in preparation for TRMM. The present model retrieved distributions of the mass density of the precipitation.

Observation of convective storms will be a problem at those times when the TRMM satellite is not overhead. Very Low Frequency (VLF) radio noise, called sferics, can continuously observe the distribution of lightning from ground based receivers. Preliminary observations of the distribution of sferics are compared to the distribution of rainfall derived from the Special Sensor Microwave / Imager (SSM/I) on the Defense Meteorological Satellite Program (DMSP) satellite to provide a proxy for continuous rain observations.

KEYWORDS
Radar, rainfall retrieval, microwave radiometry, radar-radiometry algorithm, melting band, hydrometeor mass profile, Maxwell-Garnet mixing rule, Monte-Carlo radiative transfer model, SSM/I, DMSP, sferics, lightning.

I. Summary of the TRMM program

I.a Scientific background

Latent heating, attributable to the condensation of water vapor into rain, accounts for most of the energy in the atmosphere. Two thirds of that precipitation is concentrated in the tropics. Unfortunately, the distribution of tropical rainfall has been difficult to measure

NATO ASI Series, Vol. I 26
Global Precipitations and Climate Change
Edited by M. Desbois and F. Désalmand

because much of the tropics are oceanic or underdeveloped, and because rainfall is highly variable both temporally and spatially. Because of these considerations, the Tropical Rainfall Measuring Mission (TRMM) has been initiated by the Japanese and American space agencies. The TRMM satellite is designed to measure tropical precipitation and its variation from a low-inclination orbit combining a suite of sensors that will overcome many of the limitations that were previously encountered in the acquisition of rainfall measurements. With the inclusion of a rain radar, TRMM will provide opportunities to estimate vertical profiles of the latent heat released through condensation. TRMM rainfall data will be important for studies of the global hydrological cycle and for testing the realism of climate models.

I.b Characteristics of the TRMM satellite system

TRMM is a comprehensive program that consists of research on cloud models, rain retrieval algorithms for space sensors, the use of TRMM measurements with other satellite data to improve sampling, a surface-based verification system, and a TRMM Science Data and Information System (TSDIS). The TRMM satellite (Fig. 1) described by Theon et.al. (1992), will be launched on a Japanese H-II rocket in 1997 with a designed lifetime of three years.

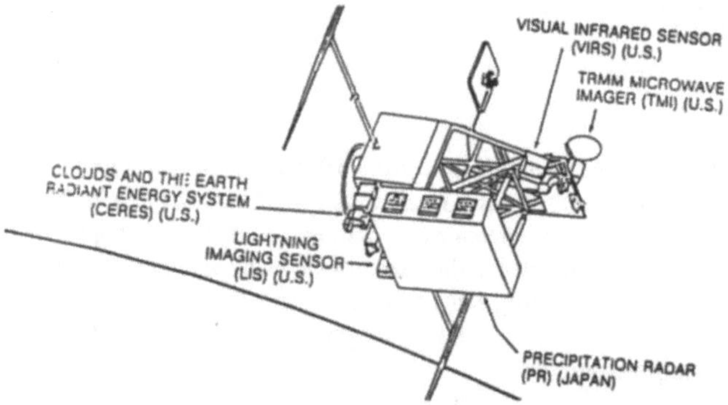

Fig . 1 The TRMM satellite with the Precipitation Radar (PR), the Visible and Infrared Radiometer System (VIRS), and the TRMM Microwave Imager (TMI) are illustrated along with the radiation and lightning sensors (CERES and LIS) that are located behind the 2 m PR antenna.

The TRMM satellite will carry a precipitation radar (provided by Japan), a multi-frequency microwave radiometer and a visible and infrared radiometer.The satellite orbit of 365 km altitude and inclination of 35°, will fulfill the stated science objectives of the mission. For related precipitation and climate observations, a lightning imager and an Earth radiation budget sensor have been added. A brief description of each of the five instruments follows.

The Precipitation Radar (PR) will measure the three dimensional rainfall distribution. A unique feature is that it will permit the measurement of rain over land where passive microwave channels have more difficulty. The PR is an electronically scanning rain radar operating at 13.8 GHz with horizontal polarization that will provide 4.3 km horizontal resolution at nadir, the range resolution is 250 m and the scanning swath width is 220 km.

The TRMM Microwave Imager (TMI), is designed to provide information on the integrated column precipitation content, its areal distribution, and its intensity. The TMI will operate at 5 frequencies, of which 4 will have dual polarization thus providing 9 channels of data. The 5 frequencies are 10.65 GHz, 19.35 GHz, 22.235 GHz (single polarization), 37.0 GHz and 85.5 GHz. The horizontal resolution of the TMI will range from 5 km at 85.5 GHz to 45 km at 10.65 GHz. A scan angle of 45° will provide a swath width of 680 km.

The Lightning Imaging Sensor (LIS) is a calibrated optical sensor operating at 0.7774 μm that is optimized to detect the lightning location, mark the time of occurrence, and measure the radiant energy to provide the distribution of lightning. The horizontal resolution at nadir will be 5 km and the swath width will be 660 km. LIS will provide a link between optically observed lightning flash rates and the radio noise from lightning that can be measured continuously from the Earth's surface.

The Visible Infrared Scanner (VIRS) will provide high resolution information on cloud coverage, type, and cloud top temperatures and also serve as the link between these data and the long and virtually continuous coverage by geosynchronous meteorological satellites. The VIRS is a 5 channel cross-track scanning radiometer operating at 0.63, 1.6, 3.75, 10.80, and 12.0 μm. The VIRS swath width of 720 km will provide cloud distributions by type and height and rain estimates that can be correlated with data from the PR, TMI and LIS.

The Clouds and Earth's Radiant Energy System (CERES) is a visible and infrared sensor designed to measure emitted and reflected radiative energy from the surface of the Earth and the atmosphere and its constituents (e.g., clouds, aerosols). The CERES is a broad band scanning radiometer that will operate at 0.3 to 5.0 μm in the shortwave range and at 8.0 - 12.0 μm in the long wave range.

I.c TRMM support program

The TRMM validation program, described by Thiele et.al (1993), will calibrate and validate the spaceborne measurements and algorithms used to derive rainfall. In the pre-mission phase, the program includes: 1) establishing and/or supporting rain gauge networks in company with existing radars at various tropical sites; 2) preparing climatologies for various tropical sites based upon available radar observations and gauge networks; 3) conducting a broad range of precipitation research to develop both surface and airborne methods and algorithms for the improved use of radar to measure rainfall; 4) developing operational algorithms for the TSDIS; 5) conducting specialized experiments to test instruments and methods such as new optical rain gauges, microwave and millimeter-wave path attenua-tion measurements for rainfall estimates and underwater acoustic sensors to estimate rainfall impacting on the sea; and 6) investigating advanced surface and aircraft radar techniques for more accurate areal coverage and for observations in remote areas.

Routine collection of radar and rain gauge data is already underway from sites shown in Fig. 2. Although there is considerable variability in the type and quality of these radars and rain gauges, rainfall climatologies are being generated with state-of-the-art estimation and analysis techniques.

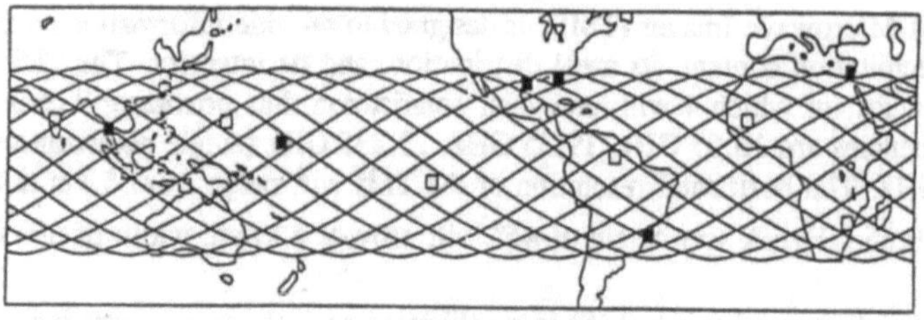

Fig. 2 Current and future validation sites providing radar and rain gauge data. A TRMM 16 orbit (one day) ground track (35° inclination) is shown for location reference.

The operational processing of validation products is a responsibility of the TSDIS that is currently under development. Approximately 6 months before launch routine processing for representative tropical climatologies will be taken on by TSDIS. Beside data from the mission validation sites, specialized data products from other sources of interest to TRMM scientists will also be obtained from the TRMM Project Office.

To examine new surface measurement techniques, a small scale experimental rain measurement research facility is being developed at the NASA Goddard Wallops Flight Facility, Virginia. A polarized multifrequency microwave attenuation link will be the key component of that facility. Attenuation of horizontally propagating microwave radiation is strongly related to the intervening precipitating mass and is less sensitive than radar reflectivity-rainfall (Z-R) relationships to rain drop size distributions. The six frequencies to be employed are two near 8.7 GHz and vertically, horizontally polarized 24, 35, 82, and 245 GHz and 82 and 245 GHz (furnished by the Communications Research Laboratory of Japan). For purposes of evaluating new systems/calibrations and intercomparisons, a variety of state-of-the-art rain gauges and disdrometers will be located along the link path.

Over the ocean, surface-based techniques, e.g., optical devices for measuring rainfall from ship and buoy platforms, and hydrophones for the acoustical detection of rainfall have been employed to calibrate surface based radars. Optical rain gauges have been deployed on a NOAA boat buoy off the east coast of Florida. Hydrophones have also been included in the oceanic rainfall measuring program; these can either be tethered below moored or drifting buoys or anchored on the ocean floor.

II. Airborne radar and radiometric measurements

An ensemble of airborne multi-frequency radiometers and a single frequency radar comprise a powerful tool to measure distributions of hydrometeor characteristics. The radiometric measurements yield information on the phase of the hydrometeors and they provide a constraint on the retrieval of the vertical profiles of hydrometeor distributions from radar observations. While the radiometric measurements have poor vertical resolution, the radar provides information regarding the vertical hydrometeor structure.

In order to simulate the TRMM measurements a major effort has been initiated to measure precipitation from airborne radars and microwave radiometers. Data from airborne 10 GHz radar and radiometric measurements were combined by Weinman et.al. (1990) to derive rainfall rate profiles. The rain rate profiles compared favorably with profiles retrieved from dual frequency radar measurements obtained on the same flights for light to moderate rainfall rates. However no attempt was made to probe the ice hydrometeors that may have existed in those clouds. Schols and Weinman (1994) addressed that problem and they also retrieved several profiles of the hydrometeor mass density, the average particle density and effective diameter throughout the entire depth of the precipitating clouds. The following presentation describes recent developments in the analysis of extensive precipitation fields derived from combined airborne radar and microwave radiometric data.

A field campaign, described by Wang et. al. (1994), obtained data from which an algorithm was developed to retrieve distributions of hydrometeor mass from simultaneous measurements provided by a nadir directed 10 GHz radar and 10, 19, 34 and 92 GHz radiometers carried aboard the NASA DC-8 aircraft. Those measurements, similar to those that the TRMM will provide, were obtained on September 17, 1990 between 4:30 and 6:00 UMT from typhoon Flo in the western Pacific.

Because the radar profiles were obtained at horizontal intervals of 0.5 km and the footprints of the radar and radiometers were of the order of 1 km in diameter, the radar profiles were preprocessed by applying a median filter across five profiles. Occasional missing radiometric data were filled in by scaling missing brightness temperatures with those available from the nearest frequencies or using interpolating spline fits to the data stream of brightness temperatures.

II.a Hydrometeor distribution retrieval algorithm

The precipitating clouds were assumed to consist of three layers, each containing a different type of hydrometeors. The lowest layer was assumed to consist of only spherical rain drops. A cloud physics model was used to describe the melting process of the hydrometeors below the freezing level in the middle layer. The upper layer contained only frozen hydrometeors.

II.a.1 Lower layer hydrometeor model

Precipitation in the lower layer was assumed to consist of spherical rain drops. The particle size distribution, PSD, $n_l(D)$, of the raindrop diameters, D, was assumed to be:

$$n_l(D) = n_{l,o}(\Lambda_l) \cdot D^{\mu} \cdot \exp(-\Lambda_l \, D) \tag{1}$$

The parameters Λ_l and $n_{l,0}(\Lambda_l)$ are a function of the rain mass, and $\mu = 2.16$, as described by Willis and Tattleman, W-T, (1989). The particular representation of the PSD is not critical because the 10 GHz radar reflectivity factor is proportional to the sixth moment of the PSD, while the extinction coefficient is proportional to the third moment of the PSD in the Rayleigh limit. We found that the radar reflectivity factor and the attenuation coefficient were related nearly the same for the W-T PSD as they were for the widely cited Marshall-Palmer PSD.

In the first approximation, the brightness temperature measured by a nadir directed 10 GHz radiometer, T_{b10}, was found to depend only on the total path attenuation in the lower layer at 10 GHz, τ_{10}. The results obtained from a simple plane parallel radiative transfer model were empirically fitted to:

$$\tau_{10} = 1.6 - 0.3\ln(263 - T_{b10}) \qquad (2)$$
$$110\text{ K} < T_{b10} < 257\text{ K}$$

where

$$\tau_{10} = \int_0^{z^*} k_{t,10}(z'').dz'' \qquad (3)$$

$k_{t,10}(z)$ is the total extinction coefficient as a function of height, z, which is the sum of the hydrometeor extinction coefficient, $k_{10}(z)$, and the absorption coefficients for liquid cloud water, $k_{c,10}(z)$, and water vapor, $k_{v,10}(z)$. z^* is the height of the top of the rain layer.

Meneghini and Kozu (1990) show that the radar reflectivity factor, $Z_{10}(z)$, of the rain is related to the returned radar power, $P_r(z)$, by:

$$Z_{10}(z) = \frac{P_r(z).(H-z)^2}{c_R.|K_w|^2}.e^{2.[\int_z^{z^*} k_{t,10}(z'').dz'']} \qquad (4)$$

H is the altitude of the aircraft, and $|K_w|^2$ depends on the refractive index of water and is equal to 0.93 at 10 GHz. c_R is the radar constant at 10 GHz, determined from radiometric measurements by Schols and Weinman (1993).

The extinction by rain, $k_{10}(z)$, is assumed to be related to the radar reflectivity factor, $Z_{10}(z)$, by:

$$k_{10}(z) = \xi \cdot Z_{10}^{\zeta}(z) \qquad (5)$$

ξ and ζ are constants determined from Mie theory applied to the W-T PSD.

We can define a normalized radar signal power:

$$S_{10}(z) = \frac{P_r(z).(H-z)^2}{c_R.|K_w|^2}.e^{2\int_z^{z^*}[k_{c,10}(z'')+k_{v,10}(z'')].dz''}$$
$$= Z_{10}(z).e^{-2.[\int_z^{z^*} k_{10}(z'').dz'']} \qquad (6)$$

The quantity, $S_{10}(z)$, in Eq. (6) depends only on the scattering and extinction characteristics of the rain drops.

The total path attenuation, derived from Eq.(2), was applied as a constraint to the radar equation of Hitschfeld-Bordan (1954), as modified by Weinman et al. (1990), to yield the vertical profile of the hydrometeor extinction coefficient :

$$k_{10}(z) = \frac{S_{10}^{\zeta}(z)\,[1 - e^{-2\zeta\tau_{10}}]}{2\zeta[\int_0^z S_{10}^{\zeta}(z'')dz'' + e^{-2\zeta\tau_{10}} \int_z^{z^*} S_{10}^{\zeta}(z'')dz'']} \tag{7}$$

Eq. (7) is valid only when ξ and ζ are constant throughout the layer. $S_{10}(0)$, closest to the surface, was estimated from Eq. (6) using $Z_{10}(0)$ and the total path attenuation τ_{10} in Eq.(2) with the total path absorption by assumed profiles of liquid cloud water and water vapor removed. $Z_{10}(0)$ was obtained from Eq. (4) using the measured value of the radar power closest to the surface corrected for τ_{10}. The value of $k_{10}(0)$ was then computed, using the estimates of $S_{10}(0)$ and τ_{10} in Eq. (7). The remainder of the first guess of the vertical profile of $k_{10}(z)$ was computed proceeding upwards towards the radar, repeating these steps, while the decrease in path attenuation in Eqs. (4) and (6) was accounted for by using the values of $k_{10}(z)$ between the present range and the surface. The profile of $Z_{10}(z)$ was obtained using Eq.(5), while the profile of rain mass was calculated from

$$M(z) = \gamma . Z_{10}^{\delta}(z) \tag{8}$$

where the values of the constants γ and δ were determined from Mie theory applied to the W-T PSD. The profile of rain mass was used to compute the profile of backscatter at 10 GHz, $\eta_{10}(z)$, from Mie theory. The modeled effective reflectivity profile was defined as

$$Z_{m,10}(z) = \frac{\eta_{10}(z).\lambda^4}{\pi^5|K_w|^2}.e^{-2\int_z^{z^*} k_{t,10}(z'').dz''} \tag{9}$$

where λ is the wavelength of the microwave. The profile of $Z_{m,10}(z)$ was compared with the profile of the measured radar reflectivity factor

$$Z_{e,10}(z) = Z_{10}(z).e^{-2\int_z^{z^*} k_{t,10}(z'').dz''} \tag{10}$$

which was readily derived from Eq. (4).

While the radar parameters defined in Eqs. (3-7) were initially derived as functions of z alone, the ultimate solutions of $M(z,x)$ are functions of the horizontal coordinate as well. The reason for this is that the radiation that produces the brightness temperature at any horizontal position, x, may have been scattered from another horizontal position. The brightness temperatures from clouds with variable $k_{10}(z,x)$ were computed with the backward Monte-Carlo radiative transfer model of Roberti et.al. (1994) for a horizontally and vertically structured cloud with horizontal dimensions of 3 by 10 km.

Nonprecipitating cloud water was assumed to exist only between a height of 1 km and the

263 K isotherm. The cloud water profile was assumed triangular with its maximum value, to be optimized, at the freezing level, i.e the height of the 273 K isotherm. The vertical profiles of temperature were obtained from sondes dropped from the NASA DC-8. The vertical distribution of water vapor, taken from tropical climatology data, assumed that the relative humidity varied as a smooth one-parameter function of height. The emissivity of the underlying sea surface was initially assumed to be wind roughened and foam covered, as described by Schols and Weinman (1994). Each iteration of the backward Monte-Carlo radiative transfer model and reflectivity retrieval sought to obtain $T_{b10}(x)$ and $Z_{m,10}(z)$ distributions from the model which were consistent with the measured values. New total path attenuations, τ_{10}, which no longer satisfied Eq. (2), were then calculated from Eq. (3). The Marquardt (1963) nonlinear least squares optimization procedure was applied to the measured $T_{b10}(x)$, $T_{b19}(x)$ and the $Z_{m,10}(z)$ to obtain optimal mass density profile employed in the W-T PSD and the total path absorption attributable to non-precipitating cloud water.

II.a.2 Upper layer hydrometeor model

The hydrometeors in the upper layer were assumed to be spheres with a PSD that was characterized by the same μ that was used in Eq.(1). The density of the frozen particles was given by:

$$\rho_f = \frac{0.91}{D_{fl}^{e}} \tag{11}$$

subject to:

$$0.05 < \rho_f < 0.91 \quad (g/cm^3)$$

D_{fl} is the melted diameter in *mm* of the frozen particles. e is a density factor, ranging from 0 (hail) to 10 (an upper limit to describe light fluffy snow). Eq. (11) is similar to a density parameterization employed by Klaassen (1988).

The frozen hydrometeors were assumed to consist of a two-component mixture with an average dielectric constant, ϵ_{av}, that consisted of contributions by a matrix with ϵ_m and inclusions with ϵ_i. For homogeneous substances the dielectric constants are only functions of frequency and temperature. The theory of Maxwell-Garnet for elliptical inclusions in a matrix that was described by Bohren and Battan (1982) yields:

$$\epsilon_{av} = \frac{(1-f)\cdot\epsilon_m + \beta\cdot f\cdot\epsilon_i}{1 - f + \beta\cdot f} \tag{12a}$$

where β is given by:

$$\beta = \frac{2\epsilon_m}{\epsilon_i - \epsilon_m}\left[\frac{\epsilon_i}{\epsilon_i - \epsilon_m}\ln\left(\frac{\epsilon_i}{\epsilon_m}\right) - 1\right] \tag{12b}$$

f is the volume fraction of the inclusions and the ln(..) is the principle value of the natural log. In the upper layer the matrix is assumed to be ice and the inclusions are air.

The 92 GHz brightness temperature and the vertical profile of the 10 GHz radar reflectivity factor were used to find the density factor e in Eq. (11) and the vertical profile of M.

II.a.3 Melting layer hydrometeor model

The presence of a bright band was derived from an algorithm that required a maximum of $Z_{10}(z)$ by at least 3 dBZ in a relatively narrow region 0.5 to 2 kilometers wide, below the freezing level. The adjacent local minima were used to define the thickness of the meting layer. If no maximum could be defined, the melting was assumed to take place over a 1 km thickness.

The melting hydrometeors were assumed spherical and to contain a mixture of ice, air and water. The precipitation rate in the melting layer was assumed to be independent of height, while matching the precipitation rates at both the bottom of the upper layer and the top of the lower layer, i.e. the PSDs of the melting particles were matched to those of the liquid and the frozen hydrometeors at the base and the top of the melting layer respectively. In the melting layer, the cloud microphysics model originally developed by of Yokoyama and Tanaka (1984) was used to describe the liquid mass fraction and PSD in the melting layer.

The refractive indices of the melting particles were determined by the liquid volume fraction and the way in which the ice, air and water mixture were distributed inside the melting particles. These refractive indices were modeled as a Maxwell-Garnet mixture of ice inclusions in a water matrix that was then treated as an ice-water matrix containing air inclusions as shown in Eqs. (12), see Bohren and Battan (1982). The measured Tb_{34} contributed mainly to the determination of the precipitation rate in the melting layer.

II.a.4 Optimization algorithm

In the previous section, models for the PSD and refractive indices of the hydrometeors in the upper layer and melting layer were described to define the radiative characteristics of the hydrometeors. The models were used in Mie scattering computations and a backward Monte-Carlo horizontally and vertically structured cloud radiative transfer model. Marquardt non-linear least squares optimization calculations were applied over a complete cloud configuration to match the model parameters to the observed distributions of the 10 GHz radar reflectivity factor within the noise characteristics of radar (~ 2 dBZ) and the 10, 19, 34 and 92 GHz brightness temperatures within the noise range of the radiometers (~4 K).

II.b Results

As a result of the optimization, distributions of precipitable hydrometeor mass were retrieved:

$$M = \frac{4}{3}\pi \int_0^\infty \rho(D)\cdot\left(\frac{D}{2}\right)^3 n(D)\,dD \tag{13}$$

$\rho(D)$ is the density of the hydrometeors. Because the hydrometeor mass is the third moment of the PSD, we can make no claim about the uniqueness of the PSD that appears in Eq. (13).

The measured and optimized brightness temperatures at 10, 19, 34 and 92 GHz are compared in Fig. 3 (upper). The agreement is within the measurement uncertainty, ~4K. Cross sections of the measured and optimized effective 10 GHz radar reflectivity factors, $Z_{e,10}(z,x)$, are shown in Figs. 3 (upper middle) and (lower middle) respectively as a function of the distance, x, from the center of the typhoon. The gray scale is contrast equalized, rather than uniformly graduated, to provide the best view of the features in the typhoon transect with the limited grey scale that was available. The distribution of $Z_{e,10}(z,x)$ shows a fairly extensive bright band that extends beyond 30 km from the typhoon center in both directions. Fig. 3 (lower) shows the retrieved profiles of the hydrometeor mass density, $M(z,x)$, $[g/m^3]$ that permit the computation of latent heating profiles. The grey scale is also nonlinear to emphasize the features with the limited dynamic range afforded by the grey scale. Most of the transect, with the exception of a the rain bands had light rainfall.

II.c Validation

Because no ground truth was available, the retrieved hydrometeor model was validated by comparing the computed 192 GHz brightness temperature and the distributions of the effective 35 GHz radar reflectivity factor, $Z_{e,35}(z,x)$, with measurements of those quantities that were not used in the retrieval algorithm. The model parameters that were found by the optimization were applied to model computations for these parameters. Transects of the modeled 192 Ghz brightness temperatures shown at the top of Fig. 4 were found within 4K of the measured values, and Fig. 4 (lower) and (middle) show that the model produced cross sections of $Z_{e,35}(z,x)$ agreed within ~ 2 $dBZ_{e,35}$ with the corresponding measured values. Note that this agreement was not the result of any optimization of those quantities. Some of the measured $Z_{e,35}(z,x)$ data were too small to be displayed when the rainfall rates were large and attenuation below the melting layer was significant. This effect is evident for x ~ 20 km and z < 3 km. These checks suggest that the retrieval algorithm yielded reasonably valid inferences about the characteristics of the hydrometeors near the cloud tops.

II.d Future algorithm development plans

Future algorithm development will employ rainfall measurements obtained from the Tropical Oceans and Global Atmosphere Coupled Ocean-Atmosphere Response Experiment (TOGA-COARE). This field experiment took place around an Intensive Flux Array (IFA) located at *156°E. x 2°S.* in the western Pacific. The Intensive Operating Period (IOP) of that experiment extended from November 1, 1992, through February 28, 1993. The IFA included an extensive network of moorings for current and surface flux measurements with 13 ships and 7 instrumented aircraft operating in the area. These platforms were heavily instrumented to obtain the most comprehensive observations of the ocean and atmosphere ever attempted.

NASA contributed its aircraft to the TOGA-COARE campaign to measure rainfall and micro physical parameters in convective system, e.g., velocity fields, phase changes, etc. Both the NASA DC-8 and ER-2 were equipped with microwave radiometers with frequencies similar to those of TRMM. The DC-8 was also equipped with a TRMM radar simulator, e.g., a 14

GHz multi-parameter radar. One of the two shipboard Doppler radars in the IFA region was also supported by NASA. During the IOP, Doppler radar data was continuous, with three 20-day periods in which both radars were on station. Experiments which included the aircraft in combination with the ship radars provided many observational data for research and precipitation processes studies. The NASA aircraft overflew the NOAA, NCAR, and UK-C-130 turboprop aircraft to provide data from the surface to an altitude near 21 km. Additionally, significant contributions to TRMM "ground truth" studies were provided by Japan with its J-COARE program.

III. Expansion of convection measurements beyond those provided by the TRMM satellite

III.a Sampling strategy

Tropical rainfall exhibits variability over many time scales: diurnal, 4-5 and 10-15 day periods, intraseasonal oscillations over time scales of 30-60 days, and interannual. Many of these features have been derived from the measurement of a proxy variable, such as Outgoing Long wave IR Radiation,(OLR). It is typical to find that only about 10-15 percent of the days with rain account for half of the total monthly rain, while the remaining 85-90 percent of the rainy days contribute the other half. This clearly represents a sampling challenge, and emphasizes the need for rainfall measurements with high spatial and temporal resolution.

The TRMM will observe the tropics from orbits chosen to observe a given region at all hours of the day in the course of a month. The monthly sampling of $5° x\ 5°$ boxes by TRMM, analyzed by Bell et.al. (1990), show that 40% or more of the area of those boxes are observed by TRMM at least eight times during a given month. However gaps of the order of 10 to 20 days may occur between a given visitation hour so that periodicities in the rainfall distribution may alias the retrieved rainfall. North et. al. (1993) addressed this problem and they outlined statistical techniques that may enable data from several satellites to reduce this sampling limitation. It is not evident at this time whether those efforts will be sufficient to permit the TRMM program to meet all of its goals. Other sensors may be needed to fill the temporal and spatial gaps between the TRMM overpasses. The TRMM will however provide excellent calibration for coincident measurements of those other measurements.

Because cold and thick clouds were known to be often associated with precipitation, particularly in convective systems, one of the earliest uses of satellite imagery was to infer rainfall. Although geostationary IR and visible imagery provide only a proxy measure of rainfall, these observations are nearly continuous, see Arkin and Ardanuy (1989). Unfortunately IR scenes are often obscured by cirrus anvils that can persist long after nearby rainfall has ceased, thereby rendering IR imagery somewhat unreliable as a proxy for rainfall measurements. An additional technique is needed to overcome the cirrus IR obscuration effect.

Wilheit et.al. (1991) showed that microwave radiometry provides excellent rainfall measurements over the ocean. However, microwave brightness temperatures are only measured for a few minutes as the satellite passes over a 200 x 200 km sampled region. Moreover, the SSM/I on the DMSP satellite, described by Hollinger et. al. (1990), can view

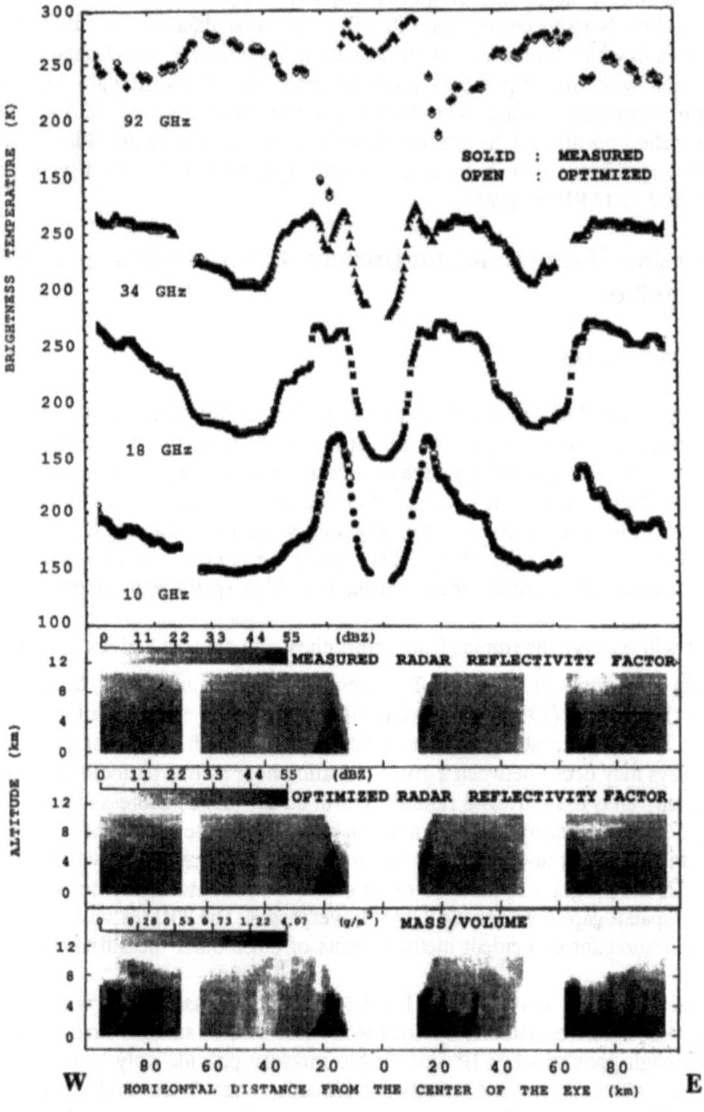

Fig. 3 Upper: Transects of measured and optimized 10,19,34 and 92 GHz brightness temperatures observed along a 200 km long flight path that crossed over the eye of typhoon Flo on 9/17/90. Note that the brightness temperatures agree within ~4 K on the average. Upper middle and lower middle: The distributions of the measured and optimized effective radar reflectivity factors, $Z_{e,10}(z,x)$. Lower: Distributions of hydrometeor mass density averaged over the PSD, $M(z,x)$, defined in Eq. (13). Radar measurements were available at the apparent gaps around the eye and at a range around 60 km from the center of the eye, however no rain was attributed to those regions because $dBZ_{e,10}(z,x) < 15$.

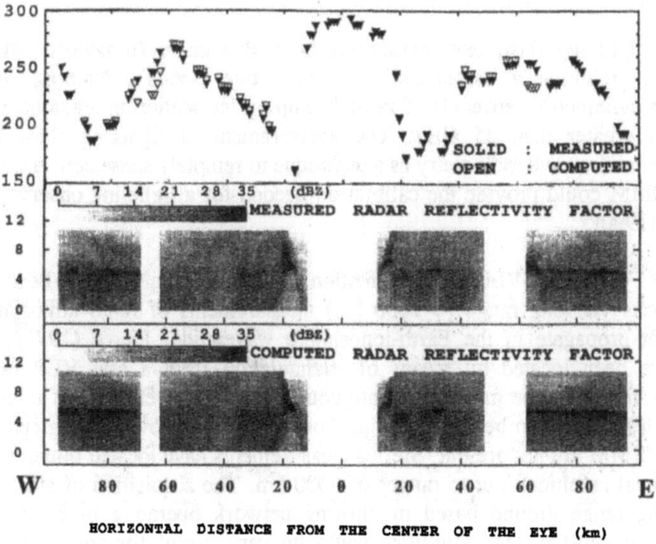

Fig. 4 (Upper) Comparison of the measured and computed 192 GHz brightness temperatures. Note that the 192 GHz brightness temperatures were not optimized, rather they were computed from the retrieved model precipitation distributions. Comparison of the measured (Middle) and computed (Lower) 35 GHz radar reflectivity factor, $Z_{e,35}(z,x)$. This comparison serves as partial validation of the model.

a given area only twice a day at best. The temporal sampling rate and spatial coverage of the SSM/I is inadequate to determine large-scale rain fields from a single DMSP satellite. Two DMSP satellites are now in operation so that somewhat better, albeit still sparse, sampling is now available.

IR imagery from geostationary satellites tuned with coincident SSM/I microwave retrievals has been used to fill the temporal gaps between SSM/I overpasses. Kummerow and Giglio (1992) filled the temporal gaps between SSM/I overpasses near Darwin Australia with rainfall estimates derived from IR imagery obtained from the GMS geostationary satellite. Those retrievals were tuned to the microwave retrievals obtained at times when the two systems provided coincident data. Microwave tuning reduced the accumulated monthly rainfall discrepancy between radar and spaceborne measurements from 50% to 15% in an area that was somewhat smaller than a TRMM sample area. Adler et. al. (1993) recently described a microwave tuned IR retrieval that they applied on a regional scale. Techniques based on IR imagery will no doubt be improved by the time that the TRMM satellite is launched; however other continuous proxy data also will need to be considered to assure that rainfall is adequately measured.

III.b Comparison between distributions of sferics and rainfall derived from microwave radiometry

Radar and lightning flash rate measurements in Britain by Illingworth and Lees (1991) suggest that lightning is associated with convective clouds containing wet hail. Some microwave radiometric retrievals of rainfall employ ice scattering signatures at microwave frequencies greater than 35 GHz. The measurement of lightning flash rate may thus complement microwave radiometry as a technique to remotely sense convective activity. The LIS on TRMM could provide the calibration needed for a lightning observation technique, Goodman (1990).

Very Low Frequency (VLF) radio direction finding techniques have been used to track thunderstorms via long range (>2000 km) measurements of radio atmospheric noise, or sferics, that propagate in the Earth-ionosphere waveguide. Pierce (1977) describes how sferics have been located by means of triangulation from a few VLF radio receivers. Moreover, sferics can be measured continuously from receivers so that intense convection that generates sferics can be monitored at those times when low orbiting spacecraft such as DMSP or TRMM are not present. Sferics measurements have located thunderstorms with ~ 200 km spatial resolution out to ranges of 9000 km. The acquisition of sferics observations from a long range ground based monitoring network operating in conjunction with IR imagery from geostationary satellites would be very useful for continuously monitoring convective activity because the IR imagery can locate cold clouds within a few tens of km while the sferics could determine whether those clouds were convectively active.

Sferics are now measured operationally from the United Kingdom, U.K., Meteorological Office sferics observation network, described by Lee (1986). Triangulation from a few stations permits coverage of large regions by means of an Arrival Time Difference (ATD) radio-location technique. The ATD system consists of a network of seven outstations dispersed over the United Kingdom and the Mediterranean. A single lightning stroke produces sferic wave forms which are detected through their vertical electric field between 8.1 and 11.7 kHz at the unmanned stations. Each outstation stores all of the data that it observes above a threshold, and forwards selected data on request to the control station. The service area covered by this system exceeds 30° to 70°N and 40°W to 40°E, with flash location accuracy of 2-20 km; however the sources of sferics can be located from regions well outside this area out to ranges of several thousand km. The ATD system detects both ground strokes and cloud-to-cloud strokes. The latter are detectable in less severely convective systems. It is estimated that the current system detects about a quarter of all strokes, the limitations being caused by computer processing speeds and the present communications links between the outstations and the control station. The high time-accuracy results in a low rate of miscorrelated signals of an estimated 0.004%. The ATD system started automatic international dissemination of lightning locations on 17 June, 1988.

We conducted a preliminary comparison of the sferics distributions measured from the U.K. Met. Office sferics observation network and the rainfall rate distribution in $.5^o \times .5^o$ boxes obtained from SSM/I data. We applied a scattering index rainfall retrieval procedure defined by Grody (1991) as the difference between the 85 and the 22 GHz brightness temperatures, to locate ice hydrometeors over land. Brightness temperatures measured at 19 and 22 GHz were employed over the oceans to estimate the total path attenuation of water vapor and cloud liquid water in a manner similar to that shown in Eq.(2). Departures of the ratios of the total path attenuations at those two frequencies by more than 10% from those found in rain free areas were used to identify raining regions. The SSM/I measurements obtained at 1800 Local Standard Time (LST) on 11/22/92 is shown in Fig. 5(left) and the corresponding sferics distributions observed within ± 3 h of the SSM/I overpass is shown in Fig. 5(right).

The sferics data were binned in $1^o \ x \ 1^o$ boxes. The maximum flash rate within such a box was 87 during the 6 h interval. There is a noteworthy resemblance between the distributions of flash rates and regions classified as raining by the SSM/I observations. This is particularly evident in the north Atlantic, and off the coasts of southern Brazil and West Africa. The temporal and spatial gaps in the SSM/I data are a vexing limitation. This is especially distressing near the frontal rainband in the middle of the north Atlantic ocean and in the Middle East. There are also a few areas of discrepancy. For example, the areas of intense lightning in the US mid west appear to the west of the SSM/I ice hydrometeor observations. The Gulf Stream boundary appears in the microwave retrieval and it is absent in the sferics data. Grody (private communication) suggests that his criteria for identifying scattering pixels may be too stringent. It is also possible that some electrically active clouds may be too small to be resolved by the microwave retrieval algorithm.

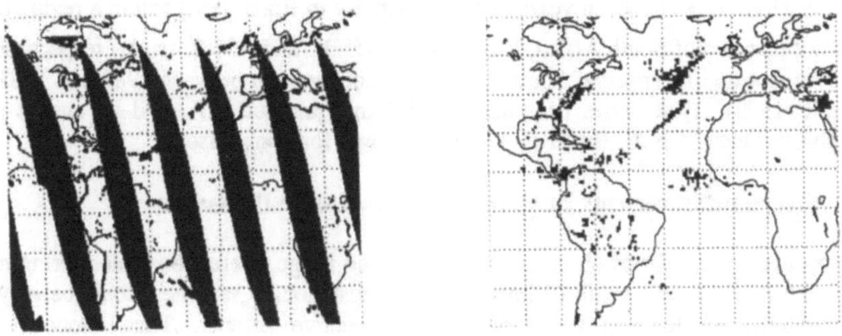

Fig. 5 A preliminary comparison of the distribution of rainfall inferred from SSM/I observations, left, and sferics obtained during a 6 h window centered at 1800 LST, right, on 11/22/1992. The precipitating areas over land were located by means of Grody's (1991) algorithm and they were located over the ocean from a combination of $T_{b,19}$ and $T_{b,22}$. Sferics measurements were kindly provided by Dr. A.C. Lee of the U.K. Met. Office.

The long term measurent of sferics may have considerable value for monitoring climate change because VLF sferics observations can continuously monitor the cumulus convection. Figs. 6 show the interannual variations of sferics rates over Brazil and the Agulhas current off the southeast coast of South Africa measured by Grandt (1992) from a single VLF receiver located in Pretoria, South Africa. (Regrettably, his receiver was not functioning during the 1982-83 El Nino year.) Figs. 6 also show the total annual rainfall in the Transvaal, South Africa, and a drought variability index, defined as 100 x the ratio of the root mean square departure from the annual mean rainfall divided by the mean annual rainfall, measured in the Northeast of Brazil. This superficial comparison suggests that sferics may be useful as a proxy for rainfall measurements. It also demonstrates that interannual variations in the distribution of sferics could be observed long after the TRMM mission has ended. Long time series of such measurements may be relatively cheap indicators of global climate variations. This may become especially important when we seek to monitor large scale convection over an entire El Nino cycle.

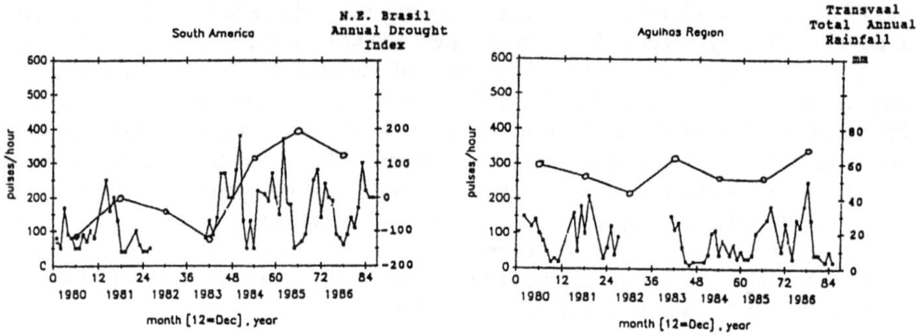

Figs. 6 Time series of seven years of sferics activity in Brazil (left) and in a region bounded by the east coast of South Africa, Madagascar, 45° E. and 40° S. (right) respectively. The receiver, located in Pretoria, South Africa, did not function between 1982 and 1983. From Grandt (1992). Time series of annual rainfall in Transvaal S.A. and an annual drought variability index in N.E. Brazil (---O---) are shown for comparison.

III.c Sferics measurements combined with spaceborne IR imagery

The acquisition of sferics observations from a long range international monitoring network operating in conjunction with IR imagery from geostationary satellites would be very useful for continuously locating and monitoring convective activity. The European science community, with its access to geographically widely dispersed confederated territories, could contribute significantly to such a program.

Acknowledgments: The authors are very thankful to Bob Meneghini and Jim Wang who provided the Typhoon-1 data used in this study.The authors also wish to thank Chris Kummerow and Laura Roberti, who provided a copy of the backward Monte Carlo radiative transfer code.

REFERENCES

Adler, R.F., A.J. Negri, P.R. Keehn and I.M. Hakkarinen, 1993: Estimation of Monthly Rainfall over Japan and Surrounding Waters from a Combination of Low-Orbit Microwave and Geosynchronous IR Data. J. Appl. Met., 32, 335-356.

Arkin, P.A. and P.E. Ardanuy, 1989: Estimating Climatic-Scale Precipitation from Space: A review. J. Climate, 2, 1229-1238.

Bell, T.L., A. Abdullah, R.L. Martin and G.R. North, 1990: Sampling Errors for Satellite-Derived Tropical Rainfall: Monte-Carlo Study Using Space-Time Stochastic Model. J. Geophys. Res., 95, 2195-2206.

Bohren, C.F. and L.J. Battan, 1982: Radar Backscattering of Microwaves by Spongy Ice Spheres. J. Atmos. Sci., 39, 2623-2628.

Goodman, S.J., 1990: Predicting Thunderstorm Evolution Using Ground-Based Lightning Detection Networks. NASA Tech. Memo. TM-103521.

Grandt, C., 1992: Thunderstorm Monitoring in South Africa and Europe by Means of Very Low Frequency Sferics. J. Geophys. Res., 97, 18215-18226

Grody, N.C., 1991: Classification of Snow Cover and Precipitation using the Special Sensor Microwave Imager. J. Geophys. Res., 96, 7423-7435.

Hitschfeld W. and J. Bordan, 1954: Errors Inherent in The Radar Measurement of Rainfall at Attenuating Wavelengths. J. Meteorol., 11, 58-67.

Hollinger, J., J. Peirce and G. Poe, 1990: Validation for the Special Sensor Micro wave/Imager. IEEE Trans. Geosci. Remote Sens., 28,781-790.

Illingworth, A.J. and M.I. Lees, 1991: Comparison of Lightning Location Data and Polarisation Radar Observations of Clouds. 1991 International Aerospace and Ground Conference on Lightning and Static Electricity, April 16-19, J.F. Kennedy Space Center, FL, NASA Conf. Pub. 3106, 85-1 -85-10.

Klaassen, W., 1988: Radar Observations and Simulation of the Melting Layer of Precipitation. J. Atmos. Sci., 45, 3741-3753.

Kummerow, C.D. and L. Giglio, 1992: A Passive Microwave/Infrared Technique for Overcoming Sampling Limitation of TRMM. Presented at the International Workshop on the Processing and Utilization of Rainfall Data Measured from Space, Tokyo Japan, 3/16-20. Also submitted to J. Appl. Met.

Lee, A. C. L., 1986: An Operational System for the Remote Location of Lightning Flashes using a VLF Arrival Time Difference Technique. J. Atmos. Oceanic Technol., 3, 630-642.

Marquardt, D.W., 1963: An Algorithm for Least Squares Estimation of Nonlinear Parameters. Jour. S.I.A.M., 11, 431-441.

Meneghini, R. and T. Kozu, 1990: Spaceborne Weather Radar. Artech House, pp 190.

North, G.R., S.S.P. Shen and R. Upson, 1993: Evaluation of Sampling Errors of Precipitation from Spaceborne and Ground Sensors. J. Appl. Met., 32, 399-410.

Pierce, E.T., 1977: Atmospherics and Radio Noise, in Lightning. Ed. R.H. Golde,

Academic Press, London, 351-384.

Roberti, L., J. Haferman and C. Kummerow, 1994: Microwave Brightness Temperatures from Horizontally and Vertically Structured Precipitating Clouds. Accepted for publication by J. Geophys. Res.

Schols, J.L. and J.A. Weinman, 1993: Microwave Radiometric Calibration of Airborne Weather Radars Under Rain Free Conditions Over the Ocean. Int'l Geosci. and Rem. Sens. (IGARSS '93), Tokyo, Japan.

Schols, J.L. and J.A. Weinman, 1994: Retrieval of Hydrometeor Distributions over the Ocean from Airborne Single-frequency Radar and Multi-frequency Radiometric Measurements. Accepted for publication in J. Atm. Res.

Theon, J.S., T. Matsuno, T. Sakata and N. Fugono, 1992: The Global Role of Tropical Rainfall, A. Deepak Pub. Co., Hampton, VA, pp 280.

Thiele, O.W., D.A. Short and J.C. Wilkerson, 1993: Oceanic Rainfall Observations. Proc. of the Int'n'tl Symp. on Precip. and Evaporation, ed. B. Sevruk & M. Lapin, Bratislava, Slovakia, 1, 23-29.

Wang, J.R., R. Meneghini, H. Kumagai, T.T. Wiheit, W.C. Boncyk, P. Racette, J. Tesmer and B. Maves, 1994: Airborne Active and Passive Microwave Observations of Typhoon Flo. Submitted to IEEE Trans. Geosci. and Remote Sensing.

Weinman, J.A., R. Meneghini and K. Nakamura, 1990: Retrieval of Precipitation Profiles from Airborne Radar and Passive Radiometer Measurements: Comparison with Dual-Frequency Radar Measurements. J. Appl. Meteorol., 29, 981-993.

Wilheit, T. T., A. T. C. Chang and L. S. Chiu, 1991: Retrieval of Monthly Rainfall Indices from Microwave Radiometric Measurements Using Probability Distribution Functions. J. Atmos. Ocean. Tech., 3, 118-136.

Willis, P.T. and P. Tattleman, 1989: Drop Size Distributions associated with Intense Rainfall. J. Appl. Met., 28, 3-15.

Yokoyama, T. and H. Tanaka, 1984: Microphysical Processes of Melting Snowflakes Detected by Two-Wavelength Radar. Part I. Principle of Measurement Based on Model Calculation. J. Meteorol. Soc. Japan, 62, 650-666.

RECOMMENDATIONS OF WORKING GROUPS AT THE END OF THE WORKSHOP

General statements

1. Precipitation is the main input to water cycle which govern the climate and the life on the Earth.

2. However, precipitation processes are not sufficiently known to obtain the amount of water falling to and to be stored on the Earth with required accuracy for many practical purposes.

3. Precipitation measurements and estimates have to be adjusted for errors and spatial and temporal inhomogeneities if to be used for hydrological and climatological computations including the assessment of climate change.

4. To achieve this goal a strong interdisciplinary and international cooperation should be promoted, in particular between paleoclimatologists, physicists, ...
   This includes the data exchange, which should be encouraged by international bodies, and the establishment of international data centers and intercomparisons.

Working Group I
"Precipitations over land"

1. Why are *measured* data needed :

Precipitation = source for agriculture, drinking water, other water resources, energy ...
= important component of *climate*.

Measured data are needed for :

- analysis of climate *variability*, detection of *changes, trends* ;
- understanding, determination of *global energy and water cycle* (and budgets) ;
- validation of *remote* sensing ;
- initialisation, *verification of GCMs* ;
- development of *down-scaling* methods.

2. Conventional "rain-gauge" measurements are the only *direct* measure of *fallen* precipitation.

- But correction of systematic errors is necessary before application (also needed : quality control, metadata).

  Time series going back to last centuries are only available from conventional measurements (rain gauge, snow depth).
- Homogeneous series *must* continue (parallel use of old and new instruments).

3. For different purposes, different ***temporal*** resolution is necessary (sub-hourly, hourly, daily, monthly, long-term means).

4. Most applications require ***area-integrated data.***
   - ***Aspects*** :
     - high data density required (high spatial variability of precipitation, ***events***),
     - application of advanced objective analysis methods (***orography*** !),
     - siting of new gauges in data poor regions !!
     - filling of remaining gaps by remote sensing data,
     - RADAR can provide regional scale information of high resolution in time and space ; but has to be calibrated by reliable ground measurements (online, dense networks, qualified instruments).

5. Existing international activities

   to be supported and to continue into next century.

GHCN	Long-term series of monthly data, precipitation, temperature and others.
GPCC	Global gridded monthly precipitation data including error estimates → analysis ; → monitoring.
GPCP/cal-val	Data of highest accuracy and resolution for limited areas and periods (including radar).
GRDC	River runoff data from many rivers, world-wide.

Summarized recommendations

* Forcing and regulation of international data exchange (international data centres).
* International funds for international data centres (in the framework of GCOS).
* Providing reliable measured data (quality control, correction of systematic errors, but hold also original data, maintain homogeneous series).
* Enhance research on data analysis (methods intercomparison project).

Working Group II "Oceanic Working Group"

General statements

Global precipitation climatologies have their largest uncertainties over the oceans. This is due to the historic and continuing lack of direct measurements of rainfall over most of these regions. Fortunately, satellites are well placed to help solve this problem in the future for in some spectral regions evidence of rainfall or rainfall-related clouds is most direct over such surfaces. The amount and variety of satellite data generally available for oceanic satellite rainfall estimation are both increasing rapidly. Particular progress has been made in the last five years with the use of passive microwave radiometry for rainfall estimation of the world's major water bodies; However, the algorithms are not yet mature and further development of them is required. A key related problem is the sparseness of surface observation data suitable for calibration and validation of satellite rainfall algorithms : this situation must be improved at least in selected regions if the algorithm error statistics are to be improved.

Recommendations (in logical, not rank order)

1. Intensification of the collection of coastal geological indications of paleoclimates, in order to obtain more information on :
   a) ocean level changes and global SSTs,
   b) reactions thereto of atmospheric patterns (including precipitation changes) in areas not dominated by continentality.
   c) coastal zones directly affected by important ocean currents (Canaries, Benguela, Humboldt, etc.) and their variations through time.
2. New long-term time-dependent, climatologies or rainfall over surfaces.
3. More intensive efforts to analyse ship observations of weather related to rainfall, e.g. in terms of rain/no-rain statistics, to help validate satellite rainfall algorithms in areas of sparse instrumental data.
4. Programmes to continue existing long-term climate data sets of rainfall etimated from satellite data over the world's oceans (e.g. the infrared-based GPI).
5. Supplementation of existing data sets of rainfall estimated from satellite data sets with results from newer radiometric sources, particularly in the passive microwave : SSMR and SSM/I (imaging) data records already extend over more than 16 years, and MSU

(profiling) data records over more than 24 years. This should be undertaken immediately using existing algorithms.

6. Organisation of new intercomparison campaigns for existing satellite algorithms to learn more about their behaviour in different regions an under different synoptic situations, and to effect related improvements in them.

7. Co-ordinated efforts to improve the satellite algorithms with special emphasis on :
   a) Further development of visible/infrared algorithms especially fur use over shorter time-periods, and to help elucidate growth cycles, diurnal patterns, and physical processes of cloud and associated rain cells.
   b) Development of improved passive microwave procedures, including composite algorithms involving surface properties, cloud systems, synoptic situations as well as the levels of performance of individual algorithms in different surface/atmospheric environments.
   c) Investigations of the potential benefits of inverse modelling for rainfall algorithm performance, using products of numerical weather prediction models.
   d) Increased attention to more physically-based passive microwave rainfall algorithm.
   e) Experimentation with combined visible/infrared/passive microwave/surface data approaches in the search for optimum global oceanic rainfall products.

8. Immediate expansion of the present very sparse global network of radars on research vessels, coasts, and even more importantly on islands to significantly improve surface calibration/validation data sets as soon as possible.

9. Accelerated devevopment and deployment of new surface-based rainfall measurement devices for use on small islands, ships and buoys, e.g. optical raingauges, improved mechanical instruments, and small, simple (vertically-pointed) radars.

10. Installation of small radio frequency receivers on ships and buoys to estimate rainfall through related attenuation of signals from communication satellites.

11. Accelerated testing of the value of satellite data products (especially water vapour, cloud liquid water, other cloud properties, and rainfall) to numerical weather prediction models, to reduce their spin-up problems and improve their rainfall outputs.

12. Research into possible uses of satellite observations in relatively raw forms (e.g. brightness temperatures) as new inputs of potential value to synoptic analyses and the performance of NWP models.

13. Consideration of the potential value of satellite and combined global oceanic (and continental) rainfall products for the development and use of global circulation models (GCMs).

14. Special attention to rainfall processes and climatologies in the global areas of heaviest precipitation, i.e. in the Intertropical Convergence Zone, and the mid-latitude depression belts of both northern and southern hemispheres.

15. Encouragement of pre-launch studies in support related to the Japanese/US Tropical Rainfall Monitoring Mission (TRMM) due for launch in 1997.

16. New and expanded preparatory studies in support of the forthcoming Multichannel Imaging Microwave Radiometer (MIMR) taking special note of the needs and aspirations of the global climate change community.

17. Further design studies for a geostationary-based passive microwave radiometer taking account of recent advances in space technology and understanding of the scientific problems involved in passive microwave and infrared estimation of rainfall from satellite altitudes.

Of the above recommendations, those of highest urgence include : (1) in respect of the more distant past, (2) in respect of the more recent past, and (6) and (7) in relation to the present and near-term future.

Working Group III
"Process studies of long- and short-term variations in precipitation regimes"

- Studies to understand mechanisms
  - Process that maintain present climate may be different from processes that cause climatic changes.
  - Models are typically tuned to simulate present climate.
  - Model results can be "right" for the wrong reasons.

- We need data to verify model results, if possible at the process level.

- There is a need to identify physically meaningful space and time scales in relation to precipitation climatology and variability.

- Paleoclimatic reconstructions are important to studies of climate in order to understand the range of sensitivities to variable external and internal forcings.
  - → Glacial / Interglacial
  - → Holocene changes.

Working group IV
"Modelling group"

• Model developments on precipitation process data requirements :

- Best estimates of daily and monthly mean areally averaged precipitation data. 1°x1° grid with error bounds ; as many years as possible. Tropical coverage, global if possible.
- For model evaluation of diurnal change : 3 hourly precipitation at some evenly distributed land sites (100).
- Daily rainfall data be made available on GTS for real time usage. The current coverage needs enhancement.
- Access of real time OLR, DMSP/SSMI along satellite swaths for initialization of models is recommended, inverse modelling.
- Some modelling groups would like to analyze cloud liquid water and cloud amounts and cloud heights. Real time availability of such data from satellites needs to be explored.

Above data are required for data assimilation and model validation.

Additional requirements

Snowdepth data is desirable for hydrological balance studies and for estimating river runoff.

Variance of precipitation

On a 1°x1° grid for as many years as possible.

This would be the variance of daily precipitation evaluated on a month by month basis.

Composite data

For the further development of rain-rate strategies, a data set covering at least a month (preferably during TOGA-COARE) should be prepared that might include :

- precipitation from rain gauge,
- OLR observations at high resolution,
- SSM/I observations at highest resolution.

INDEX

ALPHABETICAL LIST OF AUTHORS

Dr Phillip A. ARKIN
U.S. Department of Commerce - NOAA
National Weather Service
W/NMC52, 5200 Auth Road
Washington, DC 20233
UNITED STATES

Dr Klaus ARPE
Max Planck Institute für Meteorologie
Bundesstrasse 55
20146 Hamburg
GERMANY

Dr Edouard BARD
Université Aix-Marseille III, Case 431
Laboratoire de Géologie du Quaternaire
Avenue Escadrille Normandie-Niemen
13397 Marseille, Cedex 20
FRANCE

Dr Franck C. BASSINOT
Université Aix-Marseille III, Case 431
Laboratoire de Géologie du Quaternaire
Avenue Escadrille Normandie-Niemen
13397 Marseille, Cedex 20
FRANCE

Dr Luc BEAUFORT
Université Aix-Marseille III, Case 431
Laboratoire de Géologie du Quaternaire
Avenue Escadrille Normandie-Niemen
13397 Marseille, Cedex 20
FRANCE

Dr. H.S. BEDI
Department of Meteorology B-161
The Florida State University
Tallahassee, Fl. 32306-3034
UNITED STATES

Dr L. BENGTSSON
Max Planck Institute für Meteorologie
Bundesstrasse 55
20146 Hamburg
GERMANY

Dr Patrick BENICHOU
Centre National de Recherches Météorologiques
Météo-France
42, Avenue Coriolis
31057 Toulouse, Cedex
FRANCE

Dr O. BOUCHER
Laboratoire de Météorologie Dynamique
Ecole Normale Supérieure
24, rue Lhomond
75231 Paris, Cedex 05
FRANCE

Dr. Long S. CHIU
General Sciences Corporation
Laurel, MD 20707
UNITED STATES

Prof. Kerry H. COOK
Atmospheric Sciences Program
1110 Bradfield Hall
Cornell University
Ithaca, NY 14853
UNITED STATES

Dr Michel DEQUE
Centre National de Recherches Météorologiques
Météo-France
42, Avenue Coriolis
31057 Toulouse, Cedex
FRANCE

Dr Françoise DESALMAND
Laboratoire de Météorologie Dynamique
Ecole Polytechnique
91128 Palaiseau, Cedex
FRANCE

Dr Michel DESBOIS
Laboratoire de Météorologie Dynamique
Ecole Polytechnique
91128 Palaiseau, Cedex
FRANCE

Dr Henry DIAZ
U. S. Department of Commerce - NOAA
Environmental Research Laboratories
325 Broadway
Boulder, Colorado 80303-3328
UNITED STATES

Prof. George DUGDALE
Department of Meteorology
University of Reading
2 Earley Gate
Reading RG6 2AU
UNITED KINGDOM

Dr L. DÜMENIL
Max Planck Institute für Meteorologie
Bundesstrasse 55
20146 Hamburg
GERMANY

Dr David R. EASTERLING
NOAA - National Climatic Data Center
Federal Building
23 Second Line Federal Building
Asheville, NC 28801-2696
UNITED STATES

Dr Christophe GENTHON
Laboratoire de Glaciologie et Géophysique
de l'Environnement
54, Rue Molière, DU BP 96
38402 Saint-Martin d'Hères
FRANCE

Dr Pavel Ya. GROISMAN
State Hydrological Institute
23 Second Line
194018 St - Petersburg
RUSSIA

Dr H. HAUSCHILD
Global Precipitation Climatology Center
Deutscher Wetterdienst
Postfach 10 04 65
63004 Offenbach am Main
GERMANY

Dr Mike HULME
Climatic Research Unit
School of Environmental Sciences
University of East Anglia
Norwich NR4 7TJ
UNITED KINGDOM

Dr Serge JANICOT
Météo-France
Laboratoire de Météorologie Dynamique
Ecole Polytechnique
91128 Palaiseau, Cedex
FRANCE

Dr Isabelle JOBARD
Laboratoire de Météorologie Dynamique
Ecole Polytechnique
91128 Palaiseau, Cedex
FRANCE

Dr Jean JOUZEL
Laboratoire de Modélisation du Climat
et de l'Environnement
CEA CDM, CE l'Orme des Merisiers
91191 Gif sur Yvette, Cedex
FRANCE

Dr Graeme KELLY
European Center for Medium-Range
Weather Forecasts
Shinfield Park
Reading RG2 9AX
UNITED KINGDOM

Dr Khamil KHAIRULLIN
Voeikov Main Geophysical Observatory
7, Karbyshev Ulitsa
194018 St - Petersburg
RUSSIA

Dr Elena KITAIGORODSKAYA
Voeikov Main Geophysical Observatory
7, Karbyshev Ulitsa
194018 St - Petersburg
RUSSIA

Prof. T.N. KRISHNAMURTI
Department of Meteorology B-161
The Florida State University
Tallahassee, Fl. 32306-3034
UNITED STATES

Dr Milan LAPIN
Slovak Hydrometeorological Institute
Jeséniova 17
833 15 Bratislava
SLOVAKIA

Dr Hervé LE TREUT
Laboratoire de Météorologie Dynamique
Ecole Normale Supérieure
24, rue Lhomond
75231 Paris, Cedex 05
FRANCE

Dr Z.X. LI
Laboratoire de Météorologie Dynamique
Ecole Normale Supérieure
24, rue Lhomond
75231 Paris, Cedex 05
FRANCE

Dr Michaël J. MANTON
Bureau of Meteorology Research Center
GPO Box 1289K
150 Lonsdale House
Melbourne VIC 3001
AUSTRALIA

Dr Martin MILLER
European Center for Medium-Range
Weather Forecasts
Shinfield Park
Reading RG2 9AX
UNITED KINGDOM

Dr Jean-Jacques MORCRETTE
European Center for Medium-Range
Weather Forecasts
Shinfield Park
Reading RG2 9AX
UNITED KINGDOM

Dr Peter J. MÜLLER
FB Geowissenschaften
Universität Bremen
GERMANY

Dr Vladislav NESPOR
Swiss Federal Institute of Technology
Geographisches Institut ETH
Winterthurer Strasse 190
8092 Zürich
SWITZERLAND

Prof. Sharon NICHOLSON
Department of Meteorology B-161
The Florida State University
C.P. Box 6.70.12
Tallahassee, FL 32306-3034
UNITED STATES

Prof. Nicole PETIT-MAIRE
Laboratoire de Géologie du Quaternaire
Case 907 Luminy
13288 Marseille, Cedex 9
FRANCE

Dr M. REISS
Global Precipitation Climatology Center
Deutscher Wetterdienst
Postfach 10 04 65
63004 Offenbach am Main
GERMANY

Dr E. ROECKNER
Max Planck Institute für Meteorologie
Bundesstrasse 55
20146 Hamburg
GERMANY

Dr. G.D. ROHALY
Department of Meteorology B-161
The Florida State University
Tallahassee, Fl. 32306-3034
UNITED STATES

Dr Frauke ROSTEK
Université Aix-Marseille III, Case 431
Laboratoire de Géologie du Quaternaire
Avenue Escadrille Normandie-Niemen
13397 Marseille, Cedex 20
FRANCE

Dr Bruno RUDOLF
Global Precipitation Climatology Center
Deutscher Wetterdienst
Postfach 10 04 65
63004 Offenbach am Main
GERMANY

Dr W. RUETH
Global Precipitation Climatology Center
Deutscher Wetterdienst
Postfach 10 04 65
63004 Offenbach am Main
GERMANY

Dr Götz RUHLAND
FB Geowissenschaften
Universität Bremen
GERMANY

Prof. Eberhard RUPRECHT
Institut für Meereskunde
Universität Kiel
Düsternbrooker Weg 20
24105 Kiel
GERMANY

Dr U. SCHNEIDER
Global Precipitation Climatology Center
Deutscher Wetterdienst
Postfach 10 04 65
63004 Offenbach am Main
GERMANY

Dr. Jack L. SCHOLS
General Sciences Corporation
Laurel, MD 20707
UNITED STATES

Dr Andréi SELIVANOV
Water Problem Institute
Russian Academy of Sciences
Novobasmannaya St. 10, P.O. Box 524
107078 Moscow
RUSSIA

Dr Boris SEVRUK
Swiss Federal Institute of Technology
Geographisches Institut ETH
Winterthurer Strasse 190
8092 Zürich
SWITZERLAND

Dr Marina SHABALOVA
Royal Netherlands Meteorological Institute
P.O. Box 201
3730 AE De Bilt
THE NETHERLANDS

Prof. Georgy SHCHUKIN
A.I. Voeikov Main Geophysical Observatory
7, Karbyshev Ulitsa
194018 St - Petersburg
RUSSIA

Dr. Christian THOMAS
Institut für Meereskunde
University Kiel
Düsternbrooker Weg 20
24105 Kiel
GERMANY

Dr Michael TIEDTKE
European Center for Medium-Range
Weather Forecasts
Shinfield Park
Reading RG2 9AX
UNITED KINGDOM

Prof. James A. WEINMAN
NASA
Goddard Space Flight Center
Greenbelt, MD 20771
UNITED STATES

The ASI Series Books Published as a Result of Activities of the Special Programme on Global Environmental Change

This book contains the proceedings of a NATO Advanced Research Workshop held within the activities of the NATO Special Programme on Global Environmental Change, which started in 1991 under the auspices of the NATO Science Committee.

The volumes published as a result of the activities of the Special Programme are:

Vol. 1: **Global Environmental Change.**
Edited by R. W. Corell and P. A. Anderson. 1991.
Vol. 2: **The Last Deglaciation: Absolute and Radiocarbon Chronologies.**
Edited by E. Bard and W. S. Broecker. 1992.
Vol. 3: **Start of a Glacial.** Edited by G. J. Kukla and E. Went. 1992.
Vol. 4: **Interactions of C, N, P and S Biogeochemical Cycles and Global Change.** Edited by R. Wollast, F. T. Mackenzie and L. Chou. 1993.
Vol. 5: **Energy and Water Cycles in the Climate System.**
Edited by E. Raschke and D. Jacob. 1993.
Vol. 6: **Prediction of Interannual Climate Variations.**
Edited by J. Shukla. 1993.
Vol. 7: **The Tropospheric Chemistry of Ozone in the Polar Regions.**
Edited by H. Niki and K. H. Becker. 1993.
Vol. 8: **The Role of the Stratosphere in Global Change.**
Edited by M.-L. Chanin. 1993.
Vol. 9: **High Spectral Resolution Infrared Remote Sensing for Earth's Weather and Climate Studies.**
Edited by A. Chedin, M.T. Chahine and N.A. Scott. 1993.
Vol. 10: **Towards a Model of Ocean Biogeochemical Processes.**
Edited by G. T. Evans and M.J.R. Fasham. 1993.
Vol. 11: **Modelling Oceanic Climate Interactions.**
Edited by J. Willebrand and D.L.T. Anderson. 1993.
Vol. 12: **Ice in the Climate System.** Edited by W. Richard Peltier. 1993.
Vol. 13: **Atmospheric Methane: Sources, Sinks, and Role in Global Change.** Edited by M. A. K. Khalil. 1993.
Vol. 14: **The Role of Regional Organizations in the Context of Climate Change.** Edited by M. H. Glantz. 1993.
Vol. 15: **The Global Carbon Cycle.**
Edited by M. Heimann. 1993.
Vol. 16: **Interacting Stresses on Plants in a Changing Climate.**
Edited by M. B. Jackson and C. R. Black. 1993.
Vol. 17: **Carbon Cycling in the Glacial Ocean: Constraints on the Ocean's Role in Global Change.** Edited by R. Zahn, T. F. Pedersen, M. A. Kaminski and L. Labeyrie. 1994.

Vol. 18: **Stratospheric Ozone Depletion/UV-B Radiation in the Biosphere.** Edited by R. H. Biggs and M. E. B. Joyner. 1994.

Vol. 19: **Data Assimilation: Tools for Modelling the Ocean in a Global Change Perspective.** Edited by P. O. Brasseur and J. Nihoul. 1994.

Vol. 20: **Biodiversity, Temperate Ecosystems, and Global Change.** Edited by T. J. B. Boyle and C. E. B. Boyle. 1994.

Vol. 21: **Low-Temperature Chemistry of the Atmosphere.** Edited by G. K. Moortgat, A. J. Barnes, G. Le Bras and J. R. Sodeau. 1994.

Vol. 22: **Long-Term Climatic Variations – Data and Modelling.** Edited by J.-C. Duplessy and M.-T. Spyridakis. 1994.

Vol. 23: **Soil Responses to Climate Change.** Edited by M. D. A. Rounsevell and P. J. Loveland. 1994.

Vol. 24: **Remote Sensing and Global Climate Change.** Edited by R. A. Vaughan and A. P. Cracknell. 1994.

Vol. 25: **The Solar Engine and Its Influence on Terrestrial Atmosphere and Climate.** Edited by E. Nesme-Ribes. 1994.

Vol. 26: **Global Precipitations and Climate Change.** Edited by M. Desbois and F. Désalmand. 1994.

Vol. 27: **Cenozoic Plants and Climates of the Arctic.** Edited by M. C. Boulter and H. C. Fisher. 1994.

MIX
Papier aus verantwortungsvollen Quellen
Paper from responsible sources
FSC® C105338

If you have any concerns about our products,
you can contact us on
ProductSafety@springernature.com

In case Publisher is established outside the EU,
the EU authorized representative is:
Springer Nature Customer Service Center GmbH
Europaplatz 3, 69115 Heidelberg, Germany

Printed by Libri Plureos GmbH
in Hamburg, Germany